D0753325

35

BUCHSBAUM'S COMPLETE HANDBOOK OF PRACTICAL ELECTRONIC REFERENCE DATA

Third Edition

BUCHSBAUM'S COMPLETE HANDBOOK OF PRACTICAL ELECTRONIC REFERENCE DATA

Third Edition

by
Walter H. Buchsbaum

as revised by
Robert C. Genn, Jr.

Prentice-Hall, Inc.
Englewood Cliffs, New Jersey

Prentice-Hall International, Inc., *London*
Prentice-Hall of Australia, Pty. Ltd., *Sydney*
Prentice-Hall Canada, Inc., *Toronto*
Prentice-Hall of India Private Ltd., *New Delhi*
Prentice-Hall of Japan, Inc., *Tokyo*
Prentice-Hall of Southeast Asia Pte. Ltd., *Singapore*
Editora Prentice-Hall do Brasil Ltda., *Rio de Janeiro*
Prentice-Hall Hispanoamericana, S.A., *Mexico*

© 1987 by

PRENTICE-HALL, INC.

Englewood Cliffs, N.J.

All rights reserved. No part of this
book may be reproduced in any form or
by any means, without permission in
writing from the publisher.

Library of Congress Cataloging-in-Publication Data

Buchsbaum, Walter H.
    Buchsbaum's complete handbook of practical
electronic reference data.

    Includes bibliographies and index.
    1. Electronics—Handbooks, manuals, etc.
I. Genn, Robert C.   II. Title.
III. Title: Complete handbook of practical
electronic reference data.
TK7825.B8   1987   621.381   86-30284

ISBN 0-13-084633-3

Printed in the United States of America

ACKNOWLEDGMENTS

Each of the following made significant contributions to this third edition.

Mrs. Ann Buchsbaum
Mr. W. M. Flanagan
Mrs. E. Louise Genn
Mr. Morris Grossman
Mr. Paul Palko
Mr. G. Rosenthal
Dr. Samuel Seely
Mrs. Inge Seymour
Mr. W. Ralph Siena

ABOUT THIS BOOK

The first and second editions of this book enjoyed many printings and were offered by two of the largest electronic book clubs. Comments from a wide cross-section of electronics professionals have been overwhelmingly favorable, especially concerning the practical approach to every technical topic. It was this, the strictly practical application of electronics, that earned the first and second editions such a large readership; the same no-nonsense, practical approach has been followed in this completely revised and updated *Third Edition*.

During the revision for this third edition, we modified and updated many topics. Among these are solid state devices, computer principles, radio/TV broadcasting, multiple-beam CRTs, and memory designs. In all, eighteen chapters have been modified and new schematics and tables inserted in key places.

We have supplied new material in place of the original in numerous areas, to improve the usefulness of this practical handbook. For example, satellite and space communications signal propagation, very large scale integrated circuits (VLSI), microwave relay stations, practical tuned circuits, multichannel television sound (stereo TV), optical communications, and television receive only satellite (TVRO) systems are now included.

Other significant changes are: (1) References to the literature available to the reader have been supplemented and brought up to date; (2) common carrier satellite communication and optical communications have been included; (3) in the material pertaining to the field of digital electronics, we have added digital modulation techniques and digital television.

The first nine chapters of this book contain the factual foundations of all electronics work. Starting with the physical constants, the frequency spectrum and contemporary electrical parameters, this part provides complete reference data for all types of state-of-the-art active and passive components and networks.

These original chapters have been expanded to include electronic tuning components, multiple resistor arrays in dual in-line and single in-line packages, solid state filters, special rectifiers and diodes, very large scale integration (VLSI), and ultra large scale integration (ULSI). In each chapter the basic characteristics are presented, together with practical application data and typical circuits.

Chapters 10 through 23 deal with specialized areas of electronics and feature the fundamentals of each area as well as the most widely used applications in today's electronic industry. Subjects such as microwave power transistors (2 to 4 GHz), new microwave diodes (hot carrier, PIN types), stereo TV, satellite TV, optical computing technology, telecommunications, AM stereo broadcasting, microwave relay systems, and robotics technology are now included in these chapters.

Another expanding area of electronics is optoelectronic devices and applications. These include optoisolators, optocouplers, phototransistors, photo Darlington amplifiers, and optical transmission systems. Chapter 19 covers all aspects of this growing field, ranging from the fundamentals of optics through the latest in fiber optic systems.

The updated material covered in Chapters 10 through 23 will be especially valuable to those electronics specialists whose work involves going outside their own area of specialization. A digital logic designer, for example, will be able to use Chapter 19 to understand just how the CRT or liquid crystal-type display works (since his equipment usually interfaces with one or the other). Similarly, a TV service technician can learn the fundamentals of digital logic needed to service digital television systems by referring to Chapter 16. Whatever area of specialization you work in, the probability is high that every one of the topics covered in the third edition of this reference handbook will be useful at one time or another.

No reference handbook of electronics would be complete without the essential mathematical formulas and application examples in Chapter 23. Here we find all essential formulas, tables, and graphs one is likely to need in electronics.

The detailed index at the end of the book will help you find the particular topic, component, device, circuit, or system you want to know about. Then, after you have read the portion that interests you, check the completely updated bibliography in case you want more information.

The best recommendation for this third edition is, of course, the success of the first and second editions. We have Mr. Buchsbaum's second edition on our bookshelf; it has been in constant use since its original publication. We find that it contains just about all the electronics data needed on a day-to-day basis. We hope you find that this updated third edition contains all the essential information anyone interested in electronics needs.

Robert C. Genn, Jr.

TABLE OF CONTENTS

Chapter 1

Units, Constants, Frequency Spectrum, and Conversion Factors

Chapter 2

Properties of Materials and Wire Tables

Chapter 3

Components and Parts

Chapter 4

Fundamentals of Networks

Chapter 5

Filters and Attenuators

Chapter 6

Ferromagnetic Core Transformers and Reactors

Chapter 7

Practical Tuned Circuits

Chapter 8

Semiconductors, Transistors, and Circuits

Chapter 9

Integrated Circuits

Chapter 10

Antennas and Transmission Lines

Chapter 11

RF and Microwave Fundamentals

Chapter 12

Radio, Broadcasting, and Recording

Chapter 13

Radar and Navigation Fundamentals

Chapter 14

Communications Systems

Chapter 15

Television Systems

Chapter 16

Digital Logic

Chapter 17

Computer Principles

Chapter 18

Programming Systems and Languages

Chapter 19

Opto-Electronic Devices

Chapter 20

Power Supplies

Chapter 21

Control Devices and Transducers

Chapter 22

Industrial Electronics

Chapter 23

Mathematical Formulas

TABLE OF ILLUSTRATIONS

Chapter 1

Chapter 2

Chapter 3

Chapter 4

Chapter 5

Chapter 6

Chapter 7

Chapter 8

Chapter 9

Chapter 10

Chapter 11

Chapter 12

Chapter 13

Chapter 14

Chapter 15

Chapter 16

Chapter 17

Chapter 21

Chapter 22

Chapter 23

BUCHSBAUM'S COMPLETE HANDBOOK OF PRACTICAL ELECTRONIC REFERENCE DATA

Third Edition

CHAPTER 1

UNITS, CONSTANTS, FREQUENCY SPECTRUM, AND CONVERSION FACTORS

1.0 INTRODUCTION

Like any technical field, electronics is based on standard units by which different characteristics are measured. Some of these units are common to other branches of physics and some are unique to electronics, even though they are named after the physicists who first measured and described the particular quantities.

1.1 UNITS

	Electron:	*Proton:*
Mass	$m_e = 9.1 \times 10^{-31}$ kg	$m_p = 1.67 \times 10^{-27}$ kg
Radius	$r_e = 1.4 \times 10^{-15}$ m	$r_p = 10^{-15}$ m
Charge	$-e = 1.602 \times 10^{-19}$ coulombs	$+e = 1.602 \times 10^{-19}$ coulombs

Current: 1 Ampere = 1 coulomb/second
(Andre Marie Ampere, 1775–1836)   Symbol: A

Charge: 1 Coulomb = 6.25×10^{18} electrons
(Charles Augustin de Coulomb, 1736–1806)   Symbol: C

Potential: 1 Volt = 1 joule/coulomb (MKS)
                                                           Voltage = potential energy per unit
                                                           charge
(Alessandro Volta, 1745–1827)   Symbol: V

Energy: $1 \text{ Joule} = \dfrac{1 \text{ kilogram-meter}^2}{\text{second}^2}$

(James Prescott Joule, 1818–1889)   Symbol: J

Power: 1 Watt = 1 volt × 1 ampere
(James Watt, 1736–1819)   Symbol: W

Resistance: $1 \text{ Ohm} = \dfrac{1 \text{ volt}}{1 \text{ ampere}}$

(George Simon Ohm, 1787–1854)   Symbol: Ω

Capacitance: 1 Farad = 1 coulomb/volt
(Michael Faraday, 1791–1867)   Symbol: F

Inductance: 1 Henry = 1 volt-second/ampere
(Joseph Henry, 1797–1878)   Symbol: H

Magnetic Flux Density: $1 \text{ Tesla} = \dfrac{1 \text{ weber}}{\text{meter}^2}$

(Nikola Tesla, 1857–1943)   Symbol: T

Magnetic Field Intensity: 1 Oersted = 79.6 amperes/meter
(Hans Christian Oersted, 1777–1851)   Symbol: Oe

Magnetic Flux: 1 Weber = 1 volt × 1 second
(Wilhelm Eduard Weber, 1804–1891)   Symbol: Wb

Electrical Field: E = newton/coulomb

The Metric System

The metric system is based on the meter as the unit of length, the gram as the unit of mass and the liter, and the volume of 1 kilogram of water as the unit of volume. The metric temperature scale, named after Celsius, or sometimes referred to as Centigrade, bases its 0-degree point upon the temperature of melting ice and its 100-degree point on the temperature of boiling water.

The trend in scientific measurements in the United States and England, the only two major holdouts from the metric system, is now strongly toward the use of this system. In accordance with the recommendations accepted by the IEEE, the MKS (meter-kilogram-second)

system is to be used in preference to the CGS (centimeter-gram-second) version of the metric system. For this reason, although both systems will be described in this chapter, in subsequent chapters the MKS system will be used exclusively.

The metric system uses a series of multipliers, all powers of ten, which, together with Greek and Latin terminology, indicate the actual size of its units. A kilogram, for example, is 10^3 or 1000 grams. The same multipliers, prefixes, and symbols used for the metric system in measurement are also applied to electrical quantities. Table 1-1 lists multiples and sub-multiples, with their prefixes and symbols.

Multiple	Prefixes	Symbols
10^{18}	exa	E
10^{15}	peta	P
10^{12}	tera	T
10^{9}	giga	G
10^{6}	mega	M
10^{3}	kilo	k
10^{2}	hecto	h
10	deka	dk
10^{-1}	deci	d
10^{-2}	centi	c
10^{-3}	milli	m
10^{-6}	micro	μ
10^{-9}	nano	n
10^{-12}	pico	p
10^{-15}	femto	f
10^{-18}	atto	a

Table 1-1.   Prefixes and multiples.

Systems of Measurement

In physics, and in particular in electronics, three basic systems of measurement are in common use. The English system, used in the United States as well, is based on the ancient foot, pound, and second.

The other two systems, the MKS and the CGS systems, are both based on the metric system of meters, grams, and seconds, in different multiplier arrangements. A comparison of the three systems is presented in Table 1-2. A conversion table to facilitate conversion between the three systems follows at the end of this chapter (see Table 1-7).

Decibels

The ratios of power, voltage, and current are frequently expressed in terms of the decibel. By definition, decibels are equal to ten times the

Quantity	Symbol	Units		
		MKS	**Cgs**	**English**
1. Length	l	Meter	Centimeter	Foot
2. Mass	m, M	Kilogram	Gram	Slug
3. Time	t	Second	Second	Second
4. Charge	Q, q	Coulomb	Statcoulomb, abcoulomb	—
5. Linear velocity	v	Meter/second	Centimeter/second	Foot/second
6. Angular velocity	ω	Radian/second	Radian/second	Radian/second
7. Momentum	p	Kilogram-meter/ second	Gram-centimeter/ second	Slug-foot/ second
8. Linear acceleration	a	Meter/second^2	Centimeter/second^2	Foot/second^2
9. Angular acceleration	α	Radian/second^2	Radian/second^2	Radian/second^2
10. Force	F	Newton	Dyne	Pound
11. Work	W	Newton-meter	Dyne-centimeter	Foot-pound
12. Energy	E, T, H, V	Joule	Erg	Foot-pound
13. Power	P	Watt	Erg/second	Horsepower
14. Gravitational field	G	Newton/kilogram	Dyne/gram	Pound/slug
15. Current	I, i	Ampere	Statampere, abampere	—
16. Electric field	$\mathscr{E}$	Volt/meter	Statvolt/centimeter	—
17. Potential	U	Volt	Statvolt	—
18. Voltage	V	Volt	Statvolt	—
19. Frequency	f	1/second	1/second	1/second
20. Period	T	Second	Second	Second
21. Permittivity of space	ϵ_0	Coulomb^2/ newton-meter^2	—	—
22. Conductivity	σ	Mho/meter	—	—
23. Resistivity	ρ	Ohm-meter	Statohm-centimeter	—
24. Resistance	R	Ohm	Statohm	
25. Conductance	G	Mho	—	—
26. Capacitance	C	Farad	—	—
27. Electric susceptibility	η	Couolmb/ volt-meter	—	—
28. Permittivity	ϵ	Farad/meter	—	—

Table 1-2.   MKS, CGS, and English systems.

(Lane K. Branson, *Introduction to Electronics*, © 1967. Reprinted by permission of Prentice-Hall, Inc.)

Quantity	Symbol	Units		
		MKS	**Cgs**	**English**
29. Pole strength	P, p	Ampere-meter	Abampere-centimeter	—
30. Magnetic induction	B	Weber/meter^2	Gauss	—
31. Magnetic-field intensity	H	Ampere/meter	Oersted	—
32. Permeability	μ	Kilogram-meter/coulomb^2	—	—
33. Magnetic flux	Φ	Weber	Maxwell	—
34. Relative permeability	K_m	(Unitless)	—	—
35. Magnetic susceptibility	χ	(Unitless)	—	—
36. Inductance	L	Henry	—	—
37. Mutual inductance	M	Henry	—	—
38. Mobility	μ	Meter^2/volt-second	—	—

Table 1-2. Continued

log of the ratio of power:

$$dB = 10 \log \frac{P_2}{P_1}$$

where P_2 is the output power and P_1 is the input power and the log is taken to the base ten. Logarithms are treated in Chapter 23.

Similarly, decibels are equal to 20 times the log of the voltage or current:

$$dB = 20 \log \frac{V_2}{V_1} = 20 \log \frac{I_2}{I_1}$$

where V_2 and V_1 or I_2 and I_1 are the respective voltage and current ratios.

An example for power gain would be a power ratio of 10:1, meaning that P_2 is ten times as great as P_1. In this case:

$$10 \log \frac{10}{1} = 10 \text{ dB}$$

Power ratio	Voltage and current ratio	Decibels	Power ratio	Voltage and current ratio	Decibels	Power ratio	Voltage and current ratio	Decibels
1.0233	1.0116	0.1	2.2387	1.4962	3.5	158.49	12.589	22.0
1.0471	1.0233	0.2	2.5119	1.5849	4.0	251.19	15.849	24.0
1.0715	1.0351	0.3	2.8184	1.6788	4.5	398.11	19.953	26.0
1.0965	1.0471	0.4	3.1623	1.7783	5.0	630.96	25.119	28.0
						1000.0	31.623	30.0
1.1220	1.0593	0.5	3.5481	1.8836	5.5	1584.9	39.811	32.0
1.1482	1.0715	0.6	3.9811	1.9953	6.0	2511.9	50.119	34.0
1.1749	1.0839	0.7	5.0119	2.2387	7.0	3981.1	63.096	36.0
1.2023	1.0956	0.8	6.3096	2.5119	8.0	6309.6	79.433	38.0
						10^4	100.000	40.0
1.2303	1.1092	0.9	7.9433	2.8184	9.0	$10^4 \times 1.5849$	125.89	42.0
1.2589	1.1220	1.0	10.0000	3.1623	10.0	$10^4 \times 2.5119$	158.49	44.0
1.3183	1.1482	1.2	12.589	3.5481	11.0	$10^4 \times 3.9811$	199.53	46.0
1.3804	1.1749	1.4	15.849	3.9811	12.0	$10^4 \times 6.3096$	251.19	48.0
						10^5	316.23	50.0
1.4454	1.2023	1.6	19.953	4.4668	13.0	$10^5 \times 1.5849$	398.11	52.0
1.5136	1.2303	1.8	26.119	5.0119	14.0	$10^5 \times 2.5119$	501.19	54.0
1.5849	1.2589	2.0	31.623	5.6234	15.0	$10^5 \times 3.9811$	630.96	56.0
1.6595	1.2882	2.2	39.811	6.3096	16.0	$10^5 \times 6.3096$	794.33	58.0
						10^6	1,000.00	60.0
1.7378	1.3183	2.4	50.119	7.0795	17.0	10^7	3,162.3	70.0
1.8197	1.3490	2.6	63.096	7.9433	18.0	10^8	10,000.0	80.0
1.9055	1.3804	2.8	79.433	8.9125	19.0	10^9	31,623	90.0
1.9953	1.4125	3.0	100.00	10.0000	20.0	10^{10}	100,000	100.0

Table 1-3. Power, voltage, and decibels.

(Harry E. Thomas, *Handbook for Electronic Engineers and Technicians*, © 1965. Reprinted by permission of Prentice-Hall, Inc.)

For a voltage gain of ten, meaning that V_2 is ten times as great as V_1:

$$20 \log \frac{10}{1} = 20 \text{ dB}$$

Table 1-3 permits conversion of power and voltage ratios into decibels, and vice versa.

1.2 CONSTANTS USED IN ELECTRONICS

Table 1-4 lists a number of frequently used constants that occur in a variety of electronics formulas.

Two charts providing useful numerical data and the properties of free space follow.

Usual Symbol	Denomination	Value and Units
$F'=Ne/c$	Faraday's constant (physical scale)	9652.19 ± 0.11 emu (g mole)$^{-1}$
N	Avogadro's constant (physical scale)	$(6.02486\pm0.00016)\times10^{23}$ (g mole)$^{-1}$
h	Planck's constant	$(6.62517\pm0.00023)\times10^{-27}$ erg second
m	Electron rest mass	$(9.1083\pm0.0003)\times10^{-28}$ g.
e	Electronic charge	$(4.80286\pm0.00009)\times10^{-10}$ esu
$e'=e/c$		$(1.60206\pm0.00003)\times10^{-20}$ emu
e/m	Charge-to-mass ratio of electron	$(5.27305\pm0.00007)\times10^{17}$ esu g^{-1}
$e'/m=e/(mc)$		$(1.75890\pm0.00002)\times10^{7}$ emu g^{-1}
c	Velocity of light in vacuum†	$299\,793.0\pm0.3$ km second^{-1}
$h/(mc)$	Compton wavelength of electron	$(24.2626\pm0.0002)\times10^{-11}$ cm
$a_0=h^2/(4\pi^2me^2)$	First Bohr electron-orbit radius	$(5.29172\pm0.00002)\times10^{-9}$ cm
$\sigma=\dfrac{\pi^2\,k^4\,8\pi^3}{60\;c^2\;h^3}$	Stefan-Boltzmann constant	$(0.56687\pm0.00010)\times10^{-4}$ erg cm^{-2} deg^{-4} second^{-1}
$\lambda_{\max}\,T$	Wien displacement-law constant	(0.289782 ± 0.000013) cm deg
$\mu_0=he/(4\pi mc)$	Bohr magneton	$(0.92731\pm0.00002)\times10^{-20}$ erg gauss^{-1}
Nm	Atomic mass of the electron (physical scale)	$(5.48763\pm0.00006)\times10^{-4}$
M_p/Nm	Ratio, proton mass to electron mass	1836.12 ± 0.02
$E_0=e\cdot10^8/c$	Energy associated with 1 eV	$(1.60206\pm0.00003)\times10^{-12}$ erg
$(mc^2/E_0)\times10^{-6}$	Energy equivalent of electron mass	(0.510976 ± 0.000007) MeV
$k=R_0/N$	Boltzmann's constant	$(1.38044\pm0.00007)\times10^{-16}$ erg deg^{-1}
R_∞	Rydberg wave number for infinite mass	$(109\,737.309\pm0.012)$ cm^{-1}
H	Hydrogen atomic mass (physical scale)	1.008142 ± 0.000003
R_0	Gas constant per mole (physical scale)	$(8.31696\pm0.00034)\times10^{7}$ erg mole^{-1} deg^{-1}
V_0	Standard volume of perfect gas (physical scale)	$(22\,420.7\pm0.6)$ cm^3 atm mole^{-1}

Table 1-4. Constants used in electronic formulas.

(From *Reference Data for Radio Engineers*, 5th Edition, by H. P. Westman, © 1968. Used with permission of Howard W. Sams & Co., Inc.)

USEFUL NUMERICAL DATA

1 cubic foot of water at 4°C (weight)	62.43 lb
1 foot of water at 4°C (pressure)	0.4335 lb/in.^{2}

Velocity of light in vacuum, c

186 280 mi/second = 2.998×10^{10} cm/second

Velocity of sound in dry air at 20°C, 760 mm Hg

1127 ft/second

Degree of longitude at equator

68.703 statute miles, 59.661 nautical miles

Acceleration due to gravity at sea level, 40° latitude, g

32.1578 ft/second^2

$(2g)^{1/2}$

8.020

1 inch of mercury at 4°C

1.132 ft water = 0.4908 lb/in^2

Base of natural logs ϵ

2.718

1 radian

$180° \div \pi = 57.3°$

360 degrees

2π radians

π

3.1416

Sin 1'

0.00029089

Arc 1°

0.01745 radian

Side of square

$0.707 \times$ (diagonal of square)

(From *Reference Data for Radio Engineers*, 5th Edition, by H. P. Westman, © 1968. Used with permission of Howard W. Sams & Co., Inc.)

PROPERTIES OF FREE SPACE

Velocity of light = c = $1/(\mu_v \epsilon_v)^{1/2}$ = 2.998×10^8 meters per second

= 186,280 miles per second

= 984×10^6 feet per second.

Permeability = μ_v = $4\pi \times 10^{-7}$ = 1.257×10^{-6} henry per meter.

Permittivity = ϵ_v = 8.85×10^{-12} $\approx$ $(36\pi \times 10^9)^{-1}$ farad per meter.

Characteristic impedance = Z_0 = $(\mu_v/\epsilon_v)^{1/2}$ = 376.7 $\approx$ 120π ohms.

(From *Reference Data for Radio Engineers*, 5th Edition, by H. P. Westman, © 1968. Used with permission of Howard W. Sams & Co., Inc.)

1.3 FREQUENCY SPECTRUM

Frequency is defined as the variation per unit time. Thus, if a particular process varies ten times within a second, its frequency is said to be ten cycles per second or, in the preferred electrical nomenclature, ten Hertz. The Hertz, Hz, was named after the German physicist Heinrich Hertz, who proved the existence of electromagnetic waves and the concept of their alternation or frequency. In the electronics field, two types of time-variant energies are generally considered: the acoustical and the electromagnetic. Acoustical frequencies are those in which air vibrates, and this vibration is detected by the human ear. Electrical signals, radio-waves, visible light, ultraviolet, and X rays are considered electromagnetic energy.

The frequency spectrum can be broadly divided into the following bands:

1. Acoustical band, up to about 20,000 Hertz or 20 kilohertz, 20 kHz

2. The ultrasonic band, from 20 kHz to about 100 kHz. Medical ultrasonic equipment operates up to 25 MHz.

3. Radio frequencies, from 100 kHz up to about 1 terahertz or about 10^{12} Hertz. This band is divided into many subgroups, ranging from the so-called very low frequency (vlf) to the so-called extremely high frequencies (ehf).

4. The infrared band is no longer considered in terms of Hertz but rather in terms of wavelengths, which range from 300 to 3 microns or micrometers in wavelength.
(Frequency-wavelength conversion figures can be found in Table 1-6).

5. The visible spectrum, ranging in wavelength from 0.7 to 0.4 microns.

6. The ultraviolet, X-ray, and gamma-ray range, which is considered to go from 0.4 to 3×10^{-6} microns.

Table 1-5 lists the frequency bands by band number and nomenclature.

Band Number*	Frequency Range	Metric Subdivision	Adjectival Designation	
2	30 to 300 hertz	Megametric waves	ELF	Extremely low frequency
3	300 to 3000 hertz	————	VF	Voice frequency
4	3 to 30 kilohertz	Myriametric waves	VLF	Very-low frequency
5	30 to 300 kilohertz	Kilometric waves	LF	Low frequency
6	300 to 3000 kilohertz	Hectometric waves	MF	Medium frequency
7	3 to 30 megahertz	Decametric waves	HF	High frequency
8	30 to 300 megahertz	Metric waves	VHF	Very-high frequency
9	300 to 3000 megahertz	Decimetric waves	UHF	Ultra-high frequency
10	3 to 30 gigahertz	Centimetric waves	SHF	Super-high frequency
11	30 to 300 gigahertz	Millimetric waves	EHF	Extremely high frequency
12	300 to 3000 gigahertz or 3 terahertz	Decimillimetric waves	—	—

* "Band Number N" extends from 0.3×10^N to 3×10^N hertz. The upper limit is included in each band; the lower limit is excluded.

Table 1-5.   Frequency bands.

(From *Reference Data for Radio Engineers*, 5th Edition, by H. P. Westman, © 1968. Used with permission of Howard W. Sams & Co., Inc.)

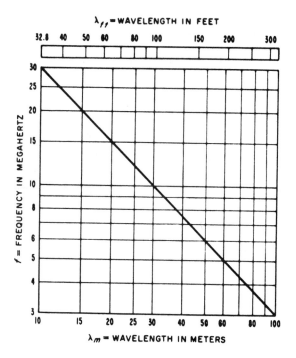

Table 1-6. Wavelength frequency conversion.

For Frequencies in Megahertz from		Multiply f by	Multiply λ by
0.03–	0.3	0.01	100
0.3 –	3.0	0.1	10
3.0 –	30	1.0	1.0
30 –	300	10	0.1
300 –	3 000	100	0.01
3 000 –	30 000	1 000	0.001
30 000 –	300 000	10 000	0.0001

(From *Reference Data for Radio Engineers*, 5th Edition, by H. P. Westman, © 1968. Used with permission of Howard W. Sams & Co., Inc.)

1.4 CONVERSION TABLE (TABLE 1-7)

To use this conversion table, find the original units in either the first or second column, then find the new units in the other column and use the multiplier as indicated.

To convert	Into	Multiply by	Conversely multiply by
Ampere-hours	Coulombs	3,600	2.778×10^{-4}
Amperes per sq. cm	Amperes per sq. inch	6.452	.155
Ampere turns	Gilberts	1.257	.7958
Ampere turns per cm	Ampere turns per inch	2.54	.3937
Btu (British thermal unit)	Foot-pounds	778.3	1.285×10^{-3}
Btu	Joules	1,054.8	9.48×10^{-4}
Btu	Kilogram-calories	.252	3.969
Btu	Horsepower-hours	3.929×10^{-4}	2,545
Centigrade	Fahrenheit	$(C^\circ \times 9/5) + 32$	$(F^\circ - 32) \times 5/9.$
Circular mils	Square centimeters	5.067×10^{-6}	1.973×10^{5}
Circular mils	Square mils	.7854	1.273.
Cubic inches	Cubic centimeters	16.39	6.102×10^{-2}
Cubic inches	Cubic feet	5.785×10^{-4}	1,728
Cubic inches	Cubic meters	1.639×10^{-5}	6.102×10^{4}
Cubic meters	Cubic feet	35.31	2.832×10^{-2}
Cubic meters	Cubic yards	1.308	.7646.
Degrees (angle)	Radians	1.745×10^{-2}	57.3
Dynes	Pounds	2.248×10^{-6}	4.448×10^{5}
Ergs	Foot-pounds	7.367×10^{-8}	1.356×10^{7}
Feet	Centimeters	30.48	3.281×10^{-2}
Foot-pounds	Horsepower-hours	5.05×10^{-7}	1.98×10^{6}
Foot-pounds	Kilogram-meters	.1383	7.233
Foot-pounds	Kilowatt-hours	3.766×10^{-7}	2.655×10^{6}
Gauss	Lines per sq. inch	6.452	.155
Grams	Dynes	980.7	1.02×10^{-3}
Grams	Ounces (avoirdupois)	3.527×10^{-2}	28.35
Grams per cm	Pounds per inch	5.6×10^{-3}	178.6
Grams per cubic cm	Pounds per cu. inch	3.613×10^{-2}	27.68
Grams per sq. cm	Pounds per sq. foot	2.0481	.4883
Horsepower (550 ft.-lb. per sec.)	Foot-lb. per minute	3.3×10^{4}	3.03×10^{-5}
Horsepower (550 ft.-lb. per sec.)	Btu per minute	42.41	2.357×10^{-2}
Horsepower (550 ft.-lb. per sec.)	Kg-calories per minute	10.69	9.355×10^{-2}
Horsepower (Metric) (542.5 ft.-lb. per sec.)	Horsepower (550 ft.-lb. per sec.)	.9863	1.014
Inches	Centimeters	2.54	.3937
Inches	Mils	1,000	.001
Joules	Foot-pounds	.7376	1.356.
Joules	Ergs	10^{7}	10^{-7}
Kilogram-calories	Kilojoules	4.186	.2389
Kilograms	Pounds (avoirdupois)	2.205	.4536
Kg per sq. meter	Pounds per sq. foot	.2048	4.882
Kilometers	Feet	3,281	3.048×10^{-4}
Kilowatt-hours	Btu	3,413	2.93×10^{-4}
Kilowatt-hours	Foot-pounds	2.655×10^{6}	3.766×10^{-7}
Kilowatt-hours	Joules	3.6×10^{6}	2.778×10^{-7}
Kilowatt-hours	Kilogram-calories	860	1.163×10^{-3}

Table 1-7.   Conversion factors.

(Harry E. Thomas, *Handbook for Electronic Engineers and Technicians,* © 1965. Reprinted by permission of Prentice-Hall, Inc.)

To convert	Into	Multiply by	Conversely multiply by
Kilowatt-hours	Kilogram-meters	3.671×10^5	2.724×10^{-6}
Liters	Cubic meters	.001	1,000
Liters	Cubic inches	61.02	1.639×10^{-2}
Liters	Gallons (liq. US)	.2642	3.785
Liters	Pints (liq. US)	2.113	.4732
Meters	Yards	1.094	.9144
Meters per min	Feet per min	3.281	.3048
Meters per min	Kilometers per hr	.06	16.67
Miles (nautical)	Kilometers	1.853	.5396
Miles (statute)	Kilometers	1.609	.6214
Miles per hr	Kilometers per min	2.682×10^{-2}	37.28
Miles per hr	Feet per minute	88	1.136×10^{-2}
Miles per hr	Kilometers per hr	1.609	.6214
Poundals	Dynes	1.383×10^4	7.233×10^{-5}
Poundals	Pounds (avoirdupois)	3.108×10^{-2}	32.17
Sq inches	Circular mils	1.273×10^6	7.854×10^{-7}
Sq inches	Sq centimeters	6.452	.155
Sq feet	Sq meters	9.29×10^{-2}	10.76
Sq miles	Sq yards	3.098×10^6	3.228×10^{-7}
Sq miles	Sq kilometers	2.59	.3861
Sq millimeters	Circular mils	1,973	5.067×10^{-4}
Tons, short (avoir 2,000 lb.)	Tonnes (1,000 Kg.)	.9072	1.102
Tons, long (avoir 2,240 lb.)	Tonnes (1,000 Kg.)	1.016	.9842
Tons, long (avoir 2,240 lb.)	Tons, short (avoir 2,000 lb)	1.120	.8929
Watts	Btu per min	5.689×10^{-2}	17.58
Watts	Ergs per sec	10^7	10^{-7}
Watts	Ft-lb per minute	44.26	2.26×10^{-2}
Watts	Horsepower (550 ft-lb per sec.)	1.341×10^{-3}	745.7
Watts	Horsepower (metric) (542.5 ft-lb per sec.)	1.36×10^{-3}	735.5
Watts	Kg-calories per min	1.433×10^{-2}	69.77

Table 1-7.   Continued

Example 1:   Convert 200 cubic inches into cubic centimeters.
    Step 1: Cubic inches appear in column 1 and cubic centimeters in column 2.
    Step 2: Multiply 200 by 16.39 as directed by column 3.
    Result: 200 cubic inches = 3278 cubic centimeters.
Example 2:   Convert 75° Fahrenheit into Centigrade.
    Step 1: Fahrenheit appears in column 2 and Centigrade in column 1.
    Step 2: Column 4 directs to subtract 32 from the °F and multiply by 5/9.
    Result: 75° Fahrenheit = 43 × 5/9 = 23.9° C.

BIBLIOGRAPHY—CHAPTER 1

DEEM, W., *Electronics Math*. Englewood Cliffs, NJ, Prentice-Hall, Inc., 1981.

GOTHMAN, W., *Contemporary Mathematics for Electronics*. Englewood Cliffs, NJ, Prentice-Hall, Inc., 1981.

HEISLER, S. I., *The Wiley Engineer's Desk Reference*. New York, NY, John Wiley & Sons, Inc., 1984.

IEEE Standard Dictionary of Electrical and Electronics Terms. New York, NY, Institute of Electrical and Electronics Engineers, 1984.

Index of EIA and JEDEC Standards and Engineering Publications. EIA, Eng. Dept., Washington DC.

Reference Data for Engineers: Radio, Electronics, Computer, and Communications. Indianapolis, IN, Howard W. Sams & Co., Inc., 1985.

CHAPTER 2

PROPERTIES OF MATERIALS AND WIRE TABLES

2.1 THE ELEMENTS

Table 2-1 of the basic elements contains the name, symbol, and a number of other important characteristics of each element. The atomic number Z indicates the number of protons (positively charged particles) per atom. The mass number $Z + N$ describes the number of protons plus the number of neutrons, and is therefore an indication of the total mass of the nucleus. The atomic weight of each element is, as apparent from the table, closely related to the mass number but differs from it in that it is calculated as the weight per atom at sea level. The thermionic, photoelectric, and contact electron work functions are given in electron volts and describe the energy that must be supplied to move an electron across the surface barriers of the element. This energy may be in the form of heat, light, or by contact with a dissimilar metal. The next-to-

14

	Symbol	Atomic Number Z	Mass Number $Z+N$	Atomic Weight	Electron Work Function			Electrochemical Equivalent	
					Thermionic	Photoelectric	Contact	Valence* Involved	Amp-Hours per Gram
Actinium	Ac	89	227	227				3	0.35
Aluminum	Al	13	27	26.98		4.08	3.38	3	2.98
Americium	Am	95	243						
Antimony	Sb	51	121	121.76		4.1	4.14	5	1.1
Argon	Ar or A	18	40	39.948				n	0.67
Arsenic	As	33	75	74.92		5.11		5	1.79
Astatine	At	85	210						
Barium	Ba	56	138	137.34	2.11	2.48	1.73	2	0.39
Berkelium	Bk	97	249						
Beryllium	Be	4	9	9.012		3.92	3.10	2	5.94
Bismuth	Bi	83	209	208.98		4.25	4.17	5	0.64
Boron	B	5	11	10.81		4.5		3	7.43
Bromine	Br	35	79–81	79.91				1	0.335
Cadmium	Cd	48	114–112	112.40		4.07	4.0	2	0.477
Calcium	Ca	20	40	40.08	2.24	2.706	3.33	2	1.337
Californium	Cf	98	251						
Carbon	C	6	12	12.011	4.34	4.81		4	8.93
Cerium	Ce	58	140	140.12	2.6	2.84		3	0.574
Cesium	Cs	55	133	132.905	1.81	1.92		1	0.2
Chlorine	Cl	17	35	35.453				1	0.756
Chromium	Cr	24	52	51.996	4.60	4.37	4.38	3	1.546
Cobalt	Co	27	59	58.93	4.40	4.20	4.21	2	0.91
Copper	Cu	29	63	63.54	4.26	4.18	4.46	2	0.84
Curium	Cm	96	247						
Dysprosium	Dy	66	164–162–163	162.50				3	0.495
Einsteinium	Es or E	99	254	254					
Erbium	Er	68	166–168–167	167.26				3	0.48
Europium	Eu	63	153–151	151.96				3	0.53
Fermium	Fm	100	253	253					
Fluorine	F	9	19	18.998				1	1.41
Francium	Fr	87	223	223					
Gadolinium	Gd	64	158–160–156	157.25				3	0.513
Gallium	Ga	31	69–71	69.72	4.12		3.80	3	1.15
Germanium	Ge	32	74–72–70	72.59		4.5	4.5	4	1.48
Gold	Au	79	197	196.967	4.32	4.82	4.46	3	0.41
Hafnium	Hf	72	180–178	178.49	3.53			4	0.600
Helium	He	2	4	4.003				n	6.698
Holmium	Ho	67	165	164.93				3	0.488
Hydrogen	H	1	1	1.0080				1	26.59
Indium	In	49	115	114.82				3	0.700
Iodine	I	53	127	126.904		6.8		1	0.211
Iridium	Ir	77	193	192.2	5.3		4.57	4	0.555
Iron	Fe	26	56	55.847	4.25	4.33	4.40	3	1.440
Krypton	Kr	36	84	83.80				n	0.32
Lanthanum	La	57	139	138.91	3.3			3	0.579
Lead	Pb	82	208–206	207.21		4.05	3.94	4	0.517
Lithium	Li	3	7	6.940		2.35	2.49	1	3.862
Lutetium	Lu	71	175	174.99				3	0.46
Magnesium	Mg	12	24	24.32		3.68	3.63	2	2.204
Manganese	Mn	25	55	54.94	3.83	3.76	4.14	4	1.952
Mendelevium	Md or Mv	101	256	256					
Mercury	Hg	80	202–200–119	200.59		4.53	4.50	2	0.267
Molybdenum	Mo	42	98–96–95	95.94	4.20	4.25	4.28	6	1.67

* n = nonvalent.

Table 2-1. Basic elements.

(From *Reference Data for Radio Engineers*, 5th Edition, by H. P. Westman, © 1968. Used with permission of Howard W. Sams & Co., Inc.)

	Symbol	Atomic Number Z	Mass Number $Z+N$	Atomic Weight	Electron Work Function			Electrochemical Equivalent	
					Thermionic	Photoelectric	Contact	Valence* Involved	Amp-Hours per Gram
Neodymium	Nd	60	142–144–146	144.24	3.3			3	0.557
Neon	Ne	10	20	20.183				n	1.33
Neptunium	Np	93	237	237					
Nickel	Ni	28	58	58.71	5.03	5.01	4.96	2	0.913
Niobium	Nb	41	93	92.91					
Nitrogen	N	7	14	14.007				5	9.57
Osmium	Os	76	192–190	190.2			4.55	4	0.56
Oxygen	O	8	16	16.00				2	3.35
Palladium	Pd	46	108–106–105	106.4	4.99	4.97	4.49	4	1.005
Phosphorus	P	15	31	30.974				5	4.33
Platinum	Pt	78	195–194	195.09	5.32	5.22	5.36	4	0.549
Plutonium	Pu	94	242	242					
Polonium	Po	84	209	210				6	0.766
Potassium	K	19	39	39.102		2.24	1.60	1	0.685
Praseodymium	Pr	59	141	140.907		2.7		3	0.571
Promethium	Pm	61	145	145					
Protactinium	Pa	91	231	231				5	0.580
Radium	Ra	88	226	226.05				2	0.237
Radon	Rn	86	222	222				n	0.121
Rhenium	Re	75	187	186.2		5.1	5.0	7	1.007
Rhodium	Rh	45	103	102.905	4.57	4.80	4.52	4	1.042
Rubidium	Rb	37	85	85.47		2.09		1	0.314
Ruthenium	Ru	44	102–104–101	101.07			4.52	4	1.054
Samarium	Sm	62	152–154–147	150.35	3.2			3	0.535
Scandium	Sc	21	45	44.956				3	1.783
Selenium	Se	34	80–78	78.96		4.8	4.42	6	2.037
Silicon	Si	14	28	28.086	3.59	4.52	4.2	4	3.821
Silver	Ag	47	107–109	107.870	3.56	4.73	4.44	1	0.248
Sodium	Na	11	23	22.99		2.28	1.9	1	1.166
Strontium	Sr	38	88	87.62		2.74		2	0.612
Sulfur	S	16	32	32.064				6	5.01
Tantalum	Ta	73	181	180.95	4.19	4.14	4.1	5	0.741
Technetium	Tc	43	99	99					
Tellurium	Te	52	130–128	127.60		4.76	4.70	6	1.260
Terbium	Tb	65	159	158.924				3	0.505
Thallium	Tl	81	205	204.37		3.68	3.84	3	0.393
Thorium	Th	90	232	232.038				4	0.462
Thulium	Tm	69	169	168.934				3	0.475
Tin	Sn	50	120–118	118.69		4.38	4.09	4	0.903
Titanium	Ti	22	48	47.90	3.95	4.06	4.14	4	2.238
Tungsten	W	74	184–186–182	183.85	4.52	4.49	4.38	6	0.874
Uranium	U	92	238	238.03	3.27	3.63	4.32	6	0.676
Vanadium	V	23	51	50.94	4.12	3.77	4.44	5	2.63
Xenon	Xe	54	132–129	131.30				n	0.204
Ytterbium	Yb	70	174–172	173.04				3	0.465
Yttrium	Y	39	88–91	88.905				3	0.904
Zinc	Zn	30	64–68	65.87		3.73	3.78	2	0.820
Zirconium	Zr	40	90–94–92	91.22	4.21	3.82	3.60	4	1.175

Table 2-1. Continued

last column, "Valence," is used in chemistry to express the relative electrochemical values of the particular element.

2.2 METALS AND ALLOYS

One of the important electrical characteristics of materials is their resistance to the flow of electric current. This resistance is a property of the particular material and its geometry. It is generally expressed as:

$$R = \rho \frac{L}{A} \text{ ohms}$$

The resistance R equals the resistivity (Greek letter rho, ρ) multiplied by the geometry of the material; that is, length over the area in centimeters. From this consideration, it is apparent that the resistivity (rho) is expressed in ohms per centimeter. Depending on the resistivity of a material, it can be classified into conductors, semiconductors, or insulators, as illustrated in Table 2-2.

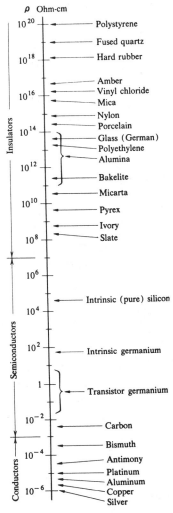

Table 2-2.   Relative resistivity scale.

(Lane K. Branson, *Introduction to Electronics*, © 1967. Reprinted by permission of Prentice-Hall, Inc.)

The resistivity for commonly used conducting and insulating materials is listed in Table 2-3.

Material	Resistivity, ohm/centimeter
Aluminum	2.83×10^{-6}
Antimony	4.17×10^{-5}
Brass	7×10^{-6}
Calcium	4.6×10^{-6}
Carbon	3.5×10^{-3}
Copper, annealed	1.72×10^{-6}
Copper, hard-drawn	1.77×10^{-6}
Copper, pure	1.70×10^{-6}
Gold	2.44×10^{-6}
Graphite	8×10^{-4}
Iron	1×10^{-5}
Lead	2.2×10^{-5}
Magnesium	4.6×10^{-6}
Mercury	9.67×10^{-5}
Nickel	6.8×10^{-6}
Platinum	1×10^{-5}
Silver	1.63×10^{-6}
Tin	1.15×10^{-5}
Titanium	3.2×10^{-6}
Tungsten	5.51×10^{-6}
Zinc	5.8×10^{-6}
Amber	5×10^{16}
Alumina	$10^{11} \longleftrightarrow 10^{14}$
Mica	9×10^{15}
Nylon	4×10^{14}
Paraffin	$10^{15} \longleftrightarrow 10^{19}$
Plate glass	2×10^{13}
Porcelain	$10^{12} \longleftrightarrow 10^{14}$
Rubber	2×10^{15}
Shellac	2×10^{9}
Steatite	$10^{13} \longleftrightarrow 10^{15}$
Vinyl chloride	$10^{11} \longleftrightarrow 10^{16}$

Table 2-3. Resistivity table.

(Lane K. Branson, *Introduction to Electronics*, © 1967. Reprinted by permission of Prentice-Hall, Inc.)

Temperature Effects of Metals

All metals have a number of important thermal characteristics, such as thermal conductivity, a thermal coefficient of expansion, and a melting point. The thermal conductivity of a metal describes the amount of power in watts per centimeter, per degree Centigrade that can be conducted in this metal. This information is important in evaluating metals for uses as heat sinks and similar applications. The coefficient of thermal expansion per degree Centigrade refers to the percentage of linear increase in size as the temperature is increased by 1° Centigrade. As the name indicates, the melting point is that temperature in Centigrade at which a particular metal changes from the solid to the liquid state. Table 2-4 shows these characteristics for most common metals.

Metal	Melting Point, °C	Thermal Conductivity	Linear Expansion
Aluminum	659	2.18	22.9
Antimony	271	0.19	8.5-10.8
Bismuth	271	0.084	13.3
Cadmium	321	0.91	29.8
Cobalt	1480	0.69	12.3
Copper	1083	3.94	16.5
Gold	1063	2.96	14.2
Iron	1530	0.79	11.7
Lead	327	0.35	28.7
Magnesium	651	1.55	25.2
Mercury	−38.9	0.084	——
Molybdenum	2500	1.46	4.9
Nickel	1452	0.90	13.3
Palladium	1550	0.70	11.8
Platinum	1755	0.69	8.9
Silver	960	4.08	18.9
Tantalum	2850	0.54	6.6
Tin	232	0.64	23.0
Tungsten	3400	1.99	4.3
Titanium	1800	——	8.5
Zinc	419	1.10	17-39

Table 2-4.   Temperature effects of metals.

(Lane K. Branson, *Introduction to Electronics*, © 1967. Reprinted by permission of Prentice-Hall, Inc.)

2.3 PROPERTIES OF PRACTICAL INSULATING MATERIAL

Two major characteristics are important in the use of insulating materials—the dielectric constant and the dielectric strength. The dielectric constant of air is nominally 1.00, and that of all other solid insulators is considerably greater. The dielectric constant is used in various electrical formulas concerning transmission lines and antennas, where electromagnetic waves are propagated in a medium other than free air.

The dielectric strength describes its insulating property and is expressed in volts per 0.001 inch or volts/mil. The dielectric strength of hard rubber, for example, is given as 470, which means that a 0.001-inch thick insulation of hard rubber will withstand 470 volts. A hard rubber strip of 0.003-inch thickness will not break down at less than 1410 volts. Table 2-5 contains the dielectric constants and strengths for most commonly used insulating materials.

2.4 PROPERTIES OF MAGNETIC MATERIAL

Iron, steel, and its many alloys have magnetic characteristics that are analogous to the electrical properties of conductors and insulators. The relative permeability "mu" (μ_r) of magnetic material is defined as:

$$\mu_r = 1 + \frac{M}{\mu_o H}$$

where M is the flux density, H the magnetic field strength, and μ_o the permeability of free space. As in the case of electrical conduction, three types of magnetic permeability are possible: those materials commonly used in magnetic applications are called *ferromagnetic*, those that do not react to magnetic fields *paramagnetic*, and those that are analogous to insulators *diamagnetic*. The values of permeability are expressed in Henry per meter in the MKS system. In ferromagnetic materials, the permeability varies with temperature, pressure, signal frequency, and relative magnetic strength. Relative permeability figures at small magnetization and at maximum for representative materials are listed in Table 2-6.

2.5 WIRE TABLES

Almost all wire used in electronics is copper wire. Table 2-7 shows the important mechanical and electrical characteristics of bare copper wire

Material	Dielectric Constant, K_e	Dielectric Strength, S_D, volts/mil
Mica	2.5–7.0	125–5500
Nylon	3.5	470
Neoprene	4.1	300
Polyethylene	2.3	460
Polyester	4.1–5.2	250–500
Polystyrene moulding	2.4–2.5	500–700
Rubber, hard	2.8	470
Shellac	—	200–600
Vinyl chloride	3.5–4.5	800–1000
Alumina	4.5–8.4	40–160
Porcelain	6.0–8.0	40–400
Porcelain, zircon	7.1–10.5	250–400
Titanates (Bo, Sr, Ca, Mg, Pb)	15–12,000	50–300
Paraffin	2.0–2.5	250
Gases (STP)		
Air	1.000590	
Ammonia	1.0072	
Argon	1.000545	
Carbon dioxide	1.000985	
Helium	1.0000684	
Hydrogen	1.000764	
Neon	1.000127	
Oxygen	1.000523	
Liquids, 25°C		
Ammonia (liquid)	22	
Methanol	32.63	
Benzene	2.283	
Carbon tetrachloride	2.24	
Oil, petroleum	2.13	
Oil, transformer	2.24	
Oil, turpentine	2.23	
Water	78.54	
Solids		
Diamond	16.5	
Quartz	4.34	
Ruby	13.27	
Zircon	12	
Selenium	6.6	

Table 2-5. Dielectric constants.

(Lane K. Branson, *Introduction to Electronics*, © 1967. Reprinted by permission of Prentice-Hall, Inc.)

according to diameter size. The mechanical dimensions of corresponding insulated copper wire is presented in Table 2-8.

	μ_r Maximum	μ_r at Small Magnetization
Ferromagnetic:		
Cobalt	60	60
Nickel	50	50
Cast iron	90	60
Silicon iron	7 000	3 500
Transformer iron	5 500	3 000
Very pure iron	8 000	4 000
Machine steel	450	300
Paramagnetic:		
Aluminum	1.000 000 65	
Beryllium	1.000 000 79	
$MnSO_4$	1.000 100	
NiCl	1.000 040	
Diamagnetic:		
Bismuth	0.999 998 600	
Paraffin	0.999 999 42	
Silver	0.999 999 81	
Wood	0.999 999 50	

Table 2-6.   Permeability.

(From *Reference Data for Radio Engineers*, 5th Edition, by H. P. Westman, © 1968. Used with permission of Howard W. Sams & Co., Inc.)

Gage (AWG) or (B&S)	Diameter, inches			Area	Weight	Length	Resistance at 68° F			Gage (AWG) or (B&S)
	Min.	Nom.	Max.	Circular mils	Pounds per M'	Feet per lb.	Ohms per M'	Feet per ohm	Ohms per lb.	
0000	.4554	.4600	.4646	211600.	640.5	1.561	.04901	20400.	.00007652	0000
000	.4055	.4096	.4137	167800.	507.9	1.968	.06180	16180.	.0001217	000
00	.3612	.3648	.3684	133100.	402.8	2.482	.07793	12830.	.0001935	00
0	.3217	.3249	.3281	105500.	319.5	3.130	.09827	10180.	.0003076	0
1	.2864	.2893	.2922	83690.	253.3	3.947	.1239	8070.	.0004891	1
2	.2550	.2576	.2602	66370.	200.9	4.977	.1563	6400.	.0007778	2
3	.2271	.2294	.2317	52640.	159.3	6.276	.1970	5075.	.001237	3
4	.2023	.2043	.2063	41740.	126.4	7.914	.2485	4025.	.001966	4
5	.1801	.1819	.1837	33100.	100.2	9.980	.3133	3192.	.003127	5
6	.1604	.1620	.1636	26250.	79.46	12.58	.3951	2531.	.004972	6
7	.1429	.1443	.1457	20820.	63.02	15.87	.4982	2007.	.007905	7
8	.1272	.1285	.1298	16510.	49.98	20.01	.6282	1592.	.01257	8
9	.1133	.1144	.1155	13090.	39.63	25.23	.7921	1262.	.01999	9
10	.1009	.1019	.1029	10380.	31.43	31.82	.9989	1001.	.03178	10
11	.08983	.09074	.09165	8234.	24.92	40.12	1.260	794.	.05053	11
12	.08000	.08081	.08162	6530.	19.77	50.59	1.588	629.6	.08035	12
13	.07124	.07196	.07268	5178.	15.68	63.80	2.003	499.3	.1278	13
14	.06344	.06408	.06472	4107.	12.43	80.44	2.525	396.0	.2032	14
15	.05650	.05707	05764	3257.	9.858	101.4	3.184	314.0	.3230	15
16	.05031	.05082	05133	2583.	7.818	127.9	4.016	249.0	.5136	16
17	.04481	.04526	04571	2048.	6.200	161.3	5.064	197.5	.8167	17
18	.03990	.04030	04070	1624.	4.917	203.4	6.385	156.5	1.299	18
19	.03553	.03589	.03625	1288.	3.899	256.5	8.051	124.2	2.065	19
20	.03164	.03196	.03228	1022.	3.092	323.4	10.15	98.5	3.283	20
21	.02818	.02846	.02874	810.1	2.452	407.8	12.80	78.11	5.221	21
22	.02510	.02535	.02560	642.4	1.945	514.2	16.14	61.95	8.301	22
23	.02234	.02257	.02280	509.5	1.542	648.4	20.36	49.13	13.20	23
24	.01990	.02010	.02030	404.0	1.223	817.7	25.67	38.96	20.99	24
25	.01770	.01790	.01810	320.4	.9699	1031.	32.37	30.90	33.37	25
26	.01578	.01594	.01610	254.1	.7692	1300.	40.81	24.50	53.06	26
27	.01406	.01420	.01434	201.5	.6100	1639.	51.47	19.43	84.37	27
28	.01251	.01264	.01277	159.8	.4837	2067.	64.90	15.41	134.2	28
29	.01115	.01126	.01137	126.7	.3836	2607.	81.83	12.22	213.3	29
30	.00993	.01003	.01013	100.5	.3042	3287.	103.2	9.691	339.2	30
31	.008828	.008928	.009028	79.7	.2413	4145.	130.1	7.685	539.3	31
32	.007850	.007950	.008050	63.21	.1913	5227.	164.1	6.095	857.6	32
33	.006980	.007080	.007180	50.13	.1517	6591.	206.9	4.833	1364.	33
34	.006205	.006305	.006405	39.75	.1203	8310.	260.9	3.833	2168.	34
35	.005515	.005615	.005715	31.52	.09542	10480.	329.0	3.040	3448.	35
36	.004900	005000	.005100	25.00	.07568	13210.	414.8	2.411	5482.	36
37	.004353	.004453	.004553	19.83	.06001	16660.	523.1	1.912	8717.	37
38	.003865	.003965	.004065	15.72	.04759	21010.	659.6	1.516	13860.	38
39	.003431	.003531	.003631	12.47	.03774	26500.	831.8	1.202	22040.	39
40	.003045	.003145	.003245	9.888	.02993	33410.	1049.	0.9534	35040.	40
41	.00270	.00280	.00290	7.8400	.02373	42140.	1323.	.7559	55750.	41
42	.00239	.00249	.00259	6.2001	.01877	53270.	1673.	.5977	89120.	42
43	.00212	.00222	.00232	4.9284	.01492	67020.	2104.	.4753	141000.	43
44	.00187	.00197	.00207	3.8809	.01175	85100.	2672.	.3743	227380.	44
45	.00166	.00176	.00186	3.0976	.00938	106600.	3348.	.2987	356890.	45
46	.00147	.00157	.00167	2.4649	.00746	134040.	4207.	.2377	563900.	46

Table 2-7. Bare copper wire.

(Harry E. Thomas, *Handbook for Electronic Engineers and Technicians*, © 1965. Reprinted by permission of Prentice-Hall, Inc.)

AWG of bare copper conductor	Diameter of bare copper conductor		Insulation									
			Enamel & single cotton		Enamel & single silk		Enamel & nylon		Resin & single cotton		Resin & nylon	
			+Diameter over insulation, inches									
	Min.	Max.	Min.	Max.	Min.	Max.	Min.	Max.	Min.	Max.	Min.	Max.
4	.2023	.2063	.2110	.2173	—	—	—	—	.2110	.2173	—	—
5	.1801	.1837	.1888	.1947	—	—	—	—	.1888	.1947	—	—
6	.1604	.1636	.1690	.1745	—	—	—	—	.1690	.1745	—	—
7	.1429	.1457	.1514	.1565	—	—	—	—	.1514	.1565	—	—
8	.1272	.1298	.1356	.1404	*	*	*	*	.1356	.1404	—	—
9	.1133	.1155	.1209	.1251	*	*	*	*	.1208	.1251	—	—
10	.1009	.1029	.1075	.1114	*	*	*	*	.1075	.1114	—	—
11	.0898	.0916	.0960	.0996	*	*	*	*	.0956	.0991	—	—
12	.0800	.0816	.0861	.0895	*	*	*	*	.0857	.0890	—	—
13	.0713	.0727	.0774	.0805	*	*	*	*	.0770	.0800	—	—
14	.0635	.0647	.0696	.0725	*	*	*	*	.0692	.0720	—	—
15	.0565	.0577	.0625	.0654	.0590	.0619	.0591	.0621	.0621	.0649	.0596	.0628
16	.0503	.0513	.0562	.0589	.0527	.0554	.0528	.0556	.0558	.0584	.0533	.0563
17	.0448	.0458	.0507	.0533	.0472	.0498	.0473	.0500	.0503	.0528	.0478	.0507
18	.0399	.0407	.0457	.0481	.0422	.0446	.0423	.0448	.0453	.0476	.0428	.0455
19	.0355	.0363	.0413	.0437	.0378	.0402	.0379	.0404	.0409	.0432	.0384	.0411
20	.0317	.0323	.0374	.0396	.0339	.0361	.0340	.0363	.0370	.0391	.0345	.0370
21	.0282	.0288	.0339	.0361	.0304	.0326	.0305	.0328	.0335	.0356	.0310	.0335
22	.0250	.0256	.0303	.0323	.0272	.0293	.0273	.0295	.0303	.0323	.0278	.0302
23	.0224	.0228	.0276	.0294	.0245	.0264	.0246	.0266	.0276	.0294	.0251	.0273
24	.0199	.0203	.0251	.0268	.0220	.0238	.0221	.0240	.0251	.0268	.0226	.0247
25	.0177	.0181	.0224	.0240	.0198	.0215	.0199	.0217	.0224	.0240	.0204	.0224
26	.0157	.0161	.0203	.0219	.0177	.0194	.0178	.0196	.0203	.0219	.0183	.0203
27	.0141	.0143	.0187	.0201	.0161	.0176	.0162	.0178	.0187	.0201	.0167	.0185
28	.0125	.0127	.0170	.0184	.0144	.0159	.0145	.0161	.0170	.0184	.0150	.0168
29	.0112	.0114	.0157	.0171	.0131	.0146	.0132	.0148	.0157	.0171	.0137	.0155
30	.0099	.0101	.0143	.0157	.0117	.0132	.0118	.0134	.0143	.0157	.0123	.0141
31	.0088	.0090	.0132	.0145	.0106	.0120	.0107	.0122	.0132	.0146	.0112	.0130
32	.0079	.0081	.0123	.0136	.0097	.0111	.0098	.0113	.0123	.0136	.0103	.0120
33	.0070	.0072	.0113	.0126	.0087	.0101	.0088	.0103	.0114	.0127	.0094	.0111
34	.0062	.0064	.0105	.0117	.0079	.0092	.0080	.0094	.0106	.0118	.0086	.0102
35	.0055	.0057	.0097	.0109	.0071	.0084	.0072	.0086	.0098	.0110	.0078	.0094
36	.0049	.0051	.0090	.0101	.0065	.0078	.0066	.0080	.0088	.0099	.0072	.0088
37	.0044	.0046	.0084	.0095	.0059	.0072	.0060	.0074	.0083	.0093	.0067	.0082
38	.0039	.0041	.0079	.0090	.0054	.0067	.0055	.0069	.0077	.0087	.0061	.0076
39	.0034	.0036	.0073	.0084	.0048	.0061	.0049	.0063	.0072	.0082	.0056	.0071
40	.0030	.0032	.0069	.0080	.0044	.0057	.0045	.0059	.0068	.0078	.0052	.0067

*Special construction–not standardized.

Table 2-8.   Insulated copper wire.

(Harry E. Thomas, *Handbook for Electronic Engineers and Technicians*, © 1965. Reprinted by permission of Prentice-Hall, Inc.)

BIBLIOGRAPHY—CHAPTER 2

Index of EIA and JEDEC Standards and Engineering Publications. EIA, Eng. Dept., Washington, DC.

Handbook of Engineering Fundamentals (3rd Ed.). New York, NY, John Wiley & Sons, Inc., 1975.

JACKSON, H., *Introduction to Electric Circuits* (3rd Ed.). Englewood Cliffs, NJ, Prentice-Hall, Inc., 1981.

LANDEE, R., *Electronics Designers' Handbook.* New York, NY, McGraw-Hill, 1977.

LENK, J., *Handbook of Electronic Charts, Graphs and Tables.* Englewood Cliffs, NJ, Prentice-Hall, Inc., 1970.

LUDWIG, R. H., *Illustrated Handbook of Electronic Tables, Symbols, Measurements and Values.* Englewood Cliffs, NJ, Prentice-Hall, Inc., 1977.

MORRIS, N. M., *Essential Formulae for Electrical Engineers.* New York, NY, John Wiley & Sons, Inc., 1974.

1975 Index to IEEE Publications. Institute of Electrical and Electronics Engineers, New York, NY, 1975.

POPOV, E., *Mechanics of Materials* (2nd Ed.). Englewood Cliffs, NJ, Prentice-Hall, Inc., 1976.

Reference Data For Engineers: Radio, Electronics, Computer, and Communications (7th Ed.). Indianapolis, IN, Howard W. Sams & Co., Inc., 1985.

CHAPTER 3

COMPONENTS AND PARTS

3.1 STANDARDS

This chapter deals with passive electrical components and the most frequently used hardware. These components and parts are produced in large quantities, and it is therefore necessary that their electrical and mechanical dimensions follow certain fixed standards. Military standards, which are also sometimes used for commercial work, are set either by Joint Services committees or else by the individual services, such as the Army, Navy, or Air Force, responsible for a particular MIL standard. In the civilian area, the Electronic Industries Association (EIA) and the Institute of Electrical and Electronics Engineers (IEEE) publish the largest number of pertinent standards. Most of the mechanical parts used in electronics equipment are specified by the American Standards Association (ASA) or by the American Society of Mechanical Engineers (ASME). Again, military standards exist for all of the hardware. In the bibliography of this chapter we list the name and address of a standards organization which should be contacted for detailed publications of specific component parts standards.

The standards applied to all components and parts determine the nominal electrical or mechanical dimensions of the particular item, the tolerance over which this dimension may vary, as well as temperature, voltage, power dissipation, and other characteristics. In each case, the tolerance stated applies only to the primary value of the component and, unless specifically stated, does not apply to the other characteristics. For example, a 1500-ohm 5% resistor may have an actual resistance anywhere from 1425 to 1575 ohms. If this resistor has a rating of 0.5 watts, there is no implied tolerance. The power dissipation rating is usually related to a specific temperature, such as 25°C, and this means that it is safe to dissipate 0.5 watts in the particular resistor when the ambient temperature is 25°C.

Since it is not always practicable to indicate component values by numerals or letters, particularly on small cylindrical components, a standard color code is used throughout the industry. This color code is shown in Table 3-1 and applies to resistors, capacitors, inductors, and, in many instances, even to the wire number in a particular cable.

Mass-produced electrical components are available in a series of preferred values. In this system, each value represents a constant percentage increase over the next lower value. By using tolerances of 5, 10, and 20%, a different number of component values is required for

Color	Digit	Multiplier
Black	0	$10^0 = 1$
Brown	1	$10^1 = 10$
Red	2	$10^2 = 100$
Orange	3	$10^3 = 1000$
Yellow	4	$10^4 = 10,000$
Green	5	$10^5 = 100,000$
Blue	6	$10^6 = 1,000,000$
Violet	7	$10^7 = 10,000,000$
Grey	8	$10^8 = 100,000,000$
White	9	$10^9 = 1,000,000,000$
Gold	–	$10^{-1} = .1$
Silver	–	$10^{-2} = .01$

Table 3-1.   The standard color code.

(Lane K. Branson, *Introduction to Electronics*, © 1967. Reprinted by permission of Prentice-Hall, Inc.)

20% tolerance	10% tolerance	5% tolerance
10	10	10
	12	12
15	15	15
		16
	18	18
		20
22	22	22
		24
	27	27
		30
33	33	33
		36
	39	39
		43
47	47	47
		51
	56	56
		62
68	68	68
		75
	82	82
		91
100	100	100

Table 3-2.   Tolerance and component values.

each tolerance level. As indicated in Table 3-2, the range from 10 to 100 is covered in seven values for the 20% component and in 23 values for the 5% component. If a precision greater than 5% is required in a component, the usual procedure is either to select a special component or have one custom-made.

3.2 RESISTORS, FIXED

Fixed resistors are available in several basic types. Multiple resistors and resistors in single or dual in-line packages are available. Such arrays are very useful in a variety of systems and can mate with IC sockets.

Carbon composition molded resistors are widely used in applications of less than 2 watts. Also, resistor families are often surface mounted; in many commercial applications they have contacts rather than leads. For higher power ratings, or where great precision and stability are required, wire-wound resistors are used. For applications where high frequency and great stability are essential characteristics, film resistors are available.

Figure 3-1 is an example of various types of resistor packages. (A) shows multiple resistors and arrays in dual in-line and single in-line packages, and (B) is a composite resistor with color code.

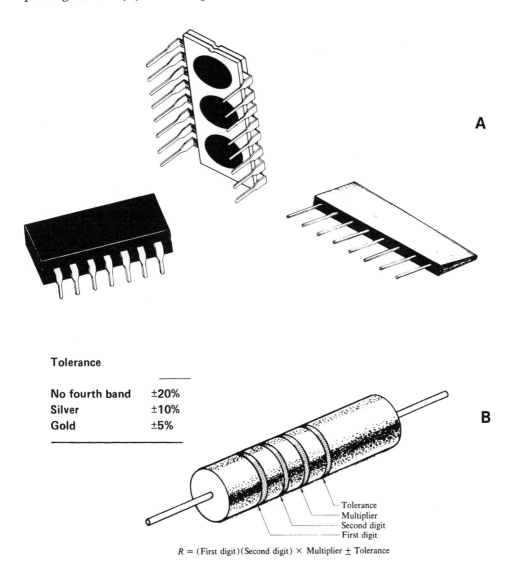

Tolerance	
No fourth band	±20%
Silver	±10%
Gold	±5%

Tolerance
Multiplier
Second digit
First digit

$R = (\text{First digit})(\text{Second digit}) \times \text{Multiplier} \pm \text{Tolerance}$

Figure 3-1.   (A) Multiple resistor and arrays in dual (DIP) and single (SIP) in-line packages. (B) Composite resistor color code and bands.

Carbon Composition Resistors

Commercially available carbon composition resistors are usually produced in 5, 10, and 20% tolerance values. The actual resistance values
follow those shown in Table 3-2. Figure 3-1 shows how the color code
is applied to a typical resistor, including the tolerance-indicating band.
If the first digit is yellow, the second digit violet, and the third digit
yellow, with a gold band indicating the tolerance, a reference to Table
3-1 will show that this resistor is 470,000 ohms, plus or minus 5%. If
the first band were green, the second band brown, and the third band
yellow, its value would be 510,000 ohms. From Table 3-2, we know
that this latter value is only available in 5% tolerance and would therefore expect to find a gold band.

　　Carbon composition resistors are available in five different power
dissipation ranges, with the maximum RMS voltage applied and the
range of resistance values as shown in Table 3-3.

Max, RMS volts

1/8 watt	150 V	2.7 Ω to 22 Meg
1/4 watt	250 V	2.7 Ω to 22 Meg
1/2 watt	350 V	1.0 Ω to 22 Meg
1　watt	500 V	2.7 Ω to 22 Meg
2　watt	750 V	10.0 Ω to 22 Meg

Table 3-3.　Maximum voltages for carbon composition resistors.

　　Although the power dissipation rating of the resistor is clearly
stated, this merely means that the resistor will be able to dissipate that
much power but does not indicate the effect of this power dissipation
on the actual resistance value, the temperature, or the life of the resistor.
The nominal resistance values shown by the color code on the composition resistor body can only be expected to be accurate at room
temperature. As the temperature increases, resistance decreases. The
actual temperature of a resistor is dependent, to a great extent, on its
surroundings. If, for example, a stream of cooling air passes a resistor,
its temperature rise will be much less than if a resistor is located in a
thermal hot spot where heat continues to build up. For safe, reliable
operation of typical electronics circuitry, Table 3-4 indicating power
dissipation derating is recommended. This derating table is for use at
25°C. If the ambient temperature is higher, additional derating is recommended.

Nominal (Watts)	Derated (Watts)
0.125	0.05
0.25	0.10
0.50	0.20
1.0	0.50
2.0	1.25

Table 3-4.   Power derating for carbon composition resistors.

Wire-Wound Resistors

Wire-wound resistors are generally available at power dissipation ratings from 3 up to several hundred watts, with tolerances ranging from 0.05 up to 10%. Because such resistors are in essence merely a layer of resistance wire wound on a mandrel, they exhibit inductive properties, but it is possible to purchase noninductive, wire-wound resistors. These devices use parallel windings of opposite polarity to cancel out the inductive effect and are considerably more expensive than the regular wire-wound types.

Wire-wound resistors are generally available in values ranging from 0.1 up to 50,000 ohms. Their maximum voltage ratings depend mostly on the insulation of the particular resistor type. Four different types of insulation are in common use:

1. A simple coating of vitreous enamel.
2. The entire device molded in a silicone compound.
3. The entire device molded in a ceramic material.
4. The molded resistor further encased in a metal heat sink with radiating fins.

A variation of standard wire-wound resistors is available in which a portion of the resistance wire is exposed and a movable, pressure-type contact provides the adjustment. These types of resistors are available in power ranges from 5 to 250 watts, in tolerances of 1 to 10% and in resistances from 1 to 10,000 ohms. Almost all of them use a vitreous enamel coating. Because of the pressure contact, these devices are only recommended for experimental work and should not be used where great stability and long-term reliability is required.

Power dissipation derating of wire-wound resistors is usually unnecessary. If sufficient heat dissipation capability is available, i.e., if there is no likelihood of heat buildup in a particular hot spot, wire-wound resistors may be safely used up to 100% of their rated power. At ambient temperatures considerably above 25°C, however, derating will be necessary for wire-wound resistors. However, power wire-wound resistors can be operated with a body temperature of 275°C. Aluminum and water-cooled housing are also available.

Film Resistors

In general, metal-film resistors are used in applications requiring higher stability and precision than is possible using carbon composition types. In addition, they perform well in dc circuits, from dc to the MHz part of the RF spectrum.

Film Resistor Types

   Thin-film resistors   are highly stable, have low-noise character-istics, and have a very low temperature coefficient, making them ex-cellent for precision applications such as digital multimeters and sim-ilar instrument circuits. Thin-film resistor networks are available in DIPs and SIPs (see Figure 3-1).

   Thick-film resistors   are usually used in applications requiring high power density and/or rough use where occasional overloads and power spikes are expected.

   Carbon-film resistors   perform the same basic functions as the carbon composition resistors described previously under the heading "Carbon Composition Resistors." The carbon-film and carbon compo-sition both have poor temperature coefficients and do not stand up well during transient voltage spikes. They are available in the same resis-tance values as the carbon composition type but usually cost less.

3.3 POTENTIOMETERS (VARIABLE RESISTORS)

Almost all variable resistors are available in the form of a potentiometer, which is defined as a three-terminal device with a constant resistance between two of the terminals, but the resistance between the third ter-minal and either of the other two is variable. A rheostat we generally consider a device that varies the resistance between two points. It is possible to connect a potentiometer "as rheostat" by connecting the variable terminal to either one of the fixed terminals.

   Potentiometers are available in three basic types. For resistances of 2 watts or less, a carbon composition molded element with a movable contact on it is usually employed. For resistances ranging from 2 up to 1000 watts, wire-wound elements are used, usually arranged on a circular form with a sliding contact, rotating on a shaft, to provide the variable contact. One of the limitations of the wire-wound potentiometer is its resolution, since the movable contact usually connects only between discrete points on the spiral-wound resistance element. Where infinite resolution and great precision are required, film potentiometers are used, which are available in power ratings from 1 to 5 watts.

The basic characteristics of the three types of potentiometers correspond closely to the basic characteristics of the corresponding types of fixed resistors, with the addition of a new characteristic, the taper. The taper of a potentiometer is the relation between resistance change and motion of the variable contact. A linear taper, for example, would be a straight-line plot between resistance change and contact motion. In some applications, a fixed tap is added to the potentiometer, which permits an abrupt change when the moving contact reaches a certain point. A typical example is the tone compensation circuit used on the volume control in a hi-fidelity audio system. Table 3-5 contains three taper charts, with an explanation of the individual taper curves and typical applications in industry.

Carbon Composition Potentiometers

This type of potentiometer is available in many different physical models, from the familiar rotary, single-turn potentiometer used in volume controls, tone control circuits, etc., to the micro-miniature, rectilinear motion, screwdriver-adjustable trimmers found in print wiring boards of sophisticated equipment. Resistance ranges are available from 50 ohms to 50 megohms; with tolerances of 5, 10, and 20%; in power ratings of 0.25 to 2 watts; with maximum voltage ratings of 200 to 400 volts. Carbon composition potentiometers have infinite resolution; i.e., it is possible to adjust the movable tap to infinitely fine resistance increments.

Power dissipation specifications for potentiometers are always stated for the entire resistance element. It is important to consider the power dissipation between the variable tap and either end of the potentiometer, particularly when only a small portion of the total resistance is involved.

Example: A 2-watt carbon compositon potentiometer of 1000 ohms is used in a circuit so that the variable tap and one end provide 100 ohms resistance. The power dissipated in this 100-ohm segment of the total resistance must be less than 0.2 watts. For reliable operation, a derating factor of 50% should be applied to reduce the likelihood of a

Design Parameters

Electrical Specifications

1. **RESISTANCE RANGE** — 100 ohms to 10 meg ohms

   Nonlinear tapers — normally supplied with resistance values of 200 ohms to 1 megohm at midpoint of rotation

2. **TOLERANCE**

   ± 30% standard
   ± 20% or ± 10% upon reques

3. **WATTAGE RATING**

   ½ watt at +40° C. standard for linear taper

   Nonlinear tapers should be derated by ½

   1 watt at +70° C. linear taper available upon request

4. **MAXIMUM OPERATING TEMPERATURE**

   Derated to 0 watts at +105 C. for ½ watt unit

5. **ELECTRICAL ROTATION**

   Plain type — 300°

   Switch type with throw at CCW end 300°

   Switch type with throw at CW end 280°. All above include 12° terminal flat at each end.

   Fixed end resistance 250° available at either left or right end.

   Mechanical tolerance of ± 3% on intermediate test points.

6. **TAP LOCATIONS**

   Single tap at 37½%, 50% or 62½% of rotation

   Double taps at 37½% and 62½% of rotation only

   15° resistance flat at taps

7. **TAPERS (see taper charts)**

   These charts represent only typical curves, other tapers are available upon request.

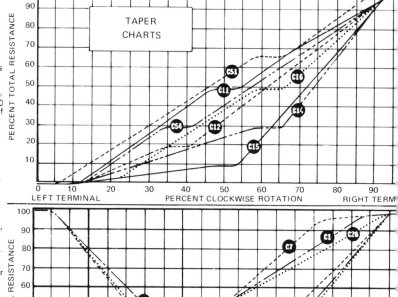

8. **CONTACT RESISTANCE (PEAK TO PEAK)**

   Special requirements as low as 0.5% of the nominal maximum resistance are available upon request.

9. **ATTENUATOR**

10. **MATCHED TWINS**

    One check point:

    5 db at −55 db attenuation

    Checked over the entire range:

    4 db tracking from 0 to −50 db attenuation

    6 db tracking from −50 db to −60 db attenuation

3 db tracking from 0 to −50 db attenuation

5 db tracking from −50 db to −60 db attenuation

2 db tracking from 0 to −50 db attenuation

4 db tracking from −50 db to −60 db attenuation

The standard tap load for a single tapped unit is 7½% of the nominal total resistance.

The standard tap loads for double tapped units are 5% of the nominal total resistance on the lower tap and 15% on the higher tap.

11. **DIELECTRIC STRENGTH**

    900 VRMS between all terminals and mounting parts.

Table 3-5. Potentiometer taper curves.

(Illustration courtesy of Centralab Electronics Division.)

Resistance Curves

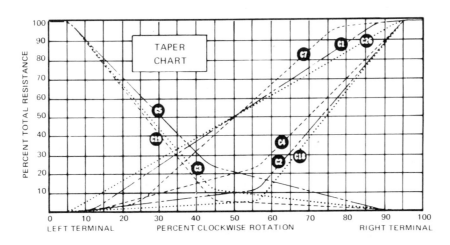

C-1 & C-26   Linear has uniform resistance change from either end. Used for Hor. & Vert. Centering; Hor. & Vert. Hold; Hor. Drive; Vert. Linearity; Focus; Brightness; Height; Contrast. Volume or Tone.

C-2   Semi-log (audio) 10% center. Used for Volume or Tone.

C-3   Right-hand semi-log. Reverse of C-2. Decreases with clockwise rotation.

C-4   Modified log with 20% center. Used principally as Volume Control, also as Tone Control

C-5   Right-hand modified log with 20% center. Reverse of C-4. Decreases with clockwise rotation

C-7   Symmetrical-straight line with slow resistance change at either end. Used principally as Tone Control and Balance Control.

C-11*   Taper tapped at 50% rotation. 50% resistance. Used as Tone Control or in TV for Vert. and Hor. Centering

C-12*   Taper tapped at 37½% rotation, 20% resistance. Used principally as Volume Control with Tone compensating tap.

C-14*   Taper tapped at 62½% rotation, 30% resistance. Used principally as Volume Control with Tone compensating tap.

C-15*   Taper tapped at 50% rotation, 10% resistance. Used principally as Volume Control with Tone compensating tap.

C-16*   Taper tapped at 62½% rotation, 50% resistance. Used principally as Volume Control with Tone compensating tap.

C-18   Modified log with 5% center. Used principally as Volume or Tone Control

C-19   Right-hand modified log with 5% center. Reverse of C-18.

C-51*   L.H. linear taper, tap at 62½% cw rot. from L.T., with 62⅔% of res. between L.T. & tap. Used as TV contrast control.

C-54*   L.H. semi-linear taper, tap at 37½% cw rot. from L.T., with 30% of res. between L.T. & tap. Used as loudness control.

*Available with Model 2 Only

Table 3-5. Continued

35

defect forming at the point where the variable contact touches the carbon element. This derating can be neglected for short periods, but if the potentiometer is permanently set at the 100-ohm point, a 200-ohm potentiometer should be used in series with an 800-ohm fixed resistor.

Table 3-4 should be used as a reference for derating carbon composition potentiometers.

Wire-Wound Potentiometers

These are generally available in the range from 1 to 10,000 ohms, with tolerances of 0.5 to 10%, and in power ratings from 2 to 1000 watts. Voltage ratings are in the order of 500 to 1000 volts, with higher ratings available where larger power dissipation and larger physical size are used. The resolution of wire-wound potentiometers depends on the size of the wire and its spacing. A 1000-ohm potentiometer having 1000 turns and a moving contact that is simultaneously connected to two turns will have a resolution of 2 ohms. In actual practice, the exact position of the moving contact cannot be fixed that precisely, and a coarser resolution must be expected.

Wire-wound potentiometers are available as precision components for servo systems and analog computers. For these applications, special tapers may be designed.

The power dissipation rating for wire-wound resistors is essentially the same as for wire-wound potentiometers, with the exception of precision components used in servo systems and industrial control applications. For these components, manufacturers' data must always be consulted concerning power dissipation rating. Precision wire-wound potentiometers have resistance ranges from 0.1 through 800 k-ohms, power ratings to 15 watts, and tolerances of 0.01 to 1%.

Film Potentiometers

These components are available in resistance ranges from 100 ohms to 2 megohms at tolerances of 0.5 to 5%, with power dissipation of 1 to 5 watts and maximum voltages ranging up to 500 volts. Film potentiometers combine the precision of wire-wound potentiometers with the infinite resolution of the carbon composition type. Many applications of film potentiometers require that the total resistance as well as the taper curve remain precisely constant over a relatively large range of temperatures. Since film potentiometers are usually used in precision instruments, servo systems, and other special applications, the power dissipation derating should be determined from the manufacturer's literature.

Film potentiometers are frequently used in a multiple-turn variation as well as in a concentric shaft or stacked arrangement.

Conductive Plastic Potentiometers: Resistance ranges of these potentiometers can vary from 150 ohms up to 5 megohms. However, their power rating is typically no greater than 1 watt. Also, derating factors are 50% for power and 80% for voltage. Plus, internally generated noise may cause problems.

Conductive Plastic Potentiometers—General Purpose: Resistance ranges on these plastic potentiometers depends on the power rating, which can be as high as 1000 watts. In general, the resistance range can be anywhere from 1 ohm to 15 k-ohms.

Conductive Plastic Potentiometers—Precision: Resistance ranges are usually from 100 to 500 k-ohms and the power rating is 7 watts. Tolerance is often listed as 3%; as a general rule, conductive plastic potentiometers have a long life expectancy (in some cases, over 2 million rotations).

Hybrid potentiometers have a longer life expectancy—10 million rotations with a tolerance of 5%, a power rating of 7 watts. Incidentally, hybrid potentiometers are wire-wound devices with a conductive plastic track attached to the contact path of the resistive element.

3.4 CAPACITORS, FIXED

The same basic color code and preferred standard values as applied to resistors apply to capacitors. In practice, the color code is used only on micacapacitors and those ceramic capacitors which are relatively small, as shown in Table 3-6. On larger capacitors, the capacitance and other important characteristics are printed directly on the body of the capacitor. Fixed capacitors are available in values from as small as 1 picofarad (pF) to values as high as 1000 microfarads (μF). This wide range of capacitor values is covered by a number of different types of construction. In general, capacitors with values below 0.01 μF are either of the mica or ceramic construction type. Capacitors ranging from 0.01 to approximately 1.0 μF are generally of the paper, ceramic or plastic dielectric type of construction. Capacitors larger than 1 μF and ranging all the way up to 1000 μF are usually of the electrolytic type, using such different materials as zinc, aluminum, and tantalum.

EIA #RS 153A AND MIL-C-5A COLOR CODE

Color	Capacitance		Capacitance Tolerance in Percent	Charac-teristic
	Significant Figure	Decimal Multiplier		
Black	0	1	20 (M)*	—
Brown	1	10	1 (F)	B
Red	2	100	2 (G)	C
Orange	3	1,000	—	D
Yellow	4	10,000*	—	E
Green	5	—	5 (J)	F
Blue	6	—	—	—
Violet	7	—	—	—
Gray	8	—	—	—
White	9	—	—	—
Gold	—	0.1	½ (E)*	—
Silver	—	0.01*	10 (K)	—

*Not Included in the MIL-C-5 Specification

WHITE DOT DESIGNATES EIA, WHILE BLACK DOT DESIGNATES MIL

ADDITIONAL TYPE DESIGNATION INFORMATION

EIA #RS 153A & MIL-C-5C TEMPERATURE RANGE

Symbol	Degree Centigrade
N	−55 TO −85
O	−55 TO −125

EIA #RS 153A DC WORKING VOLTAGE

Symbol	Voltage	Symbol	Voltage
1	100	5	500
3	300	10	1000

MIL-C-5C DC WORKING VOLTAGE

Symbol	Voltage	Symbol	Voltage
C	300	F	1000
D	500	H	1500

MIL-C-5C VIBRATION GRADE

Symbol	Cycles Per Second
1	10 TO 55
3	10 TO 2000

THE TYPE DESIGNATION FOR FIXED MICA-DIELECTRIC CAPACITORS IS FORMED AS FOLLOWS:

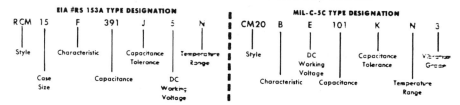

EIA #RS 153A TYPE DESIGNATION

R CM   15   F   391   J   5   N

Style — Characteristic — Capacitance Tolerance — Temperature Range
Case Size — Capacitance — DC Working Voltage

MIL-C-5C TYPE DESIGNATION

CM20   B   E   101   K   N   3

Style — DC Working Voltage — Capacitance Tolerance — Vibration Grade
Characteristic — Capacitance — Temperature Range

CHARACTERISTIC LIMITS

Symbol	Temperature Coefficient Parts/Million/-Degree C.	Capacitance Drift
B	Not specified	Not specified
C	±200	±(0.5% + 0.1 MMF)
D	±100	±(0.3% + 0.1 MMF)
E	−20 to +100	±(0.1% + 0.1 MMF)
F	0 to +70	±(0.05% + 0.1 MMF)

Table 3-6.   Mica-capacitor table.

(Illustration courtesy of Electro Motive Mfg. Co.)

At higher frequencies, capacitors also exhibit resistance and inductance effects. These latter two characteristics become inportant when either reactive losses or self-resonance is a factor. The resistance of a capacitor becomes important when the dissipation factor, always given together with the frequency, must be considered.

Capacitors are subject to voltage breakdown, and the voltage ratings are therefore of prime importance. Power ratings are really a description of the dissipation, caused by the inherent resistance in the capacitor structure.

Capacitance varies with temperature, and this change may be either in positive or negative direction. Ceramic capacitors in particular are available in a wide range of positive and negative temperature coefficients. Table 3-6 indicates some of these temperature characteristics.

A summary of the basic types of capacitors is presented in Table 3-7. This summary is arranged, as seen in the left-hand column, according to the dielectric material used in the capacitor. Under the heading of "paper" we include all those capacitors that use a plastic film or a plastic-coated paper or similar material. Under the heading of "mica" we include all those capacitors that use mica as a dielectric, regardless of whether they are molded, potted, or in a metal can. Similarly, under the heading of "ceramic," all those capacitors that depend on a ceramic insulator material are listed. A variety of electrolytic capacitors and special vacuum and gas capacitors are shown. The latter two types are used quite rarely and only in very unique applications. Note that among the characteristics, the operating temperature range is given for each type. This does not mean that a particular capacitor will retain its capacity without change over the entire temperature range. It only means that a particular type of capacitor is generally capable of working in that temperature range.

Capacitor package styles are also important to the user. High lead inductance, usually associated with tubular units, degrades high-frequency performance. Nevertheless, tubular ceramic capacitors are one of the most stable forms of capacitors available.

Dipped or molded radial lead packages are often used because they help reduce interconnecting circuit impedance. This is accomplished by mounting the capacitor as close to the PC board as practical.

To further reduce interconnecting circuit impedance, chip capacitors may be used. These capacitors have contacts rather than leads. Another technique to reduce this type of impedance when fabricating PC boards is surface mounting, which is even better than chip capacitors thin enough to mount beneath unsocketed ICs.

Table 3-7. Fixed capacitor characteristics.

Dielectric classification	Capacity range (MF)	Voltage range	Leads, mounting	Operating temp. range (°C)	Application frequency and range	Marking
PAPER:						
Tubular	.0005-6.0	200-400	Axial	-55-+150	General purpose, up to 1 MC	EIA standard color-coded bands
Plastic film		200-600				
Special: bathtub, rectangular case	.1-2.0	100-12,500	Terminals			Direct MIL-STD labelling
MICA:						
Molded	.00001-.015	300-3000	Axial	-60-+85	General purpose, up to 10 KMC	EIA or MIL-STD color-dot coding
Potted plastic (ceramic case)	.00005-.1	250-5000	Terminals, ears, stud mounting			
Button	.00001-.0039	300	Tab and mounting stud			
CERAMIC:						
Tubular	.00001-.03	0 to 500 or 600	Radial or parallel wire		General purpose, 50-1000 MC	
Disc						
Special: standoff feedthrough compensating	0-.00001 0-.00005		Body or stud mounting	-55-+125	UHF bypass, critically tuned circuits, 50-1000 MC	EIA or MIL-STD color-dot or band marking MMF stamped on body
high voltage	.0005	10,000-20,000	Stud and terminal mounting		High-voltage TV filter 0-20 KC	
ELECTROLYTIC:						
Tubular	0-100	150-450	Axial leads			Body stamping
Aluminum foil (plug-in)	6-2000	6-500	End mounting	0-+65	0-20 KC	Body stamping or coded indentations
Non-polarized (screw)	450-50					
Wet or dry (base)						
Cylindrical (twist, prong)						
Tantalum: foil	.5-60	.5-150.0	Lead mounting	-55-+85		Body stamping
tubular slug	7-300	.5-600.0	Body mounting	55-+200		
GLASS:						
Rectangular	.00005-.005	0-300	Axial	-60-+200	0-100 MC	EIA color-dot
VITREOUS ENAMEL:						
Rectangular	.00001-.01	0-300	Axial	-60-+200	0-100 MC	EIA color-dot
SPECIAL:						
Vacuum	.0001-.0005	0-50KV	Terminals	-60-+400	Blocking and decoupling	Body stamping
Gas	.00001-.00005		Terminals	-60-+200	Blocking and decoupling	Name plate

(Harry E. Thomas, *Handbook for Electronic Engineers and Technicians*, © 1965. Reprinted by permission of Prentice-Hall, Inc.)

3.5 CAPACITORS, VARIABLE

Variable capacitors come in a variety of styles, and several materials are used during the manufacturing process. For example, materials used in manufacturing a trimmer capacitor may be glass/quartz, sapphire, plastic, ceramic, air, mica, or vacuum/gas.

In addition, trimmer capacitors are available in three types: multi-turn, single turn, and compression. The dielectric used in multi-turn may be either glass, quartz, sapphire, plastic, or aid dielectric, depending on the application and frequency requirements. Glass and quartz devices can be used at frequencies up to 300 MHz. All others should not be used above 1 GHz. The exceptions are ceramic and high-grade plastic, which are generally usable up to about 2 GHz.

Air-Variable Capacitors

These capacitors are available in a variety of capacitance and voltage ranges and generally consist of one set of stationary metal blades, the stator, and a set of rotating blades, the rotor, which mesh with the stator according to the shaft rotation. If it is desired to obtain a linear variation of capacitance with shaft rotation, the rotor blades must have a special shape. If it is desired to have a linear change of frequency with rotation, the shape of the rotor blades must be different again. Air variable with minimum capacitance of 3 pF and maximum capacitance of several hundred pF. They are available as small as ½ inch in diameter and as large as 8 to 10 inches in blade diameter, the latter being used for high-powered transmitters. Voltage ranges depend upon the size and spacing of the stator and rotor blades and the insulating frame. General voltage ranges vary from 100 volts for the smallest capacitors to several thousand volts for large, transmitter-type units.

Mica Compression Capacitors

These capacitors, usually called trimmers, are screwdriver adjustable and use mica sheets placed between spring-loaded metal plates, which are compressed by a central screw. Typical values range from 1.5 to 15 pF for the smallest size and 390 to 1400 pF for the largest mica compression capacitor. Voltage ranges go from 175 to 500 volts. This type of capacitor is found as a trimmer capacitor in radios, the horizontal oscillator of TV receivers, and in similar applications.

Ceramic Variable Capacitors

Screwdriver adjustable, this type of capacitor is available either in the disc or tubular form. Disc variable capacitors consist of two ceramic discs with appropriate silver coating, and capacitance is varied by changing the position of the two silver-coated areas with respect to each other. Depending upon the type of ceramic used, different temperature characteristics are available. Typical capacitance ranges go from 2 to 60 pF. In the tubular ceramic capacitor, a metal piston moves inside a ceramic tube with an outer conductive coating. Capacity depends on the proximity of the inner metal piston and the outer ceramic coating. Voltage ranges for ceramic capacitors vary from 200 to 500 volts.

Glass Variable Capacitors

This type of capacitor is similar to a tubular ceramic capacitor, except that the material of the tube providing the dielectric is glass. Typical capacity ranges go from 1 to 16 pF; voltages vary from 200 to 400 volts.

Voltage Variable Diodes

Different manufacturers provide a whole range of special purpose diodes which change their inherent capacity according to the voltage applied across them. The general range of capacity variation is approximately 3:1, and voltage variable diodes are available with minimum capacities ranging from 1.3 to 330 pF. These capacitors are available under certain trade names such as "Varactor," "tuning diodes," etc. Voltage variable diodes are used in automatic frequency control circuits, in TV receivers, FM receivers, and a host of other applications where the frequency of an oscillator is controlled by a voltage. See Chapter 8 for principles of operation and Chapter 11 for microwave applications.

3.6 COILS, FIXED AND VARIABLE

Commercially available coils are produced in a wide variety of types, sizes, and electrical performance characteristics. The standard color code described at the beginning of this chapter is used to indicate the value of the inductance, where applicable, and in some cases is used to identify terminal numbers or wires. Inductance values are available according to the series of preferred values described in Table 3-2.

Basic Characteristics

- Inductance—stated in Henries, milli-henries, and micro-henries.
- Q (quality factor)-describes the ratio of the reactance of a coil and its dc resistance.

$$Q = \frac{X_L}{R_{dc}} \text{ (} X_L \text{ depends on frequency)}$$

- Saturation—applies only to coils using ferromagnetic cores and describes the amount of direct current which will substantially change the magnetic characteristics of the core material and therefore the inductance of the entire coil.
- dc resistance—the actual dc resistance of the conductors.
- Breakdown voltage—describes the maximum voltage that can be applied across a coil before breakdown of the insulation occurs, either across terminals or between turns.

Fixed Coils

Fixed coils can be classified into the five basic types listed below, together with applications and useful frequency range.

1. RF chokes and peaking coils—small fixed inductances, ranging from 0.1 to 10,000 micro-henries. They are used to filter radio frequencies, extend the bandwidth of amplifiers, etc. These coils may be wound on plastic, ceramic, powdered iron, or ferrite. When used as RF chokes, the powdered iron core material is also frequently used as parallel resistance to increase the losses.

2. Air coils—these may be either self-supported or wound on insulating forms. Available in ranges from 0.1 to 2000 microhenries, these coils are used for RF work, covering a frequency band from approximately 100 kHz to 1000 MHz. At the higher frequencies, where fewer turns are used, most air coils are custom-made. Particularly in the radio amateur field, the design and construction of air coils is widespread. For this reason, the following formulas are presented:

$$L = \frac{r^2 n^2}{9r + 10l} \text{ or } n = \sqrt{\frac{L(9r + 10l)}{r^2}}$$

where: L = microhenries
            r = radius (½ diameter) of the coil, in inches
            l = length of coil, in inches
            n = number of turns

Example: $n = 20$
            $r = 0.5''$ $L = \dfrac{0.25 \times 400}{4.5 + 10} = \dfrac{100}{14.5} = 6.9$ microhenries
            $l = 1''$

3. Iron-core coils (See Chapter 6 for detailed data.)—These coils are generally used for frequencies below 10 kHz and are available in inductance values from 1 millihenry to 10 henries. The iron core is available either in the form of a powdered iron mold or else as sheets of iron or steel, laminated in a suitable configuration to form the core.

4. Ferrite-core coils (See Chapter 6 for detailed data.)—Coils having a ferrite material as a core are generally used in the frequency range from 10 to 100 kHz and are available in 0.5 to 10,000 microhenry inductances.

5. Toroids (See Chapter 6 for detailed data.)—Coils wound on a ring instead of a cylinder are called toroids and are generally used in the frequency range from 10 kHz to 100 kHz. They are available in inductances from 1 millihenry to 10 henries, and their special characteristic is the relatively high Q and special shape of the magnetic field.

Variable Coil Types

These are generally coils wound on a cylinder with a ferrite, powdered iron, or brass core, which can be moved inside the cylinder by means of a screw. Increasing the amount of powdered iron or ferrite material in the center of the coil increases the inductance. Increasing the amount of brass core in the center of the coil reduces the inductance. The presence of the core reduces the inductance. The presence of the core material has a great effect on the Q of the coil as well. Variable coils are used in the IF amplifiers of all types of receivers, in the horizontal deflection circuits of television sets and in many other applications where the inductance of a coil is variable over a 2:1 range. The core material determines the frequency range over which a given Q can be obtained.

3.7 TRANSFORMERS

The subject of transformers, particularly those using ferromagnetic cores, is quite extensive and complex. For this reason Chapter 6 is devoted entirely to ferromagnetic inductors and transformers, and the reader is referred to that chapter for detailed information.

One of the few standards that apply to all transformers is the color coding of terminals and lead wires, which is shown in Table 3-8.

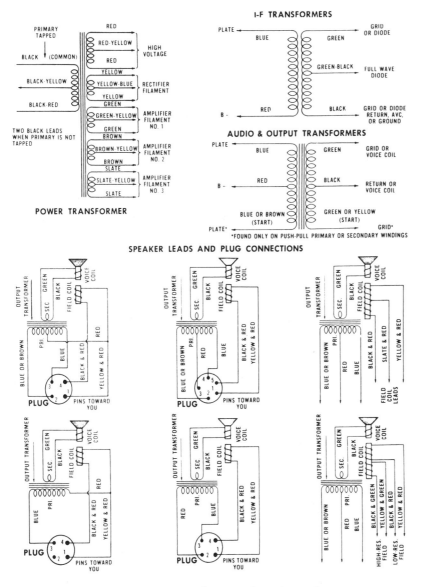

Table 3-8.   Transformer wire color code.

(Harry E. Thomas, *Handbook for Electronic Engineers and Technicians*, © 1965. Reprinted by permission of Prentice-Hall, Inc.)

Air-core transformers, like air-core coils, are used for radio-frequency work and basically consist of two or more air coils linked by electromagnetic coupling. Air-core transformers are generally custom-made and custom-designed and are not commercially available from stock. Design formulas for air-core transformers are contained in Chapter 6.

3.8 PRINTED CIRCUITS

This heading covers all the techniques in which the interconnection of components is achieved by printed or machine-deposited conductors on an insulating surface. Printed circuits or printed wiring is available in the following forms:

1. Single-sided—conductor pattern is contained on one side of the board and components are mounted on the opposite side. Component leads and connections pass through suitably placed holes and are soldered onto the conductor.

2. Double-sided printed wiring—where greater complexity is required, both sides of the insulating material contain the printed wiring pattern. Components are again mounted solely on one side. Connections between conductors of two opposite sides can be made by insertion of a metal conductor, such as eyelets, rivets, or a simple wire. For many applications, plated-through holes are used. In this process, conductive material is deposited inside the holes connecting the two printed wiring patterns, and, subsequently, plating is applied to provide a firm connection between the material in the hole and the land pattern on each side.

3. Multi-layer printed wiring—this technique uses several layers of printed wiring patterns, each mounted on a thin insulating material, and carefully aligned and molded together. This technique invariably uses plated-through holes to provide the connection of conductors that are in the internal conducting layers and those of the outside conducting printed wiring pattern.

Materials:

Although a wide variety of different materials is used by different manufacturers and for different applications, they are all basically of the following three types:

- Phenolic—usually a brown material used in commercial applications where temperature, great stability, and reliability are not too essential.
- Glass epoxy laminate—the most widely used material, it is available in green and blue colors, is considerably more translucent than phenolic material, and has excellent temperature stability and considerably greater strength than phenolic.
- Ceramic—this is usually used only for very small printed wiring applications since large ceramic pieces tend to be brittle and are not as flexible as phenolic or the glass epoxy. Ceramic material is chosen when extreme dimensional stability is required at very high temperatures.

Metal Finishes and Impregnation

Although in some commercial applications the copper pattern remains bare, in most applications the copper conductor is plated with a second metal or alloy. Following are the most commonly used conductor plating materials:

- "Solder plating"—uses a mixture of tin and lead to improve solderability and prevent corrosion.
- Gold plating—provides excellent solderability, withstands corrosion for long periods of time, and provides low-resistance surfaces particularly important in plug-in boards.
- Rhodium plating—is used because of its hardness and surface stability for extreme environmental and wear conditions. Fingers on printed circuit boards which mate with connectors and switch surfaces are usually rhodium plated for long wear.
- Nickel plating—nickel may be used as a substitute for gold or rhodium but most often is used directly over the copper, below the gold or rhodium plating, to provide a hard base for thin layers of gold or rhodium.

In many high reliability applications, an insulating material is sprayed over the entire printed wiring assembly. Three basic types of these insulating or impregnating materials are in common use:

- Special lacquers—these provide good insulation but can be melted with a soldering iron to permit repair of selected portions of an assembly.

- Polyvinyl fluoride (or equivalent plastics)—these materials provide excellent insulating capability and withstand extreme environmental conditions as well, but are more difficult to repair. The polyvinyl fluoride material must be scraped away before soldering.
- Epoxy-type coatings—these are excellent mechanical and thermal protection, as well as electrical insulators, but do not lend themselves to easy repair. Removal of the epoxy coating to gain access to a solder connection is usually complicated.

Electrical Performance

Defects in printed wiring boards of all types consist either of open circuits, shorts, or potential arc-over points. Open circuits are generally caused by a break in the conductor, while short circuits are usually caused by bridging of solder between conductors. Arc-over occurs when the gap between adjacent conductors is so small that an arc can occur at higher voltages. Table 3-9 shows the maximum permissible voltage between minimum-spaced conductors at altitudes of 10,000 feet and inside of multi-layer boards at any altitude. This latter list is very similar to properly impregnated surface conductors when the impregnation is either polyvinyl fluoride or an epoxy fluoride or an epoxy-type coating.

Performance Standards

A group of specifications for printed wiring boards of all types have been agreed on by the Institute of Printed Circuits, Evanston, IL 60202,

SURFACE (UP TO 10,000 FEET)		SURFACE (UP TO 10,000 FEET)		INTERNAL LAYERS (ANY ALTITUDE)	
VOLTAGE	INCHES (MIN.)	VOLTAGE	INCHES (MIN.)	VOLTAGE	INCHES (MIN.)
0 - 150	0.025	0 - 50	0.025	0 - 30	0.010
150 - 300	0.050	50 - 100	0.060	30 - 50	0.015
300 - 500	0.100	100 - 170	0.125	50 - 150	0.020
		170 - 250	0.250	150 - 300	0.030
		250 - 500	0.500	300 - 500	0.060

Table 3-9. PC board voltages between conductors.

and these also correspond to certain military specifications. These standards and specifications deal with acceptability of unassembled printed wiring boards, completely assembled printed wiring boards, and the performance of the entire unit over all ranges of environmental conditions.

Surface Mounting

The practice of surface mounting components and surface layers on printed circuit boards has produced some remarkable changes in the electronics industry. For example, one Japanese computer contains 419-by-280-mm logic boards that hold 72 LSI flat packs. The polyimede-glass boards have a total of 20 layers, 8 signal layers, plus 10 power and ground layers.

Components such as leadless chip resistors, capacitors, inductors, toroidal transformers, and many ICs now are available in surface-mountable packages. Today, the designer has almost 3000 discrete components in surface-mountable packages that can be used in PC board designs.

3.9 HARDWARE

Electronic equipment generally uses certain standard hardware, some of which is described in this section.

Screw Types and Sizes

Figure 3-2 shows the most commonly used screw heads, wood screw and machine screw types, and the nomenclature of frequently used nuts, bolts, and washers.

Machine screws are available with three different types of threads—NC is considered coarse when compared to NF, which is considered fine, and EF is considered extra-fine. A comparison of these three types of machine screw threads and the diameter of the machine screws in which they are used is shown in Table 3-10. This shows, for example, a size 6 screw, a very commonly used size, which has a body diameter of 0.8 inches and which is available in 32 or 40 threads to the inch. In electronics the most commonly used is the 6-32, although 6-40 can also be found occasionally. Extra-fine thread is usually used only in the largest screw sizes, about ¼-inch diameter.

The machine screws listed in Table 3-10 generally require a nut or tapped hole for mating. Self-tapping screws have a hardened thread

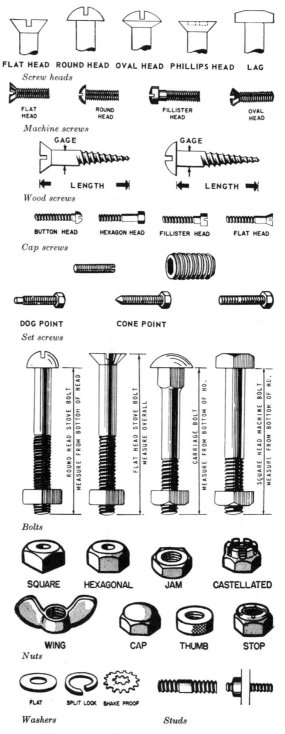

Figure 3-2. Screw and nut types.

(Harry E. Thomas, *Handbook for Electronic Engineers and Technicians,* © 1965. Reprinted by permission of Prentice-Hall, Inc.)

Diameter		Threads per inch			
No.	Inch	Decimal equiva-lent	NC (U.S.S.)	NF (S.A.E.)	EF (extra fine)
0	...	.0600	...	80	...
1	...	.0730	64	72	...
2	...	.0860	56	64	...
3	...	.0990	48	56	...
4	...	.1120	40	48	...
5	1/8	.1250	40	44	...
6	...	.1380	32	40	...
8	...	.1640	32	36	...
10	...	.1900	24	32	40
12	...	.2160	24	28	...
...	1/4	.2500	20	28	36
...	5/16	.3125	18	24	32
...	3/8	.3750	16	24	32
...	7/16	.4375	14	20	28
...	1/2	.5000	13	20	28
...	9/16	.5625	12	18	24
...	5/8	.6250	11	18	24
...	3/4	.7500	10	16	20
...	7/8	.8750	9	14	20
...	1	1.0000	8	14	20

Table 3-10.   Screw thread sizes.

(Harry E. Thomas, *Handbook for Electronic Engineers and Technicians*, © 1965. Reprinted by permission of Prentice-Hall, Inc.)

which permits the screw to make its own thread in sheet metal or other relatively thin or soft material. Self-tapping screws are generally available in the sizes and with the threads per inch shown in Table 3-11, and can be obtained in all of the head types shown in Figure 3-2.

Diam. (inches)	Size #	Threads/inch
.086	2	32/56
.112	4	24/40
.138	6	18/20/32
.164	8	15/18/32
.190	10	12/16/24
.216	12	11
.250	1/4	14/20

Table 3-11.   Self-tapping screw threads and sizes.

Size of tap		Outside diameter (inches)	Size of tap drill (size of hole)				Clearance drill		Clearance (inches)
NC (U.S.S.)	NF (S.A.E.)		Number drills	Letter drills	Fractional drills	Decimal equivalent	Size	Decimal equivalent	
	#0–80	0.0600			3/64	0.0469	#51	0.0670	0.0070
#1–64		0.0730	53			0.0595	#47	0.0785	0.0055
	#1–72	0.0730	53			0.0595	#47	0.0785	0.0055
#2–56		0.0860	50			0.0700	#42	0.0935	0.0075
	#2–64	0.0860	50			0.0700	#42	0.0935	0.0075
#3–48		0.0990	47			0.0785	#36	0.1065	0.0075
	#3–56	0.0990	45			0.0820	#36	0.1065	0.0075
#4–40		0.1120	43			0.0890	#31	0.1200	0.0080
	#4–48	0.1120	42			0.0935	#31	0.1200	0.0080
#5–40		0.1250	38			0.1015	#29	0.1360	0.0110
	#5–44	0.1250	37			0.1040	#29	0.1360	0.0110
#6–32		0.1380	36			0.1065	#25	0.1495	0.0115
	#6–40	0.1380	33			0.1130	#25	0.1495	0.0115
#8–32		0.1640	29			0.1360	#16	0.1770	0.0130
	#8–36	0.1640	29			0.1360	#16	0.1770	0.0130
#10–24		0.1900	25			0.1495	13/64	0.2031	0.0131
	#10–32	0.1900	21			0.1590	13/64	0.2031	0.0131
#12–24		0.2160	16			0.1770	7/32	0.2187	0.0027
	#12–28	0.2160	14			0.1820	7/32	0.2187	0.0027
1/4″–20		0.2500	7			0.2010	17/64	0.2656	0.0156
	1/4″–28	0.2500	3			0.2130	17/64	0.2656	0.0156
5/16″–18		0.3125		F		0.2570	21/64	0.3281	0.0156
	5/16″–24	0.3125		I		0.2720	21/64	0.3281	0.0156
3/8″–16		0.3750			5/16	0.3125	25/64	0.3906	0.0156
	3/8″–24	0.3750		Q		0.3320	25/64	0.3906	0.0156
7/16″–14		0.4375		U		0.3680	29/64	0.4531	0.0156
	7/16″–20	0.4375			25/64	0.3906	29/64	0.4531	0.0156
1/2″–13		0.5000			27/64	0.4219	33/64	0.5156	0.0156
	1/2″–20	0.5000			29/64	0.4531	33/64	0.5156	0.0156
9/16″–12		0.5625			31/64	0.4844	37/64	0.5781	0.0156
	9/16″–18	0.5625			33/64	0.5156	37/64	0.5781	0.0156
5/8″–11		0.6250			17/32	0.5312	41/64	0.6406	0.0156
	5/8″–18	0.6250			37/64	0.5781	41/64	0.6406	0.0156
3/4″–10		0.7500			21/32	0.6562	49/64	0.7656	0.0156
	3/4″–16	0.7500			11/16	0.6875	49/64	0.7656	0.0156
7/8″– 9		0.8750			49/64	0.7656	57/64	0.8906	0.0156
	7/8″–14	0.8750			13/16	0.8125	57/64	0.8906	0.0156
1″– 8		1.0000			7/8	0.8750	1 1/64	1.0156	0.0156
	1″–14	1.0000			15/16	0.9375	1 1/64	1.0156	0.0156

Table 3-12.   Drill and tap sizes.

(Harry E. Thomas, *Handbook for Electronic Engineers and Technicians*, © 1965. Reprinted by permission of Prentice-Hall, Inc.)

Drill and Tap Sizes

In order to insert a machine screw or self-tapping screw into a mating hole, two different drill sizes must be considered. For a machine screw, a tap must be inserted and turned in the hole to create a female thread into which the machine screw can be threaded. Table 3-12 shows the size of the hole made by the tap drill before the tap itself can be inserted for different screw threads. As an example, a 4-40 screw or tap will require a No. 43 drill to go through the hole before the tap can be inserted. If it is desired to pass a 4-40 machine screw through a hole without threading it, then the clearance drill, a No. 31 drill, must be

NAME OF GAGE	American or Browne & Sharpe	United States Standard	Manufacturers Standard ([2])
PRINCIPAL USE	aluminum	ferrous sheet and plate (480 lb / cu ft)	ferrous sheet
No ([1])	Aluminum	THICKNESS OR DIAMETER—Inches	
2	.2576	.265625	. . . . . .
4	.2043	.234375	.2242
6	.1620	.203125	.1943
8	.1285	.171875	.1644
10	.1019	.140625	.1345
12	.08081	.109375	.1046
14	.06408	.078125	.0747
16	.05082	.0625	.0598
18	.04030	.05	.0478
20	.03196	.0375	.0359
22	.02535	.03125	.0299
24	.02010	.025	.0239
26	.01594	.01875	.0179
28	.01264	.015625	.0149
30	.01003	.0125	.0120

([1])Designation of size in decimals of an inch instead of gage numbers is recommended. If gage numbers are used, the name of the gage referred to must be specified.

([2])Recently adopted by the American Iron and Steel Institute as a modification of United States Standard Gage to reflect present average unit weights of sheet steel.

Table 3-13.   Sheet metal gauges and sizes.

used. For self-tapping screws, it is good practice to use the same size drill before the tap itself can be inserted for different screw threads. For example, a 4-24 self-tapping screw will work best in a hole drilled by a No. 43 drill.

Sheet Metal Gauges and Sizes

Almost all of the sheet metal used in electronic work is either aluminum or cold rolled steel, plated to protect it against corrosion. These sheet metals are available in standard gauges and sheet sizes, as shown in Table 3-13. Aluminum, in particular, is available in different ranges of hardness which determine its stability to bend.

BIBLIOGRAPHY—CHAPTER 3

DOUGLAS-YOUNG, J., *Illustrated Encyclopedic Dictionary of Electronics.* Englewood Cliffs, NJ, Prentice-Hall, Inc., 1981.

HARPER, C. A., *Handbook of Components for Electronics.* New York, NY, McGraw-Hill, 1977.

JACKSON, H., *Introduction to Electric Circuits* (3rd Ed.). Englewood Cliffs, NJ, Prentice-Hall, Inc., 1970.

KAUFMAN, M., AND SEIDMAN, A. H., *Handbook for Electronics Engineering Technicians* (2nd Ed.). New York, NY, McGraw-Hill, 1984.

LANDEE, R., *Electronics Designers' Handbook.* New York, NY, McGraw-Hill, 1977.

LENK, J., *Handbook of Electronic Charts, Graphs and Tables.* Englewood Cliffs, NJ, Prentice-Hall, Inc., 1970.

1975 Index to IEEE Publications. New York, NY, Institute of Electrical and Electronics Engineers, 1975.

Reference Data for Engineers: Radio, Electronics, Computer, and Communications (7th Ed.). Indianapolis, IN, Howard W. Sams & Co., Inc., 1985.

SCARLET, J. A., *Printed Circuit Boards for Microelectronics.* New York, NY, Van Nostrand Reinhold Co., 1970.

CHAPTER 4

FUNDAMENTALS
OF NETWORKS

4.0 SUMMARY

This section contains a summary of the important results that are discussed in greater detail in this chapter.

1. Terminal Properties of Electrical Elements (Voltage-Current Reactions):

$$\text{capacitor} \quad i = C\,\frac{dv}{dt} \qquad v = \frac{1}{C}\int i\,dt$$

$$\text{resistor} \quad i = Gv \qquad v = Ri$$

$$\text{inductor} \quad i = \frac{1}{L}\int v\,dt \qquad v = L\,\frac{di}{dt}$$

where v is the instantaneous voltage across and i is the instantaneous current through the element.

2. Kirchhoff's Laws:

   a. *Current:* The algebraic sum of the currents toward a node is zero.

$$\sum_{\text{node}} i_b = 0$$

   b. *Voltage:* The algebraic sum of the voltages around any closed loop of a network is zero.

$$\sum_{\text{loop}} v_b = 0$$

3. Circuit Equilibrium Laws:

   a. Node pair equations (y-equations)

$$i_1 = y_{11}(p)\, v_1 + y_{12}(p)\, v_2 + \ldots + y_{1n}(p)\, v_n$$

$$\begin{matrix} \cdot & \cdot & \cdot & \cdot \\ \cdot & \cdot & \cdot & \cdot \\ \cdot & \cdot & \cdot & \cdot \end{matrix}$$

$$i_n = y_{n1}(p)\, v_1 + y_{n2}(p)\, v_2 + \ldots + y_{nn}(p)\, v_n$$

where: y_{jj} = sum of admittance functions of all elements connected to node j.

   y_{ij} = negative sum of admittance functions of elements connecting nodes i and j.

   i_j = algebraic sum of current drivers to node j.

   b. Loop equations (z-equations)

$$v_1 = z_{11}(p)\, i_1 + z_{12}(p)\, i_2 + \ldots + z_{1n}(p)\, i_n$$

$$\begin{matrix} \cdot & \cdot & \cdot & \cdot \\ \cdot & \cdot & \cdot & \cdot \\ \cdot & \cdot & \cdot & \cdot \end{matrix}$$

$$v_n = z_{n1}(p)\, i_1 + z_{n2}(p)\, i_2 + \ldots + z_{nn}(p)\, i_n$$

where: z_{jj} = sum of impedance functions of all elements on contour of loop j.

   z_{ij} = negative sum of impedance functions of elements common to loops i and j.

   v_j = algebraic sum of voltage drivers on contour of loop j.

4. Sinusoidal Excitation:

   a. Waveform properties

   Peak value = I_m

   Full cycle average $I_{av} = \langle i \rangle = 0$

   RMS value $I_{RMS} = I = 0.707\, I_m$

   b. Circuit element properties

   resistor: resistance R/O

$$\text{capacitor: capacitive reactance } X_c = \frac{1}{\omega C} \; \underline{/-90^\circ}$$

$$\text{inductor: inductive reactance } X_L = \omega L /90^\circ$$

The angle specifies the time phase between the current through and the voltage across the element.

  5. Properties of Networks:

    a. Electrical impedances in series

$$Z_t = Z_1 + Z_2 + \ldots = \sum_j Z_j$$

    b. Electrical elements in parallel

$$Y_t = Y_1 + Y_2 + \ldots = \sum_j Y_j$$

  6. Superposition Principle: The voltage or current response of a linear circuit to a number of sources applied at the same time at any point in the circuit is equal to the sum of the individual responses due to the sources applied separately.

  7. Reciprocity Theorem: If a voltage or current source is placed at one location in a linear circuit and the current or voltage is measured at another location in the circuit, then if the locations of the source and measured points are interchanged, the measured response is the same.

  8. Thevenin and Norton Theorems:

    a. Thevenin Theorem: Relative to a pair of terminals of an extensive network with sources, the total circuit can be replaced by a single voltage source and an equivalent single series impedance.

    b. Norton Theorem: Relative to a pair of terminals of an extensive network with sources, the total circuit can be replaced by a single current source and an equivalent single shunting impedance.

  9. Source Transformation: A voltage source and a series impedance in a network can be replaced by a current source and a shunting impedance.

  10. Twoport Networks: Y- and Z-System Description: A complicated network relative to one pair of terminals as input port and a second pair of terminals as output port can be described by the equations:

$$I_1 = y_{11}V_1 + y_{12}V_2$$

$$I_2 = y_{21}V_1 + y_{22}V_2$$

and

$$V_1 = z_{11} I_1 + z_{12} I_2$$

$$V_2 = z_{21} I_1 + z_{22} I_2$$

with y_{ij} and z_{ij} being defined through proper interpretation of these equations.

11. T and π (Y and Δ) Equivalent Networks: Relative to an input and an output port, every linear network can be replaced by an equivalent T and π network.

12. Series RLC and Parallel GLC Networks:

   a. Series RLC circuit. Important in discussing such networks are:

$$\text{series resonant frequency } \omega_n = \frac{1}{\sqrt{LC}}$$

$$Q \text{ (quality) of circuit} \qquad Q = \frac{\omega_n L}{R}$$

$$\text{Bandwidth} \qquad B\omega = \frac{\omega_n}{Q}$$

   b. Parallel GLC circuit. For parallel circuits:

$$\text{parallel resonance when } \omega_n = \frac{1}{\sqrt{LC}}$$

$$Q \qquad\qquad\qquad Q = \frac{1}{\omega_n LG}$$

$$\text{Bandwidth} \qquad B\omega = \frac{\omega_n}{Q}$$

13. Transformer Networks: When two inductors interact with each other, the mutual inductance is

$$M = k\sqrt{L_1 L_2}$$

where k is the coefficient of coupling, $0 \leqslant k \leqslant 1$. For tightly coupled coils,

$$v_2 = \frac{n_2}{n_1} v_1 \quad i_2 = \frac{n_1}{n_2} i_1 \quad Z_1 = \left(\frac{n_1}{n_2}\right)^2 Z_2$$

14. Transient Response of Networks:

   a. Series RL circuit, step input excitation. The current is

$$i = \frac{V}{R}\left(1 - \epsilon^{-Rt/L}\right)$$

The quality $L/R = \tau$ is the time constant of the circuit.

   b. Series RC circuit to step voltage excitation. The current is

$$i = \frac{V}{R}\,\epsilon^{-t/RC}$$

The quantity $RC = \tau$ is the time constant of the circuit.

15. Filters: Usually the cascade of frequency-sensitive twoport networks. Simple RL and RC networks as filters; cascade of LC sections to provide low pass, high pass, band pass, and band rejection circuits. Elementary considerations of constant-K filter sections.

4.1 BASIC NETWORK PRINCIPLES

The properties of the electrical parts that are used in an electrical circuit were considered at some length in Chapter 3. We now address ourselves to an interconnection of such parts to which one or more voltage or current sources are applied. The problem of network analysis is to determine the voltages and currents in the several branches of a network. A branch is defined as having a single passive element.

As mentioned in Chapter 3, the electrical parts—resistors, capacitors, inductors—can be discussed in physical terms, such as dimensions, areas, dielectric materials, and magnetic materials. When one considers the electrical behavior, these elements are described by electrical quantities, voltage, and current, as measured respectively across and through the terminals of the element in question. Depending on which electrical quantity is considered to be independent, one can relate the voltage and current for linear elements by the sets of relations given in Table 4-1.

These relations are valid for all independent variable wave shapes. Figure 4-1a shows a schematic designation of the interrelationships. Actually the reference polarity and the reference current directions may be chosen arbitrarily, but for the selection shown, all the relations in Table 4-1 carry a + sign.

The initial studies relating the voltage and the current for the resistor were conducted by Ohm, and the relation is known as Ohm's law. The other i, v and v, i relationships ordinarily carry no name, but

Element		independent variable	
		voltage v	current i
capacitor	c ——⊣(——	$i = C\dfrac{dv}{dt}$	$v = \dfrac{1}{C}\displaystyle\int i\, dt$
resistor	R ——\/\/\/——	$i = \dfrac{v}{R}$	$v = Ri$
inductor	L ——⦙⦙⦙⦙——	$i = \dfrac{1}{L}\displaystyle\int v\, dt$	$v = L\dfrac{di}{dt}$

Table 4-1.   Terminal properties of electrical elements.

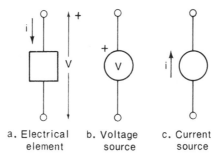

Figure 4-1. Relation between voltage and current.

may be referred to as generalized Ohm's law relationships. This generalization will actually be extended in our development of network theory.

4.2 KIRCHHOFF'S LAWS

Refer to Figure 4-2, which shows two circuits that are composed of a number of elements plus, in one case, a voltage source, and in the second case, a current source. We address ourselves to each of these networks in turn. In Figure 4-2a there is a current driver, and this establishes the voltage v as the dependent variable. Correspondingly, in Figure 4-2b, there is a voltage driver, which thus establishes the current as the dependent variable.

The symbolism introduced in these figures is to distinguish the functional relationships that exist between the current and voltage, written $y(p)$, and that between voltage and current, written $z(p)$, in

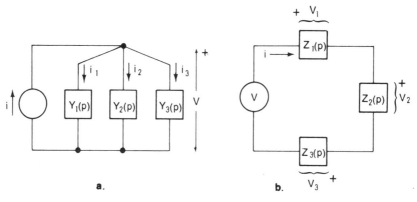

Figure 4-2. Two representative networks.

accordance with the entries in Table 4-1. Specifically,

$$v = z(p) \, i$$
$$i = y(p) \, v \qquad \textbf{(4-1)}$$

where, by definition, p denotes the derivative operator d/dt, and correspondingly, $1/p$ denotes the integral operator $\int dt$. The forms for $y(p)$ appropriate to the first column in Table 4-1 are: Cp, $1/R$, $1/Lp$. Note that p must be identified with a variable; it is an operator and not an algebraic quantity.

There are two equilibrium laws that provide the basis for writing the dependent-independent equilibrium relationships. Appropriate to the first is the Kirchhoff current law, and appropriate to the second is the Kirchhoff voltage law. The Kirchhoff current law, KCL, is essentially a statement of conservation of charge, and states that at any point in the circuit (and usually it is chosen at the junction of several branches), the rate at which the charge reaches the junction or node is equal to the rate at which it leaves the node. More often it is written, "the sum of the currents toward a node is equal to the sum of the currents away from the node," or equivalently, "the algebraic sum of the currents toward a node is zero." Symbolically, this law is written

$$\sum_{node} i_b = 0 \qquad \textbf{(4-2)}$$

where i_b denotes the branch currents. When applied to Figure 4-2a, this law requires that

$$i = i_1 + i_2 + i_3 \qquad \textbf{(4-3)}$$

More extensive networks would require that the KCL be applied to each node.

The Kirchhoff voltage law, KVL, is essentially a statement of conservation of energy when applied to a closed loop in a circuit. This law states that "the algebraic sum of the voltages around any closed loop of a network is zero," or alternatively, "the sum of the voltage rises (in going from $-$ to $+$) equals the sum of the voltage drops (in going from $+$ to $-$) when traversing a closed path in a network." Symbolically, this law is written

$$\sum_{loop} v_b = 0 \qquad \textbf{(4-4)}$$

where v_b denotes the branch voltages. When applied to Figure 4-2b, this law specifies that

Single Node Example

$$v = v_1 + v_2 + v_3 \tag{4-5}$$

Write the node equilibrium equation for the network shown in Figure 4-3.

Solution: By a direct application of the KCL, we write

$$i = i_1 + i_2 + i_3$$

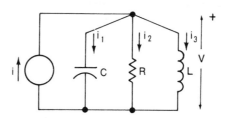

Figure 4-3. Network for a single node.

Now we use the appropriate v, i relations from Table 4-1; the resulting expression is

$$i = Cpv + \frac{v}{R} + \frac{1}{Lp}v \tag{4-6}$$

This is an integrodifferential equation that relates the node voltage v with the source current driver i. Later we shall discuss the solution of this integrodifferential equation.

It is usual to write this expression in the operational form

$$i = \left(Cp + \frac{1}{R} + \frac{1}{Lp} \right) v$$

Now we extend the concept of $y(p)$ by writing

$$i = Y(p)\, v$$

where

$$Y(p) = y_1(p) + y_2(p) + y_3(p) \tag{4-7}$$

This latter is a very important generalization that relates $Y(p)$, called the input or driving point admittance operator, with the admittance operators of the separate branches $y(p)$.

Two-Node Example

Write the node equations for the two-node network shown in Figure 4-4.

Solution: The KCL is applied to each node. Then

$$i = i_1 + i_2 + i_3$$

$$0 = -i_3 + i_4 + i_5$$

Using the v, i relationships from Table 4-1, these equations can be written

$$i = \frac{v_1}{R_1} + C_1 p v_1 + \frac{1}{L_1 p}(v_1 - v_2)$$

$$0 = -\frac{1}{L_1 p}(v_1 - v_2) + \frac{v_2}{R_2} + \frac{1}{L_2 p} v_2$$

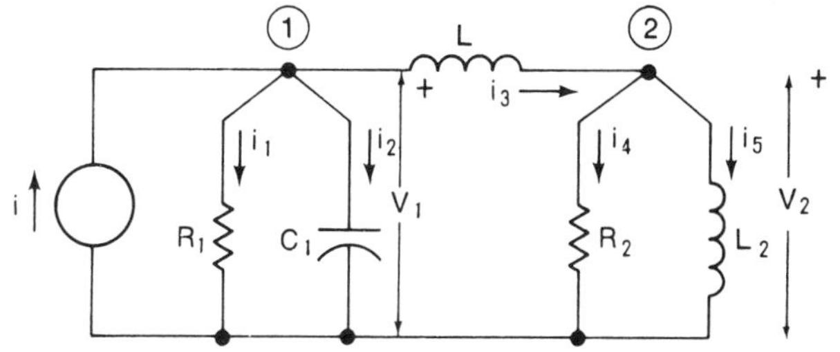

Figure 4-4.   Network for two nodes.

In determining the response variables v_1 and v_2, these simultaneous integrodifferential equations must be solved.

Suppose that we regroup these terms in the following manner:

$$i = \left(\frac{1}{R_1} + C_1 p + \frac{1}{L_1 p} \right) v_1 - \frac{1}{L_1 p} v_2$$

$$0 = -\frac{1}{L_1 p} v_1 + \left(\frac{1}{R_2} + \frac{1}{L_1 p} + \frac{1}{L_2 p} \right) v_2$$

(**4-8**)

These equations can now be written in the following form:

$$i = y_{11}(p) v_1 + y_{12}(p) v_2$$

$$0 = y_{21}(p) v_1 + y_{22}(p) v_2$$

(**4-9**)

where, by examining the contributing terms in each $y(p)$, we see that

$y_{11}(p)$ = the sum of all $y(p)$'s that connect to node 1

$y_{22}(p)$ = the sum of all $y(p)$'s that connect to node 2

$y_{12}(p) = y_{21}(p)$ = the negative sum of all $y(p)$'s that are common to nodes 1 and 2

$\quad\quad i$ = the algebraic sum of all current sources connected to node 1

This is a general approach to writing network equations and is known as the *node-pair* formulation. The form is general and can be extended to a network of any number of nodes and any number of current drivers.

Single-Loop Example

Write the loop equilibrium equation for the network shown in Figure 4-5.

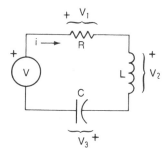

Figure 4-5. Network for a single loop.

Solution: the KVL is applied around the loop to get

$$v = v_1 + v_2 + v_3$$

Now use the appropriate i, v relations from Table 4-1. We then write

$$v = Ri + Lpi + \frac{1}{Cp}i$$

This is the integrodifferential equation that relates the loop current with the source voltage for the series network shown. The solution of this type of differential equation will be considered later.

We now extend the concept of the impedance operator $z(p)$. Let us write this dynamic equation in the form

$$v = \left(R + Lp + \frac{1}{Cp} \right) i$$

which is now written as

$$v = Z(p) \, i$$

where

$$Z(p) = z_1(p) + z_2(p) + z_3(p) \qquad \text{(4-10)}$$

where each $z(p)$ corresponds to a term in the above equation. This latter generalization is very important in that it relates the impedance function $Z(p)$ with the impedance functions of the separate branches of the loop.

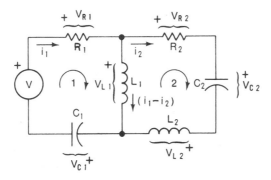

Figure 4-6.   Network for two loops.

Two-Loop Example

Write the loop equilibrium equations for the two-loop network shown in Figure 4-6.

Solution: The KVL is applied to the left- and right-hand loops, with the resulting equations

$$v = v_{R1} + v_{L1} + v_{C1}$$

$$0 = -v_{L1} + v_{R2} + v_{C2} + v_{L2}$$

Using the i, v relationships from Table 4-1, these equations can be written in the form

$$v = R_1 i_1 + L_1 p (i_1 - i_2) + \frac{1}{C_1 p} i_1$$

$$0 = -L_1 p (i_1 - i_2) + R_2 i_2 + \frac{1}{C_2 p} i_2 + L_2 p i_2 \qquad \text{(4-11)}$$

The response variables i_1 and i_2 would be obtained from a solution of these simultaneous integrodifferential equations.

This set of equations is regrouped as follows:

$$v = \left(R_1 + L_1 p + \frac{1}{C_1 p} \right) i_1 - L_1 p i_2$$

$$0 = -L_1 p i_1 + \left(R_2 + \frac{1}{C_2 p} + L_1 p + L_2 p \right) i_2 \qquad (4\text{-}12)$$

These equations are written

$$v = z_{11}(p)\, i_1 + z_{12}(p)\, i_2$$

$$0 = z_{21}(p)\, i_1 + z_{22}(p)\, i_2 \qquad (4\text{-}13)$$

where, by examining the contributing terms in each $z(p)$, we see that

$z_{11}(p) =$ the sum of all $z(p)$'s on the contour of loop 1

$z_{22}(p) =$ the sum of all $z(p)$'s on the contour of loop 2

$z_{12}(p) = z_{21}(p) =$ the negative sum of all $z(p)$'s that are common to loops 1 and 2

$v =$ the algebraic sum of all voltage drivers in loop 1

This is a general approach to writing network equations on a *loop* basis. It can be extended to a network of any number of loops and any number of voltage drivers.

4.3 SINUSOIDAL EXCITATION FUNCTIONS

The properties of system response to sinusoidal excitation functions are of considerable importance in practical problems. There are several reasons for this: (a) sinusoidal sources over wide frequency ranges and appropriate instrumentation are readily available; (b) many important areas—e.g., communication systems—operate largely with sinusoidal wave forms; (c) the basic sources of electrical power are sinusoidal.

Sinusoidal wave forms are illustrated in Figure 4-7. As indicated, the waves are described analytically as $I_{m1} \sin \omega t$ and $I_{m2} \sin(\omega t + \theta)$. I_{m1} and I_{m2} denote the peak amplitudes of the sinusoids, ω is the angular frequency and is related to the source frequency f by $f = \dfrac{\omega}{2\pi}$, and θ denotes the phase angle of the sinusoid with respect to $\omega t = 0$. In this figure, i_2 is said to lead i_1 by the phase angle θ, or correspondingly, i_1 lags i_2 by the phase angle θ.

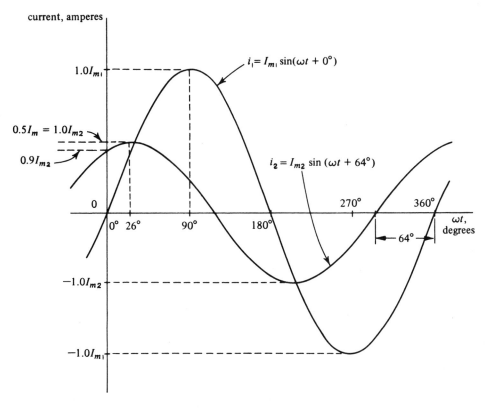

Figure 4-7. Properties of the sinusoidal time function.

(A. Pike, *Fundamentals of Electronic Circuits,* © 1971. Reprinted by permission of Prentice-Hall, Inc.)

Important in the discussion of circuit properties involving sinusoidal signals (voltage or current) are the time average (or full cycle average) and the rms (root mean squared) value. The full cycle average is given by

$$I_{avg} = \langle i \rangle = \frac{1}{T} \int_0^T i \, dt = \frac{1}{2\pi} \int_0^{2\pi} i \, d(\omega t) \qquad (4\text{-}14)$$

This leads to zero for the sinusoid since

$$\langle i \rangle = \frac{1}{2\pi} \int_0^{2\pi} I_m \sin \omega t \, d(\omega t) = \frac{1}{2\pi} [-\cos \omega t]_0^{2\pi} = 0$$

The rms value is defined by the expression

$$I_{RMS} = I = \sqrt{\frac{1}{2\pi} \int_0^{2\pi} i^2 \, d(\omega t)} \qquad (4\text{-}15)$$

For the sinusoid, this yields

$$I_{RMS} = \sqrt{\frac{1}{2\pi} \int_0^{2\pi} I_m^2 \sin^2 \omega t \, d(\omega t)}$$

$$= \sqrt{\frac{I_m^2}{2\pi} \left(\frac{\omega t}{2} + \frac{\sin 2\omega t}{4}\right) \Big|_0^{2\pi}} = \frac{I_m}{\sqrt{2}} = 0.707 \, I_m$$

These calculations show that the full cycle average of a sinusoid is zero, whereas the RMS value is a constant equal to 0.707 of the peak value of the sinusoid. Moreover, these results are independent of any time phase displacement.

4.4 STEADY STATE SYSTEM RESPONSE TO SINUSOIDAL EXCITATION FUNCTIONS

Before considering connected networks, let us develop a table like Table 4-1, but for the case when the independent variables are sinusoids with zero phase angle. See Table 4-2.

It is seen from these entries that:

For the capacitor, the current leads the voltage by 90 degrees.
For the resistor, the current is in phase with the voltage.
For the inductor, the current lags the voltage by 90 degrees.

It is customary to define the following quantities that are associated

Element	independent variable	
	voltage = $V_m \sin\omega t$	current = $I_m \sin\omega t$
capacitor	$i = \omega C V_m \sin(\omega t + \pi/2)$	$v = \dfrac{I_m}{\omega C} \sin(\omega t - \pi/2)$
resistor	$i = \dfrac{V_m}{R} \sin \omega t$	$v = R I_m \sin \omega t$
inductor	$i = \dfrac{V_m}{\omega L} \sin (\omega t - \pi/2)$	$v = \omega L I_m \sin (\omega t + \pi/2)$

Table 4-2. Properties of electrical elements for sinusoidal independent variables.

with electrical elements when in sinusoidal environments:

capacitor: capacitive reactance $X_C = \dfrac{I}{\omega C\underline{/90}}$; susceptance $B_C = \omega C\underline{/90}$

resistor: resistance $R = R\underline{/0}$; conductance $\dfrac{I}{R} = G = G\underline{/0}$

inductor; inductive reactance $X_L = \omega L\underline{/90}$; susceptance $B_L = \dfrac{I}{\omega L}\underline{/90°}$

We shall also write these quantities in a different, though equivalent, form. Here we use the Euler identity

$$\epsilon^{+jx} = \cos x + j \sin x \tag{4-16}$$

from which we see that

$$\cos x = \text{Re}\ [\epsilon^{jx}]; \quad \sin x = \text{Im}\ [\epsilon^{jx}] \tag{4-17}$$

where Re and Im are contractions of *real* and *imaginary*, and specify which term in the general expansion is to be used. We observe now that

$$\frac{d}{dx}\epsilon^{jx} = j\epsilon^{jx} \quad \epsilon^{jx}\ dx = \frac{\epsilon^{jx}}{j} = -j\epsilon^{jx}$$

so that differentiation introduces j and integration introduces $-j$. But by direct substitution in Equation (4-16).

$$\epsilon^{j\pi/2} = 0 + j1$$

Hence, j denotes a phase shift of $+90$ degrees; correspondingly, $-j$ denotes a phase shift of -90 degrees. Thus, equivalent to the above tabulation for $\epsilon^{j\omega t}$ are:

resistor: resistance $R = R$; conductance $= G = \dfrac{1}{R}$

capacitor: capacitive reactance $X_C = \dfrac{1}{j\omega C}$; susceptance $B_C = j\omega C$

inductor: inductive reactance $X_L = j\omega L$; susceptance $B_L = \dfrac{1}{j\omega L}$

Now let us apply these ideas to the Single Node Example of Figure 4-3. Here, if we assume that a sinusoidal excitation function $i = \text{Re}\ I_m\epsilon^{j\omega t}$ is applied, then, because differentiation and integration do not alter the form of the exponential function, we expect that the voltage v

will be $v = \text{Re } V_m \epsilon^{j\omega t}$. For the network shown, Equation (4-6) becomes

$$\text{Re }[I_m \epsilon^{j\omega t}] = \text{Re }\left[\left(j\omega C + \frac{1}{R} + \frac{1}{j\omega L}\right) V_m \epsilon^{j\omega t}\right]$$

from which we write the admittance function

$$Y(j\omega) = \frac{I_m}{V_m} = \left(j\omega C + \frac{1}{R} + \frac{1}{j\omega L}\right) \tag{4-18}$$

Attention is called to the fact that, for sinusoidal excitation in the steady state, the time derivative operator $p \rightarrow$ complex quantity $j\omega$, and the function

$$Y(p) \rightarrow Y(j\omega)$$

Correspondingly, for the series RLC circuit of the Single Loop Example, we shall find that

$$Z(j\omega) = \left(R + j\omega L + \frac{1}{j\omega C}\right) \tag{4-19}$$

Note that, in general, $Y(j\omega)$ for any circuit will be a complex quantity, and there is both magnitude and angle associated with these quantities. Specifically, we can write

$$Y(j\omega) = \sqrt{\left(\frac{1}{R}\right)^2 + \left(\omega C - \frac{1}{\omega L}\right)^2} \underline{\bigg/ \tan^{-1} \frac{\omega C - 1/\omega L}{1/R}} \tag{4-20}$$

with a similar form for $Z(j\omega)$.

4.5 PROPERTIES OF NETWORKS

There are a number of general network properties that can be developed from the material presented above. We shall consider some of the more important properties.

a. *Electrical elements in series.* As an extension of the Single Loop Example, the impedance of a number of impedances in series is

$$Z = Z_1 + Z_2 + Z_3 + \ldots \tag{4-21}$$

that is, for series circuits, impedances add directly (but with due account of the phase angles). From Equation (4-21)

resistors: $R = R_1 + R_2 + R_3 + \ldots$ add resistors in series inductors: $L = L_1 + L_2 + L_3 + \ldots$ add inductors in series

capacitors: $\dfrac{1}{C} = \dfrac{1}{C_1} + \dfrac{1}{C_2} + \dfrac{1}{C_3} + \cdot$

b. *Electrical elements in parallel.* As an application of Equation (4-7), the admittance of a number of admittances in parallel is

$$Y = Y_1 + Y_2 + Y_3 + \ldots \qquad (4\text{-}22)$$

It follows from this equation that for parallel

resistors: $G = \left(\dfrac{1}{R}\right) = G_1 + G_2 + G_3 + \ldots$

inductors: $\dfrac{1}{L} = \dfrac{1}{L_1} + \dfrac{1}{L_2} + \dfrac{1}{L_3} + \ldots$

capacitors: $C = C_1 + C_2 + C_3 + \ldots$ add capacitors in parallel

For the special case of two impedances in parallel, $Z_1 \| Z_2$,

$$Y = Y_1 + Y_2$$

or equivalently

$$\frac{1}{Z} = \frac{1}{Z_1} + \frac{1}{Z_2}$$

so that

$$Z = \frac{Z_1 Z_2}{Z_1 + Z_2}$$

c. *Series-parallel combination of electrical elements.* We discuss this by means of two examples.

Series-Parallel Example

Refer to the diagram of Figure 4-8 and find the total effective R, L, C, as appropriate, if all Z's are one class of elements.

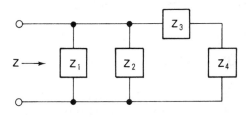

Figure 4-8. Series-parallel network.

Solution: We observe that this network consists of three parallel paths: Z_1, Z_2, and $Z_3 + Z_4$. We write directly that

$$Y = Y_1 + Y_2 + Y_{34}$$

where $Y_{34} = 1/Z_{34} = 1/(Z_3 + Z_4)$. We now write for the case when all Z's are

resistors: $\quad \dfrac{1}{R} = \dfrac{1}{R_1} + \dfrac{1}{R_2} + \dfrac{1}{R_3 + R_4}$

inductors: $\quad \dfrac{1}{L} = \dfrac{1}{L_1} + \dfrac{1}{L_2} + \dfrac{1}{L_3 + L_4}$

capacitors: $\quad C = C_1 + C_2 + \dfrac{1}{\dfrac{1}{C_3} + \dfrac{1}{C_4}} = C_1 + C_2 + \dfrac{C_3 C_4}{C_3 + C_4}$

Repeat the above example where

$Z_1 = R_1 + j\omega L_1$; series combination of R_1 and L_1

$Z_2 = 1/j\omega C_2$; a pure capacitor C_2

$$Z_3 = R_3 \| C_3 = \dfrac{R_3 \dfrac{1}{j\omega C_3}}{R_3 + \dfrac{1}{j\omega C_3}} = \dfrac{R_3}{1 + j\omega R_3 C_3}; \quad R_3 \text{ and } C_3 \text{ in parallel}$$

$$Z_4 = L_4 \| C_4 = \dfrac{j\omega L_4 \dfrac{1}{j\omega C_4}}{j\omega L_4 + \dfrac{1}{j\omega C_4}} = \dfrac{L_4/C_4}{1 - \omega^2 L_4 C_4}; \quad L_4 \text{ and } C_4 \text{ in parallel}$$

Solution: Using the general expression for combining impedances in parallel, we have

$$Y = \dfrac{1}{R_1 + j\omega L_1} + j\omega C_2 + \dfrac{1}{\dfrac{R_3}{1 + j\omega R_3 C_3} + \dfrac{L_4/C_4}{1 - \omega^2 L_4 C_4}}$$

and the input impedance $Z = 1/Y$. We shall not carry out the algebra involved in combining the terms, which is straightforward but quite lengthy.

4.6 SUPERPOSITION PRINCIPLE

In all our past examples, a single driving source has been assumed. We now extend this and examine the consequences of having a network with multiple sources. This is done by reference to an example.

Two-Source Example

Consider the two-loop circuit shown in Figure 4-9. Find expressions for the currents I_1 and I_2.

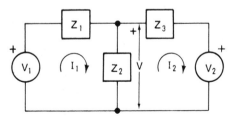

Figure 4-9.   Two-source network.

Solution: By means of the loop method of analysis [Equation (4-13)], the network equations are:

$$V_1 = (Z_1 + Z_2)\, I_1 - Z_2 I_2$$

$$-V_2 = -Z_2 I_1 + (Z_2 + Z_3)\, I_2$$

The currents are obtained by solving this pair of simultaneous equations. This is done by Cramer's rule, which yields

$$I_1 = \frac{\begin{vmatrix} V_1 & -Z_2 \\ -V_2 & Z_2 + Z_3 \end{vmatrix}}{\begin{vmatrix} Z_1 + Z_2 & -Z_2 \\ -Z_2 & Z_2 + Z_3 \end{vmatrix}} = \frac{V_1(Z_2 + Z_3) - V_2 Z_2}{Z_1 Z_2 + Z_2 Z_3 + Z_3 Z_1}$$

$$I_2 = \frac{\begin{vmatrix} Z_1 + Z_2 & V_1 \\ -Z_2 & -V_2 \end{vmatrix}}{\begin{vmatrix} Z_1 + Z_2 & -Z_2 \\ -Z_2 & Z_2 + Z_3 \end{vmatrix}} = \frac{-V_2(Z_1 + Z_2) + V_1 Z_1}{Z_1 Z_2 + Z_2 Z_3 + Z_3 Z_1}$$

These expressions for the currents show that each voltage source contributes separately to the currents. We write the expressions for current as

$$I_1 = I_1^{(1)} + I_1^{(2)}$$

$$I_2 = I_2^{(1)} + I_2^{(2)}$$

where (1) denotes the contribution due to source V_1 and (2) denotes the contribution due to source V_2. These results can be interpreted and generalized to show that "the voltage or current response of a *linear* circuit to a number of sources applied at the same time at any number of points in the circuit is equal to the sum of the individual responses due to the sources applied separately.'' This is the superposition principle; it is valid only for linear circuits.

4.7 RECIPROCITY THEOREM

We now wish to demonstrate that linear circuits obey the reciprocity theorem. It says, in effect, if a voltage or current source is placed at one location in a reciprocal circuit and if the current or voltage is measured at another location in the circuit, then if the locations of the driving source and measured points are interchanged, the measured response remains the same. As an example, consider a slight modification of the Two-Source Example.

Reciprocity Example

Two circuits are shown in Figure 4-10. Find expressions for I_a for each circuit.

Solution: By a straightforward application of the loop equations, we find for circuit a:

$$V = (Z_1 + Z_2) I_1 - Z_2 I_a$$

$$0 = -Z_2 I_1 + (Z_2 + Z_3 + Z_4) I_a$$

From these we calculate I_a:

$$I_a = \frac{\begin{vmatrix} Z_1 + Z_2 & V \\ -Z_2 & 0 \end{vmatrix}}{\begin{vmatrix} Z_1 + Z_2 & -Z_2 \\ -Z_2 & Z_2 + Z_3 + Z_4 \end{vmatrix}} = \frac{VZ_2}{\text{Denominator}}$$

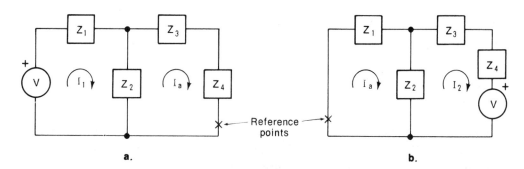

Figure 4-10.   Reciprocity of voltage sources.

Now refer to Figure 4-10b. The loop equations are

$$0 = (Z_1 + Z_2) I_a - Z_2 I_2$$

$$-V = -Z_2 I_a + (Z_2 + Z_3 + Z_4) I_2$$

From these we calculate I_a:

$$I_a = \frac{\begin{vmatrix} 0 & -Z_2 \\ -V\, Z_2 + Z_3\ + Z_4 \\ Z_1 + Z_2 & -Z_2 \\ -Z_2\ Z_2 + Z_3 + Z_4 \end{vmatrix}}{} = \frac{-VZ_2}{\text{Denominator}}$$

If we now take into account the relative positive polarity of the source
and the current directions, I_a being essentially out of the $+$ marked
terminal in Figure 4-10a and into the $+$ marked terminal in Figure
4-10b, then we account for this reversal in sense by introducing a
$-$ sign. Thus, we see that our example is in accord with the reciprocity
theorem.

4.8 THEVENIN THEOREM; NORTON THEOREM

Suppose that we have a complicated network with a number of voltage
and current sources of the same frequency, and suppose also that we
are interested in the current through a particular element in the circuit.
We could proceed by writing the loop equations (or node pair equations) and then solving for the current through the element in question.
The Thevenin theorem addresses itself to the entire network, less the
element in question, and when viewed from the specified terminals,
replaces this by an equivalent network containing an equivalent single
source and an equivalent single series impedance. Figure 4-11a shows
a general network containing sources and the element of interest; Figure 4-11b is the Thevenin equivalent circuit.

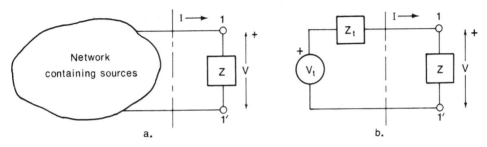

Figure 4-11.   Networks containing sources.

The Thevenin source V_t is the voltage that appears across the terminals 1-1' when Z is open circuited (i.e., Z is set equal to infinity); the Thevenin impedance Z_t is the impedance looking back into the network through terminals 1-1' when all independent source voltages are reduced to zero (internal source impedances must be retained in the circuit).

The Norton theorem is the current counterpart of the Thevenin theorem and states that the general network containing sources, as shown in Figure 4-11a, can be replaced by a current source I_n and a shunt impedance Y_n, as shown in Figure 4-12. In this network, I_n is the current that appears in a short circuit across the terminals 1-1', and Y_n is the inverse of the Thevenin impedance.

A proof of the Norton theorem is readily effected by a simple transformation of the Thevenin circuit. To do this, we write from an inspection of Figure 4-11b:

$$V = V_t - IZ_t$$

Rearrange this as follows:

$$\frac{V_t}{Z_t} - I = \frac{V}{Z_t}$$

Now write this

$$I = I_n - VY_n$$

where $Y_n = 1/Z_t$. This equation defines Figure 4-12. Observe that when the terminals 1-1' are shorted, $I = I_n$, as specified.

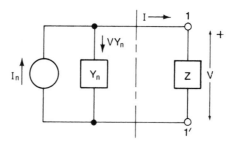

Figure 4-12.   Norton equivalent network.

Three-Impedance Example

Deduce the Thevenin and the Norton circuits relative to element Z in the network shown in Figure 4-13.

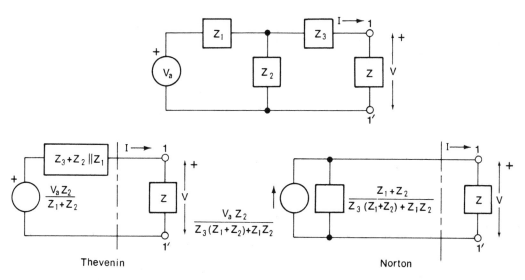

Figure 4-13. Three-impedance network, Thevenin and Norton equivalent.

Solution: Relative to terminals 1-1', the Thevenin impedance is

$$Zt = Z_3 + Z_2 \| Z_1$$

$$= Z_3 + \frac{Z_1 Z_2}{Z_1 + Z_2}$$

Also, for V_t, with 1-1' open circuited,

$$V_t = \frac{V_a Z_2}{Z_1 + Z_2}$$

From these

$$I = \frac{V_t}{Z_t + Z} = \frac{V_a Z_2/(Z_1 + Z_2)}{\dfrac{Z_3 + Z_1 Z_2 + Z}{Z_1 + Z_2}} = \frac{V_a Z_2}{(Z_3 + Z)(Z_1 + Z_2) + Z_1 Z_2}$$

For the Norton equivalent, we require

$$I_n = \frac{V_t}{Z_t} = \frac{V_a Z_2}{Z_3(Z_1 + Z_2) + Z_1 Z_2}$$

The equivalent circuits are given in the accompanying figures.

The Thevenin and Norton theorems seldom save many steps in the actual solution of a particular problem. However, the equivalent networks often provide an important step in general circuit analysis.

4.9 SOURCE TRANSFORMATIONS

A very useful concept is that of source transformation, which allows a voltage source and a series impedance to be replaced by an equivalent current source and a shunting admittance. The results bear a strong resemblance to the Thevenin-Norton transformation, but now we are concerned with single elements that might appear embedded within a larger network. The present results are shown in Figure 4-14. A proof will parallel that for the Thevenin-Norton transformation.

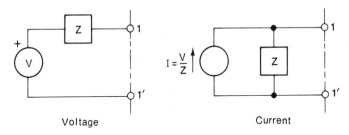

Figure 4-14.   Source transformation theories.

Source Transformation Example

Find the voltage across Z_2 in the network of Figure 4-15.

Solution: According to the methods in Figure 4-9, the voltage across Z_2 is $V = Z_2(I_1 - I_2)$ and both currents have been evaluated.

Now apply the source transformations discussed here and redraw the circuit as shown. We now write the solution directly:

$$V = \left(\frac{V_1}{Z_1} + \frac{V_2}{Z_3}\right)(Z_1\|Z_2\|Z_3) = \frac{V_1Y_1 + V_2Y_3}{Y_1 + Y_2 + Y_3}$$

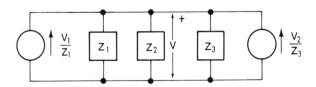

Figure 4-15.   Three-impedance source transformation.

4.10 TWO PORTS: Y- AND Z-SYSTEM DESCRIPTION

A problem of considerable interest is the response of passive networks at one pair of terminals when an excitation is applied to another pair of terminals. In such a case, we are really discussing twoport networks. Here, as in the discussion of the Thevenin and Norton theorems there is the implication that the total network, except for the input and output ports, can be represented by a relatively simple equivalent network. The situation is illustrated in Figure 4-16.

We begin with the general network in Figure 4-16a. If we assume

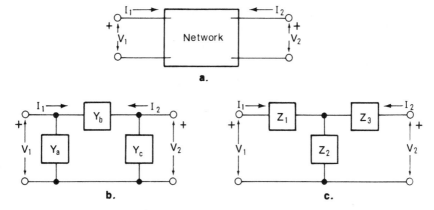

Figure 4-16.   Twoport network analysis

that the currents are independent variables, then we would apply the KCL to all nodes of the network. Calling the input 1 and the output 2, we can then solve the resulting set of simultaneous equations by Cramer's rule to find

$$I_1 = \frac{\Delta_{11}V_1}{\Delta} + \frac{\Delta_{12}V_2}{\Delta}$$

$$I_2 = \frac{\Delta_{21}V_1}{\Delta} + \frac{\Delta_{22}V_2}{\Delta}$$

(4-23)

where Δ is the network determinant, and $\Delta_{12} = \Delta_{21}$ are the first cofactors. We now introduce the quantities

$$y_{11} = \frac{\Delta_{11}}{\Delta} \quad y_{12} = y_{21} = \frac{\Delta_{12}}{\Delta} = \frac{\Delta_{21}}{\Delta} \quad y_{22} = \frac{\Delta_{22}}{\Delta}$$

Equations (4-23) become

$$I_1 = y_{11}V_1 + y_{12}V_2$$

$$I_2 = y_{21}V_1 + y_{22}V_2$$

(4-24)

These equations are used to interpret the coefficients y_{ij}. It is seen that

$$y_{11} = \frac{I_1}{V_1}\bigg|V_2 = 0 = \begin{array}{l}\text{driving point admittance at port 1 when port 2 is}\\ \text{short circuited}\end{array}$$

$$y_{22} = \frac{I_2}{V_2}\bigg|V_1 = 0 = \begin{array}{l}\text{driving point admittance at port 2 when port 1 is}\\ \text{short circuited.}\end{array}$$

$$y_{12} = \frac{I_1}{V_2}\bigg|V_1 = 0 = \text{transfer admittance when input is short circuited}$$

$$y_{21} = \frac{I_2}{V_1}\bigg|V_2 = 0 = \text{transfer admittance when output is short circuited}$$

Because of the form of these coefficients, the y-system description is often referred to as the short-circuit description of the network.

If the V's were the independent variables, then the network would be described by applying the KVL to the loops. The resulting set of equations, when solved for the input and output, yields

$$V_1 = \frac{\Delta_{11}}{\Delta} I_1 + \frac{\Delta_{12}}{\Delta} I_2$$

$$V_2 = \frac{\Delta_{21}}{\Delta} I_1 + \frac{\Delta_{22}}{\Delta} I_2 \tag{4-25}$$

where Δ is the network determinant and $\Delta_{12} = \Delta_{21}$ are the first cofactors. We now introduce the quantities

$$z_{11} = \frac{\Delta_{11}}{\Delta} \quad z_{12} = z_{21} = \frac{\Delta_{12}}{\Delta} = \frac{\Delta_{21}}{\Delta} \quad z_{22} = \frac{\Delta_{22}}{\Delta}$$

and Equation (4-25) becomes

$$V_1 = z_{11}I_1 + z_{12}I_2$$

$$V_2 = z_{21}I_1 + z_{22}I_2 \tag{4-26}$$

The Z_{ij} coefficients are obtained from these equations. They are:

$z_{11} = \dfrac{V_1}{I_1}\bigg|_{I_2 = 0} =$ driving point impedance at port 1 when port 2 is open circuited

$z_{22} = \dfrac{V_2}{I_2}\bigg|_{I_1 = 0} =$ driving point impedance at port 2 when port 1 is open circuited

$z_{12} = \dfrac{V_1}{I_2}\bigg|_{I_1 = 0} =$ transfer impedance when input is open circuited (also called the reverse transfer impedance)

$z_{21} = \dfrac{V_2}{I_1}\bigg|_{I_2 = 0} =$ transfer impedance when output is open circuited (also called the forward transfer impedance)

Since both sets of Equations (4-24) and (4-26) refer to the same network, then clearly, one can invert one set of equations to obtain the second set. If this is done, it will be found that

$$z_{11} = \frac{y_{22}}{\Delta y} \quad z_{12} = z_{21} = -\frac{y_{12}}{\Delta y} \quad z_{22} = \frac{y_{11}}{\Delta y} \tag{4-27}$$

$$\Delta y = y_{11} y_{22} - y_{12} y_{21}$$

4.11 T AND Π (Y AND Δ) EQUIVALENT NETWORKS

We shall now show that the general network descriptions of the twoport allow simple equivalent network representations. These are the Π and the T networks shown in Figure 4-16b and 4-16c. In fact, by applying the descriptions of the coefficients Y_{ij} and Z_{ij} to these networks, we shall find the following:

Parameter	T (or Y) Section	Π (or Δ) Section
y_{11}	$\dfrac{y_1(y_2 + y_3)}{y_1 + y_2 + y_3}$	$y_a + y_b$
y_{22}	$\dfrac{y_3(y_1 + y_2)}{y_1 + y_2 + y_3}$	$y_b + y_c$
$y_{12} = y_{21}$	$\dfrac{y_1 y_3}{y_1 + y_2 + y_3}$	y_b
z_{11}	$z_1 + z_2$	$\dfrac{z_a(z_b + z_c)}{z_a + z_b + z_c}$
z_{22}	$z_2 + z_3$	$\dfrac{z_c(z_a + z_b)}{z_a + z_b + z_c}$
$z_{12} = z_{21}$	z_2	$\dfrac{z_a z_c}{z_a + z_b + z_c}$

Table 4-3.   System parameters for T and Π networks.

By equating the quantities on the separate lines, sets of equations are obtained. When solved, these relate the y's with the z's in Figure 4-16. The results are:

$$y_a = \frac{z_3}{Z^2} \quad y_b = \frac{z_2}{Z^2} \quad y_c = \frac{z_1}{Z^2} \qquad (4\text{-}28)$$

where $Z^2 = z_1 z_2 + z_2 z_3 + z_3 z_1$. The corresponding expressions for the z's are

$$z_1 = \frac{y_c}{Y^2} \quad z_2 = \frac{y_b}{Y^2} \quad z_3 = \frac{y_a}{Y^2} \qquad (4\text{-}29)$$

where $Y^2 = y_a y_b + y_b y_c + y_c y_a$.

4.12 SERIES RLC AND PARALLEL GLC NETWORKS

Series RLC Circuit

We wish to discuss the important properties of the series RLC circuit, and also of the parallel GLC circuit. We begin our considerations with Equation (4-19) for the driving point impedance of the series RLC circuit, which we write

$$Z = \left(R + j\omega L + \frac{1}{j\omega C}\right) = \sqrt{R^2 + \left(\omega L - \frac{1}{\omega C}\right)^2} \Big/ \underline{\tan^{-1} \frac{\omega L - 1/\omega C}{R}}$$

(4-30)

Owing to the different variation with frequency of the capacitive and inductive reactances, both the magnitude and phase angle are frequency dependent. A sketch of the variation of the inverse of $Z(j\omega)$, the driving point admittance $Y(j\omega)$, is given in Figure 4-17. We observe that the magnitude function

$$Y(j\omega) = \frac{1}{\sqrt{R^2 + (\omega L - 1/\omega C)^2}}$$

(4-31)

has a maximum value $= 1/R$ when the quantity

$$\omega_{nL} = \frac{1}{\omega_n C}$$

or when

$$\omega_n = \frac{1}{\sqrt{LC}}$$

(4-32)

ω_n is defined as the *resonant* angular frequency, and the circuit is in series resonance when the capacitive reactance equals the inductive reactance. Also, at this frequency, the phase angle is zero. For frequencies less than ω_n, the phase angle is positive or leading, and for frequencies higher than ω_n, the phase angle is negative or lagging.

Additional properties of the series RLC circuit follow from the expression for $Y(j\omega)$ or $Z(j\omega)$. Let us consider the case when the net reactance is equal to the resistance; i.e., when

$$\omega_p L - \frac{1}{\omega p C} = \mp R$$

(4-33)

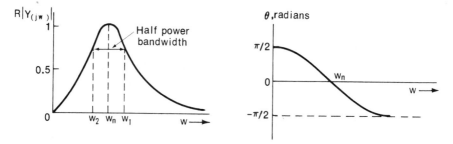

Figure 4-17.   Frequency response of series RLC circuit.

Under these conditions, the phase angle of the function will be ± 45 degrees and the amplitude has a magnitude $Z(j\omega_p) = \sqrt{2}\,R$ and $Y(j\omega_p) = 0.707/R$. From this equation, which is written

$$\omega_p{}^2 \pm \frac{R\omega}{L}\,p - \frac{1}{LC} = 0$$

we find that

$$\omega_p = \mp\frac{R}{2L} \pm \sqrt{\left(\frac{R}{2L}\right)^2 + \frac{1}{LC}}$$

which we write as

$$\omega_p = \omega_n(\mp\zeta \pm \sqrt{\zeta^2 + 1}) \tag{4-34}$$

where, by definition,

$$\zeta\omega_n = \frac{R}{2L} \qquad \zeta = \frac{R}{2}\sqrt{\frac{L}{C}}$$

For the case when ζ is small,

$$\omega_p = \omega_n(1 \mp \zeta) \tag{4-35}$$

We designate the upper value

$$\omega_1 = \omega_n(1 + \zeta)$$

and the lower value

$$\omega_2 = \omega_n(1 - \zeta) \tag{4-36}$$

The total frequency spread, $\omega_1 - \omega_2$, appropriate to these 0.707 points (at these points, the current is 0.707 of the maximum value, and the power, which is I^2R, is down to half the maximum value) is called the half-power or 3-dB bandwidth of the circuit. Thus

$$B_w = \omega_1 - \omega_2 = 2\zeta\omega_n \tag{4-37}$$

Resonant circuits are often discussed in terms of the circuit Q, sometimes called the "quality" of the circuit. The Q of a series RLC circuit is defined by the relation

$$Q = \frac{\omega_n L}{R} = \frac{\frac{1}{2}\omega_n}{R/2L} = \frac{\frac{1}{2}\omega_n}{\zeta\omega_n} = \frac{1}{2\zeta} \tag{4-38}$$

In terms of the circuit Q, the bandwidth is

$$B\omega = \frac{\omega_n}{Q} \tag{4-39}$$

This expression shows that the bandwidth is narrow for high Q circuits, and is broad for low Q circuits. This means that to increase the bandwidth of a series circuit, the Q must be reduced, or correspondingly, R of the circuit must be increased.

As a final feature of the series RLC circuit under resonant conditions, we note that at resonance

$$V_L = V_C \quad V = V_R$$

from which it follows that

$$V_L = \omega_n LI = \frac{\omega_n L}{R} RI = QV \tag{4-40}$$

This shows that the voltage across the inductor and also across the capacitor can be many times the input voltage. In this sense, the series RLC circuit under resonant or near resonant conditions will act as a voltage step-up transformer with a step-up ratio equal to the circuit Q.

Parallel GLC Circuit

Many of the foregoing ideas that were developed for the series RLC circuit apply to the parallel GLC circuit. We now begin with the admittance function $Y(j\omega)$ of the parallel circuit and then work with the inverse of this function, which is

$$Z(j\omega) = \frac{1}{G + j\omega C + \dfrac{1}{j\omega l}} \tag{4-41}$$

Now the admittance is a minimum at parallel resonance (often called anti-resonance), and this occurs when

$$\omega_n = \frac{1}{\sqrt{LC}} \quad \zeta = \frac{G}{2}\sqrt{\frac{L}{C}} \tag{4-42}$$

The impedance at parallel resonance is the *shunt* resistance $= 1/G$. The parallel circuit Q is given by

$$Q = \frac{1}{\omega_n LG} = \frac{1}{G}\sqrt{\frac{C}{L}} = \frac{1}{2\zeta} \tag{4-43}$$

The bandwidth considerations for the series RLC circuit apply equally to the parallel circuit, and Equation (4-39) remains valid for the parallel circuit.

As a final feature of the parallel resonant circuit, we note that I_L is equal and opposite to I_C; hence,

$$I_L = I_C \quad I = I_G$$

from which it follows that

$$I_L = \frac{V}{\omega_n L} = \frac{VG}{\omega_n LG} = QI_G \qquad \textbf{(4-44)}$$

This shows that the current through the inductor and capacitor may be many times the input current, and in this sense the GLC circuit at or near resonance acts as a current transformer with a step-up ratio of Q.

4.13 TRANSFORMER NETWORKS

In our foregoing considerations, we have assumed direct coupling; that is, R, L, C were connected together conductively. A situation does exist when more than a single inductor is present in a given region and mutual interaction of the magnetic fields of the inductors can exist. To examine this matter in some detail, refer to Figure 4-18, which shows two coils on a common core. This is just a schematic representation, since in many cases no clearly discernible core exists (e.g., when the two coils are in the air), but it does aid in visualizing the mutual magnetic flux that links both coils.

    An intimate relation of the winding distributions on the core of Figure 4-18a and the dots on Figure 4-18b exists. The meaning of the dots is the following: When progressing along the winding from the dotted terminal, each winding encircles the core in the same sense. This means that if the currents enter the dotted terminals, the component fluxes in the core will be in the same direction, and will add. Clearly, the algebraic sign to be assigned to the mutual flux, and so to the mutual inductance, is intimately associated with the assumed current directions. Consider the case of *unity* coupled coils, which means

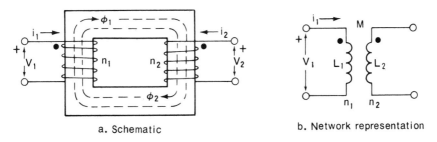

a. Schematic   b. Network representation

Figure 4-18.   Two coupled coils.

that all of the flux ϕ_1 produced by the current in coil 1 links coil 2, and correspondingly, all of the flux ϕ_2 produced by current i_2 in coil 2 links coil 1. For the coupled inductors in Figure 4-18a, we write, by Faraday's law of induction

$$v_1 = n_1 \frac{d}{dt}(\phi_{11} + \phi_{12})$$

$$v_2 = n_2 \frac{d}{dt}(\phi_{22} + \phi_{21})$$

(4-45)

where ϕ_{jk} denotes the flux coupling coil j that arises from a current in coil k. By definition of inductance (total flux linkages per ampere),

$$L_1 = \frac{n_1\phi_{11}}{i_1} \quad L_2 = \frac{n_2\phi_{22}}{i_2}$$

$$M = \frac{n_1\phi_{12}}{i_2} = \frac{n_2\phi_{21}}{i_1}$$

(4-46)

It follows from these that

$$M = \sqrt{L_1 L_2}$$

(4-47)

Achieving unity coupling is very difficult but can be closely approximated in tightly wound coils on a high permeability core. Ordinarily, not all the flux produced by coil 1 links coil 2. In this case, one separates the flux into components—that which is mutual to the two coils and the leakage flux that does not link the two coils. In this case, it can be shown that

$$M = k \sqrt{L_1 L_2}$$

(4-48)

where k, the coefficient of coupling, may range from zero to unity.

By combining Equations (4-45) and (4-46), we find that

$$v_1 = L_1 \frac{di_1}{dt} + M \frac{di_2}{dt}$$

$$v_2 = M \frac{di_1}{dt} + L \frac{di_2}{dt}$$

(4-49)

These specify the manner of taking the mutual effects of one circuit on a second magnetically coupled circuit into account.

For tightly coupled coils, we have, by Equation (4-45),

$$\frac{v_1}{n_1} = \frac{v_2}{n_2}$$

from which

$$v_2 = \frac{n_2}{n_1} v_1 \tag{4-50}$$

This shows that the output voltage v_2 can be greater or smaller than v_1, depending on the turns ratio. This is also evident, because for zero power losses in the transformer, the total power to the system is

$$v_1 i_1 + v_2 i_2 = 0 \tag{4-51}$$

where due account is taken of the reference conditions for voltage and current. This expression is combined with Equation (4-50) to find

$$-\frac{i_2}{i_1} = \frac{v_1}{v_2} = \frac{n_1}{n_2} \tag{4-52}$$

which relates the current in winding 2 with that in winding 1. If we take the ratio of the appropriate v to i, we find that

$$\frac{v_1}{i_1} = \left(\frac{n_1}{n_2}\right)^2 \frac{v_2}{i_2}$$

or

$$Z_1 = \left(\frac{n_1}{n_2}\right)^2 Z_2 \tag{4-53}$$

This shows that the impedance transforms as the square of the number of turns. Clearly, a properly chosen transformer can be used to effect an impedance match between a load, such as a loudspeaker, and the internal resistance of the driving source, such as an amplifier. Transformers are used extensively for sound equipment involving vacuum tubes and for certain transistor amplifiers.

The windings on the transformer need not be isolated from each other. They can have a connection in common. A tapped coil, which is an autotransformer, has many applications, as shown in Figure 4-19.

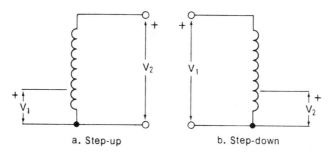

a. Step-up   b. Step-down

Figure 4-19.   Autotransformer.

For less than unity coupling, and this is the case for all air-core coils and some iron-core coils, Figure 4-18 must be modified to take the leakage inductance into account. The equivalent network is that shown in Figure 4-20.

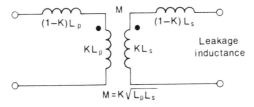

Figure 4-20. Self and leakage inductances.

An important network configuration involves a capacitor across the output terminals of a coupled coil and a series capacitor in the primary; a shunt capacitor across the primary is also used extensively. The analysis of such "double-tuned circuits" becomes rather involved. Of particular importance in amplifiers is such a circuit with both primary and secondary coils being shunt-tuned to the same resonant frequency. A feature of this configuration is that, when the coupling is loose, the response is essentially that of either circuit. As the coupling is increased, by bringing the inductors into closer proximity, interaction occurs, and there is, in fact, a displacement in resonant frequency upward for one circuit and down for the other. With this interaction, the overall bandwidth becomes greater than that of either circuit alone. However, once a critical coupling is exceeded, then the response breaks into two peaks. The results are illustrated in Figure 4-21.

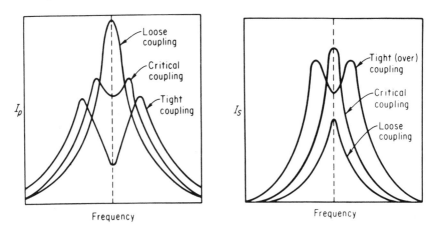

Figure 4-21. Coupling changes for double-tuned transformer.

(M. Mandl, *Fundamentals of Electric and Electronic Circuits*, © 1964. Reprinted by permission of Prentice-Hall, Inc.)

4.14 TRANSIENT RESPONSE OF NETWORKS

Our studies to this point have been concerned with methods for writing the equations that describe the dynamics of the interconnected system (applications of KVL and KCL). This gives rise in the general case to a function of the form

$$v(t) = Z(p)\, i(t) \tag{4-54}$$

where $Z(p)$ is the general impedance operator. We have found that when the excitation is sinusoidal, the KVL leads to the steady state response which is given by

$$V = Z(j\omega)\, I \tag{4-55}$$

where V and I are the rms values of the sinusoidal function, and $Z(j\omega)$ is the appropriate frequency-dependent description of the system. There is an implied sin ωt or cos ωt with Equation (4-55) if the steady state time response is desired.

   If there is a switching operation in a circuit, as occurs when a voltage or current source is switched into a circuit, whether relaxed (no initial currents in inductors or voltages across capacitors) or with specified initial conditions, then we must solve the differential equation of the circuit (see Chapter 20 for details). If we examine several simple, though important, cases, these establish the general pattern for handling such problems.

RL Networks

We shall consider the series RL circuit and its dual, as shown in Figure 4-22. Closing the switch introduces, instantaneously, a constant voltage

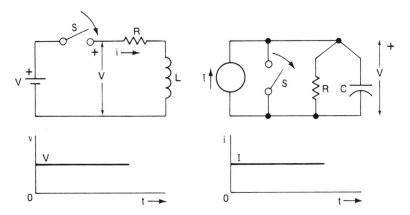

Figure 4-22.   RL circuit with step function excitation.

V into the circuit (or a constant current in the dual circuit). The excitation function is called a *step function* and is conveniently written $Vu_{-1}(t)$, where the symbol $u_{-1}(t)$ denotes the unit step function, which is defined as

$$u_{-1}(t) = \begin{cases} 0 & t < 0 \\ 1 & t \geq 0 \end{cases} \tag{4-56}$$

The dynamic equation for Figure 4-22 is

$$L\frac{di}{dt} + Ri = v(t) \tag{4-57}$$

We note that, except for a change of symbols, the equation applies for the given circuit and its dual.

The complete solution to this differential equation consists of two parts—the complementary function, transient term, or free response, and the particular solution, steady state term, or forced response. To find the complementary function, we must solve the differential equation with the right-hand side set to zero; that is

$$L\frac{di}{dt} + Ri = 0 \tag{4-58}$$

Clearly, this portion of the solution is a function of the network alone and not of the excitation, hence the designation "free response." A trial solution is chosen; it is standard practice to choose an exponential function of the form

$$i_t = B\epsilon_{st} \tag{4-59}$$

This is inserted into the complementary equation, Equation (4-58), to find

$$sLB\epsilon^{st} = RB\epsilon^{st} = 0$$

from which we find that the trial solution is valid if

$$s = -\frac{R}{L} \tag{4-60}$$

The particular solution is readily found since, in the steady state with an applied dc excitation, the inductor has no effect; i.e., $L\,di/dt = 0$ when i is constant. Thus,

$$i_s = \frac{V}{R}$$

The complete solution is then

$$i = i_s + i_t = \frac{V}{R} + B\epsilon^{-Rt/L} \qquad (4\text{-}61)$$

To evaluate the constant B in this equation, we require a knowledge of one initial condition. Here we shall assume that the initial current, the current through the inductor just prior to the switching instant $i(0-)$ is zero; i.e., $i(0-) = 0$; that is, we assume that the switch in the series circuit had been open for a long time prior to $t = 0$. But the current in an inductor cannot change instantaneously; this is so since $di/dt = \infty$ for an instantaneous change in i, a possibility only if an impulse voltage has been applied. Since no impulse voltage is provided, then $i(0-) = i(0+)$; i.e., the current is unchanged during the switching interval, and in this case is 0. Then we have at $t = 0$

$$0 = \frac{V}{R} + B$$

from which

$$B = -\frac{V}{R} \qquad (4\text{-}62)$$

The final solution for the current is thus

$$i = \frac{V}{R}(1 - \epsilon^{-Rt/L}) \qquad (4\text{-}63)$$

The initial slope is an important feature of the exponential function, as indicated in Figure 4-23. The slope of the curve at any point is

$$\frac{di}{dt} = \frac{V}{L}\epsilon^{-Rt/L} \qquad (4\text{-}64)$$

This has the value V/L at $t = 0$. We use this initial slope to calculate

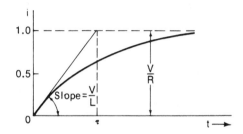

Figure 4-23.   Step response of RL circuit.

the intercept with the final value of the current V/R. This is

$$\tau = \frac{V/R}{\text{slope}} = \frac{L}{R} \tag{4-65}$$

The quantity $\tau = L/R$ is known as the *time constant* of the circuit, and is given in seconds when L is in Henries and R is in ohms. The time constant can be interpreted as the time it would take the current to reach the steady state value if it were to continue at its original rate of increase. τ may also be interpreted from Equation (4-63) to show that for $t = \tau$, the current will reach 0.632 of its final value. Another interesting figure is that the current will reach 99% of its final value in five time constants.

RC Networks

The general procedure in finding the response of the series RC circuit and its dual exactly parallels that discussed above. Now refer to Figure 4-24, which shows the circuits being studied. The controlling equilibrium equation for the circuit after switching is

$$Ri + \frac{1}{C} \int i \, dt = v(t) \tag{4-66}$$

As before, a simple exponential form is assumed for the complementary solution. This leads to a requirement on s that

$$s = \frac{-1}{RC} \tag{4-67}$$

We note that since a capacitor cannot support a dc current, the steady state or particular solution is zero. Hence, the complete solution is

$$i = B\epsilon^{-t/RC} \tag{4-68}$$

To find the value of B requires an appropriate initial condition $i(0+)$. This can be found from the fact that the voltage across the ca-

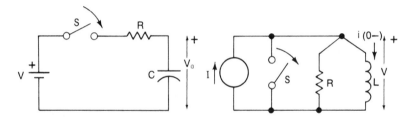

Figure 4-24. Series RC circuit with step function excitation.

pacitor cannot change instantaneously; hence, $v_c(0-) = v_c(0+) = V_o$, and the initial current is then

$$i_o = i(0+) = \frac{V - V_o}{R} \tag{4-69}$$

This value is combined with Equation (4-68) to yield

$$i = \frac{V - V_o}{R} \epsilon^{-t/RC} \tag{4-70}$$

The nature of this function is shown in Figure 4-25. The time constant is $\tau = RC$ (τ is in seconds when C is is farads and R is in ohms). It can be related to both the initial slope of the curve and the 0.632 value of the change in current, or the time to reach $1/\epsilon$-th of its final value.

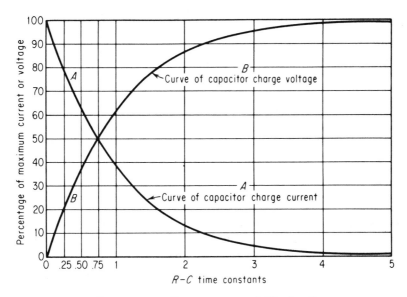

Figure 4-25.   Step response of RC circuit.

(M. Mandl, *Fundamentals of Electric and Electronic Circuits*, © 1964. Reprinted by permission of Prentice-Hall, Inc.)

4.15 FILTERS

Our discussion in Section 4.12 showed that series or parallel circuits involving L's and C's are frequency sensitive; that is, they respond more vigorously at some frequencies than at others. In this sense, they may be called *filters*. However, as ordinarily used, the term *filter* is applied

to twoport networks of appropriate characteristics, which are the subject of Chapter 5.

To examine the situation, refer to Figure 4-26, which shows an RL
and an RC section. The describing input-output equation for the RL
section is ($\tau = L/R$)

$$\frac{V_2}{V_1} = \frac{R}{R + j\omega L} = \frac{1}{1 + j\tau\omega} \tag{4-71}$$

Likewise, the input-output relation for the RC section is ($\tau = RC$)

$$\frac{V_2}{V_1} = \frac{1/j\omega C}{R + 1/j\omega C} = \frac{1}{1 + j\tau\omega} \tag{4-72}$$

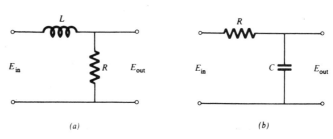

(a)   (b)

Figure 4-26.   RL and RC low pass sections.

(Lane K. Branson, *Introduction to Electronics*, © 1967. Reprinted by permission of Prentice-Hall, Inc.)

Clearly, both sections have the same frequency response characteristic,
and this is shown in Figure 4-27. It appears entirely reasonable, if one
wants a sharper cutoff filter, to cascade a number of identical sections.
While this is true, RL or RC cascaded filters are not often used, owing
to the dissipation in the R's.

Let us examine the effects of interchanging the elements, as shown

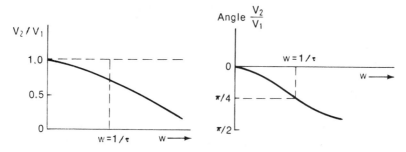

Figure 4-27.   Frequency response of RL and RC low pass
sections.

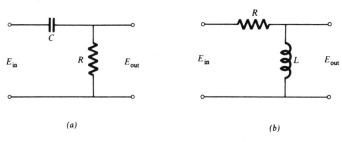

(a)   (b)

Figure 4-28.   RL and RC high pass sections.

(Lane K. Branson, *Introduction to Electronics*, © 1967. Reprinted by permission of Prentice-Hall, Inc.)

in Figure 4-28. The describing input-output equations are, respectively,

$$\text{For (a)} \quad \frac{V_2}{V_1} = \frac{j\omega L}{R + j\omega L} = \frac{j\tau\omega}{1 + j\tau\omega}$$

$$\text{For (b)} \quad \frac{V_2}{V_1} = \frac{R}{R + 1/j\omega c} = \frac{j\tau\omega}{1 + j\tau\omega}$$

(4-73)

The frequency response characteristics of these are given in Figure 4-29. Here too we note that we could cascade identical sections for improved response properties, but other methods are preferred.

Filter Example

Examine the effect of cascading n parallel RLC tuned circuits, assuming that loading of successive circuits does not occur (this assumes that an ideal coupling network having an infinite input impedance and a zero output impedance is used between stages, a situation that is approximated by a vacuum tube, FET, or transistor circuit).

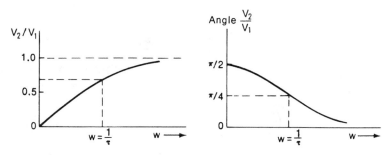

Figure 4-29.   Frequency response of RL and RC high pass sections.

Solution: Refer to Equation (4-41), which we write as:

$$Z(j\omega) = \frac{1}{G + j\omega C + \dfrac{1}{j\omega L}} = \frac{1}{G + j\omega C\left(1 - \dfrac{1}{\omega^2 LC}\right)} \qquad \textbf{(4-74)}$$

In terms of ωn and Q, this becomes

$$Z(j\omega) = \frac{1}{G\left[1 + \dfrac{j\omega C}{G}\left(1 - \dfrac{\omega_n^2}{\omega^2}\right)\right]}$$

$$= \frac{1}{G\left[1 + jQ\left(\dfrac{\omega}{\omega_n} - \dfrac{\omega_n}{\omega}\right)\right]} \qquad \textbf{(4-75)}$$

Now define

$$\delta = \frac{\omega}{\omega_n} - 1; \quad \frac{\omega}{\omega_n} = 1 + \delta \qquad \textbf{(4-76)}$$

and so

$$Z(j\omega) = \frac{1}{G\left[1 + jQ\left[(1 + \delta) - \dfrac{1}{(1 + \delta)}\right]\right]}$$

$$= \frac{1}{G\left[1 + jQ\left(\dfrac{\delta^2 + 2\delta}{1 + \delta}\right)\right]} \qquad \textbf{(4-77)}$$

In the neighborhood of resonance, $\omega \simeq \omega_n$ and δ is small. Approximately, therefore,

$$Z(j\omega) \simeq \frac{1}{1 + j\delta^2 Q} \qquad \textbf{(4-78)}$$

For n such stages in cascade, we write, using the magnitude function, with $R_s = 1/G$,

$$\left(\frac{Z}{R_s}\right)^n = \frac{1}{[1 + (2\delta Q)^2] \, n/2} \qquad \textbf{(4-79)}$$

To deduce the corresponding bandwidth of this cascaded circuit, we use the fact that the bandwidth is defined by the 0.707 points, so that

$$[1 + (2\delta Q)^2]^{n/2} = \sqrt{2}$$

from which

$$1 + (2\delta Q)^2 = 2^{1/n}$$

But the bandwidth is

$$B_1 = \frac{\omega_n}{Q} = 2\zeta\omega_n \tag{4-80}$$

However, from Equation (4-36), we observe that δ is precisely ζ when ω is ω_1, the upper bandwidth frequency. Hence, the bandwidth B_n in terms of the bandwidth of the single stage B_1 is

$$B_n = B_1 \sqrt{2^{1/n} - 1} \tag{4-81}$$

This shows that the resulting cascade becomes progressively sharper with an increase in the number of stages. For example, a two-stage cascade has a bandwidth that is only 0.64 times that of a single stage.

In the more important cases, filters are made up of L's and C's, usually with the minimum possible R, except possibly for the termination. The networks of Figure 4-30 show the essential features of filters having the pass bands indicated, but practical designs may differ in detail from those shown. There are many techniques for designing filters, and all techniques are rather involved. We shall here give a sketchy review of the subject indicating something of what determines the pass band of a filter. This approach illustrates principles, but does not represent more advanced design.

The networks of Figure 4-30 are known as ladder networks. It is possible by repeated applications of the T-Π and the Π-T transformations discussed earlier to reduce a ladder network to the form of Figure 4-31. It is here assumed that there is no dissipation and that all branches of the equivalent T are nondissipative.

We now assume that the source will supply the same power to the output load, here chosen to be a resistor, as it would be if there were no network between the source and the resistor (zero insertion loss). This will be possible if the impedance Z looking into the left-hand terminals is R. For what value of R can this be true? We write from Figure 4-31

$$Z = R = jX_a + \frac{jX_b(R + jX_a)}{R + j(X_a + X_b)} \tag{4-82}$$

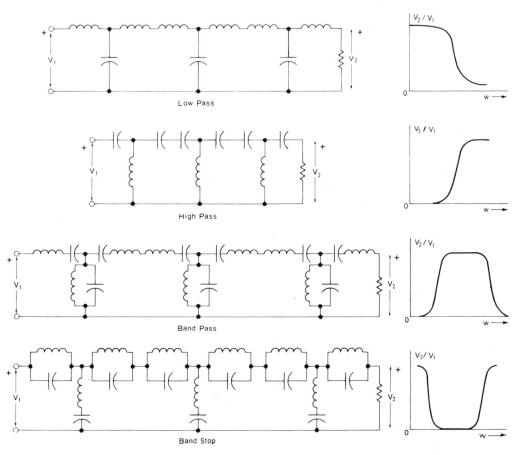

Figure 4-30.   Filters and their frequency response.

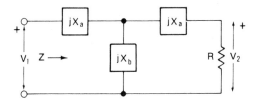

Figure 4-31.   T-equivalent of ladder network.

Solve for R to find

$$R_o = j \sqrt{X_a^2 + 2X_aX_b} \qquad (4\text{-}83)$$

where R_o is used for the particular value of R which is a solution of Equation (4-82). R_o is called the *characteristic impedance*.

In our further analysis, we shall use the conditions illustrated in

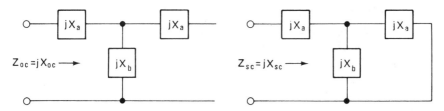

Figure 4-32. Open and short-circuit terminations.

Figure 4-32. From these figures we see that

$$Z_{oc} = jX_{oc} = j(X_a + X_b)$$

$$(4\text{-}84)$$

$$Z_{sc} = jX_{sc} = jX_a - \frac{X_a X_b}{j(X_a + X_b)} = j\left(\frac{X_a^2 + 2X_a X_b}{X_a + X_b}\right)$$

Therefore,

$$Z_{oc} Z_{sc} = -(X_a^2 + 2X_a X_b) \tag{4-85}$$

Comparing this result with Equation (4-83) allows us to write

$$R_o = \sqrt{Z_{oc} Z_{sc}} = \sqrt{-X_{oc} X_{sc}} \tag{4-86}$$

However, since we require that power must be transmitted to the system R_o must be real. For this condition to be fulfilled, X_{oc} and X_{sc} must have opposite signs. Whenever X_{oc} and X_{sc} have the same signs, the network cannot transmit power without attenuation. Frequencies for which the network can transmit power without attenuation constitute the pass bands.

Refer to Figure 4-30 and suppose that the total filter can be regarded as a cascade connection of T sections, as illustrated. Consider a single section and call the series element X_1 and the shunt element X_2. It is then postulated that

$$X_1 X_2 = -K^2 \tag{4-87}$$

where K is a constant real number and is a design parameter of the filter. Such filters are called constant-K filters. Each of the filters in Figure 4-30 can be built up of constant-K sections by proper choice of the L and C parameters. A similar design can be based on the cascade connection of Π sections.

A feature of constant-K filters is that R_o changes rapidly near the band edge, hence this proves to be the poorest part of the response curve. It is possible to improve the situation by using one or more corrective networks in cascade connection at each end of the network.

These corrective networks are not constant-K type; they have the property of making the sides of the pass band steeper than they would be if only the constant-K sections were used. For details, Chapter 5 and the references listed there should be consulted.

BIBLIOGRAPHY—CHAPTER 4

BRUTON, P., *RC-Active Circuits: Theory and Design.* Englewood Cliffs, NJ, Prentice-Hall, Inc., 1980.

GHAUSI, M., and K. LAKER, *Modern Filter Design; Active RC and Switched Capacitor.* Englewood Cliffs, NJ, Prentice-Hall, Inc., 1980.

McGraw-Hill Encyclopedia of Electronics and Computers. New York, NY, McGraw-Hill, 1984.

Reference Data for Engineers: Radio, Electronics, Computer, and Communications (7th Ed.). Indianapolis, IN, Howard W. Sams & Co., Inc., 1985.

VAN VALKENBURG, M. E., *Network Analysis*, (3rd Ed.). Englewood Cliffs, NJ, Prentice-Hall, Inc., 1974.

CHAPTER 5

FILTERS AND ATTENUATORS

5.1 BASIC FILTER AND ATTENUATOR PRINCIPLES

Attenuators are generally resistive networks that reduce the input signal by the desired amount, whether the input signal is dc, a sine wave, square wave, or some other voltage or current variant electrical phenomenon. The simplest attenuator consists of two resistors, with the input signal applied across both resistors and the output taken off across one. This so-called voltage divider reduces or attenuates the signal output according to the ratio of resistance. Resistors in parallel provide current dividers. Filters are also attenuators but they are frequency selective and attenuate some frequencies more than others. A low pass filter, for example, will allow the lower frequencies to pass and will attenuate the higher frequencies. Similarly, a high pass filter will allow the higher frequencies to pass and will reduce the lower frequencies. A band pass filter attenuates higher and lower frequencies and passes only a selected band of frequencies. Conversely, a band-stop filter acts as an attenuator for a certain band of frequencies but allows others to pass.

Network theory, covered in Chapter 4, governs attenuators as well as filters and contains the network equations for both. The actual design of practical filters can generally be approached by either of two different methods.

The simplest filter design method is called the *image parameter method* and is based on the characteristic impedance at the input and output of the filter network. In actual circuits it is difficult to obtain characteristic impedances that remain constant over a range of frequencies and pure inductances and capacitors. For these reasons the image parameter method is usually employed to design filters for experimental networks and for limited bandwith applications. The advantage of this method lies in the relatively simple design formulas which require simple arithmetic. A pocket calculator or the old-fashioned slide rule is a help, but you really need only a pencil and paper.

Almost all commercially designed filters are based on the *modern network theory* approach. In this method it is possible to account for the resistive, capacitive, and inductive elements of each component, using the exact impedance formulas given in Chapter 4. The calculations involve complex numbers and are usually performed by computer.

With the availability of integrated circuit operational amplifiers (IC Op Amp) it has become feasible to design so-called "active filters." An Op Amp with capacitive feedback can be designed to act like an inductance or a complete L-C network. The main advantages of active filters are that they can be designed to have no insertion loss—they can even have some gain factor—and that they can be constructed in much smaller volumes than their passive counterparts.

For example, there are filters available on a single chip that can simplify the construction of any type filter. Each filter IC, together with two to five resistors, produces various second-order filter functions such as low pass, band pass, high pass notch, and all-pass. Plus, the user can tune the center frequency of these functions by an external clock and resistor ratio. Also, by cascading three IC (LTC1061) filters, any of the classical filter functions such as Butterworth, Chebyshev, Bessel, and Cauer may be implemented.

This chapter will first deal with basic filter types, using the image parameter design method, including a typical sample filter design. Because of the complexity of the modern network theory method and its reliance on computer design, its treatment in this book will be limited to fundamentals. The same approach is applied here to active filter principles.

The last part of this chapter is devoted to resistive attenuators. Crystal filters, digital filters, surface wave filters, and mechanical filters are commercially available items; their design and fabrication are too specialized to be included in this book. The bibliography at the end of this chapter, however, includes references to these specialty areas.

5.2 IMAGE PARAMETER FILTER DESIGN

All filters can be realized in three basic forms, which can then be assembled in sections and added together for more effective operation. The basic section is a half-section, with one impedance in series and one in shunt. When the series impedance is divided so that one-half of it is inserted in the line on either side of the shunt impedance, a T section results. When the shunt impedance is split up and arranged in shunt at either side of the series impedance, a π section results. Following the discussion of the basic filters below, a design example will be presented, in which different sections are added to form a complete multisection filter.

Low Pass, Constant-K Filter (Figure 5-1)

This family of filters receives the name *constant K* from the fact that the product of the series and shunting impedances remains constant regardless of frequency, as indicated in the formulas below.

$$Z_o = \sqrt{Z_1 Z_2} = \sqrt{\frac{L}{C}}; \quad Z_1 Z_2 = K^2$$

$$Z_1 = X_{L1} \qquad K = \text{constant for all frequencies } (f)$$

$$Z_2 = X_{C1} \qquad f_c = \text{cutoff frequency}$$

$$L = \frac{R}{\pi f_c}; \quad C = \frac{1}{\pi f_c R}; \quad f_c = \frac{1}{\pi \sqrt{LC}}$$

Low Pass, M-Derived Filter (Figure 5-2)

While the constant-K filter response gradually drops off beyond the cutoff frequency, it is sometimes desired to assure maximum attenuation at a particular frequency, beyond cutoff. A filter designed for two

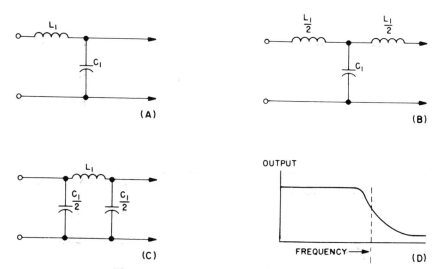

Figure 5-1. Low pass, constant-K filter.

(Matthew Mandl, *Directory of Electronic Circuits (with a Glossary of Terms)*, © 1966. Reprinted by permission of Prentice-Hall, Inc.)

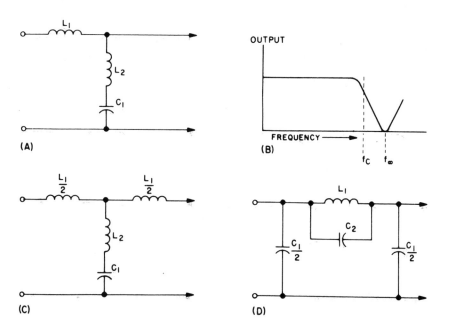

Figure 5-2. Low pass, m-derived filter.

(Matthew Mandl, *Directory of Electronic Circuits (with a Glossary of Terms)*, © 1966. Reprinted by permission of Prentice-Hall, Inc.)

frequencies, one a cutoff frequency and the other an "infinite attenuation" frequency, is called m-derived. The value of m is given in the ratios of the cutoff frequency and the infinite attenuation frequency and generally ranges between 1 and 0.5. Most commercially available m-derived filters have an m equal to 0.6. Note that in the equations below, the series and shunt impedances will depend on where the resonant circuit is located. In Figure 5-2A the shunt element contains the resonant circuit, while in Figure 5-2D the series element contains the resonant circuit. As indicated in the formulas below, these differences will affect the characteristic impedance.

$$Z_o = \sqrt{Z_1 Z_2}; \qquad m = \sqrt{1 - \left(\frac{f_c}{f_\infty}\right)^2}$$

for A: $Z_1 = X_{L1}$ 　　　　　　　f_c = cutoff frequency

　　　　$Z_2 = X_{L2} + X_{C1}$ 　　　f_∞ = infinite attenuation frequency

for D: $Z_1 = \dfrac{X_{L1} X_{C2}}{X_{L1} + X_{C2}}$ 　　　$L_1 = \dfrac{mR}{\pi f_c}; \quad C_1 = \dfrac{m}{\pi f_c R}$

　　　　$Z_2 = X_{C1}$ 　　　　　$L_2 = \dfrac{(1 - \omega^2)R}{4m\pi f_c}; \quad C_2 = \dfrac{(1 - \omega^2)}{4m\pi f_c R}$

High Pass, Constant-K Filter (Figure 5-3)

In the high pass, constant-K filter, the same basic principles apply as for the low pass, constant-K filter, except that the position of the capacitor and inductor are interchanged. The cutoff frequency, f_c, is now at the low end instead of at the high end. The characteristic impedance is the same as for a low pass, constant-K filter, and the other values are as presented below.

$$Z_o = \sqrt{Z_1 Z_2} = \sqrt{\frac{L}{C}} \qquad\qquad Z_1 Z_2 = K_2$$

$Z_1 = X_{C1}$ 　　　　　　　K = constant for all frequencies (f)

$Z_2 = X_{L1}$ 　　　　　　　f_c = cutoff frequency

$L = \dfrac{R}{4\pi f_c}; \quad C = \dfrac{1}{4\pi f_c R}; \quad f_c = \dfrac{1}{4\pi \sqrt{LC}}$

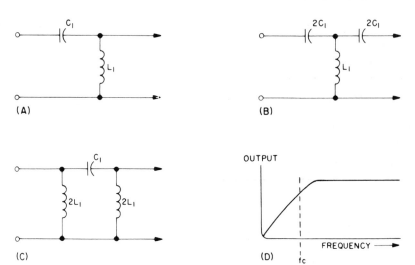

Figure 5-3. High pass, constant-K filter.

(Matthew Mandl, *Directory of Electronic Circuits (with a Glossary of Terms),* © 1966. Reprinted by permission of Prentice-Hall, Inc.)

High Pass, *M*-Derived Filter (Figure 5-4)

The high pass, m-derived filter follows the same reasoning as described for the low pass, m-derived filter, except that now the infinite attenuation frequency is directed toward the low frequency, and the series arm is now a capacitor, where it was an inductor for the low pass, m-derived filter. The detailed formulas are as presented below.

$$Z_o = \sqrt{Z_1 Z_2}; \qquad m = \sqrt{1 - \left(\frac{f_c}{f_\infty}\right)^2}$$

for A : $Z_1 = X_{C1}$ $\qquad f_c = $ cutoff frequency

$\qquad Z_2 = X_{L1} + X_{C2}$ $\qquad f_\infty = $ infinite attenuation frequency

for D : $Z_1 = \dfrac{X_{L2} X_{C1}}{X_{L2} + X_{C1}}$ $\qquad L_1 = \dfrac{R}{4m\pi f_c}; \quad C_1 = \dfrac{1}{4m\pi f_c R}$

$\qquad Z_2 = X_{L1}$ $\qquad L_2 = \dfrac{mR}{(1 - m^2)\pi f_c}; \quad C_2 = \dfrac{m}{(1 - m^2)\pi f_c R}$

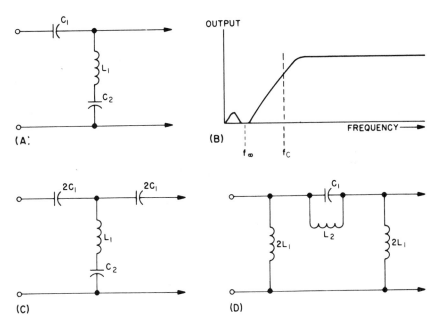

Figure 5-4.   High pass, m-derived filter.

(Matthew Mandl, *Directory of Electronic Circuits (with a Glossary of Terms)*, © 1966. Reprinted by permission of Prentice-Hall, Inc.)

Band Pass, Constant-K Filter (Figure 5-5)

A band pass filter is a combination of low pass and high pass filters, so arranged that frequencies above and below the desired frequency are attenuated. The frequency band that is not attenuated is considered the pass band and is defined by the difference between the two frequencies, which are 0.707 in amplitude of the center frequency, which corresponds to a −3-dB point. A parallel resonant circuit provides maximum impedance at its resonant frequency, while a series resonant circuit provides minimum impedance at its resonant frequency. These two characteristics are generally used in band pass filters to provide minimum attenuation over the pass band and maximum attenuation to frequencies above and below the cutoff. As illustrated in Figure 5-5, the series and parallel resonant circuits of the basic filter can be arranged in both a T and a π (pi) arrangement. Detailed formulas for basic constant-K, band pass filters are presented below.

$$Z_o = \sqrt{Z_1 Z_2} \qquad f_r = \text{resonant frequency}$$

$$f_2 - f_1 = \text{bandwidth at 0.707 peak}$$

$$\text{or } -3\text{-dB points}$$

$$Z_1 = X_{L1} + X_{C1} \qquad Q = \frac{f_r}{f_2 - f_1} \quad (\text{see Chapter 4 for other}$$

$$\text{definitions of } Q)$$

$$Z_2 = \frac{X_{L2}X_{C2}}{X_{L2} + X_{C2}} \qquad L_1 = \frac{R}{\pi(f_2 - f_1)}; \quad C_1 = \frac{f_2 - f_1}{4\pi f_1 f_2 R}$$

$$L_2 = \frac{(f_2 - f_1)R}{4\pi f_1 f_2}; \quad C_2 = \frac{1}{\pi(f_2 - f_1)R}$$

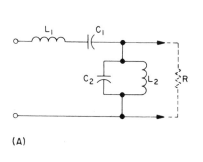

(A)

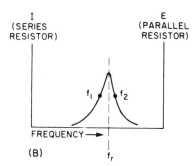

(B)

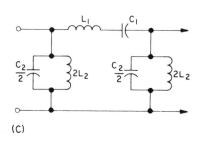

(C)

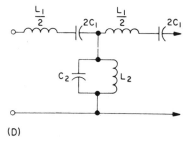

(D)

Figure 5-5.   Band pass, constant-K filter.

(Matthew Mandl, *Directory of Electronic Circuits (with a Glossary of Terms),* © 1966. Reprinted by permission of Prentice-Hall, Inc.)

Band-Stop, Constant-K Filters (Figure 5-6)

If it is desired to provide maximum attenuation over a certain frequency band, the required filter is called a band-stop filter. By interchanging series and parallel resonant elements from the band pass filter, atten-

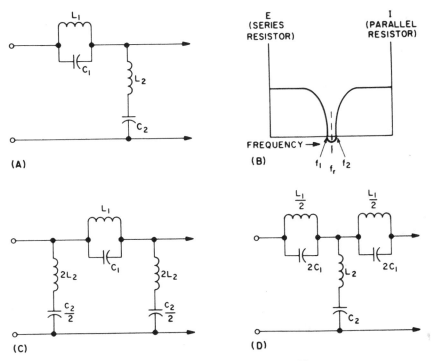

Figure 5-6.   Band-stop, constant-K filter.

(Matthew Mandl, *Directory of Electronic Circuits (with a Glossary of Terms),* © 1966. Reprinted by permission of Prentice-Hall, Inc.)

uation instead of the pass band is achieved. The center frequency and the two frequencies at either side, as well as the Q of the filter, are the same as described above, for the band pass filter. Detailed values are as given below.

$$Z_o = \sqrt{Z_1 Z_2}$$

fr, f_1, f_2, and Q are the same as for the constant-K, band pass filter.

$$Z_1 = \frac{X_{L1} X_{C1}}{X_{L1} + X_{C1}} \qquad L_1 = \frac{(f_2 - f_1) R}{\pi f_1 f_2}; \qquad C_1 = \frac{1}{4\pi R (f_2 - f_1)}$$

$$Z_2 = X_{L2} + X_{C2} \qquad L_2 = \frac{R}{4\pi (f_1 - f_2)}; \qquad C_2 = \frac{f_2 - f_1}{\pi R f_1 f_2}$$

Connection of Filters

As was shown for each of the filters discussed in the preceding paragraphs, it is possible to convert a basic two element L-filter into a T or π configuration. If it is desired to insert such a filter into a balanced

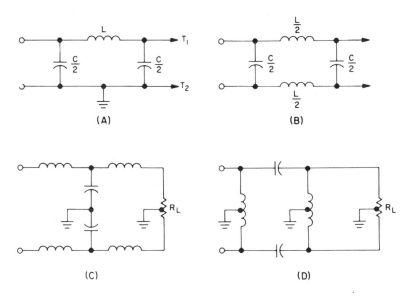

Figure 5-7.   Filter balancing.

(Matthew Mandl, *Directory of Electronic Circuits (with a Glossary of Terms),* © 1966. Reprinted by permission of Prentice-Hall, Inc.)

line, it is possible to remove the ground from the location shown and arrange the filter in a manner so that the impedances to the ground from either side are balanced. Figure 5-7 shows the circuit arrangements for balancing typical filter sections.

In order to obtain sharper attenuation at the cutoff frequency, or to obtain band pass filters with a flatter response over a wider band, it is possible to add filter sections, in series. One method is to add identical filter sections in series, which will result in a greater attenuation of the undesired frequencies, as evidenced by a steeper slope of the response curve, as described in Chapter 4. If a wider band pass or several different, sharp cutoff points are desired, filter sections of somewhat different frequency characteristics can also be connected in series. It was pointed out in the beginning of this chapter that the method using characteristic impedance as a design parameter resulted in some inaccuracies.

When several filter sections designed by this method are connected in series, the inaccuracies will be increased accordingly. One method of improving the accuracy of a multiple section filter is the use of half section, m-derived filters with $m = 0.6$ as terminations. This method is used in the example filter design shown in the following paragraph and Figure 5–8.

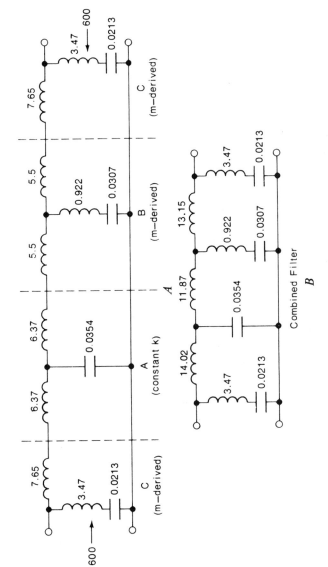

Figure 5-8. Examples of filter design.

111

Example Filter Design (Characteristic Impedance)

Using a constant-K section and an m-derived midsection, design a filter for passing frequencies below 15 kHz, with maximum attenuation at 30 kHz, matched to 600 ohms at each end.

$$600 \;-\; \boxed{\text{C}} \;-\; \boxed{\text{A}} \;-\; \boxed{\text{B}} \;-\; \boxed{\text{C}} \;-\; 600$$

$$Z_o = R = 600\Omega$$

$$f_c = 15 \text{ kHz}$$

$$f_\infty = 30 \text{ kHz}$$

Step 1.   Constant-K filter (See Figure 5-1)

$$L_1 = \frac{R}{\pi f_c} = \frac{600}{3.14 \times 15 \times 10^3} = 12.74 \text{ mh}$$

$$C_1 = \frac{1}{\pi f_c R} = \frac{1}{3.14 \times 15 \times 10^3 \times 600} = 0.0354 \text{ mfd}$$

For T-section $\dfrac{L_1}{2}$ is used twice.

Step 2. m-derived filter (See Figure 5-2):

$$m = \sqrt{1 - \left(\frac{f_c}{f_\infty}\right)^2} = \sqrt{1 - \left(\frac{15}{30}\right)^2} = \sqrt{0.75} = 0.866$$

$$L_1 = \frac{mR}{\pi f_c} = \frac{0.866 \times 600}{3.14 \times 15 \times 10^3} = 11.04 \text{ mh}$$

For T-section $\dfrac{L_1}{2}$ is used twice.

$$L_2 = \frac{(1 - m^2)\,R}{4mf_c\pi} = \frac{(1 - 0.75)\,600}{4 \times 0.866 \times 3.14 \times 15 \times 10^3} = 0.922 \text{ mh}$$

$$C_1 = \frac{m}{\pi f_c R} = \frac{0.866}{3.14 \times 15 \times 10^3 \times 600} = 0.0307 \text{ mfd}$$

Step 3. m-derived filter (See Figure 5-2): Same as in step 2, but $m = 0.6$ and L-sections are used for each end of the filter.

$$L_1 = 3.82 \text{ mh}$$

$$L_2 = 6.80 \text{ mh}$$

$$C_1 = 0.016 \text{ mfd}$$

Sections C-A-B-C are shown in series in Figure 5-8a. (In practice, two L's in series are combined, resulting in the combined filter of Figure 5-8b.)

5.3 MODERN NETWORK THEORY FILTER DESIGN

This method of filter design is based on the use of accurate solutions to the complex network equations discussed in Chapter 4, Section 4.15, including the resistive losses due to each filter element. By selecting filter chips that are now available from firms such as Linear Technology, construction of the following type filters is simplified. Four basic types of filters are used for most applications; their characteristics are described below.

Butterworth Filters

This type of filter is closely related to the filters obtained by the image parameter method because the performance at zero frequency is optimized. Figure 5-9 illustrates a low pass Butterworth filter frequency response. At the low frequency pass band the response is flat, with uniform insertion loss. The steepness of the slope beyond the 3-dB cutoff frequency f_c depends on the number of poles or the number of filter

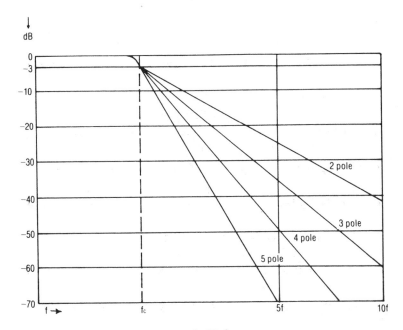

Figure 5-9.   Butterworth LP frequency response.

sections. One of the drawbacks of Butterworth filters is their poor response to transient or pulse signals. Such signals will suffer from overshoot, which increases with the number of poles.

The design of a Butterworth filter is based on the equation:

$$\left(\frac{V_p}{V}\right)^2 = 1 + \left(\frac{x}{x_{3dB}}\right)^{2n}$$

where

V_p = peak output voltage in the passband
V = output voltage at x
n = number of poles (reactances in low or high pass filters or resonators in band pass filters)
x = variable selected in tables
x_{3dB} = x at point on slope where attenuation is 3 dB below V_p

The key variable, x, is defined differently for different types of filters. For low pass filters $x = 2\pi f = \omega$. For high pass filters $x = \dfrac{-1}{\omega} = \dfrac{-1}{2\pi f}$. For symmetrical band pass filters $x = \dfrac{\text{bandwidth}}{f_o}$ and for symmetrical band reject filters $x = \dfrac{-f_o}{\text{bandwidth}}$.

Working charts for different values of n, the number of poles, plot $\left(\dfrac{V_p}{V}\right)_{dB}$ versus $\dfrac{X}{X_{3dB}}$, for low pass and band pass filters. For high pass and band reject filters the charts plot the same ordinate versus the abscissa $\dfrac{\text{3dB bandwidth}}{\text{bandwidth}}$. These curves show the actual optimum attenuation characteristics for the given number of poles, based on rigorous calculations of the ladder network.

Figure 5-10 shows the comparison between a typical low pass filter derived by the image parameter method and its equivalent derived by the network theory method. Note that the resistive components for both capacitors and inductors are included in the latter.

Chebyshev Filters

This type of filter design is based on the assumption that all frequencies in the pass band are equally important and that maximum and constant attenuation in the stop band, the skirts of the response, is desired. Typical Chebyshev frequency responses for low pass filters of different

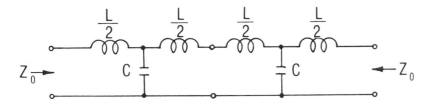

a) IMAGE PARAMETER LP FILTER

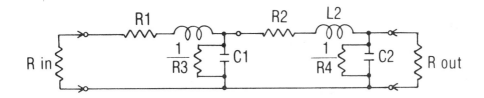

b) BUTTERWORTH LP FILTER

Figure 5-10.   Comparison between filter networks.

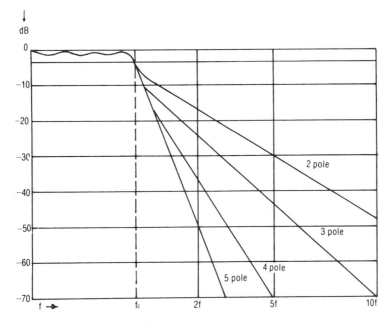

Figure 5-11.   Chebyshev LP frequency response.

numbers of poles are shown in Figure 5-11. Like its Butterworth equivalent, the Chebyshev filter also has considerable overshoot when pulse or transient signals are applied.

The design of the Chebyshev filter is based on the equation:

$$\left(\frac{V_p}{V}\right)^2 = 1 + \left[\left(\frac{V_p}{V}\right)^2 - 1\right] \cosh^2\left[\eta \cosh^{-1}\left(\frac{x}{x_v}\right)\right]$$

The significance of all parameters is identical to those used in the Butterworth equations, except that V_v and x_v refer to the valley or ripple shown in the frequency response of Figure 5-11. Working charts for different numbers of poles are derived for Chebyshev filters in the same manner described above for Butterworth filters.

Bessel or Thompson Filters

This type of filter is used primarily for pulse circuits because its group delay characteristics are uniform over the pass band and they therefore exhibit very little overshoot. Unfortunately this is achieved at the expense of stop band attenuation, which is more gradual than for an equivalent Butterworth filter as illustrated in the response curves of Figure 5-12. Design curves for this type of filter are closely related to the Butterworth curves, with additional tables and charts providing the desired pulse signal response.

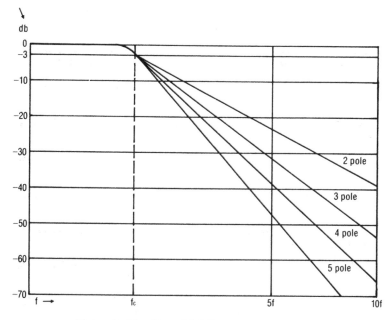

Figure 5-12. Bessel LP frequency response.

Elliptic Filters

In the previous three filter types only poles have appeared in the transfer function. In the elliptic or, as they are also called, inverse hyperbolic, type of filter zeros as well as poles are used. This results in response curves in which the attenuation between the pass band and the stop band is maximized. Figure 5-13 illustrates the response curve of a typical elliptic filter. Note the correspondence between poles and zeros and their location on the response curve. This is reminiscent of the effect of the sound IF trap in the video IF response of TV receivers. In actual circuitry the elliptic filter uses parallel or series resonant networks to produce these effects.

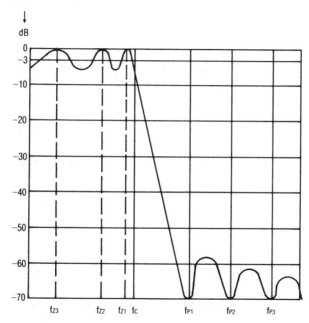

Figure 5-13.   Elliptic LP frequency response.

The design of the elliptic filter is based on the equation:

$$\left(\frac{V_p}{V}\right)^2 = 1 + \left[\left(\frac{V_p}{V}\right)^2 - 1\right] \cdot cd_v^2 \cdot \left[n\left(\frac{K_v}{K_f}\right) \cdot cd_f^{-1}\left(\frac{x}{x_v}\right)\right]$$

where

$cd = \dfrac{cn}{dn}$ = the ratio of the two elliptic functions cn and dn

(Smithsonian Elliptic Function Tables).

n = number of poles (see Butterworth)

x = variable selected in tables (see Butterworth)

v = refers to ripple or valley (see Chebyshev)

K = constant found in tables (elliptic integrals evaluated for the value of the respective subscript)

Charts and tables for frequently used elliptic filters must be consulted to correlate the elements of the design equation.

Practical Filter Design

The most efficient application of the Modern Network Theory of filter design requires access to a computer and a suitable filter design program. The designer must first establish his requirements in terms of input and output circuit, band pass characteristics, phase shift, group delay, overshoot, signal parameters, and power levels. These data are then input to the computer according to the program sequence. In many programs it is possible to limit the size of some circuit elements, such as inductors or capacitors. The computer program will then process the data and output the circuit element values and their connection in the actual filter network.

If a computer program is not available, the design is based on a series of precomputed tables and graphs. The list of references at the end of this chapter includes books containing such tables and graphs. There are two different methods of obtaining circuit element values from precomputed data. One method uses ladder network coefficients, or singly loaded Q's, and the second method uses normalized Q's, or normalized decrements. The first method is based on a definition of one or both of the terminal impedances and the 3-dB bandwidth. The second method also uses the 3-dB bandwidth but relies on the relation between successive impedances or coupling coefficients. In each instance the actual filter depends on how accurately the graphs and tables are read and interpolated. To provide the required phase shift, group delay, overshoot, etc., it is usually necessary to apply correction factors to the basic design or else find a filter design that matches the requirements exactly.

5.4 ACTIVE FILTERS

The availability of integrated circuit (IC) operational amplifiers (op-amp) has made it possible to provide all the electrical characteristics of RLC networks by using only resistors and capacitors. This approach to filter design eliminates bulky inductors and ferromagnetic effects.

Inductance can be simulated because the basic electrical characteristics (see Chapter 4) can be provided by a combination of RC feedback and an op-amp.

The key element in all active filters is the op-amp. This is a high-gain amplifier which has an inverting and a noninverting input. By definition, the ideal op-amp has infinite input resistance, infinite gain, and zero output resistance. In actual ICs, these characteristics are not fully achieved, but devices are now available that provide very high input impedance, very large amplification, and very low output impedance over a useful frequency band. For more details on op-amps see Chapter 9.

Active filter design is generally based on the modern network theory method. As has been explained in Section 5.1, the filters obtained are usually either Butterworth or Chebyshev types, although elliptic filters can also be implemented. This book covers the fundamentals of active filters. For design details the reader is referred to specialized texts listed in the bibliography at the end of this chapter.

Low pass, high pass, and band pass filters are classified into degree of complexity, with second-order filters being the simplest. Of the many different circuits in use, the multiple feedback (MFB) and the voltage-controlled voltage source (VCVS) types are the most popular. The former, MFB, inverts the signal while the latter, VCVS, is noninverting. The basic transfer functions and circuit configurations are presented below.

Low Pass Filter

Figure 5-14a shows the circuit of a second-order MFB low pass filter. Its transfer function is given by:

$$\frac{V_2}{V_1} = \frac{Kb\omega_c^2}{s^2 + a\omega_c s + b\omega_c^2}$$

where

K = gain; a and b are coefficients based on Butterworth or Chebyshev design data

$$s = -j\omega$$

$$\omega_c = 2\pi f \text{ cutoff}$$

Figure 5-14b shows the circuit of a second order VCVS low pass filter with the same transfer function as above. Note the difference in input and output polarities between the MFB and the VCVS filters.

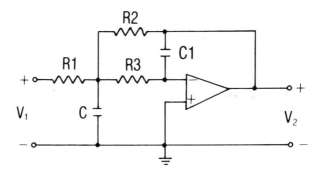

a) SECOND ORDER MFB LP FILTER

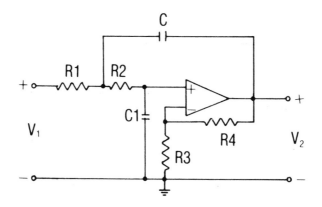

b) SECOND ORDER VCVS LP FILTER

Figure 5-14.   Active LP filters.

High Pass Filter

The second-order multiple feedback (MBF) high pass filter circuit of
Figure 5-15a, like its low pass equivalent, inverts the input signal,
while the voltage-controlled source (VCVS) circuit of Figure 5-15b does
not invert the input signal. The transfer equation shown below uses the
same symbols as the transfer equation for low pass filters.

$$\frac{V_2}{V_1} = \frac{Ks^2}{s^2 + \dfrac{a}{b\omega_c}s + \dfrac{1}{b\omega_c^2}}$$

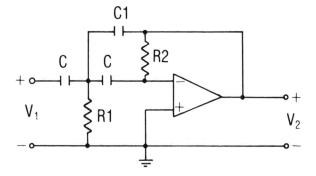

a) SECOND ORDER MBF HP FILTER

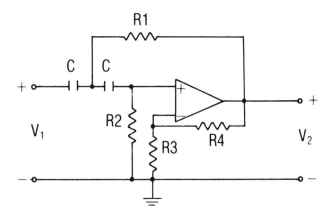

b) SECOND ORDER VCVS HP FILTER

Figure 5-15.   Active HP filters.

Band Pass Filter

The second order band pass transfer function is obtained from the previously described equations for low pass or high pass filters as shown
here.

$$\frac{V_2}{V_1} = \frac{KBs}{s^2 + Bs + \omega_o^2} = \frac{\dfrac{K\omega_o}{Q}}{s^2 + \dfrac{\omega_o}{Q}s + \omega_o^2}$$

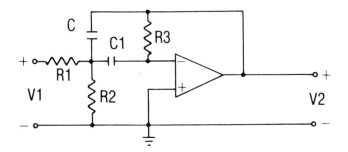

a) SECOND ORDER MFB BP FILTER

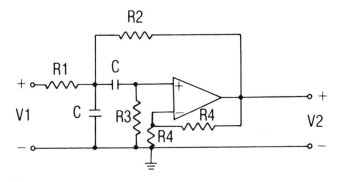

b) SECOND ORDER VCVS BP FILTER

Figure 5-16. Active BP filters.

where:

$$\omega_o = 2\pi f_o \text{ (center frequency)}$$

$$Q = \text{quality factor; } B = \frac{\omega_o}{Q}$$

A multiple feedback (MFB), second-order band pass filter circuit is shown in Figure 5-16a and its equivalent voltage controlled voltage source (VCVS) version is shown in Figure 5-16b.

Practical Filter Design

Commercially available active filters are usually composed of several op-amps with external RC networks determining the frequency characteristics. The design of such filters requires highly specialized expertise and is usually based on a sophisticated computer program. Such filters often use op-amps in series, in parallel, and in a variety of com-

binations, including op-amps as part of the feedback path. Because of
the high gain it is often necessary to shield and isolate complex, active
filters to avoid oscillation and unwanted pickup. In general, the field
of active filter design requires specialized knowledge beyond the scope
of this book. The bibliography at the end of this chapter includes some
advanced texts on this subject.

Digital Filters

All the filtering functions described in the preceding pages often may
be accomplished using digital techniques.

One necessary stage in any digital system that processes information originating in analog form is an analog-to-digital (A/D) converter. Therefore, the first stage in a digital filter is usually an A/D
converter. Following this stage is the digital processor that provides
the filter functions. The output is necessarily a digital-to-analog (D/A)
converter stage.

There are several companies that manufacture ICs that can be used
in this type application. However, the sampling rate of the A/D converter IC must be greater than the highest input signal expected, and
the digital processing stage must be compatible with the sampling rate.
The advantage of using digital filtering is that the designer can achieve
almost ideal filter characteristics.

5.5 ATTENUATORS

Unlike filters, attenuators affect all frequencies equally since they are
essentially resistive networks. The fundamentals of resistive networks
were covered in detail in Chapter 4; Chapter 5 deals with specific types
of attenuators used for specific purposes. While one group of attenuators serves specifically to provide impedance-matching networks, the
vast majority of attenuators are used to reduce the input signal a specific
amount without affecting the impedance arrangements of the circuits
to which the input and output of the attenuator are connected.

The signal reduction or attenuation of an attenuator is measured
either according to the amplitude ratio or voltage of the input and output signal, or else it is given as the power ratio N of the input and
output signals, generally in decibels (db = 10 log N)

$$N = \frac{P_{in} - P_{out}}{P_{out}}$$

$$\text{Voltage attenuation } V = \frac{V_{in}}{V_{out}}$$

The above relationships are used in the following formulas for different types of attenuators.

L-Type Attenuators

The simplest type of attenuator consists of a series and a shunt element, as shown in Figure 5-17, and is essentially a voltage divider. L-type

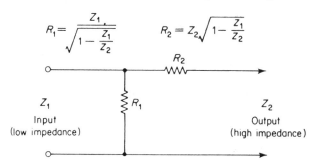

$$R_1 = \cfrac{Z_1}{\sqrt{1 - \cfrac{Z_1}{Z_2}}} \qquad R_2 = Z_2 \sqrt{1 - \frac{Z_1}{Z_2}}$$

Figure 5-17.   Basic L-type pad.

(John D. Lenk, *Handbook of Electronic Charts, Graphs and Tables,* © 1970. Reprinted by permission of Prentice-Hall, Inc.)

attenuators are used most frequently to match different impedance levels, and, as shown in Figure 5-17, the low impedance will be on the side of the shunt element as Z_1, while the higher impedance will be at the side of the series element as Z_2. A balanced arrangement of the basic L-type pad is shown in Figure 5-18. The following formulas apply to the L-type attenuator.

$$N = \left[\sqrt{\frac{Z_1}{Z_2}} + \sqrt{\frac{Z_1}{Z_2} - 1} \right]^2 = \frac{P_{in} - P_{out}}{P_{out}}$$

$$R_1 = \cfrac{Z_1}{\sqrt{1 - \cfrac{Z_1}{Z_2}}} \; ; \quad R_2 = Z_2 \sqrt{1 - \frac{Z_1}{Z_2}}$$

Example: Match 50-ohm to 300-ohm circuits

$$R_1 = \cfrac{50}{\sqrt{1 - \cfrac{50}{300}}} = \cfrac{50}{\sqrt{\cfrac{5}{6}}} \approx 56 \text{ ohms}$$

$$R_2 = 300 \sqrt{1 - \frac{50}{300}} = 300 \sqrt{\frac{5}{6}} \approx 270 \text{ ohms}$$

To get a positive N, interchange Z_1 and Z_2

$$N = \left[\sqrt{\frac{300}{50}} + \sqrt{\frac{300}{50} - 1} \right]^2 = \left[\sqrt{6} + \sqrt{5} \right]^2$$

$$= 6 + 2\sqrt{30} + 5 = 22$$

Attenuation in dB $= 10 \log 22 \approx 13.4$ *dB*

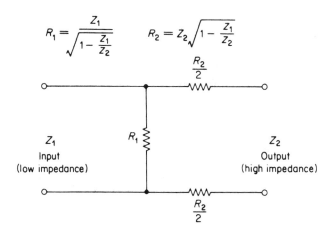

$$R_1 = \frac{Z_1}{\sqrt{1 - \frac{Z_1}{Z_2}}} \qquad R_2 = Z_2 \sqrt{1 - \frac{Z_1}{Z_2}}$$

Figure 5-18. Balanced attenuator pad.

(John D. Lenk, *Handbook of Electronic Charts, Graphs and Tables*, © 1970. Reprinted by permission of Prentice-Hall, Inc.)

T-H Attenuators

One of the most widely used types of attenuators is the T-type shown in Figure 5-19 and its balanced version, the H attenuator. T and H attenuators are generally used only to attenuate signals and not for impedance matching; therefore, the calculations shown below are based on the input and output impedance being equal.

$$Z_1 = Z_2 = Z \qquad N = \frac{P_{in} - P_{out}}{P_{out}}$$

$$R_1 = Z\left(1 - \frac{2}{\sqrt{N} + 1} \right) \quad V = \frac{V_{in}}{V_{out}} \qquad R_2 = \frac{2Z}{\sqrt{N} - \frac{1}{\sqrt{N}}}$$

$$R_1 = Z\left(\frac{V - 1}{V + 1} \right) \qquad\qquad R_2 = Z\left[\frac{2V}{(V^2 - 1)} \right]$$

$$R_1 = Z \left(\frac{V-1}{V+1}\right) \quad R_2 = Z \left[\frac{2V}{(V+1)(V-1)}\right]$$

V = input voltage divided by output voltage

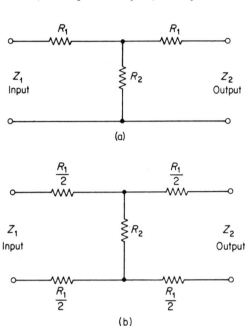

(a)

(b)

Figure 5-19. T-pad and H-pad.

(John D. Lenk, *Handbook of Electronic Charts, Graphs and Tables,* © 1970. Reprinted by permission of Prentice-Hall, Inc.)

π- and O-Attenuators

These types of attenuators are generally used in symmetrical input-output situations; i.e., $Z_1 = Z_2 = Z$. As is apparent from Figure 5-20, the π-attenuator can be converted into an O attenuator when a balanced line is desired. Shown below are the equations for each of the elements.

$$Z_1 = Z_2 = Z$$

$$N = \frac{P_{in} - P_{out}}{P_{out}}$$

$$R_2 = \frac{Z\left(\sqrt{N} - \dfrac{1}{\sqrt{N}}\right)}{2}$$

$$R_1 = Z\left(1 + \frac{2}{\sqrt{N} - 1}\right)$$

$$V = \frac{V_{in}}{V_{out}}$$

$$R_2 = Z\left(\frac{V^2 - 1}{2V}\right)$$

$$R_1 = Z\left(\frac{V + 1}{V - 1}\right)$$

$$R_1 = Z\left(\frac{V+1}{V-1}\right) \qquad R_2 = Z\left(\frac{V^2-1}{2V}\right)$$

V = input voltage divided by output voltage

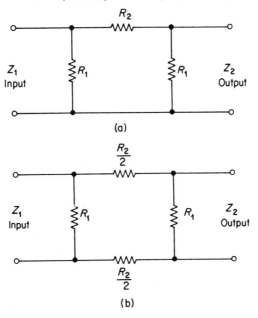

(a)

(b)

Figure 5-20. π-pad and O-pad.

(John D. Lenk, *Handbook of Electronic Charts, Graphs and Tables*, © 1970. Reprinted by permission of Prentice-Hall, Inc.)

Bridged T-H Attenuator

A very unique circuit is the bridged T attenuator, because two of the resistors are equal to the input and output impedance and all of the design formulas are extremely simple. Bridged T and H attenuators are shown in Figure 5-21; their formulas are presented below.

$$Z_1 = Z_2 = Z = R_1 = R_2 \qquad N = \frac{P_{in} - P_{out}}{P_{out}}$$

$$R_3 = \frac{2}{\sqrt{N} - 1} \qquad\qquad V = \frac{V_{in}}{V_{out}}$$

$$R_3 = \frac{Z}{V - 1} \qquad\qquad R_4 = Z(\sqrt{N} - 1)$$

$$R_4 = Z(V - 1)$$

$R_1 = R_2 = Z \qquad R_3 = \dfrac{Z}{V-1} \qquad R_4 = Z\,(V-1)$

$V =$ input voltage divided by output voltage

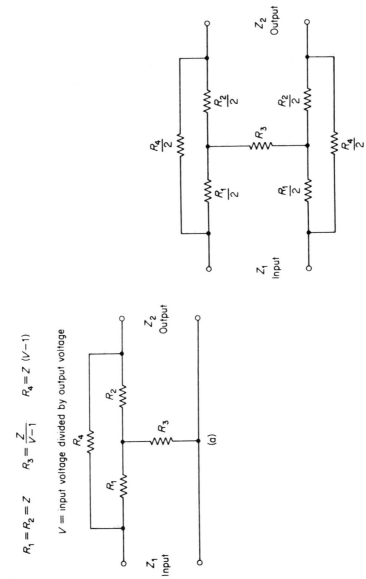

Figure 5-21. Bridged T- and H-pads.

(John D. Lenk, *Handbook of Electronic Charts, Graphs and Tables*, © 1970. Reprinted by permission of Prentice-Hall, Inc.)

Ladder Attenuators

As shown in Figure 5-22, it is possible to cascade a number of attenuator sections. By putting individual attenuator sections in series, the attenuation of each section is added to the next one, as long as the proper impedance match is provided. It is possible to consider each of the sections of the ladder arrangement of Figure 5-22 as one of the balanced L-type attenuators shown in Figure 5-18. In that case, the attenuation of each L-type section is calculated individually and the total attenuation is simply the sum of individual sections, in db (decibels). Each of the different types of attenuators described above can be arranged in a ladder attenuator. In this manner it is possible, with only a few different resistor values, to change the attenuation as well as the output impedance.

$$R_3 = \frac{Z}{V - 1} \qquad\qquad R_4 = Z(\sqrt{N} - 1)$$

$$R_4 = Z(V - 1)$$

Filter Sections in Cascade (or ladder)

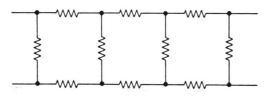

Figure 5-22.   Ladder attenuator.

(John D. Lenk, *Handbook of Electronic Charts, Graphs and Tables*, © 1970. Reprinted by permission of Prentice-Hall, Inc.)

High Frequency Attenuators

Voltage-controlled variable attenuators that have no moving parts are extremely popular with today's designers, particularly in microwave applications. For example, when used in the microwave part of the RF spectrum, a PIN diode behaves like a variable resistor with its value controlled by a dc bias current. There are several commercial microwave attenuators that are designed using PIN diodes. These devices are usually called *PIN diode attenuators.*

Another type of continuously variable attenuator is constructed using either stripline or microstrip. Typically, an energy-consuming device is moved by mechanical methods (such as a lead screw) across the conductor of the transmission line.

BIBLIOGRAPHY—CHAPTER 5

BRUTON, P., *RC-Active Circuits: Theory and Design*. Englewood Cliffs, NJ, Prentice-Hall, Inc., 1980.

GHAUSI, M., and K. LAKER, *Modern Filter Design: Active RC and Switched Capacitor*. Englewood Cliffs, NJ, Prentice-Hall, Inc., 1980.

McGraw-Hill Encyclopedia of Electronics and Computers. New York, NY, McGraw-Hill, 1984.

PHILLIPS, D., and A. GARCIA-DIAZ, *Fundamentals of Network Analysis*. Englewood Cliffs, NJ, Prentice-Hall, Inc., 1981.

Reference Data for Engineers: Radio, Electronics, Computer, and Communications (7th Ed.). Indianapolis, IN, Howard W. Sams & Co., Inc., 1985.

VAN VALKENBURG, M. E., *Network Analysis* (3rd Ed.). Englewood Cliffs, NJ, Prentice-Hall, Inc., 1974.

CHAPTER 6

FERROMAGNETIC CORE TRANSFORMERS AND REACTORS

6.0 INTRODUCTION

Transformers and reactors with ferromagnetic cores are passive inductive devices with magnetic circuits having high permeability and close coupling. Transformers are four-terminal devices that alter voltages, currents, and impedances, while reactors are two-terminal devices. The magnetic properties of transformers and reactors are similar. These devices are not controlled by industry standards, probably because the large number of variables in the specification have discouraged the classification and type reduction necessary for industrywide standardization. Because they are essentially single-application components, satisfactory performance is obtained only when they are specified, designed, and utilized with care.

The following symbols and units will be used in this chapter:

A Cross-sectional area of magnetic circuit in square centimeters

a Turns ratio

B Flux density in gauss

E RMS voltage in volts

e Instantaneous voltage in volts

f Frequency in Hertz

$£$ Magnetomotive force in gilberts

H Magnetizing force in oersteds

I RMS current in amperes

i Instantaneous current in amperes

I_{dc} Direct current in amperes

L Inductance in Henries

l Length of magnetic circuit in centimeters

N Number of turns in a winding

Q Quality factor, dimensionless

R Resistance in ohms

$\mathcal{R}$ Reluctance

t Time in seconds

Z Impedance in ohms

α Collector to emitter current gain

ϕ Flux in maxwells

μ Permeability, dimensionless

η Efficiency

Today's transformer manufacturers are producing a variety of high-frequency transformers of reduced size and, of course, weight. One of the primary uses of these devices is in video applications and ac-to-dc switching mode power supplies. These power supply high-frequency transformers usually operate using square waves ranging from 20 kHz to as high as 1 MHz. The result is very small power supplies (example, 600 watts in 260 cubic inches) partly made possible by highly efficient H7C4 ferrite cores that operate at 150 kHz in one application. See Chapter 20 for more details pertaining to switch mode power supplies, and Section 6.5 for video transformers. The following discussions are based on linear operation, that is, sine wave voltage or current, unless otherwise stated (such as pulse transformers).

6.1 MAGNETIC-CORE DEVICE PRINCIPLES

Magnetics Fundamentals

Current flowing through a conductor produces a magnetic flux perpendicular to the direction of the current. If the conductor is in the form of a coil, the flux becomes concentrated by passing through the interior of the coil. When the current varies with time, a voltage is induced across the terminals of the coil proportional to the time rate of change. This is expressed mathematically as:

$$e = L \frac{di}{dt} \tag{6-1}$$

where L is a proportionality constant called *inductance*.

A changing current causes a proportional change in flux. The induced voltage and changing flux are related by Faraday's law:

$$e = N \frac{d\phi}{dt} \times 10^{-8} \tag{6-2}$$

in which N is the number of turns being linked by the flux, ϕ.

A varying flux causes voltage to be induced in turns other than the turn that produced the flux. The percentage of the flux produced by a current in one turn which links an adjacent turn constitutes the degree of coupling between the turns. These adjacent turns may be part of the same winding or separate windings. If the turns are part of the same winding, the effect is to increase the inductance of that winding. If the turns are parts of separate windings, a voltage is induced in the second winding as a result of the varying current in the first winding. This induced voltage constitutes transformer action.

The introduction of a ferromagnetic core in the magnetic circuit of a coil increases the inductance of the coil as well as the coupling. These benefits are diminished by the introduction of dissipative losses called *core losses*. Core losses are separated into *hysteresis loss*, resulting from the reversal of magnetic domains in the core, and *eddy current loss*, resulting from conductive current flowing in the metallic core induced by the changing flux. With the introduction of the magnetic core, most of the flux becomes confined to the core. The flux is distributed almost uniformly through the cross-sectional area of that portion of the core that passes through the interior of the coil. The flux, ϕ, divided by the cross-sectional area, A, gives the flux density, B:

$$B = \frac{\phi}{A} \tag{6-3}$$

The integration of Faraday's law, Equation (6-2), for the special case of a steady state sine wave voltage, gives the familiar transformer formula used to calculate peak flux density:

$$B = \frac{E \times 10^8}{4.44 \, NAf} \qquad (6\text{-}4)$$

In this expression, E is the RMS voltage and f is the frequency. The characteristics of the core material limit the maximum value of B.

Magnetomotive force is the driving function in magnetics. Also called magnetic potential, it is analogous to voltage in an electrical circuit. Assigned the symbol £, a magnetomotive force is developed by a current flowing in a coil and is directly proportional to the ampere turns developed by that coil. In the system of units used here, the proportionality constant is 0.4π so that:

$$£ = 0.4\pi Ni \qquad (6\text{-}5)$$

Flux in a magnetic circuit is analogous to current in an electrical circuit and is proportional to the applied magnetomotive force. This is expressed in Ohm's law for magnetic circuits as:

$$£ = \Re\phi \qquad (6\text{-}6)$$

where the proportionality constant, $\Re$, is called reluctance.

The reluctance of a magnetic circuit is directly proportional to the length of the circuit and inversely proportional to the cross-sectional area. This is expressed as follows:

$$\Re = \frac{1}{\mu} \frac{l}{A} \qquad (6\text{-}7)$$

in which l is the length of the magnetic circuit and the proportionality constant, μ, is called *permeability*. Permeability is a function of the magnetic material.

The rate of change in magnetomotive force with the length of the magnetic path is of great interest in magnetics. This rate is called *magnetic potential gradient* or *magnetizing force* with the symbol H. Over a uniform magnetic path, H will be constant, allowing the following relationship:

$$H = \frac{0.4\pi Ni}{l} \qquad (6\text{-}8)$$

Substitution of Equation (6-7) into (6-6) and applying the defining relationships for B and H yields the following commonly used expression:

$$B = \mu H \qquad \qquad (6\text{-}9)$$

Characteristics of Core Materials

The characteristics of core materials of greatest interest are permeability, saturation flux density, and dissipative losses. Core material characteristics are displayed on a plot of B versus H, called a hysteresis curve. Figure 6-1 shows a typical hysteresis curve. Describing steady state sine wave conditions, this curve is closed, nonlinear, and double valued. The point B_{max} on the curve is approximately the limiting value of peak flux density allowable. This value varies from about 2 to 20 kilogauss depending on the material. The slope of the curve is the permeability. The core losses are proportional to the enclosed area of the curve.

Core materials can be classified into three groups: high silicon-iron alloys, nickel-iron alloys, and ferrites. Nickel-iron and silicon-iron alloys are available in sheets from which various shapes are punched. Nickel-iron alloys are also ground into powders, mixed with adhesives, and molded. Ferrites are molded and fired like ceramics. Silicon-iron alloys are the most widely used. Satisfactory for most noncritical and low-frequency applications, they have relatively low permeability and low cost. Nickel-iron alloys have the highest permeability of available

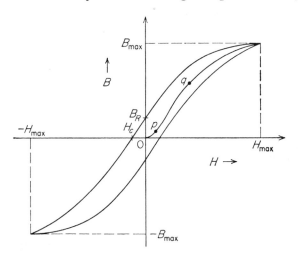

Figure 6-1. Typical hysteresis curve for a sine wave flux.

(R. F. Soohoo, *Theory and Application of Ferrites,* © 1960. Reprinted by permission of Prentice-Hall, Inc.)

materials. Ferrites have good permeability and low losses at high frequencies but low saturation flux density, and are widely used in high-frequency applications because of their small size and light weight.

Transformer Principles and Equivalent Circuits

A transformer consists of two or more windings on some type of ferromagnetic core. If all the flux links all the windings, then from Faraday's law, Equation (6-2), the voltage that is induced on any secondary winding will be equal to the voltage applied to the primary multiplied by the turns ratio:

$$e_s = \frac{N_s}{N_p} e_p = \frac{1}{a} e_p \tag{6-10}$$

in which "a" is the ratio of primary to secondary turns. If a load is connected across the secondary terminals, a current will flow as a result of the induced voltage across those terminals. The direction of the current will be such as to set up a counter magnetomotive force, £, which will oppose any change in flux. According to Faraday's law, the change in flux is established by the induced and, in this case, the applied voltage; therefore, a current must now flow in the primary to offset the opposing £ due to the secondary current. Neglecting the current that establishes the flux, the ampere turns in the primary will equal the ampere turns in the secondary:

$$i_p N_p = i_s N_s \tag{6-11}$$

Equations (6-10) and (6-11) yield the relationship that the volt-amperes in the primary will equal the volt-amperes of the secondary:

$$i_p e_p = i_s e_s \tag{6-12}$$

From Equations (6-11) and (6-12) the following impedance transformation relationship is obtained:

$$Z_p = a^2 Z_s \tag{6-13}$$

in which the impedance of the secondary load appears in the primary as the secondary impedance multiplied by the square of the turns ratio.

The departure of actual transformers from the ideal transformer can be described by means of equivalent circuits. In a practical transformer, an additional current flows in the primary, called the "no load" current. The "no load" current consists of a component in phase with the applied voltage and a quadrature component. These two components are represented in the equivalent circuit by an inductance and a resistance across the transformer terminals.

Frequently it is necessary to consider the flux generated by the current in one winding which does not link the second winding. This flux is called *leakage flux* and can be represented in the equivalent circuit by a series inductance called *leakage inductance*.

The windings are usually made of copper, which has finite resistivity. The primary and secondary winding resistance may be represented by series resistances in the equivalent circuit.

Adjacent turns of a winding and adjacent windings are maintained at different potentials. The resulting capacitive energy storage must often be considered. The capacitive effect can be approximately represented by a parallel capacitor.

Depending on conditions, one or more of the elements in the equivalent circuit may be neglected. Figure 6-2 shows a group of approximate equivalent circuits for low, middle, and high frequencies.

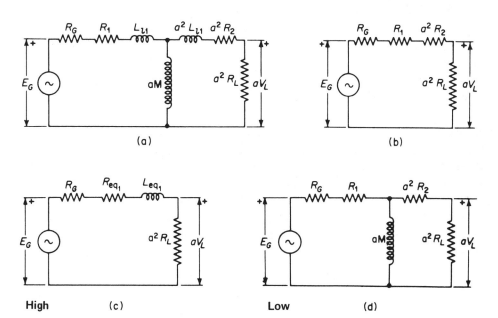

Figure 6-2. Commonly used, approximate equivalent circuits:
   a. Complete equivalent circuit, except for the parallel capacitance.
   b. Equivalent circuit for mid-band region.
   c. Equivalent circuit for high-frequency region.
   d. Equivalent circuit for low-frequency region.

(L. Matsch, *Capacitors, Magnetic Circuits and Transformers*, © 1964. Reprinted by permission of Prentice-Hall, Inc.)

Iron Core Transformer Construction

A transformer consists of a coil, core, protective case or coating, terminals and mounting facility. The coil windings are made of magnet wire, which is solid copper wire covered with thin film insulation, wound on a coil form. The layers of the winding are separated by sheet insulation. Sheet insulation is also used to insulate between windings. The core consists of thin sections of various shapes stacked together to form a three-dimensional core which is inserted in the coil. Making the core from thin sheets reduces the losses in the core. A typical small lamination is shown in Figure 6-3. Depending upon its size and weight, the transformer is equipped with brackets and other mechanical devices to provide secure mounting.

Methods are needed for protecting the coil from the environment and to bond the coil assembly. This is usually done by vacuum impregnation with resins which are forced into the interstices of the core and coil assembly. The resins are usually cured by baking, and the impregnated core and coil is then placed in a case on which feed-through terminals are mounted for the electrical connections. The spaces between the core and coil assembly and the case are filled with potting material. An alternative construction to the use of a case is to coat the core and coil with an impervious plastic material. In either case, mounting is usually accomplished by a metal bracket or threaded studs.

The film on the magnet wire, the sheet insulation, impregnating materials, potting compound, and other miscellaneous insulation constitute an insulation system. This system must be integrated into a consistent and compatible whole, and must be designed as a system to withstand the high voltage and temperature stresses to which it will be subjected. The choice of insulating materials and the electrode spacing controlled by those materials largely determine the quality of the transformer.

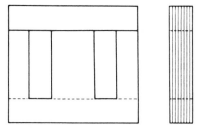

Figure 6-3. Typical transformer lamination stacked to form a core.

(L. Matsch, *Capacitors, Magnetic Circuits and Transformers,* © 1964. Reprinted by permission of Prentice-Hall, Inc.)

6.2 POWER TRANSFORMERS

As the term is used in the electronics industry, a power transformer is a device that transforms significant amounts of electrical power and is driven from a very low impedance source over a narrow frequency band. The following information is needed to specify a power transformer:

Input voltage and wave form
Input frequency
Output voltages
Output currents and wave forms
Maximum ambient temperature

Additional information is often provided, which may include:

Regulation
Efficiency
No load current
Environmental requirements
Size and weight limitations

The rated frequency of a power transformer is the minimum frequency. A power transformer may be operated at any frequency above its minimum rating up to the point where regulation due to leakage inductance becomes unacceptable. A typical rating is from 60 to 400 Hz. The voltage ratings are always maximum values, and a transformer may be operated at any input voltage below its rating. The transformer then provides proportionately lower output voltages. Output current ratings are also maximum values, and the transformer will supply any current below the rated one at an increased voltage, depending on regulation.

The sum of the products of each of the secondary voltages and currents gives the volt-ampere rating of the transformer. The transformer is fully loaded when rated current is drawn from every winding, regardless of the power factor of the load. Each winding is only capable of supplying its own rated current, regardless of the loading on the remaining windings.

Efficiency is usually defined as:

$$\eta = \frac{\text{Volt-amperes out}}{\text{Volt-amperes out} + \text{Losses in watts}} \qquad (6\text{-}14)$$

Regulation is defined as:

$$\text{Regulation} = \frac{E_{\text{No load}} - E_{\text{Full load}}}{E_{\text{Full load}}} \qquad (6\text{-}15)$$

The regulation characteristic becomes important when the load currents vary.

Insulation temperature is of vital importance in transformer operation. The insulation system of the transformer has a maximum allowable temperature rating, and a temperature greatly in excess of that rating causes immediate failure. Long-term operation at temperatures slightly above that rating reduces the life of the transformer. Momentary overloads, such as a short circuit lasting until the circuit protection functions, can usually be tolerated by transformers because of their long thermal time constant.

One of the most common applications for transformers is in power supplies that convert ac electrical power to dc. In this application, it is necessary to determine the rms value of the nonsinusoidal currents drawn by the load and to determine the correct secondary voltage on the transformer to deliver the desired dc voltage after rectification and filtering. See Chapter 5 for details on rectification and filters.

Multiphase Transformers

Multiphase transformers receive power from a multiphase source and transform voltages and currents as do single-phase transformers. In addition, phase transformation can be achieved, and especially three-to-two phase conversion is very common. The number of phases may be increased or decreased, and any desired phase displacement can be achieved by selecting the correct vector components from a multiphase source.

Inverter Transformers

The increasingly popular dc to dc inverter is a device in which transformers play a vital role. In this application, dc power is converted to an alternating square wave, transformed, and rectified. Switching transistors, often in combination with an input transformer, generate the square wave. The switching transistors drive the primary of an output transformer, where the use of high switching frequencies permits a drastic reduction in the size and weight of the transformers.

Inverter transformers have critical requirements. Careful design is needed for proper circuit operation. The switching time determines the efficiency of the circuit. If the leakage inductance of the output transformer is large, this switching time will be excessive. The switching transformer must use a core material with a square hysteresis curve. The losses of the output transformer must be kept small so that high overall efficiency is achieved; dc-to-dc efficiencies in the order of 75% are typical.

6.3 IRON-CORE REACTORS

Power Reactors

Iron-core reactors are two-terminal devices that can usually be represented by a resistance in series with a resistor-inductor parallel combination. At high frequencies, it is sometimes necessary to add a capacitor in parallel with the inductance to represent the distributed capacitance of the coil. Thus the equivalent circuit is the same as an open-circuited transformer, which in fact it physically resembles.

The iron-core reactor has an ac voltage and frequency rating. In addition, the reactor has a dc current rating. The dc current reduces the permeability of the core. When the permeability of the core and the geometry of the core and coil are known, the inductance can be calculated by means of the approximate expression:

$$L = \frac{4\pi N^2 \mu A}{l} \times 10^{-9} \qquad (6\text{-}16)$$

The central problem is the determination of permeability, which, in a ferromagnetic material, is a function of ac flux density and dc magnetizing force. Under normal circumstances, a large dc magnetizing force will saturate the core. To overcome this condition, an air gap is introduced which increases the reluctance of the magnetic circuit and reduces the dc magnetizing force. The permeability is also reduced, an undesirable but manageable limitation.

In Equation (6-6), if the reluctance, R, is considered to be two reluctances in series, the air-gap reluctance and the iron-core reluctance, an expression can be obtained which contains B and H in a linear relationship. If this line is plotted on the dc magnetization curve, the intersection of the two curves gives the values of B and H that result from the dc current. The vertical intercept of the straight line is established by the air gap. As the air gap is increased, the permeability of the iron core increases, but an increase in the air gap causes a decrease in effective permeability. For a given core, coil, and dc current, an optimum air gap exists which will result in maximum effective permeability and the largest value of inductance.

Solenoid Devices

The attractive force of magnets is well known. It is utilized in a wide variety of devices, of which the relay is typical. A relay consists of a coil around a ferromagnetic core, a spring-loaded armature, and contacts that are opened or closed when the armature is attracted to the core. This attractive force is proportional to the ampere turns in the coil and the geometry of the magnetic circuit. The rated current is determined by the temperature rise of the coil. The attractive force is at a maximum when the armature is nearest the core and approaches zero as the air gap is increased. See Chapter 21 for more details on solenoids and relays.

Audio Inductors

Audio inductors differ from filter reactors in that they usually do not have to pass dc current and they usually have a high Q requirement. The development of ferrites has greatly improved the performance of this type of component. Q's from 10 to 100, with inductances in the high millihenry range, are common with ferrites, a difficult achievement with laminated core materials. With this exception, the design of audio inductors is usually straightforward. If dc currents do exist in the winding, then the inductance will vary as the dc current varies. Appropriate use of air gaps can minimize the change in inductance but will not completely eliminate it. This can become a problem in such applications as tuned elements in a high-performance filter.

Saturable Reactors (See 6.6, Magnetic Amplifiers)

The saturating phenomenon can be utilized in control devices called saturable reactors, as shown in the circuit of Figure 6-4. The voltage appearing across R_L will be dependent upon the voltage developed

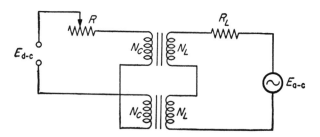

Figure 6-4. Simple saturable reactor circuit.

(S. Platt, *Magnetic Amplifiers: Theory and Application*, © 1958. Reprinted by permission of Prentice-Hall, Inc.)

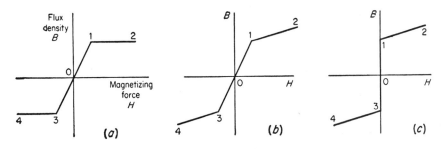

Figure 6-5. Simplified magnetization curve of saturable reactors.

(S. Platt, *Magnetic Amplifiers: Theory and Application,* © 1958. Reprinted by permission of Prentice-Hall, Inc.)

across the load windings, N_L, of the saturable reactor. Analysis of the reactor can be made by means of the simplified magnetization curve of Figure 6-5. In this curve, the magnetization is assumed to be single valued. The permeability is large and constant below saturation, and zero above saturation. An operating point on this curve is established by the dc control current flowing in the control winding. The ac flux density varies around the operating point in response to the volt-time integral of the voltage, E_{A-C}, which appears almost entirely across the reactor windings, N_L, before saturation. When saturation occurs, the voltage across N_L drops to nearly zero and the voltage, E_{A-C}, appears almost entirely across R_L. When the volt-time integral of E_{A-C} reverses, the core will be returned to the operating point.

Actual cores approximate the assumed hysteresis loop. The hysteresis loop is almost rectangular and the width of the loop is quite small. This type of hysteresis proves high gain.

A common problem in saturable reactors is the development of unwanted voltages on the control winding by transformer action. One solution to this problem is shown in Figure 6-4. Two similar reactors are connected so that the ac voltages induced in the control windings cancel out. The dc control is unaffected by this cancellation.

6.4 AUDIO TRANSFORMERS

Bandwidth Requirements and Design

Transformers impose fundamental limitations on audio circuit performance. The behavior of audio transformers can be illustrated by means of the high- and low-frequency approximate equivalent circuits of Figure 6-2c and Figure 6-2d. Elements that are insignificant at the frequency range to which the circuits apply have been omitted. In the low-frequency circuit, Figure 6-2d, the current through aM increases with decreasing frequency until the drop across R_G increases significantly and the output voltage drop becomes excessive. If the generator is matched to the load, the 3-dB point occurs where:

$$\omega a M = \tfrac{1}{2} R_G$$

Improving the low-frequency response by increasing aM without limit causes the circuit elements that affect the high-frequency response to become excessively large. In the high-frequency circuit, Figure 6-2c, if the generator is matched to the load, the 3-dB point occurs where:

$$\omega L_{eq1} = 2R_g$$

Output Transformers

Output transformers are characterized by a significant power output rating and by the fact that they usually have dc current flowing through the primary winding. The power output determines the volume which must be occupied by the windings. The higher the currents, the greater the volume, and the more difficult it is to increase aM and decrease L_{eq1}. The problem with aM is compounded when dc current flows in the primary. A single-sided transistor output stage transformer coupled to its load sends a large dc current through the primary of the transformer. This circuit will have inferior low-frequency response. A better technique is to use a push-pull arrangement where the dc magnetization of the two halves of the primary cancel out. Even here an air gap must be provided to take care of the dc magnetization resulting from imperfect cancellation. A residual magnetization of about 10% of that developed by one-half of the primary can be expected.

Impedance Matching Transformers

Impedance matching transformers are widely used where maximum power transfer from source to load is desired and the source and load impedances are unequal. Transformer coupling of source to transmission line and transmission line to load is often used when the characteristic impedance of the transmission line must be matched.

Interstage transformers are sometimes used for matching or to establish some particular voltage level. With the high impedances involved, it is sometimes difficult to avoid the loading effect the transformer presents. The highest permeability core materials are usually used for this application.

6.5 VIDEO TRANSFORMER DESIGN

Bandwidth Requirements and Design

When the frequency range of a transformer extends into the megahertz range, special consideration must be given to leakage inductance and distributed capacity. Although the representation of those parameters as lumped constants is a satisfactory approximation at audio frequencies, their distributed nature becomes significant at higher frequencies. The turn-to-turn inductance and distributed capacity resemble a transmission line or low pass filter. In fact the transformer has a well-defined characteristic impedance (see Chapters 4 and 5). The cutoff frequency is defined by the product of L_{eq1} and C. The smaller this product, the higher will be the cutoff frequency. The characteristic impedance is given by:

$$Z_o = \sqrt{\frac{L_{eq1}}{C}} \qquad (6\text{-}17)$$

This characteristic impedance is transformed by the turns ratio in the same manner as the load impedance. For proper operation, the generator impedance, the transformer characteristic impedance, and the load impedance must all be matched.

Low-frequency and impedance level considerations for video transformers are the same as for audio transformers. Video transformers have been successfully constructed which have bandwidths from 20 Hz to greater than 5 MHz.

Video Transformer Construction

It is essential that the core material used in video transformers have high permeability at low frequencies. The commonly used materials are high nickel laminations or high permeability ferrites. The high permeability requires fewer turns with consequent reduction in L_{eq1} and C. L_{eq1} can be further reduced by interleaving windings and by the use of bifilar windings, a technique in which two wires are wound simultaneously so that intimate contact exists between each turn. To control

C, it is often necessary to adjust the amount of insulation between windings and layers.

The core material used in ferrite is powdered, compressed, formed in a bonded mass of metal particles, shaped, and partially fused by pressure and heating below the melting point, that is, made into a sinter. The core is magnetic material having high resistivity, consisting mainly of ferric oxide combined with one or more other materials. Using a high resistance material such as this for the core makes eddy current losses extremely low at high frequencies.

There are numerous ferrite compositions being manufactured. These include nickel ferrite, nickel cobalt ferrite, and yttrium-ion garnet, plus several others. Ferrites are widely used in other applications such as microwave, computer, and antenna systems.

Pulse Transformers

Pulse transformers are wide-band transformers whose inputs are video pulses. The performance requirements are usually given in terms of the pulse shape. Figure 6-6 illustrates various pulse characteristics which are frequently specified as limiting values on the output pulse for a rectangular input pulse.

The performance of the transformer, as in other wide-band transformers, is intimately associated with the circuit. Source and load impedances must be defined. This often presents a problem because

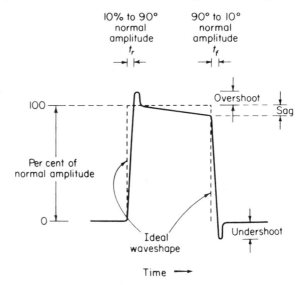

Figure 6-6. Typical transformer output pulse for a rectangular input pulse.

(J.M. Doyle, *Pulse Fundamentals*, © 1963. Reprinted by permission of Prentice-Hall, Inc.)

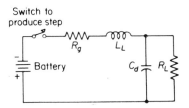

Figure 6-7. Equivalent circuit of transformers on pulse rise.

(J.M. Doyle, *Pulse Fundamentals*, © 1963. Reprinted by permission of Prentice-Hall, Inc.)

the generation of pulses involves nonlinear behavior, and the loads are often nonlinear. The theory under which pulse transformers are designed is based on linear circuit analysis, but a transformer that would work properly under the linear conditions for which it was specified and designed may perform poorly in the actual circuit. A solution to this dilemma is to specify linear approximations of load and source impedances under all four principal operating conditions: rise period, conduction period, fall period, and post pulse period.

The rise time of a pulse transformer is associated with the high-frequency response. The high-frequency equivalent circuit of Figure 6-7 is satisfactory for analysis. The smaller the quantity $\sqrt{L_LC}$, the less will be the rise time. The most satisfactory performance will be obtained when R_g and R_L are matched and:

$$R_g = \sqrt{\frac{L_L}{C}}$$

This condition provides a satisfactory compromise between a ringing and an excessively damped response.

The top of the pulse is associated with the low-frequency response. The circuit of Figure 6-2(d) is satisfactory for analysis. In this circuit, a voltage developed across the transformer will decay according to:

$$aV_L = \frac{E_g}{2\,R_g}\,\epsilon - \frac{R_g}{2_a M} \tag{6-18}$$

Analysis of the pulse decay and recovery can be made by means of the equivalent circuit of Figure 6-8. At the end of the pulse, the reactive parameters, L_o and C, contain stored energy which must be dissipated. The smaller the stored energy, the faster the recovery. Decreasing C and increasing L_o decrease the stored energy, but C must include the circuit capacitance. R_L in Figure 6-8 represents all parallel resistance during decay and recovery. If the load and source impedances are not included in this R_L, then the stored energy must be dissipated in the core, leading to slow recovery. A clipping diode and

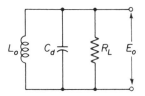

Figure 6-8. Equivalent circuit of transformers on pulse decay and recovery.

(J.M. Doyle, *Pulse Fundamentals*, © 1963. Reprinted by permission of Prentice-Hall, Inc.)

resistor placed across the transformer allowing current of opposite polarity to flow aids in dissipating the reactive energy.

The peak flux density in the core, from Faraday's law, is proportional to the volt time integral of the applied pulse. The wider the pulse, the more turns required. A transformer will function with narrower pulses, but it can only be optimized for one pulse width.

Blocking Oscillator Transformers

A blocking oscillator is an amplifier with large positive feedback provided by the feedback winding of a blocking oscillator transformer. (See Chapters 7 and 8 for details.) With a suitable transformer and circuit, each oscillation takes the form of an approximately rectangular pulse. For the collector-to-emitter feedback arrangement, the pulse width is approximately:

$$T = \frac{\left(\alpha \dfrac{N_s}{N_p} - 1 \right)}{\left(\dfrac{N_s}{N_p} \right)^2 R_e} L_p \qquad (6\text{-}19)$$

where α is the collector-to-emitter current gain of the transistor (see Chapter 8), R_e is the emitter resistance and L_p is the open circuit primary inductance of the transformer.

Pulse shape considerations in the blocking oscillator transformer are the same as for other types of pulse transformers.

Pulse Coupling Transformer Design

Pulse coupling transformers employ the same general principles of pulse transformers. Input and output impedances and times during which they are connected must be specified. The lower the impedances and the smaller the turns ratio, the better the performance obtainable.

Some winding configurations have larger effective C_d when the output is polarity reversing. Special constructions are sometimes required if it is desired to have the performance unaffected by polarity reversal.

6.6 MAGNETIC AMPLIFIERS

Magnetic amplifiers are circuits employing saturable reactors arranged to function as amplifiers. A typical amplifier may provide an ac voltage output proportional to a dc current in which the output polarity is controlled by the polarity of the input current.

To increase the gain of magnetic amplifiers, positive feedback in the form of diodes in series with the power windings is used. This technique enhances the switching action at saturation. A typical complete circuit is shown in Figure 6-9. There are four separate cores in this configuration. The four power windings are each wound on sepa-

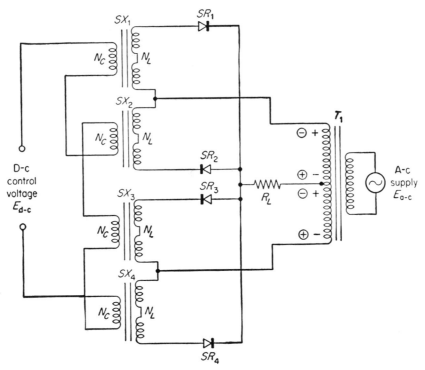

Figure 6-9. Four-reactor magnetic amplifier with feedback diodes.

(S. Platt, *Magnetic Amplifiers: Theory and Application*, © 1958. Reprinted by permission of Prentice-Hall, Inc.)

rate cores, but the two control windings are wound on two cores simultaneously. This circuit develops an ac voltage across the load, which is controlled in both magnitude and polarity by the dc control current. This circuit can also be controlled by the use of ac control voltage of the same frequency as the power source.

BIBLIOGRAPHY—CHAPTER 6

DEL TORO, V., *Electrical Engineering Fundamentals* (2nd Ed.). Englewood Cliffs, NJ, Prentice-Hall, Inc., 1972.

KAUFMAN, M., and A. SEIDMAN, *Handbook for Electronics Engineering Technicians* (2nd Ed.). New York, NY, McGraw-Hill, 1984.

KLOSS, A., *A Basic Guide to Power Electronics.* New York, NY, John Wiley & Sons, Inc., 1984.

KOSOW, I., *Electric Machinery and Transformers.* Englewood Cliffs, NJ, Prentice-Hall, Inc., 1972.

Reference Data for Engineers: Radio, Electronics, Computer, and Communications (7th Ed.). Indianapolis, IN, Howard W. Sams a Co., Inc., 1985.

CHAPTER 7

PRACTICAL TUNED CIRCUITS

7.1 FUNDAMENTALS OF TUNED CIRCUITS

In Chapter 4, the subject of reactance associated with electrical elements when in sinusoidal environments was discussed. In Section 4.4 it was noted that inductors have a quantity that can be defined as inductive reactance, the magnitude of which varies with the applied frequency. It was also noted in the same section that capacitors have a quantity that can be defined as capacitive reactance, the magnitude of which varies with the applied frequency.

Of particular importance is the introduction of the j factor in Section 4.4. It was pointed out that j denotes a phase shift of $+90°$; correspondingly, $-j$ denotes a phase shift of $-90°$. Inductive reactance is generally considered as being positive, that is, $+j$ and vectorially opposite to capacitive reactance, usually considered as being negative $(-j)$. Therefore, it follows that inductive reactance tends to cancel ca-

151

pacitive reactance. Also, in Section 4.12 it was shown that inductive and capacitive reactance behaves oppositely with respect to changes in applied frequency. For fixed values of inductance, inductive reactance increases as the applied frequency is increased. For fixed values of capacitance, it is the opposite; capacitive reactance decreases as the applied frequency is increased.

The reference to inductive reactance as being positive $(+j)$ and capacitive reactance as being negative $(-j)$ should not be misinterpreted. Using this type notation is simply a convenient way of distinguishing between the effects of the two reactances. In an inductive circuit the voltage leads the current, and in a capacitive circuit the voltage lags the current. Therefore, the effect of inductance is opposite to the effect of capacitance usually indicated by $+$ and $-$. If a plus sign is not placed before the j factor, it is assumed that it is a $+j$ factor.

The qualities of inductive reactance (X_L) and capacitive reactance (X_C) are important in any type electronic circuit—the higher the working frequency, the more important they become. The name *tuned circuits* usually refers to a combination of these two reactances in a series or parallel circuit that will produce an abrupt change of impedance at some specific frequency or narrow band of frequencies. In general, tuned circuits are made with inductors and capacitors and/or a variety of solid state integrated circuits such as tuning diodes. These voltage variable capacitance diodes are sometimes called *varactor diodes, silicon capacitor, voltage-controlled capacitor*, etc., and are a two-terminal solid state device whose capacitance varies with the applied voltage. Frequently these components are surface mounted and, in this case, are called *SMD tuning diodes*.

There are several different tuned circuit arrangements that utilize voltage variable capacitance diodes. However, tuning in these circuits is accomplished by utilizing the characteristics of the voltage variable capacitor diode. This causes them to behave as capacitors when they are reverse biased. The amount of reverse bias voltage applied to the diode will determine its capacitance and, therefore, the operating frequency of the tuned circuit.

7.2 TUNED CIRCUIT APPLICATIONS

Tuned circuits are used to perform many functions. For example, they are used for amplification of certain desired frequencies (or bands of frequencies, as in television systems) without amplifying other unwanted frequencies (adjacent channel frequencies). On the other hand, they may be used for rejection of undesired frequencies or bands of frequencies.

Tuned circuits are utilized with integrated circuits (ICs) in radio broadcast receivers. Typically, all active portions of an AM receiver are contained on a single IC. Only the tuned circuits (resonant networks) have to be provided externally. Also, FM IF system ICs are usually designed to require external coils and at least one capacitor for a quadrature-tuned circuit. Another IC that requires an external tuned circuit is a video RF modulator. These examples are just a few of the applications that use consumer ICs and external tuned circuits.

In many applications (especially on printed circuit boards) the tuned circuits are packaged in a standard discrete transistor case (TO5). Often the case has a hole in the top to provide access to the capacitor for adjustments.

7.3 CLASSIFYING TUNED CIRCUITS

There are several different ways of classifying tuned circuits. For example, they may be listed as tunable or fixed, depending on whether one of the circuit components (such as a variable capacitance diode or variable ceramic capacitor, etc.) can be adjusted to change the frequency of the tuned circuit, or whether all components operate at the same frequency.

They may be listed as narrow band or wide band. In this case it depends on the bandwidth or shape of the characteristic curve, which is often provided on spec sheets and equipment schematics. Typically, the upper and lower frequencies of the bandwidth are measured at the 3-dB points, that is, half-power points on the characteristic curve (see Chapter 4, Section 4.12).

Another method of classifying these circuits is to refer to them as being either series-tuned or parallel-tuned (see Chapter 4, Section 4.12), according to whether their components are series- or parallel-connected (series RLC or parallel GLC networks).

Series and Parallel Tuned Circuits

In general, all RF tuned circuits (whether within an IC or designed using discrete components) are based on the use of resonant circuits

(A)   **(B)**

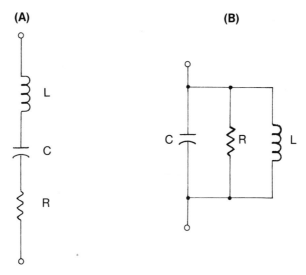

Figure 7-1. A series circuit is shown in (a). The circuit in (b) is parallel. The resonant frequencies for these circuits are listed in Chapter 4, Section 4.0.

consisting of some form of capacitance and inductance connected in series or parallel, as shown in Figure 7-1.

At the resonant frequency the inductive and capacitive reactance (X_L and X_C) are equal, and the parallel circuit presents a high input impedance, or a low input impedance in the case of the series circuit. However, in either circuit any combination of capacitance and inductance will have a resonant frequency, as the formulas in Chapter 4, Section 4.12, show.

To permit tuning of the resonant circuit over a given frequency range, either the capacitance element or the inductive element can be variable. There are several devices that have electrical characteristics similar to a capacitor or inductor. For example, a tuning diode, certain electric motors, and, in waveguide systems, tuning screws or probes are inserted into the transmission line to produce susceptance (the reciprocal of reactance) of magnitude and sign ($\pm j$) that depend on the depth of penetration of the probe. Furthermore, at extremely high frequencies any component, resistor, capacitor, or inductor can become a resonant circuit due to stray capacitance and inductance inherent in PC board construction and discrete component design.

In any case, the two basic design considerations for RF resonant circuits are resonant frequency and the Q factor discussed in Chapter 4. Figure 7-2 shows current in series-resonant circuits having different Q's (see Figure 7-2a). Relative impedance of parallel-resonant circuits with different Q's is shown in Figure 7-2b.

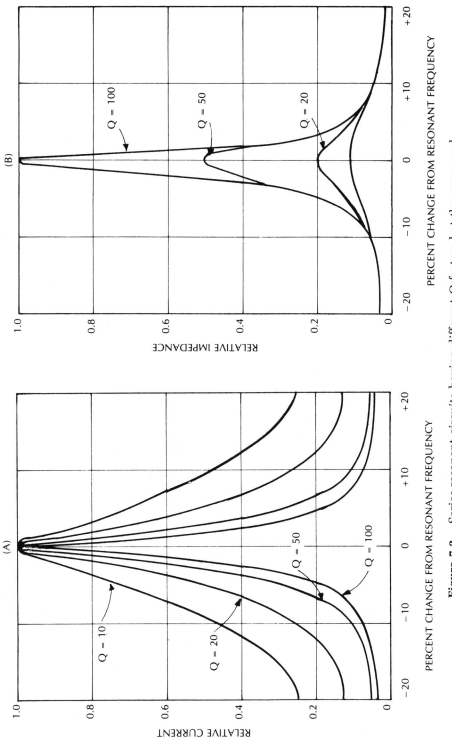

Figure 7-2. Series-resonant circuits having different Q factors but the same peak current. Note: The lower the value of Q, the wider the circuit bandwidth (shown in a). The b section shows parallel-resonant circuit impedance characteristic at different Q's.

7.4 BANDWIDTH (Q)

As Figure 7-2 shows, the bandwidth of a tuned circuit is determined by the circuit parameter called Q. From a practical standpoint, circuit Q is a very important parameter because its value indicates the performance characteristics of the tuned circuit under consideration.

Expressing Q in terms of the inductive reactance, as was done in Equation (4-38) in Chapter 4, is usually the most convenient. However, the Q of a tuned circuit is equal to the reactance of either the inductor or the capacitor at the resonant frequency divided by the effective series resistance of the components, as shown in Equation (4-38).

Another often-used way of expressing Q is in terms of the half-power bandwidth (-3-dB points) of the tuned circuit as related to its resonant frequency (see Section 4.12, Figure 4-17). A formula that is especially convenient during actual measurements of a tuned circuit's Q is:

$$Q = f_r/(f_1 - f_2) = \text{resonant frequency}/\text{bandwidth} \qquad (7\text{-}1)$$

where:

f_1 = frequency above the measured resonant frequency at the half-power point in respect to the resonant frequency maximum level.

f_2 = frequency below the measured resonant frequency at the half-power point in respect to the resonant frequency maximum level.

f_r = resonant frequency of either a series- or a parallel-tuned circuit.

It is important that the frequency measurements be made with a precision frequency counter. Typically the resolution may be selected from 0.1 Hz for frequencies below 10 MHz and 1 Hz for frequencies up to 175 MHz or better, in available precision frequency counters. Accurate measurements are especially important when working with high Q circuits because errors in measuring any one of the three frequencies (f_1, f_2, f_r) can lead to serious errors when determining the bandwidth and, of course, calculating the Q of the circuit under investigation.

Measurements based on Equation 7-1 are usually of a higher quality than the method of determining Q according to Equation (4-38) in Chapter 4. From a practical standpoint, it is very difficult to measure the true value of R (effective series resistance), especially in high-frequency applications where such things as skin effect and distributed capacitance and inductance must be taken into consideration.

7.5 PRODUCTION MEASUREMENTS OF LCR CIRCUIT PARAMETERS

There are LCR bridges that measure inductance, capacitance, resistance, and loss factor. There are also Q meters that can measure the Q of an inductor, a capacitor, or a tuned circuit directly. Some digital LCR meters have simultaneous displays of inductance and Q, capacitance and dissipation (D). An automatic zero offset function compensates for stray capacitance and residual resistance and inductance. These meters also include an equivalent circuit test (series and parallel), plus auto ranging.

7.6 PRACTICAL BANDWIDTH AND Q REQUIREMENTS

Various applications require a very narrow bandwidth and, because bandwidth and Q are inversely proportional (see Figure 7-1), the result is a very high Q circuit. In these applications the tuned circuit is designed for the highest possible degree of selectivity, that is, the lowest practical bandwidth and the highest practical Q.

There are also design problems that demand wide-band and low Q solutions. Some examples are the NTSC waveform monitors and all similar television industry equipment, including TV receiver circuits. In this case, care must be taken lest the bandwidth of the tuned circuit be too narrow (a low Q is required).

In practice, it is often desirable deliberately to introduce resistance into a circuit in order to reduce the Q and produce a desired bandwidth. For example, see the widest of the three curves (Q 20 and 50) in Figure 7-1b.

Wide-band circuits are a must in many applications, such as in the broadcasting industry. In this case the bandwidths, shapes, and amplitudes of the television composite signal waveform are dictated by FCC rules and regulations and industry standards (see Chapter 15). The standard bandwidth (6-MHz channel width) used in the television broadcast industry is comparatively narrow compared to the bandwidth used in instruments such as oscilloscopes, where the entire bandwidth is often 100 MHz and beyond.

Stagger-Tuned Circuits

As indicated in the preceding section, various applications require different bandwidths, and some must be capable of responding to a wide band of frequencies. There are several methods that can be used to

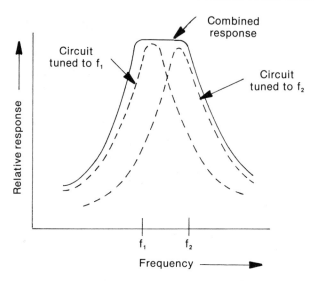

Figure 7-3. Producing a desired bandwidth by tuning two or more tuned circuits to slightly different frequencies. The combined response is the final result (the overall circuit bandwidth).

produce a specific bandwidth in a given piece of equipment. One solution is to stagger-tune cascade amplifiers. Figure 7-3 shows an example of both response curves (selectivity curves of two tuned circuits) and the combined response curve (the final bandwidth; what would be seen on an oscilloscope CRT) after the cascade amplifier circuits have been stagger-tuned, and assuming a sweep generator is being used as a signal source.

In Section 7.6 it was pointed out that wide bandwidth can be achieved by deliberately introducing resistance into a single-tuned circuit to produce a low Q. Although this technique can be used, and often is, it is important to note that obtaining wide-band response by using a stagger-tuned circuit produces an overall selectivity curve that has steep sides (a sharp cutoff). This is an excellent characteristic because it helps reject unwanted frequencies.

The advantage of multiple-tuned circuits in terms of sharp cutoff response characteristics is apparent by comparing the overall response curve in Figure 7-3 with the curve shown in Chapter 4, Figure 4-17. Also see Section 4.13 and the following section for additional information pertaining to double-tuned circuits.

Coupled Tuned Circuits

The mathematical theory of coupled circuits is complex and is presented in Chapter 4, Section 4.13. However, there are certain phenom-

ena that will be described in this section that are important from a
practical standpoint.

Mutual inductance exists in any tuned transformer network that is
inductively coupled. Mutual inductance is not a physical element such
as coil wire resistance, effective capacitance of the winding, effective
capacitance between the coupled tuned circuits, etc.; nonetheless, it is
important to the actual performance of the coupled circuit and must be
considered when dealing with a practical circuit. Figure 7-4 shows a
simplified circuit (in that symbols for distributed capacitance, etc., are
not included) of two parallel-tuned circuits inductively coupled together.

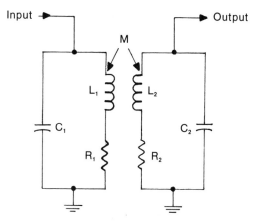

Figure 7-4.   Double-tuned transformer.

The effects of coupling changes for a double-tuned transformer are
described in Chapter 4 (see Figure 4-21). In that chapter, the mathematical definition of mutual inductance was given by the expression
(see Equation [4-48])

$$M = K \sqrt{L_1 L_2} \qquad\qquad (7\text{-}2)$$

where:

> M = mutual inductance, expressed in the same
> inductance units as the two inductors L_1 and
> L_2 (henries, millihenries, microhenries, etc.).
> L_1 = the inductance of the transformer primary.
> L_2 = the inductance of the transformer secondary.
> K = coefficient of coupling.

It is important to realize that the coefficient of coupling (the measure of how many primary winding flux lines actually link with the
secondary winding) can have any value between 0 and 1. Achieving
unity is very difficult and, in any case, the mutual inductance can never

be greater than the square root of the product of the two individual inductances L_1 and L_2—in practice less, because K seldom approaches unity coupling (1). In general, coefficients of coupling approaching unity are never found in double-tuned high-frequency transformers.

Coefficient of Coupling for Double-Tuned Circuits

From a practical standpoint, critical coefficient of coupling (K_c) is of importance.

$$K_c = 1/\sqrt{Q_1 Q_2} \qquad (7\text{-}3)$$

where:

K_c = critical coefficient of coupling.

Q_1 = Q factor of the transformer primary circuit.

Q_2 = Q factor of the transformer secondary circuit.

The critical coefficient of coupling was discussed in Chapter 4 (see Figure 4-21) and is that value of K which will produce peak response when measurements are taken across the transformer secondary winding. In other words, either loose coupling or overcoupling will produce less than maximum current in the secondary winding of the transformer. However, once a circuit is overcoupled (critical coupling is exceeded), the response breaks into two peaks, as shown in Chapter 4, Figure 4-21.

In some applications of coupled-tuned circuits, the coefficient of coupling is purposely set at some value below the critical point. A typical example is the tuned interstage transformers used in consumer products such as broadcast radio receivers.

In designs that require an even wider bandwidth and flat frequency response than simple overcoupling of individual tuned interstage transformers can produce, one stage may be overcoupled and the following critically coupled. Use of this technique will produce a wider bandwidth and a flat response curve very similar to the stagger-tuned circuits previously discussed in this section.

7.7 THERMAL DESIGN CONSIDERATIONS FOR TUNED CIRCUITS

As explained in Chapter 3, Section 3.4, capacitance varies with temperature, and this change may be either in a positive or negative direction. Table 3-6 indicates some temperature characteristics of ceramic capacitors. However, inductors exhibit a positive temperature coefficient. Therefore, with the proper application of temperature coefficient capacitors in a tuned circuit, the designer can eliminate temperature drift, and the resonant frequency of the circuit will not change with temperature changes.

To obtain a certain temperature compensation, capacitors may be combined in a series or parallel network. Referring to Chapter 2, Table 2-4, for example, temperature characteristics, the equivalent temperature coefficient may be calculated for given combinations:

Capacitors in Parallel

$$TC_c = [C_1(TC_1) + C_2(TC_2)]/(C_1 + C_2) \qquad (7\text{-}4)$$

Capacitors in Series

$$TC_c = C_T(TC_1/C_1 + TC_2/C_2) \qquad (7\text{-}5)$$

where:

TC_c = the equivalent temperature coefficient.
TC_1 = listed temperature coefficient of capacitor C_1.
TC_2 = listed temperature coefficient of capacitor C_2.
C_T = equivalent capacitance of the circuit.

Total Capacitance
   Series

$$C = 1/(1/C_1 + 1/C_2) \qquad (7\text{-}6)$$

   Parallel

$$C = C_1 + C_2 \qquad (7\text{-}7)$$

where:

C = equivalent capacitance of the network
C_1 = actual or equivalent value of capacitor C_1.
C_2 = actual value of capacitor C_2.

Resonant Frequency (Parallel-tuned Circuit) Including the Effects of Temperature Coefficient

$$F_r = 1/2\pi \sqrt{L_T(1 + TC_L)C_T(1 + TC_c)} \qquad (7\text{-}8)$$

where:

F_r = the resonant frequency derived by including the effects of temperature coefficient.
L_T = actual or equivalent value of inductance.
TC_L = value of inductance temperature coefficient.

C_T = actual or equivalent value of capacitance.

TC_c = value of capacitance temperature coefficient.

Temperature compensation and its effects described in this section are of significance to the designer if precisely tuned circuits are of paramount importance.

7.8 ALIGNMENT

This term is used to describe the adjustments of tuned circuits to obtain the desired frequency response in a radio/TV receiver or various electronic instruments. Of prime importance, alignment in all respects must conform to rigid requirements in order for a certain receiver or instrument to function properly. Alignment procedures are usually contained in the equipment service manual or other manufacturer's literature.

BIBLIOGRAPHY—CHAPTER 7

GENN, R., *Electronic Servicing Data and Procedures. A Complete Manual and Guide.* Englewood Cliffs, NJ, Prentice-Hall, Inc., 1986.

GHAUSI, M., and K. LAKER, *Modern Filter Design: Active RC and Switched Capacitor.* Englewood Cliffs, NJ, Prentice-Hall, Inc., 1981.

KAUFMAN, M., and A. SEIDMAN, *Handbook for Electronics Engineering Technicians* (2nd Ed.). New York, NY, McGraw-Hill, 1984.

THOMAS, R., and A. ROSA, *Circuits and Signals. An Introduction to Linear and Interface Circuits.* New York, NY, John Wiley & Sons, Inc., 1984.

VAN VALKENBURG, M., *Network Analysis,* (3rd Ed.). Englewood Cliffs, NJ, Prentice-Hall, Inc., 1974.

CHAPTER 8

SEMICONDUCTORS, TRANSISTORS, AND CIRCUITS

Although semiconductors are so called because of their relatively poor conductivity, more important is their ability to conduct electricity in two different modes. Depending on the addition of controlled impurities, they conduct electricity by the drift of negative electrons; or, by the addition of different impurities, they conduct electricity by the apparent drift of electron sites; that is, places in the crystal lattice where the electron can be temporarily held. These sites are called ''holes'' and behave as if they are positive charges. Electrons moving from one such site or ''hole'' to another make it appear as if the ''holes'' are moving in the opposite direction to the net electron movement.

8.1 BASIC TYPES OF SEMICONDUCTORS

Semiconductors in which the electrical properties are not modified by the presence of impurities or imperfections in the crystal lattice are

163

Material	Group in Periodic Table	Valence	Type	Ionization energy W_I, eV In Ge	In SI
Phosphorus, P.	V	5	Donor	0.012	0.045
Arsenic, As	V	5	Donor	0.013	0.052
Antimony, Sb	V	5	Donor	0.010	0.039
Boron, B	III	3	Acceptor	0.0104	0.045
Aluminum, Al	III	3	Acceptor	0.0102	0.062
Gallium, Ga	III	3	Acceptor	0.0108	0.068
Indium, In	III	3	Acceptor	0.0112	0.16

Figure 8-1. Table of typical doping materials.

(Lane K. Branson, *Introduction to Electronics*, © 1967. Reprinted by permission of Prentice-Hall, Inc.)

called intrinsic semiconductors. Extrinsic semiconductors, those dependent upon impurities for their particular electrical characteristics, however, form the bulk of the types used in solid state devices. Extrinsic semiconductors in which the current carrier is the electron are called n-type material, because the electron is a negative carrier of charge. Conversely, material in which conduction is due in the majority to the apparent movement of holes is called p-material, because the hole acts as a positive carrier.

Doping, the carefully controlled adding of impurities such as arsenic or antimony to the basic semiconductor material germanium and silicon, produces n-type material. The addition of dopents such as aluminum, indium, or gallium produce p-type material. These impurities are called "donors" and "receptors," respectively. Figure 8-1 lists some typical doping materials.

When a p-type material is in contact with an n-type material, a pn junction is formed. A voltage applied across such a junction will conduct current freely only when the positive terminal is connected to the p-type side and the negative to the n-type side. Reversing the polarity will result in only a small leakage current flowing. Figure 8-2 shows reverse and forward biased pn junctions with the typical current-voltage characteristic.

Table 8-1 lists some of the semiconductor devices that are manufactured using p-type and n-type material. Actually, the variety of semiconductor devices available is enormous, and the ones listed in Table 8-1 are basically discrete components. It should also be pointed out that Table 8-1 lists *transistors* as simply transistors, which is often done when discussing *bipolar junction* types. However, transistors fall into

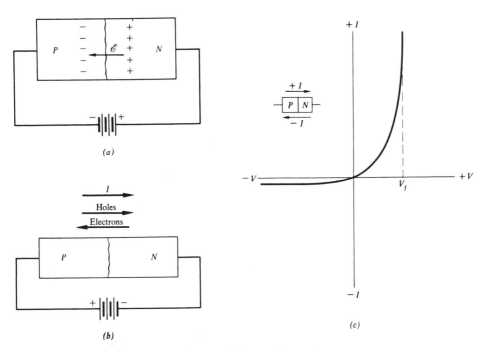

Figure 8-2.   Reverse and forward biased pn junction.

(Lane K. Branson, *Introduction to Electronics*, © 1967. Reprinted by permission of Prentice-Hall, Inc.)

two broad categories: the bipolar junction and the field effect transistor (FET). Furthermore, the FETs are further subdivided into two other basic types. Section 8.3, Fundamentals of Transistors, and the following sections provide additional information on this subject. Chapter 9 includes the various semiconductor technologies used in integrated circuit design and applications.

Effect of Temperature Upon Semiconductors

Increasing thermal energy increases the vibration of the atoms in the molecules. These vibrations are of random magnitude and, as temperature goes up, a certain percentage of the more loosely held valence electrons, in the n-type semiconductor, produce free conduction electrons. The p-type material has more tightly bound electrons and does not respond as readily to temperature increases. Conductors such as copper already have many free electrons at room temperature, and additional thermal energy only makes these electrons move faster. The resistance of a metallic conductor increases with temperature because the increased agitation of its atoms and the higher velocities of its elec-

Device	Junction Configuration or Type	Semiconductor Material	Application
Rectifiers (diodes)	pn junction	Silicon	Low fwd. voltage drop, signal and power rectification
	" "	Germanium	High rectification ratio, high inverse breakdown, high temperature, signal and power rectification
	Dry disc	Selenium	Power rectifier, low-frequency diode, self-healing
	" "	Copper oxide	Meter rectifier, low voltage power rectifier
	" "	Copper sulfide	Low voltage power rectifier
Transistors	pnp or npn	Germanium	Low saturation voltage, general purpose to 75° C
	" " "	Silicon	High temp. use to 175° C, higher voltage
Field Effect Transistors (FET)	n or p channel types	Silicon	High input impedance, resistive, bidirectional output impedance
Unijunction Transistors	n-type bar with p-junction between ends of bar	Silicon	Relaxation oscillator, timing, trigger circuits, neg. res. device
Thyristors Silicon Controlled Rectifier (SCR)	pnpn	Silicon	Phase controlled rectifier, similar to thyratron tube
Triac	two SCRs in parallel and opposite orientation	Silicon	Bidirectional control of ac, light dimmers, power tool speed control

Table 8-1. Table of semiconductor devices and applications.

Device	Junction Configuration or Type	Semiconductor Material	Application
Tunnel Diode	heavily doped pn junction	Germanium Gallium arsenide	Negative resistance, microwave amplifier oscillators, converters
Photo-diodes Photo-transistors	pn npn	Germanium Silicon	Photo-conduction and photovoltaic for control applications
Photoelectric Cells	photoresistive " photovoltaic "	Lead sulfide Lead teluride Cadmium sulfide Selenium	Infrared detector Infrared detector Light meter Light meter
Varistors	Fired Dry-disc	Silicon carbide Selenium	Surge suppressors Contact protector
Thermistors	Fired	Mixed metal oxides	Temperature sensing, control, compensating
Zener Diodes	pn junction reverse biased to Zener breakdown	Silicon	Reverse biased, voltage regulator, voltage reference
Varactor Diode (varicap)	pn junction reverse biased, no current flow	Silicon	Voltage controlled capacitor, parametric amplifier, multipliers
Light Emitting Diode (LED)	pn junction	Gallium arsenide Phosphide	Visible, infrared, light emitting displays

Table 8-1.   Continued

trons tend to scatter and interfere with the current drift of the electrons. In semiconductor devices, however, raising the temperature decreases the resistance, because of the increased number of free electron current carriers, with the resistance-increasing effect of scattering taking a backseat role.

Radiation Effects Upon Semiconductors

Semiconductor material, especially transistor grade crystals, are very sensitive to nuclear-type radiation. Neutron bombardment, gamma rays, beta rays, X rays, and all forms of particle radiation can affect semiconductor material. Radiation increases crystal lattice imperfections by knocking atoms out of position, creating voids and extra atoms at other locations, and it creates a great variety of other disturbances of a permanent nature. The conductivity and other properties of the material are altered in a complex manner, depending on the nature of the semiconductor material and the type of radiation and dosage. In general, radiation effects are detrimental, and this presents a great problem in the design of systems for use in environments of high radiation.

Photo-Effects on Semiconductors

Depending on the frequency of the light, which may extend beyond the visible range into the infrared and ultraviolet bands, different effects are observed. These phenomena may consist of absorbing sufficient energy, in the form of quanta of energy composed of photons, ejecting an electron completely from the body of a semiconductor, to merely raising the electron energy sufficient for it to become a free electron within the material and contributing to its conductivity. The first effect is called *photo-emissive*, the second *photo-conductive*. The photovoltaic effect requires a barrier-type junction which the light causes the electron to bridge, such as in photo-diodes. Photo-diodes and phototransistors will be discussed in Sections 8.2 and 8.7.

Effects of Magnetic Fields Upon Semiconductors

The basic interaction of magnetic force and the movement of electrons produces special effects in semiconductor devices. Figure 8-3 illustrates the relationships between the magnetic field, current flow, and "Hall

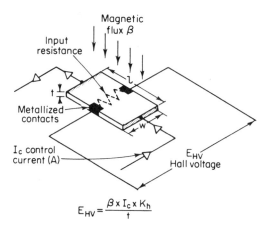

$$E_{HV} = \frac{\beta \times I_c \times K_h}{t}$$

Figure 8-3.   Magnetic effects on semiconductor devices.

(Harry E. Thomas, *Handbook of Transistors, Semiconductors, Instruments, and Microelectronics,* © 1968. Reprinted by permission of Prentice-Hall, Inc.)

effect'' voltage produced by this deflecting force. With the magnetic field β at right angles to the current, the charge carriers will be deflected toward the top surface of the crystal. This is true whether the carriers are positive (holes) or negative, because for a given current direction in the conventional sense, these carriers would move in opposite directions. A p-type semiconductor would therefore produce a positive potential on the top surface, while n-type materials give a negative potential.

Semiconductors such as antimony and bismuth exhibit large Hall effect constants.

8.2 DIODES AND CIRCUITS

Basic Diode Characteristics and Parameters

Figure 8-4 illustrates the voltage-current relationship of a typical low-power diode, and the important parameters are designated. Although in practice most diodes are used in rectifying applications, their inherent characteristics permit many uses other than simply as signal and

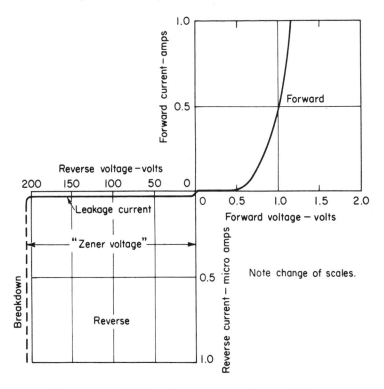

Figure 8-4.   Diode voltage-current characteristics.

	NATURE	PROPERTY	APPLICATION
TUNNEL	Guided minority carrier flow – tunnelling	Negative resistance	UHF – microwave oscillators, amplifiers
ZENER	Avalanche operation	Automatic variable resistance	Voltage reference regulation – limiting
VARACTOR	Variable depletion layer	Voltage variable capacitance	Resonant circuit and oscillator control
PIN	Variable conductance in forward bias region	Nonrectification at high frequency	UHF switching
STEP RECOVERY	Charge storage controlled carrier lifetime	Delayed current flow	Frequency multiplication
SCHOTTKY	Metal to conductor junction; Majority carrier conduction	High speed recovery	UHF switching detection
BACKWARD	Heavy, negative avalanche current flow	Negative avalanche "knee" point at zero bias	UHF, low level, low impedance detector
NOISE	Inherent zener current noise voltages	Self–generative white noise	Receiver and circuit measurement
PHOTO	Conductive or voltaic	Conductive or generative modulation	Light sensors counters

Table 8-2.   Special diode applications and characteristics.

(Harry E. Thomas, *Handbook of Transistors, Semiconductors, Instruments, and Microelectronics,*
© 1968. Reprinted by permission of Prentice-Hall, Inc.)

power rectifiers. Table 8-2 lists special diode applications and the particular characteristic of the pn junction that led to this use.

The various types listed can be broken down into other special-purpose devices such as Schottky rectifiers, which include Schottky barrier rectifiers, high-temperature Schottky rectifiers, and Schottky switch-mode rectifiers. More about this subject will be found in the following sections and chapters.

Low Power Diodes and Circuits

The fundamental "size" rating of a rectifier is the maximum average forward current, $I_{f(av)}$, which is the current of a half-wave rectifier, measured with a dc meter. While characteristics are specific measurements taken under definite conditions, ratings are limit values for safe and reliable operation. Table 8-3 lists some of the more important ratings as they apply to low-power diodes and typical ranges. It is important to refer to up-to-date manufacturers' manuals for specific ratings on particular components.

$I_{f(av)}$	Max. average fwd. current (up to 1 amp)	Average of half-wave conduction
I_{fm}	Max. peak fwd. current (up to 5 amps) repetitive	Important where large peak-to-average current exists, as in large capacitor input filters
I_{fs}	Max. peak surge current (up to 35 amps) non-repetitive	Time limit at 2 millisec or number of surges. See manufacturer's rating
$V_{f(av)}$	Max. avg. fwd. voltage drop (up to 1 volt)	Average drop across rectifier during ½ cycle conduction, usually at $I_{f(av)}$
V_f	Max. instantaneous dc voltage drop (up to 3 volts)	Voltage across rectifier under stated conditions
I_r	Max. dc reverse current (up to 1 ma)	At stated conditions
V_r	Max. peak reverse voltage (up to 1000 volts)	At stated conditions—repetitive, instantaneous, time limited, etc.
T	Max. temperature (-65 to $100°C$–silicon)	At stated conditions—continuous, momentary, free air, case, etc.

Table 8-3. Low-power rectifier ratings and definitions.

High-Power Diodes and Circuits

See Chapter 20 under rectifiers.

High-Speed Diodes and Circuits

When diodes are used as high-frequency detectors, mixers, or high-speed switches, the voltage, current, and efficiency ratings are important, as well as other factors such as low leakage, fast recovery, and low capacitance in their selection. Typical values for these are illustrated in Table 8-4. Recovery time is the time it takes to establish the potential barrier upon reversal of the diode current, or to dispel it upon applying forward bias current. It has the effect shown in Figure 8-5.

Voltage and current	50–600 volts
	1 ampere
Recovery time	150 nanoseconds
Efficiency	very good up to 250 kHz or better

Table 8-4. Typical fast-recovery rectifier parameters.

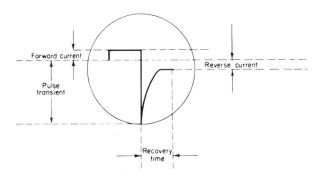

Figure 8-5. Diode recovery time.

Microwave Diodes and Circuits

See Chapter 11.

Zener Diodes and Circuits

A zener diode is a specially treated *pn* junction with a relatively low reverse breakdown voltage. When the inverse voltage applied across the diode is increased beyond the breakdown point, some electrons, injected across the potential barrier, have enough energy to ionize molecules in collisions with them so that more than one electron is knocked loose per colliding electron. This causes a so-called "avalanche" condition, resulting in a large increase in current for a small voltage change. As a result, the diode presents the effect of a very low impedance and therefore maintains a fairly constant voltage over a large range of cur-

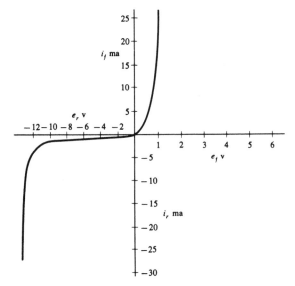

Figure 8-6. Zener diode characteristic.

rent change, as shown by the characteristic curve of Figure 8-6. This property finds application in voltage regulator circuits or as a reference voltage in regulated power supplies, as described in Chapter 20. It also is used in surge and overvoltage protection, signal clipping, etc.

Available to the designer is a wide variety of zener devices. Some of these are: 1/4 watt, 500 milliwatts, 1 watt, 10 watts, etc., silicon zener diodes; temperature-compensated zener reference diodes; constant voltage references for 120- through 200-volt applications (400 milliwatts); plus numerous wattage ratings and case designs, including anode-to-case and cathode-to-case connections. Others include zener overvoltage transient suppressors and silicon power transient suppressors.

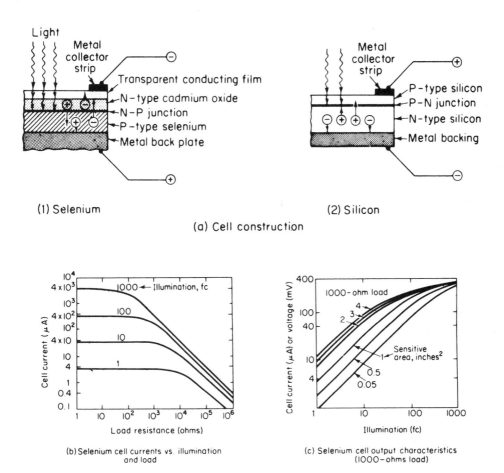

Figure 8-7. Photo-diode construction and characteristics.

(Harry E. Thomas, *Handbook of Transistors, Semiconductors, Instruments, and Microelectronics*, © 1968. Reprinted by permission of Prentice-Hall, Inc.)

Photo-Diodes and Circuits

Photo-diodes are light-sensitive pn junctions such that when light strikes near a junction, electrons are released, leaving behind positively charged holes, and a potential difference is generated between the p and n material. Figure 8-7 shows the details of construction of a silicon and selenium photo-diode and output characteristics for illumination in footcandles. More about photo-diodes in Chapter 19.

Light Emitting Diodes and Circuits

N-type gallium arsenide properly doped and joined with a p-type anode will, under forward bias, emit light in the infrared and visible ranges, depending upon the exact types of materials, treatment, and dopents used. Figure 8-8 shows the typical construction of such a diode and its characteristics. LEDs are available in a large variety of sizes and forms and are used in many types of indicators and alphanumerical displays because of their compatibility with solid state devices and long life. Chapter 19 deals in some detail with LEDs and other types of displays such as liquid crystal, etc.

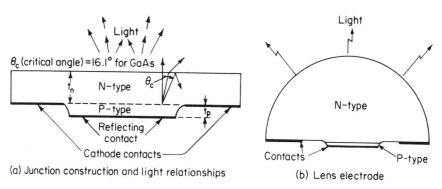

(a) Junction construction and light relationships   (b) Lens electrode

Figure 8-8.   GaAs LED.

(Harry E. Thomas, *Handbook of Transistors, Semiconductors, Instruments, and Microelectronics,* © 1968. Reprinted by permission of Prentice-Hall, Inc.)

8.3 FUNDAMENTALS OF TRANSISTORS

Basic Transistor Characteristics and Parameters

If a thin layer of p-material is sandwiched between two layers of n-material and a voltage applied as shown in Figure 8-9, very little current would flow because one of the pn junctions is reverse biased. However, if the p-material is forward biased with a small voltage, as shown,

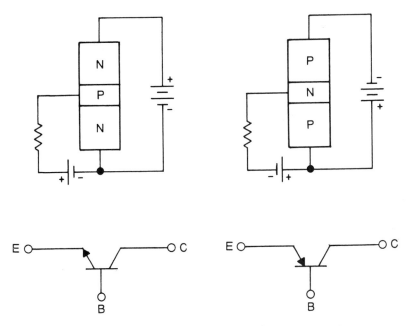

Figure 8-9.   NPN and PNP bipolar junction transistors.

electrons will flow into the p-material. Because the p-material is very thin and the voltage applied across the two outer n-type material layers is higher, electrons thus injected into the p-material will be captured by the higher voltage of the positively charged terminal. The layers of the *npn* junction are called emitter, base, and collector. Most of the electrons injected into the base (p-material) are drawn into the collector layer and a small percentage (typically 5% or less) flow out of the base. The result is that a small base current can control a much larger collector current, constituting an amplifying device.

A *pnp* junction combination of layers operates on the same principle, but now "holes" instead of electrons are involved, and the polarities on each layer are opposite that of the *npn*-transistor.

The most important parameter describing bipolar transistors is current gain, beta (β); the ratio of collector current to base current $\beta = I_c/I_b$ (also called current-transfer ratio or h_{fe}). α (alpha) is the ratio of collector to emitter current $\alpha = I_c/I_e$. For an amplifying transistor, β may range from 20 to 600, while α is somewhat less than 1. The sum of base and collector currents must equal the emitter current. The voltage and current relationships of a transistor are most easily described graphically.

Figure 8-10 shows a typical characteristic curve for a bipolar low power transistor with a $\beta = 25$. Because the emitter-to-base is forward biased, it presents a low resistance and requires less than 1 volt (typi-

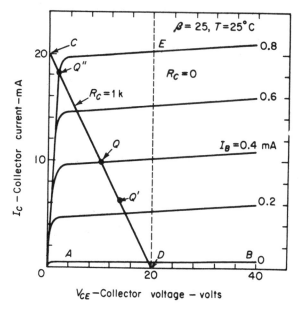

Figure 8-10.   Bipolar transistor characteristics curve.

(Laurence G. Cowles, *Transistor Circuits and Applications*, © 1968. Reprinted by permission of Prentice-Hall, Inc.)

cally about 0.5 volt for silicon and 0.25 volt for germanium) for operation. The emitter-to-collector path consists of two diodes back-to-back, with the base-to-collector diode reverse biased, thus presenting a high internal impedance, Z_C, and therefore requiring a higher voltage for operation. The flat portions of the curves in Figure 8-10 indicate this high resistance, showing a small change in current for a considerable change in collector voltage.

Small-Signal Bipolar and Field-Effect Transistors

Modern transistors are usually classified as:

1. General purpose.
2. High-speed saturated switching
3. RF/UHF/VHF amplifiers and CATV
4. Darlington
5. Low noise amplifier
6. High-voltage
7. Choppers

The typical maximum current transfer ratio of microminiature ranges from 80 to 600 for the general-purpose type, from 80 to 180 for the switching, 100 k for Darlingtons, and from about 500 to 1250 for the low-noise types.

Transistor Circuit Design Basics

Figure 8-9 shows a basic npn-transistor amplifier. Line C–D in Figure 8-10 is the load line (see Chapter 7), representing the voltage drop in the collector resistor R_c from the applied V_{cc} voltage of 20 volts, such that $V_{cc} = V_r + V_c$. To obtain a maximum swing of the voltage V_c, the bias of the circuit should be adjusted so that the base current is at point Q, somewhere midpoint of the load line. If $R_c = 1000$ ohms and $I_c = 10$ mA at Q, then

$$V_c = V_{cc} - I_c R_c = 20 - 10 \times 10^{-3} \times 1000 = 10 \text{ V}$$

With $I_b = 0.4$ ma, then

$$R_b = \frac{V_{bb}}{I_b} = 10/0.4 \times 10^{-3} = 25,000 \text{ ohms}$$

Beta (β) may be determined from the graph of Figure 8-10 by noting the change in I_c along the $R_c = 0$ line. For a change $\Delta I_b = 0.2$ mA near Q, and a corresponding change $\Delta V_c = 5v$, $\dfrac{\Delta V_c}{\Delta I_b} = \beta = 5/0.2 = 10$. "Change" or ac values, and not static dc values, are used in determining beta, because as an amplifier, the ac beta is used in most calculations. Similarly, the internal collector resistance r_c can be determined from the inverse slope of the $I_c - V_c$ characteristic curves, $\Delta V_c/\Delta I_c = r_c$. In practice, β is an approximate number that varies between transistors, even those of the same type and batch. It also varies with temperature and operating point on the characteristic curve, so that the manufacturers usually only supply a minimum β value in their ratings.

Basic Bipolar Amplifier Circuits and Biasing Methods

There are three basic bipolar transistor circuits, identified by the transistor elements, that are common to both the input and the output. The common element is frequently connected to ground, but this is not always done. Figure 8-11 shows the basic configurations for npn-transistors.

The common-base circuit provides no current gain, but couples a very low input impedance to a high output impedance. There is, however, considerable power gain because of this impedance difference between input and output, and the fact that I_e almost equals I_c, since $P = I^2 R$; therefore, $I_c^2 R_o > I_e^2 R_i$. The common-collector circuit, often

SUMMARY OF BASIC BIPOLAR TRANSISTOR AMPLIFIER CIRCUITS

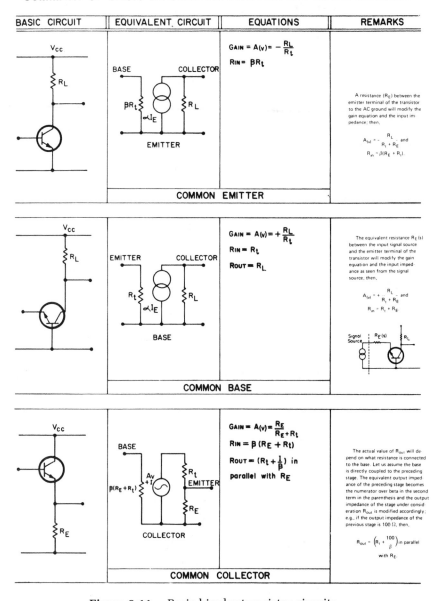

Figure 8-11. Basic bipolar transistor circuits.

(Harry E. Thomas, *Handbook of Transistors, Semiconductors, Instruments, and Microelectronics,* © 1968. Reprinted by permission of Prentice-Hall, Inc.)

called emitter-follower, provides high input impedance and low output impedance. There is very little or no voltage gain but there is considerable power gain, since $E_i \simeq E_o$ and $P = E^2/R$; therefore,

$$\frac{E_o^{\,2}}{R_o} > \frac{E_i^{\,2}}{R_i}$$

The common-emitter circuit provides current gain, power gain, and a moderately low input impedance to a moderately high output impedance. Of the three circuits, only this one inverts the signal from input to output. The simplifying assumptions listed in Table 8-5 provide approximate parameters for most practical purposes.

General simplifying assumptions: $I_e = I_c \ (\alpha \simeq 1)$
(approximations) $G_c = \beta_c$ (common emitter and collector)
$R_C \gg 0.026/I_e$ (common collector)

	Common Base Circuit	Common Emitter Circuit	Common Collector Circuit	Conditions
Input Impedance* R_i	$0.026/I_e = r_e$	$\beta\,(0.026/I_e)$	$\beta\,R_1$	$\dfrac{R_c}{r_c} < 1$
Current Gain-G_c	~ 1	$\sim\beta$	$\sim\beta$	
Voltage Gain-G_v	$\sim 38.4\,V_c$	$\sim 38.4\,V_c$	~ 1	
Output ** Impedance R_o	$\dfrac{r_c}{\beta}$ to r_c	$\dfrac{r_c}{\beta}$ (fairly constant)	$.026/I_e$ to $\dfrac{r_c}{\beta}$	$1 \gg \dfrac{R_g}{r_b} \gg 1$

Voltage gain $(G_v) = $ [Current Gain (G_c)] (R_1/R_i)

* Shockley relationship, $r_e =$ internal emitter to base resistance.

**Limits as $\dfrac{R_g}{r_b}$ varies from much less than 1 to much greater than 1. R_g is resistance of input circuit, r_b is intrinsic resistance of base element.

Table 8-5. Basic circuit formulas.

In place of individual batteries for the bias supply, it is more practical to use only a single power source. Figure 8-12 shows the biasing arrangements for each of the basic circuits. It should be noted that R_3 in each case provides a stabilizing effect by negative feedback. This will be taken up in Section 8.4.

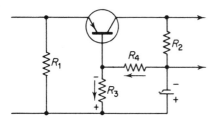

(a) Single source bias for common-base circuit.

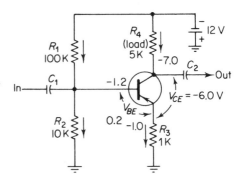

(b) Single source bias for common-emitter circuit.

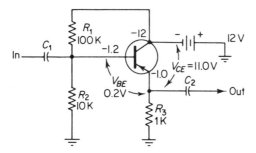

(c) Single source bias for common-collector circuit.

Figure 8-12.   Single-source bias circuits.

(John D. Lenk, *Practical Semiconductor Data Book for Electronic Engineers and Technicians,* ©
1970. Reprinted by permission of Prentice-Hall, Inc.)

Special Design Considerations of Bipolar Transistor Circuits

The *high-frequency performance* of a transistor circuit is determined
by the type of transistor used, and certain circuit configurations provide
better high-frequency performance than others. The common emitter
circuit (see Figure 8-11) can provide bandwidths anywhere from 15 kHz
to 2000 kHz, depending on the particular transistor chosen. Using a
small load resistance R_c, and with β in the neighborhood of 50, the
bandwidths specified are measured at the point where the current gain
beta drops to 70.7% of its low-frequency value. Small R_c reduces the
effect of "Miller" capacitance and any stray capacitance shunting it.
The frequency of this fall-off point is called beta cutoff frequency (f_β).
It is an important parameter describing a particular transistor's high-
frequency potential.

The *common-base* circuit (see Figure 8-11) can provide much better high-frequency performance. The same transistor types as above, in

a common-base configuration, can provide bandwidths of 700 kHz to 100 MHz. However, this impressive 50-times bandwidth improvement is at the expense of current gain; from a beta of 50 to an alpha of less than 1. In the common-base configuration, the 70.7% cutoff point is called the alpha cutoff frequency, f_α, the frequency parameter most commonly given in data sheets. It is related to f_β by the expression:

$$f\alpha = f\beta \cdot \beta$$

Driving the common-emitter configuration from a low impedance source can greatly improve its bandwidth by as much as 25 times, and is a good compromise. Putting a resistor in the emitter is also helpful.

Temperature variations affect the collector current. Even though collector-to-base current is little affected by change in collector voltage, the collector cutoff current, I_{co}, will increase by two times for every 9°C rise in temperature. In effect, the whole $I_c - V_c$ characteristic set of curves is raised, undistorted and intact, to a higher portion of the graph. If raised sufficiently, this can seriously clip and distort the signal passing through a simple common-emitter amplifier. Also, serious thermal runaway conditions can occur, destroying the transistor (refer to Section 8.7). A very important rating of transistors is their thermal resistance. It is listed in the data sheets of transistors as degrees of temperature rise per milliwatt of power fed into the unit. For example,

---

    (a) The middle letter in *PNP* or *NPN* always applies to the *base*.

    (b) The first two letters in *PNP* or *NPN* refer to the *relative bias* polarities of the *emitter* with respect to either the base or collector.

    For example, the letters *PN* (in *PNP*) indicate that the emitter is positive with respect to both the base and collector. The letters *NP* (in *NPN*) indicate that the emitter is negative with respect to both the base and collector.

    (c) The dc *electron-current flow* is always against the direction of the arrow on the emitter.

    (d) If electron flow is into the emitter, electron flow will be out from the collector.

    (e) If electron flow is out from the emitter, electron flow will be into the collector.

    (f) The collector-base junction is always reverse biased.

    (g) The emitter-base junction is always forward biased.

    (h) A *base-input* voltage that opposes or decreases the forward bias, also decreases the emitter and collector currents.

    (i) A *base-input* voltage that aids or increases the forward bias, also increases the emitter and collector currents.

Table 8-6.   List of general rules.

(John D. Lenk, *Practical Semiconductor Data Book for Electronic Engineers and Technicians*, © 1970. Reprinted by permission of Prentice-Hall, Inc.)

a thermal resistance of 0.5° C/mw means that a 10-mw power input
will cause a rise of 5°C above the ambient temperature. Maximum allowable temperatures are also usually specified. By proper circuitry,
the thermal dependence of transistors can be minimized. See Section
8.7 for discussion.

PNP-NPN Circuit Principles

The general rules of Table 8-6 can be helpful in analyzing transistor
circuits, primarily class A amplifers. Also, usually the pinout is: 1—
base, 2—emitter, 3—collector.

8.4 TRANSISTORS AS AUDIO AND VIDEO AMPLIFIERS

Coupling Circuits

Transformer coupling, as illustrated in Figure 8-13, provides close
impedance matching between the output and the input of the next stage.
This tends to preserve the power gain of each stage by insuring the best
power transfer between stages. IF and RF amplifiers and power amplifiers make use of transformer coupling, and sometimes employ special
matching techniques such as taps on the winding for optimizing performance.

Resistor-capacitor coupling is used in low-level audio signal am-

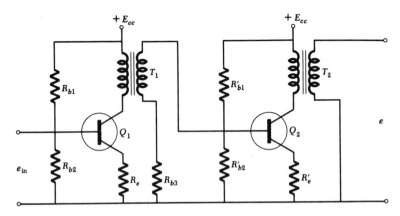

Figure 8-13. Transformer coupling.

(Lane K. Branson, *Introduction to Electronics*, © 1967. Reprinted by permission of Prentice-Hall,
Inc.)

plification, having the advantage of lower cost, smaller space requirement, and less hum pickup. Figure 8-14 shows a typical circuit. Because of the generally low impedances, coupling capacitances are usually rather large in the range of 1 to 50 mfd and, in many applications, much larger.

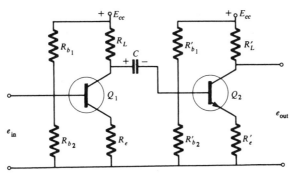

Figure 8-14.   R-C coupling.

(Lane K. Branson, *Introduction to Electronics*, © 1967. Reprinted by permission of Prentice-Hall, Inc.)

Direct coupling is used where the dc level of the signal must be preserved and amplified along with the rest of the signal, such as in operational amplifiers. Since the bias voltages generally are of the same polarity and order of magnitude as collector voltages, dc coupling is simpler in transistors than in vacuum tube circuits. Figure 8-15 illustrates a typical dc circuit.

The emitter follower provides a high input impedance and a low output impedance, permitting power matching between transistor stages without the use of transformers. The basic emitter follower circuit is shown in Figure 8-12 under the label of common collector. A

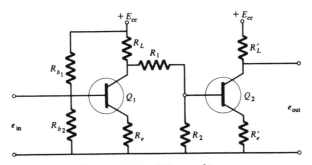

Figure 8-15.   DC coupling.

(Lane K. Branson, *Introduction to Electronics*, © 1967. Reprinted by permission of Prentice-Hall, Inc.)

very widely used emitter follower-type circuit is the Darlington connection. It consists of an emitter follower directly coupled to and driving another emitter follower. Like the simple emitter follower, the Darlington has slightly less than unity voltage gain, but a current gain of β^2 with an input resistance $R_i = \beta^2 R_1$ (see Table 8-5, common-collector circuit). A typical Darlington circuit is shown in Figure 8-16. As previously mentioned, the current gain of Darlingtons can be as high as 100 k.

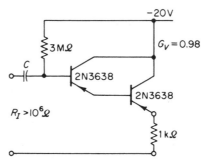

Figure 8-16. Darlington emitter follower.

(Laurence G. Cowles, *Transistor Circuits and Applications,* © 1968. Reprinted by permission of Prentice-Hall, Inc.)

Negative Feedback Circuits

As shown in Section 8.3, both β and the operating Q point of a transistor are highly sensitive to temperature and vary widely from unit to unit of the same type number. Negative feedback provides a technique to stabilize the gain of an amplifier through the values of key resistors in the circuit. Gain and impedances can be calculated readily. Negative feedback results in reduced gain in order to improve other amplifier characteristics. Typical transistor circuits using negative feedback obtain about one-fifth of the transistor gain for one stage of amplification.

Negative feedback networks extending over more than one stage result in very complex design analysis. Figure 8-17 shows a typical

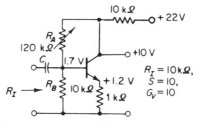

Figure 8-17. Amplifier with negative feedback.

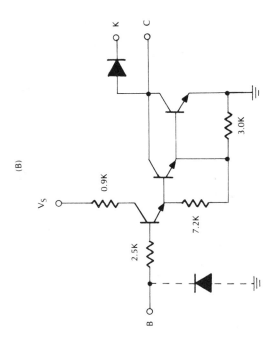

Figure 8-18. Octal Darlington array. ULN2803, with output clamp diodes.

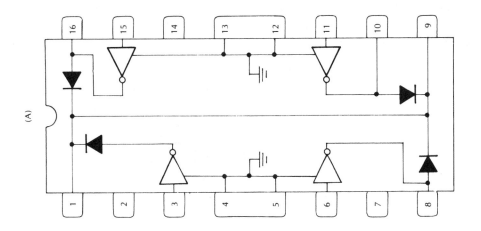

185

one-stage amplifier and the method of determining the resistor values. R_A provides the negative feedback voltage which is in opposition and subtracts from the input voltage. Variations in the Q point (or dc current) from emitter to collector are also opposed, tending to stabilize the performance (and protect the transistor from runaway heating) of the amplifier. Connecting R_A to the collector, as shown, provides additional negative feedback.

Power Amplifiers

Modern power amplifiers are constructed using both bipolar and MOS-FET transistors, usually in the form of Darlington transistors, which are very popular in the power field. See Figure 8-16 for a typical Darlington circuit.

All Darlingtons are basic emitter followers like the example shown in Figure 8-16. Many of these devices are rated for as much as 120 watts.

Darlington ICs are not limited to a single Darlington amplifier. Several Darlingtons can be included in one package. See Figure 8-18 for pin connections and a single-element Darlington schematic for this IC. Incidentally, Darlington circuits need not be limited to two transistors; three and even four transistors can be used. Also, there are common emitter as well as emitter-follower Darlingtons.

Transistor audio power amplifiers may be class A single-ended stages or class A, AB, or B push-pull stages. A simple class A single-ended power amplifier is shown in Figure 8-19 with practical approximate component values. Manufacturer's literature should be consulted for optimum component selection and design information. In a class A amplifier, the Q point of operation is selected to be roughly in the center

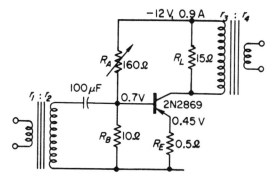

Figure 8-19.   Class A single-ended power amplifier.

(Laurence G. Cowles, *Transistor Circuits and Applications*, © 1968. Reprinted by permission of Prentice-Hall, Inc.)

of the I_c current swing, before clipping (nonlinear operation) occurs. Current flows during the complete single cycle as well as during quiescent conditions.

By adjusting the bias so that there is reduced (class AB) quiescent current, I_c, or very little (class B operation), and using a push-pull configuration, much greater efficiency and power output may be attained.

Darlington amplifiers are usually used in modern circuit design. Nevertheless, bipolar transistors are often biased to operate either class B or AB in a push-pull configuration.

8.5 TRANSISTORS IN DIGITAL CIRCUITS (SEE CHAPTER 16)

Transistor Switching Characteristics

Two-state digital logic requires that the logic device provide and be responsive to only two voltage levels. In transistor application to digital circuits, the two voltage levels correspond to cutoff and hard conduction or saturation. The transistor logic device is usually operated in the common-emitter configuration, and cutoff corresponds to zero (or very low) I_b so that I_c is also at a minimum, and the output voltage is nearly V_{cc}. At saturation, I_b is at its maximum point along the load line (see I_c-V_c characteristic curves, Figure 8-10), and the voltage output is near zero. Emitter-coupled logic does not operate to saturation and provides higher on-off switching speeds; however, the majority of logic circuits in use are saturated devices.

Whenever a transistor is turned on or off, the charge stored in the base-to-emitter capacitance must be discharged and recharged in the reverse direction. The sum of the time it takes to turn on and off is called the pair delay, and is a figure of merit for the device. Using small high-frequency transistors and careful design and construction, a high-speed saturated switching transistor time on (Ton) and time off (Toff) is typically Ton = 12 ns, Toff = 18 ns, Ton = 20 ns, Toff = 40 ns, and there are some as fast as Ton-6ns, Toff-15 ns that produce 600 MHz minimum.

Resistor-Transistor Logic

Although discrete transistors are seldom (if ever) used in modern transistor logic families, it is beneficial to examine this type of construction in order to gain an insight into the modern integrated circuits (ICs).

Figure 8-20 is the circuit of an *n* input RTL AND gate. A separate bias voltage is required to insure cutoff, under any conditions of temperature drift and variations in the transistor characteristics. Grounding of any gate input (near zero volts) will turn the transistor off and provide a "high" output. With all inputs high (near V_{cc}), the output is low (near zero). For this reason, this type of gate, a negative or inverted AND, is abbreviated as NAND. The fan-in or number of inputs of the RTL gate is limited by the interaction possibility of the inputs, and the illustrated circuit should not have more than four inputs (fan-in of four).

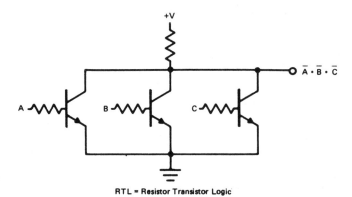

RTL = Resistor Transistor Logic

Figure 8-20.   RTL NAND logic gate.

(Gerald A. Maley, *Manual of Logic Circuits*, © 1970. Reprinted by permission of Prentice-Hall, Inc.)

Diode-Transistor Logic (DTL)

To overcome the limitation of fan-in of the RTL gate, the inputs can be isolated from each other by diodes, giving rise to DTL-type logic gates. Figure 8-21 illustrates a typical gate of this type.

Transistor—Transistor Logic (TTL) Gates

Since this combination is not usually used with discrete transistors, it is described in Chapter 9 under "Integrated Circuits" (see Figure 9-10).

Flip-Flop Circuits

The basic flip-flop circuit consists of two transistors connected to each other in such a way that when one transistor is cut off, the other conducts. When the flip-flop changes state, the cutoff transistor conducts

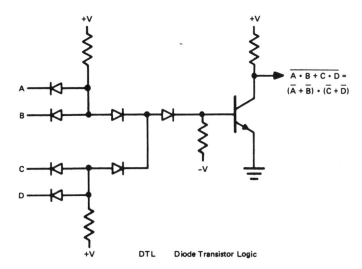

Figure 8-21.   DTL NAND logic gate.

and the conducting one is cut off. As shown in Figure 8-22, flip-flops can be made up of two gate circuits of the kind described in more detail in Chapter 16, connected back-to-back. If the output of the first gate is high (transistor is cut off), it will cause the other gate to turn on (conduct). The "on" gate has a low output, which, connected to the first gate, forces it to maintain its high output; thus a "locked-up" stable condition is obtained. By applying a negative going signal simultaneously to the second input of each gate, the "on" transistor is turned off, but the "off" transistor is turned on by the fact that it is being fed by a high signal that persists longer because of the RC time constant of the emitter-to-base connection. Thus the state of the flip-flop has been reversed.

Since the flip-flop is affected only by the negative going part of the signal, it takes two such signals to change and then restore the flip-flop to the original state. In effect, this divides the number of input signals by two. Connecting several flip-flops in series provides a binary counter chain. (See Chapter 16.)

In addition to the basic flip-flop described above, there are variations triggered by applying a pulse to both emitters, both collectors, or both bases (as in Figure 8-22). The trigger pulse may be coupled by means of diodes, capacitors, and/or resistors.

The logic diagram shown in Figure 8-22 indicates that there are two inputs (S and R) and two outputs inverted $\overline{Q}$ (the bar shows inversion), and noninverted Q.

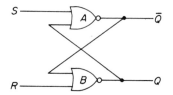

Figure 8-22. Flip-flop, two gates back to back.

(T. Sifferlen and V. Vartanian, *Digital Electronics with Engineering Application*, © 1970. Reprinted by permission of Prentice-Hall, Inc.)

One-Shot Multivibrator Circuit

The flip-flop circuit is a bi-stable device and a variation of it, the mono-stable or one-shot multivibrator, will stay in a second state only momentarily. In the bi-stable flip flop of Figure 8-22, the gate circuits were cross-coupled at both ends with dc coupling. In the one-shot, as shown in Figure 8-23, one base is coupled through capacitor C, which determines the time duration of the unstable state. The second base is dc coupled just as in Figure 8-22.

Since the base of Q_1 is returned to V_{cc}, it is biased on, and therefore keeps Q_2 off with a high output, in the quiescent state. A negative

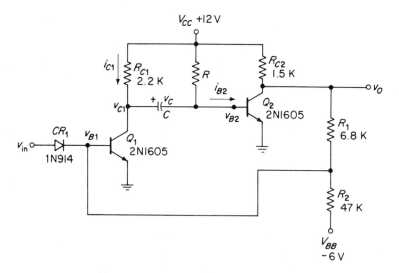

Figure 8-23. One-shot multivibrator.

(T. Sifferlen and V. Vartanian, *Digital Electronics with Engineering Applications*, © 1970. Reprinted by permission of Prentice-Hall, Inc.)

pulse, introduced into the base of Q_1, will turn it off, turning Q_2 on and reducing its output voltage to near zero. The discharge of capacitor C through the biasing resistors keeps Q_1 off temporarily, dependent upon the discharge time of C and the resistance in its discharge path (RC time constant). Upon completion of the discharge, the original state is restored, with Q_1 on and Q_2 off.

Pulse Amplifier—Schmitt Trigger Circuit

The Schmitt trigger circuit, shown in Figure 8-24, is used to convert sine waves and other wave forms into pulses. Its operation is similar to the one-shot multivibrator, except that the input signal and not a coupling capacitor determines its output voltage. Like a dc-coupled amplifier, it has one RC connection between transistors; however, a degree of positive feedback is provided through a common-emitter resistor. This speeds up the rise time and defines the input voltage level at which the output is switched from high to low and vice versa. The difference between the voltages at which the circuit will snap high and low is called the *hysteresis* or *backlash voltage*. By careful design, this can be controlled to desired levels. Also, some special Schmitt circuits are designed to trigger at specific voltage levels and act as voltage-level detectors.

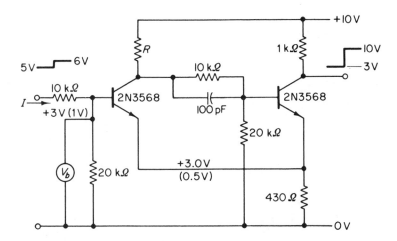

Figure 8-24. Schmitt trigger circuit.

(Laurence G. Cowles, *Transistor Circuits and Applications,* © 1968. Reprinted by permission of Prentice-Hall, Inc.)

8.6 TRANSISTORS AS OSCILLATORS

Basic Oscillator Circuits

The conditions necessary for oscillation include an amplifier with positive feedback, such that the output of the amplifier is fed back at zero (or 360°) phase with sufficient magnitude to overcome all circuit losses.

Sine wave oscillators require that the amplifier operate in the class A mode with the loop gain somewhat greater than 1, and the feedback carefully adjusted so that overloading and signal clipping does not occur. The frequency of oscillation may be determined by inductance and capacitance combinations (LC), quartz crystals, frequency-selective RC combinations, mechanically resonating devices (tuning forks, magnetostrictive, etc.), or delay lines, tuned to a frequency within the band pass of the amplifier.

Relaxation oscillators operate with the amplifier being driven hard into the saturated and cutoff modes by providing excess gain well above the loop gain of 1. The period of such an oscillator is determined by the charge and discharge time of one or more storage elements and their associated charge and discharge path resistances. There are certain types of relaxation oscillators, where the amplification is not readily apparent, such as oscillators using unijunction transistors or Shockley four-layer diodes and thyristors (See section 8.8; also, see Section 8.2 for voltage-variable capacitance diodes used in tuning circuits).

Sine Wave Transistor Oscillators

Resonant Circuit Oscillators.   Figure 8-25 illustrates some resonant circuit oscillators. The resonant circuits are either LC or equivalent electromechanical (quartz crystal) resonators. They are widely used for RF work in radio receivers and transmitters. With careful design and construction, they can be made to operate with good power efficiency, low harmonic content (almost pure sine wave), and excellent stability, especially in the case of the crystal oscillator.

Hartley Oscillator.   Part of the voltage developed across the main inductance L_1 of the main resonant circuit is fed back directly to the emitter through the coupling capacitor C. See Figure 8-25(a).

Colpitts Oscillator.   Feedback is accomplished by means of the portion of the voltage-resonant circuit divider (C_1 and C_2). The feedback appears across C_1 and is fed back to the emitter. See Figure 8-25(b).

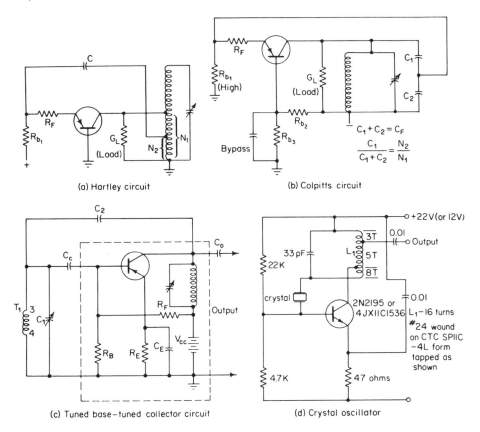

Figure 8-25.   Resonant circuit sine wave oscillators.

(Harry E. Thomas, *Handbook of Transistors, Semiconductors, Instruments, and Microelectronics,*
© 1968. Reprinted by permission of Prentice-Hall, Inc.)

Tuned Base-Tuned Collector Oscillator.   The two resonant circuits in the base and collector, respectively, of the transistor are coupled by the feedback through the collector-base capacitance. Since this capacitance is small, an external capacitor C_2 is sometimes added to provide sufficient feedback voltage. See Figure 8-25(c).

Crystal Oscillator.   This oscillator (Figure 8-25[d]) is a modification of the tuned base-tuned collector type which uses a resonant circuit in the collector. In a crystal oscillator, a piezoelectric quartz crystal is used in place of a tuned circuit in the collector circuit. The stability of the crystal frequency-determining element is not affected by supply voltage, variations, temperature, humidity, or mechanical variations. The high degree of crystal frequency stability recommends the circuit for many precision operations.

RC Circuit Sine Wave Oscillators

In the low-frequency RF and audio range, sine wave oscillators can be designed with RC networks (without inductance). The RC oscillator is usually named after the type of frequency selective network employed in its feedback circuit.

RC Phase-Shift Oscillator (a). The RC network consists of three (sometimes more) L-type RC sections in tandem, which provide a 180° phase shift at a particular frequency. As shown in Figure 8-26 (a), a single stage of transistor amplification is sufficient to overcome the attenuation of this network for a loop gain greater than 1 and provide the rest of the required 180° (phase reversal) phase shift. With a three-section network, a current gain of 56 is sufficient to insure oscillation. Each section provides the shunt impedance of the combined ladder network.

Twin-T Oscillator (b). A twin-T or bridge-T nulling network has a very sharp cusp in its transfer function, at a certain frequency. When connected in the negative feedback path of a high-gain transistor amplifier, as shown in Figure 8-26(b), it becomes a high-stability oscillator. All frequencies except the null frequency are degenerated by the negative feedback connection, and the small in-phase frequency component slightly off-null is regenerated and causes oscillation.

Wien Bridge Oscillator (c). Similar to the twin-T oscillator, the Wien bridge network shown in Figure 8-26(c) provides a sharp null at a certain frequency, oscillating in a high-gain amplifier's (OP AMP) feedback path, near its null frequency. Because it requires adjustment and tracking of only two components in changing its frequency ($C_1 = C_2$), it is suitable for use in test oscillators of wide frequency range.

Relaxation Transistor Oscillators

Oscillators that produce nonsinusoidal wave forms, such as square waves, triangular waves, sawtooth waves, etc., are known as multivibrators, blocking oscillators, and the special category corresponding to the unijunction oscillator and integrated circuits.

Multivibrators (a). The multivibrator consists of two stages of amplification coupled back-to-back as shown in Figure 8-27(a). Because the inherent frequency stability of the multivibrator is limited, it is usually used where it must be synchronized to some outside source or standard.

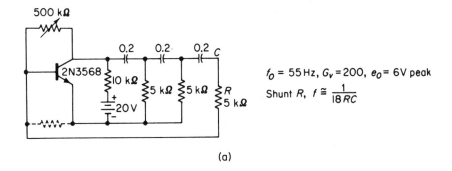

$f_0 = 55\,\text{Hz},\ G_v = 200,\ e_0 = 6\text{V peak}$

Shunt $R,\ f \cong \dfrac{1}{18\,RC}$

(a)

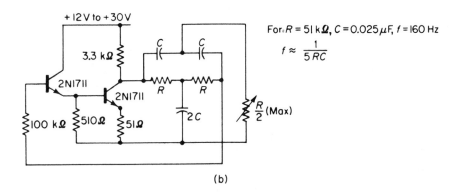

For $R = 51\,\text{k}\Omega,\ C = 0.025\,\mu\text{F},\ f = 160\,\text{Hz}$

$f \approx \dfrac{1}{5\,RC}$

(b)

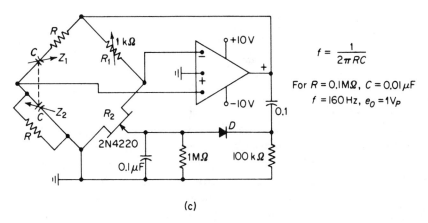

$f = \dfrac{1}{2\pi RC}$

For $R = 0.1\text{M}\Omega,\ C = 0.01\,\mu\text{F}$

$f = 160\,\text{Hz},\ e_0 = 1\text{V}_P$

(c)

Figure 8-26.   RC circuit sine wave oscillators.

(Laurence G. Cowles, *Transistor Circuits and Applications,* © 1968. Reprinted by permission of Prentice-Hall, Inc.)

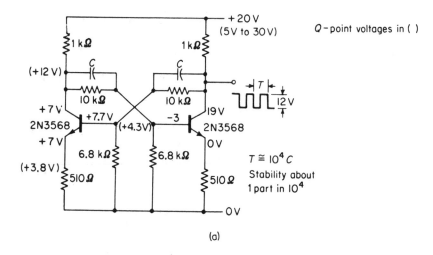

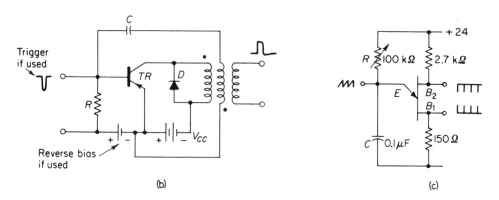

Figure 8-27.   Relaxation transistor oscillators.

(Laurence G. Cowles, *Transistor Circuits and Applications,* © 1968. Reprinted by permission of Prentice-Hall, Inc.)

Blocking Oscillators (b).   Where short, high-energy pulses are required, either driven (synchronized) or as a free running source, blocking oscillators are used. Feedback is accomplished by a tightly coupled transformer, as shown in Figure 8-27(b), that drives the single transistor hard into saturation, also saturating the transformer. The time constant of R and C determines the quiescent period between pulses. The shape and size of the pulse depend upon the transformer-transistor relationship.

Unijunction Oscillator (c).   The unijunction oscillator provides good frequency stability over a wide range of temperatures, power supply variations, and circuit parameters. The output at the emitter is a sawtooth wave form. Pulses with fast rise and fall times, as shown in Figure 8-27(c), are produced at both bases. The oscillator is readily synchronized by application of trigger pulses, and finds application in timing devices and sawtooth wave generators.

8.7 THERMAL DESIGN CONSIDERATION

Special design considerations of transistors, temperature effects, and their control are essential to semiconductor circuit operation. Wherever 1 or more watts of power are dissipated, adequate external cooling is required.

Allowable Heat Dissipation

The thermal resistance given in a transistor's data sheet may be used to compute the maximum power dissipation allowable at a given ambient temperature. A typical low-power germanium transistor may have a maximum allowable junction temperature of 85°C and a thermal resistance of 0.5°C/mw. With the ambient temperature at 25°C, it can dissipate only enough power to raise the junction temperature from 25°C to 85°C. This rise of 60°C will be produced by 120 mw in the transistor.

$$\frac{60°C}{0.5°C/mw} = 120 \text{ mw}$$

T_j = Junction temperature
T_a = Ambient temperature
θ_{ja} = Ambient temperature
        junction thermal resistance

because: $T_j = T_a + \theta_{ja} (P_d)$ P_d = Power dissipated
                          $\theta_{ja} - P_d$ = Temperature rise

If, however, the ambient were 60°C, then only a rise of 25°C would be allowable. This would limit the amount of power permissible into the transistor to 50 mw.

Cooling Methods

The simplest cooling method is a heat sink. If properly mounted, as recommended in manufacturers' manuals, using thermally conductive washers (but good electrical insulators), silicone compounds, and greases for good thermal contact, and the correct amount of bolting pressure, then the thermal resistance between transistor case and heat sink can be neglected. When mounted to a heat sink, the junction-to-case thermal resistance is used and added to the sink-to-ambient resistance to obtain the total thermal resistance.

$$\theta_{ja} = \theta_{jc} + \theta_{sa} \qquad \begin{array}{l} \theta_{ja} = \text{junction to ambient} \\ \theta_{jc} = \text{junction to case} \\ \theta_{sa} = \text{sink to ambient} \end{array}$$

Thermal Resistance

Commercial heat sinks are available in various sizes and shapes for the different size transistor cases. They are usually rated by their manufacturers in C°/W, and vary between roughly 1°C/W to 10°C/W, depen-

Fig. 1 & 2 Code No.	Shape	Surface Area (sq. in)	L (in.)	Volume Displacement		Vol. (cu. in)	Weight (grams)	Finish	Thermal Resistance °C/W
				W (in.)	H (in.)				
1	Flat-finned Extrusion	65	3.0	3.6	1.0	10.8	114	Anod. Black	2.4
2	Flat-finned Extrusion	65	3.0	3.6	1.0	10.8	114	Bright Alum.	3.0
3	Flat-finned Extrusion	65	3.0	3.6	1.0	10.8	114	Gray	2.8
4	Flat-finned Extrusion	60	3.0	4.0	.69	8.3	123	Anod. Black	2.8
5	Flat-finned Extrusion	95	3.0	4.0	1.28	15.3	189	Anod. Black	2.1
6	Flat-finned Extrusion	64	3.0	3.8	1.3	15.0	155	Black Paint	2.2
7	Flat-finned Extrusion	83	3.0	4.0	1.25	15.0	140	Anod. Black	2.2
8	Flat-finned Extrusion	44	1.5	4.0	1.25	7.5	75	Anod. Black	3.0
9	Flat-finned Extrusion	137	3.0	4.0	2.63	31.5	253	Anod. Black	1.45
10	Flat-finned Extrusion	250	5.5	4.0	2.63	58.0	461	Anod. Black	1.10
11	Flat-finned Extrusion	130	6	3.6	1.0	21.5	253	Anod. Black	1.75
12	Flat-finned Extrusion	78	3.0	3.8	1.1	12.5	190	Gray MMI	2.9
13	Flat-finned Extrusion	62	3.0	3.8	1.3	15.0	170	Gray MMI	2.2
14	Flat-finned Extrusion	78	3.0	4.5	1.0	13.5	146	Gold Alodine	3.0
15	Cylindrical Horizontal Fin, Machined Casting	30	1.75 Dia.		0.84	2.0	40	Anod. Black	8.5
16	Cylindrical Horizontal Fin, Machined Casting	50	1.75 Dia.		1.5	3.6	67	Anod. Black	7.1
17	Cylindrical Horizontal Fin, Machined Casting	37	1.75 Dia.		1.5	3.6	48	Anod. Black	6.65
18	Cylindrical Vertical Fins, Casting	7.5	1.5 Dia.		0.9	4.4	33	Anod. Black	8.1
19	Cylindrical Vertical Fins, Casting	12	1.5 Dia.		1.4	6.9	51	Anod. Black	7.0
20	Cylindrical Vertical Fins, Casting	25	1.5 Dia.		2.9	14.2	112	Anod. Black	5.6
21	Cylindrical Vertical Fins, Casting	35	1.5 Dia.		3.4	16.7	132	Anod. Black	5.1
22	Cylindrical Vertical Fins, Casting	32	2.5 Dia.		1.5	7.4	94	Anod. Black	4.5
23	Cylindrical Vertical Fins, Casting	20	2.5 Dia.		0.5	2.45	48	Anod. Black	6.6
24	Flat-Finned Casting	23	1.86	1.86	1.2	4.15	87	Anod. Black	5.06
25	Square Vertical Fin, Sheet Metal	12	1.7	1.7	1.0	2.9	19	Anod. Black	7.4
26	Cylindrical Vertical Fin, Sheet Metal	15	2.31 Dia.		0.81	3.35	18	Black	7.1
27	Cylindrical Horizontal Fin, Sheet Metal	6	1.81 Dia.		0.56	1.44	20	Anod. Black	9.15
28	Cylindrical Horizontal Fin, Sheet Metal	55	2.5 Dia.		1.1	5.4	115	Gold Irridate	7.9

Table 8-7.   Heat sink types in free air.

(Illustration courtesy of Motorola, Inc.)

dent upon their size, shape, material, surface finish, and orientation to the flow of air. The smaller the rating number, the better the heat sinking.

When natural convection air flow proves insufficient to keep the temperature within limit bounds, forced air cooling by means of a fan of proper capacity would be a next step.

Thermal Runaway And Stabilizing Suggestions

The collector cutoff current I_{co} that flows between base and collector doubles with every 9°C temperature rise of the junction. Even though this current is very sensitive to temperature, it will cause little trouble in a grounded base configuration, because I_{co} is very small in modern transistors, especially those made of silicon. When the emitter is grounded (common emitter), this cutoff current must pass through the base-emitter junction as well as the base-collector junction. It is amplified by the current amplifying mechanism of the transistor. As a result, the current that will flow in a CE configuration will be multiplied by beta, $I_c = \beta I_{co}$.

As an example, assume $\beta = 50$ and $I_{co} = 5\mu a$ at an ambient of 25°C. With a common-emitter configuration and an increase in junction temperature to 61°C, a rise of 36°C would cause I_{co} to increase 16 times, to $50 \times 5\mu a \times 16 = 4$ ma. This additional 4 ma will increase the power dissipation, raising the temperature of the transistor above 61°C, and so on. This process can continue until the destruction of the transistor and is called thermal runaway. In milder forms, it can cause clipping and distortion of the signal.

This thermal regenerative action can be limited by several techniques. The temperature-stabilizing circuit of Figure 8-28(a) used an emitter resistor, R_2, which is bypassed by C_2, as well as a thermistor feeding back current from the collector supply. This thermistor, RT_1, changes its resistance with the temperature and provides extremely stable amplifier operation over a wide temperature range.

A forward-biased diode is used to regulate the base current in the transformer-coupled amplifier circuit of Figure 8-28(b). Again, as increased base current trends to raise the junction temperature of the transistor, this current is limited by the combination of the diode CR_1 and R_1.

When a separate voltage source is available for base biasing, the reverse diode stabilization circuit of Figure 8-28(c) can be used. Here again, the value of R_1 is important. The components of collector and base current are indicated in this illustration and show the detailed temperature stabilization effect.

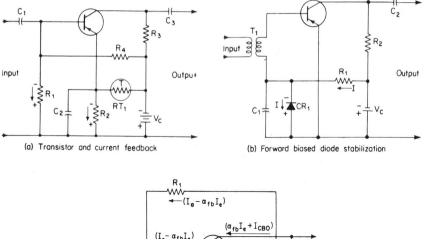

(a) Transistor and current feedback

(b) Forward biased diode stabilization

(c) Reverse biased diode stabilization

Figure 8-28. Temperature-stabilizing circuits.

(Harry E. Thomas, *Handbook of Transistors, Semiconductors, Instruments, and Microelectronics,*
© 1968. Reprinted by permission of Prentice-Hall, Inc.)

Test Precautions for Power Transistors

All transistor circuits, and especially power amplifiers, should be tested
for thermal stability. Transistors rated for 1 watt and over normally
require heat sinking. A simple temperature test is to touch the transistor
carefully with the little finger, usually the most sensitive one. At 50°C,
it can be held there indefinitely. At 60°C—a good, conservative operating temperature for germanium transistors—most fingers feel very hot
after one second. Silicon transistors can safely take at least 100°C. At
this temperature, water dropped on the transistor can will sizzle. To
test a circuit for its upper temperature limit, a hot soldering iron can
be used to provide some external heating (careful with germanium transistors). If circuit performance deteriorates with only little external
heating, the circuit is probably running too hot. At the other end of the
temperature scale, a coolant can be sprayed on the transistor. It will
quickly and temporarily reduce its temperature, while the performance
of the circuit may be observed for signs of improvement or deteriora-

tion. Precision circuits, especially high-gain dc amplifiers required to operate over wide ambient temperatures, must be carefully tested over the required temperature range in a controlled test chamber. There are several digital multimeters available that have temperature measurement capabilities.

8.8 SPECIAL SEMICONDUCTOR DEVICES

Field Effect Transistors

The field effect transistor's high input impedance and characteristics curves resemble that of a pentode vacuum tube. In circuit analysis using Thevenin and Norton equivalency diagrams, the FET is treated as a vacuum tube and the gain formulas, input and output impedance calculations, etc., derived for vacuum tubes apply here almost without variation. There are two basic types of FET transistors, known as the Junction Field Effect Transistor (JFET) and the Insulated Gate Field Effect Transistor (IGFET). The IGFET is also referred to as *physical structure*. The JFET uses as a gate two closely spaced junctions to enhance or deplete the current flow in the channel. The IGFET uses a tiny, high-quality capacitor structure as a gate over the channel between the source and the drain. Figure 8-29 provides a pictorial diagram of these structures; 8-29(a) shows the polarity required for the operation of these devices.

One important FET feature not found in other transistors is that the source-to-drain channel is a pure resistance without a diode effect. This permits FETs to act as electrically controlled resistors for variable gain control, relays (switches), modulators, and choppers, without the complication of a junction voltage offset. The gate voltage required to cut off the channel current is called the pinch-off voltage. Switching-type FETs have pinch-off voltages in the range of 2 to 8 volts, and amplifier-type FETs in the order of 0.5 volts.

The MOS (Metal Oxide Semiconductor) FETs are available in two types, characterized by whether they conduct or whether they are cut off at zero gate bias. In the depletion type, the channel conducting the current and the signal voltage on the gate can reduce it (or raise it in some types), and may cut if off completely. In the enhancement type, the channel current is normally cut off, and is turned on by applying a gate voltage.

MOS power FETs are being designed into more and more electronic equipment, offering better performance, often requiring less support circuitry, and, in the long run, costing less. The two basic types

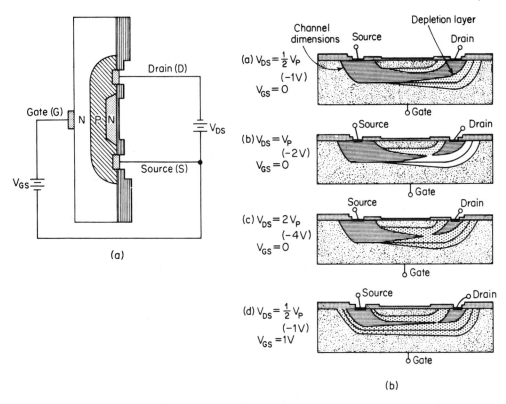

Figure 8-29. Field effect transistors, FETs.

(Harry E. Thomas, *Handbook of Transistors, Semiconductors, Instruments, and Microelectronics,*
© 1968. Reprinted by permission of Prentice-Hall, Inc.)

being used are the MOSFETs and VMOSFETs (the "V" has to do with
the FETs' physical structure—a V-shaped channel is etched in the internal material during manufacture). Both of these devices are used
widely in radio frequency applications.

One particularly interesting characteristic of the VMOSFETs is that
they have a negative temperature coefficient—there is an automatic reduction of current flow with an increase of temperature.

Unijunction Transistors (UJT)

A unijunction transistor is produced by forming a p-type junction somewhere halfway between the ends of a silicon (n-doped) filament. This
rectifying junction, called the emitter, behaves very much like a neon
tube. When a certain voltage is reached, it fires (starts conducting) and
will stay on until the voltage is removed or reduced to a very low level.
Before firing, the current drawn is very low (leakage in microamps);

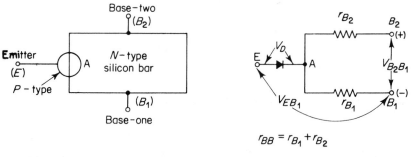

(a) A simplified bar structure   (b) Equivalent circuit valid for $I_E \leqq I_P$

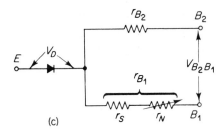

Figure 8-30.   UJT.

(John D. Lenk, *Practical Semiconductor Data Book for Electronic Engineers and Technicians,* ©
1970. Reprinted by permission of Prentice-Hall, Inc. (Figure originally courtesy of Motorola.))

after firing, the current starts out high and diminishes as the emitter
voltage increases. This is the behavior of a negative resistance.

As shown in Figure 8-30, the two ends of the silicon filament are
ohmically connected to contacts forming the base-one and base-two of
the UJT. A voltage applied between bases one and two forms a voltage
divider to the emitter point, which is about 60% of the total, up from
base-one. This voltage back-biases the emitter diode junction. When
the emitter input voltage exceeds this back-bias voltage, the UJT starts
to conduct (fires) and the current carriers injected across the junction
avalanche (as described for the zener diode), causing a large increase
in current. The increase in current is attended by a drop in resistance,
and as the current increases, the voltage regeneratively (like positive
feedback) decreases, until the so-called valley point is reached, where
the current starts to increase with voltage, as in a normal resistor.

There are six thyristors often used in today's electronic circuit
design. These are:

1. Unijunction transistors (UJT)
2. Bilateral triggers (DIAC)

3. Programmable UJTs (PUT)
4. Silicon bidirectional switch (SBS)
5. Optically coupled triac drivers (integrated circuits)
6. Optoisolators (integrated circuits)

Shockley Four-Layer Semiconductor-Thyristor

The thyristor is analogous to the thyratron tube in that it fires above a certain threshold. Figure 8-31 shows how a four-layer *pnpn* combination of semiconductor materials can be represented as two transistors connected together in a positive feedback loop with a loop gain greater than 1. When connected to polarities shown, and with gate G going to the cathode, the *npn* section is in a cutoff condition, so that the whole stack is nonconducting (except for leakage). Connecting G to the anode

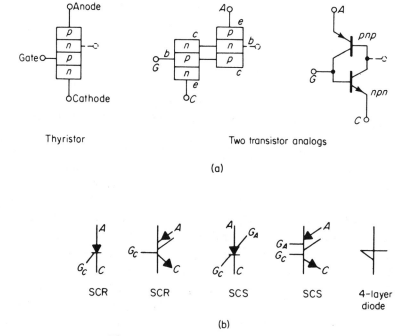

Figure 8-31.   The thyristor family.

(Laurence G. Cowles, *Transistor Circuits and Applications,* © 1968. Reprinted by permission of Prentice-Hall, Inc.)

(through a current limiting resistor), turns the npn section on, which then turns on the pnp section, and the whole stack of layers thus drives itself into saturation. The current is then limited by a load resistance.

Thyristors are manufactured as diodes, with only the cathode and anode leads presented externally (Shockley diodes). Silicon Controlled Rectifiers (SCR) are produced with the gate (G_1), anode and cathode available, and Silicon Controlled Switches (SCS), with all four semi-conductor regions accessible. Specified forward breakdown voltages are available within close tolerances, over a range of approximately 20 to 200 volts. SCRs are available in small milliampere sizes up to 100 amps turn-on current, and sizes that can block up to 1000 volts forward and reverse. The SCS is generally characterized as a sensitive SCR, designed for anode currents below 0.5 amps and for voltage ratings below 100 volts, with closely defined firing currents and voltages.

SCRs are usually turned off by removing the anode-cathode supply voltage, but they can be turned off from the gate with a relatively large negative signal pulse to the npn (G_1) base, since the current gain is much less during turn-off than turn-on. The load current must exceed a certain minimum value, called the holding current, in order that the SCR stay on upon removal of the gate signal. Capacitance coupling to the gate, due to internal capacitance, may turn an SCR on by the rate effect if the anode voltage is allowed to rise too rapidly.

Phototransistor

If the transparent emitter-base pn junction, similar to that in a photo-diode, is backed up by a collector semiconductor layer, photoconduc-tive current generated across the reverse-biased emitter-base junction is amplified, as in a transistor. The light applied to the base through a window or lens in the case is the input signal. The typical collector voltage-current characteristics for a given illumination in foot-candles are illustrated in Figure 8-32. The action of the transistor configuration amplifies the input junction current by β. The base-to-emitter resistance must be carefully chosen, because this determines to a great extent the ratio of light to dark current, and its degenerative action reduces the danger of thermal run away, since the device is quite sensitive to am-bient temperature. Figure 8-32 (c) illustrates the effect of this base-to-emitter resistance. More about phototransistors in Chapter 19.

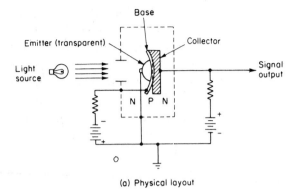

(a) Physical layout

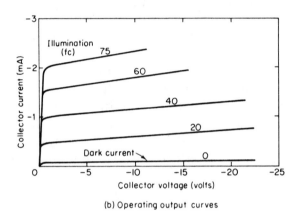

(b) Operating output curves

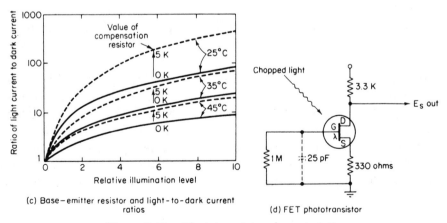

(c) Base–emitter resistor and light-to-dark current
ratios

(d) FET phototransistor

Figure 8-32.   Phototransistor characteristics.

(Harry E. Thomas, *Handbook of Transistors, Semiconductors, Instruments, and Microelectronics,*
© 1968. Reprinted by permission of Prentice-Hall, Inc.)

BIBLIOGRAPHY—CHAPTER 8

CHOMA, J., *Electrical Networks Theory and Analysis.* New York, NY, John Wiley & Sons, Inc., 1985.

GENN, R., *Illustrated Guide to Practical Solid State Circuits.* Englewood Cliffs, NJ, Prentice-Hall, Inc., 1983.

JORDON, E., ed., *Reference Data for Engineers: Radio, Electronics, Computer, and Communications* (7th Ed.). Indianapolis, IN, Howard W. Sams & Co., Inc., 1985.

MANASSE, F., *Semiconductor Electronic Design.* Englewood Cliffs, NJ, Prentice-Hall, Inc., 1977.

MILNES, A., *Semiconductor Devices and Integrated Electronics.* New York, NY, Van Nostrand Reinhold Co., 1980.

STREETMAN, B., *Solid State Electronic Devices* (2nd Ed.). Englewood Cliffs, NJ, Prentice-Hall, Inc., 1980.

SZE, S., *Semiconductor Devices: Physics and Technology.* New York, NY, John Wiley & Sons, Inc., 1985.

VASSOS, B., and G. EDWIG, *Analog and Digital Electronics for Scientists* (3rd Ed.). New York, NY, John Wiley & Sons, Inc., 1985.

CHAPTER 9

INTEGRATED CIRCUITS

9.1 CONSTRUCTION TECHNIQUES

Monolithic

In monolithic ICs all circuit elements, both active and passive, are simultaneously formed in a single chip of silicon by a planar diffusion technique. The elements are interconnected to form the required electronic circuit by metallized strips deposited on the top, protective, oxidized surface of the silicon chip, using evaporation techniques. On each silicon slice, the same circuit is repeated a large number of times.

Hybrid

Hybrid technology combines the techniques of monolithic, discrete, and thick and thin films for obtaining the "best" solution to a particular design problem. The word *best* implies a compromise between cost and technical perfection. One of the drawbacks of a straight monolithic approach to many designs is the limitations imposed by the temperature sensitivity and restricted ranges of values of passive components manufactured by this method. Many applications of ICs require closer tolerances, temperature stability, higher power handling ability, wider bandwidths, etc., than can be obtained by the monolithic technique at its present stage of development. This is especially true for linear or analog-type circuits.

Active components for hybrid packaging of circuits are most often formed just as monolithic circuits on separate wafers of silicon, using planar methods. Passive components, resistors, inductors, and capacitors are made using thick and thin film techniques. Hybrid techniques are, in a way, an extension of the older discrete-component methods, since they are assemblies of separate components that have been formed on or affixed to an insulating base material (substrate) and then interconnected.

Packaging

Of the two methods of integrated circuit design, the monolithic method has attained the greater degree of standardization in packaging because it lends itself to high quantity and automated mass production, thus permitting low cost to the consumer. A mass market was found in digital circuits because they worked with looser circuit tolerances and met the quantity and low-cost demands of the computer industry. At first the standard enclosures used for single transistors were pressed into service for housing the complete circuits of gates, flip-flops, and other digital logic circuits. Later they were replaced by the so-called dual-in-line package and the flat pack, illustrated in Figure 9-1. Special package sizes and mounting details are best obtained from up-to-date manufacturers' data sheets and manuals.

The bulk of today's hybrids are custom designs, and most hybrid packages are geared to satisfy special requirements in modest quantities. Some high-volume hybrid devices, such as operational amplifiers, etc., are found in standard packages shown in Figure 9-1.

9.2 ANALOG ICs

Hybrid circuit designs are widely used where custom requirements must be met, or where precision, high-stability, or other special performance characteristics must be met. Two major circuit types that have many

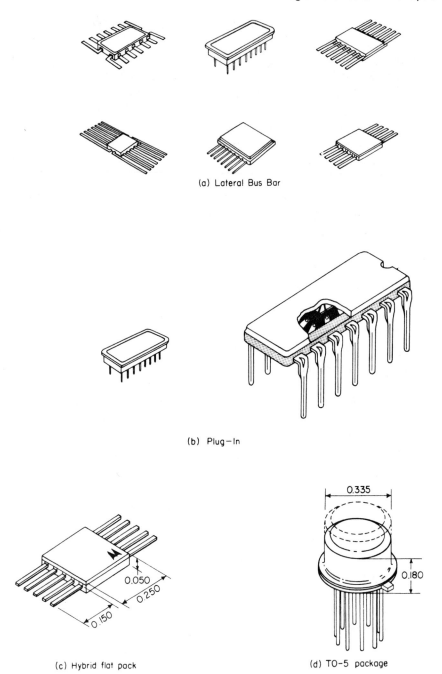

(a) Lateral Bus Bar

(b) Plug—In

(c) Hybrid flat pack

(d) TO-5 package

Figure 9-1.　Standard IC packages.

(Harry E. Thomas, *Handbook of Integrated Circuits,* © 1971. Reprinted by permission of Prentice-Hall, Inc.)

applications and readily lend themselves to monolithic techniques are the balanced differential amplifier and the operational amplifier. These types of amplifiers are also frequently constructed by hybrid methods, but where broad tolerances and other requirements permit, they make excellent and relatively inexpensive "standardized" analog ICs for mass production by the monolithic technique.

The Differential Amplifier

The differential amplifier is an ideal configuration for monolithic integrated circuit methods, because the use of capacitors and high-value resistors can be held to a minimum. Advantage is also taken of the exceptional input balance due to the inherent match between the differential transistor pairs, which are processed in exactly the same way and are located very close to each other on the same very small silicon chip. The close match in temperature coefficients of components fabricated from the same material assures stable electrical characteristics over a very broad temperature range, because ratios are held constant and absolute values are not important. The differential configuration as shown in Figure 9-2 makes excellent output-to-input isolation possible and simplifies feedback arrangements.

Another advantage of the differential amplifier is its ability to reject noise and other signals common to the two balanced inputs. It can be used to amplify only the signals connected between these inputs. This so-called common mode rejection permits amplification of low-level signals in noisy environments.

Differential amplifiers provide linear amplification from dc through the audio and video frequencies into the vhf region. They are also used for frequency multiplication, mixing, amplitude modulation, limiting, product detection, signal generation, gain control, squelch, and temperature compensation, operations that are essentially nonlinear in nature.

There are two basic categories of differential amplifiers, video and narrow band, though many in-between types have been listed in the literature variously as audio, wide-band, high-frequency, RF, and ordinary comparator circuits. The video differential amplifier usually contains all resistors as part of the IC, while the narrow-band differential amplifier depends on external components such as tuned circuits for its operation. Figures 9-2 and 9-3 show typical schematic diagrams of these two types of universal circuits. Figure 9-4 shows the typical applications, with external components connected to obtain the desired function.

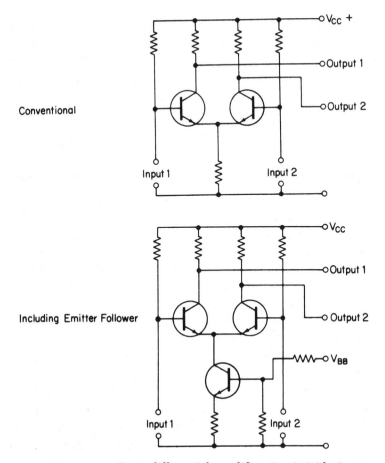

Figure 9-2. Basic differential amplifier circuit (video).

(Harry E. Thomas, *Handbook of Integrated Circuits*, © 1971. Reprinted by permission of Prentice-Hall, Inc.)

Operational Amplifiers

The term *operational amplifier* originally meant an amplifier circuit used in analog computers to perform mathematical operations, such as integration, differentiation, summation, and subtraction. The mathematical versatility of the operational amplifier is, however, not limited to these basically linear operations. By using nonlinear networks such as diodes, relays, switches, exponential semiconductors (log diode), transcendental arrays (straight-line synthesized functions), etc., in their feedback paths, operational amplifiers can perform a large variety of useful functions.

This great versatility is based on feedback, a technique discussed

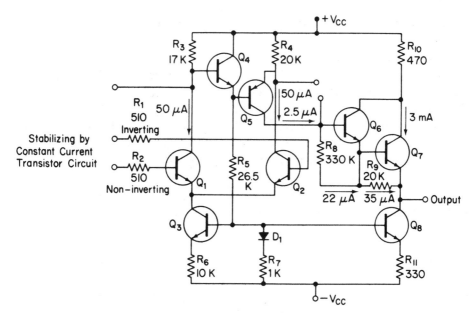

Figure 9-3.   Narrow-band differential amplifier.

(Harry E. Thomas, *Handbook of Integrated Circuits*, © 1971. Reprinted by permission of Prentice-Hall, Inc.)

in Chapter 18 under Automatic Control, which is also used in audio amplifiers to reduce distortion. Because of its widespread use in many applications, the monolithic method is suitable for this analog device and provides "standardized" operational amplifiers at relatively low cost. Hybrid circuits cover a considerable portion of the operational amplifier market, especially where external components are not per-

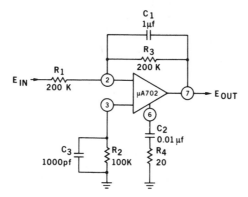

Figure 9-4.   Narrow-band amplifier applications.

(Illustration courtesy of Fairchild Semiconductor Corporation.)

mitted or where power, temperature, or other characteristics are dominant.

An ideal amplifier would have an infinite open-loop gain and bandwidth, zero noise, offset, and drift. Practical amplifiers, however, have open-loop dc gains from as low as 1000 to as high as 1 million. With very high gain, the closed-loop amplifier characteristics become a function only of the feedback components. The application potential of an operational amplifier seems limited only by the ingenuity of the circuit designer in picking and configuring the feedback network components.

Another important requirement of an operational amplifier is that it have a high input impedance and a low output impedance. In this

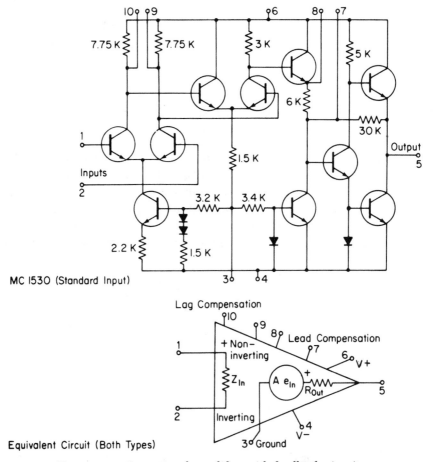

Figure 9-5. Operational amplifier with feedback circuits.

(Harry E. Thomas, *Handbook of Integrated Circuits*, © 1971. Reprinted by permission of Prentice-Hall, Inc.)

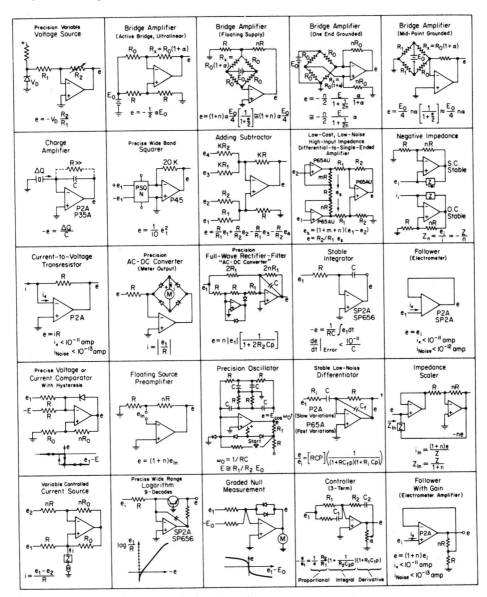

Figure 9-6. Operational amplifier applications.

(Harry E. Thomas, *Handbook of Integrated Circuits,* © 1971. Reprinted by permission of Prentice-Hall, Inc.)

way the feedback network does not load the amplifier, and the input impedance does not modify the feedback voltage.

A typical operational amplifier IC consists of two cascaded differential amplifiers, as shown in Figure 9-5, with a single-ended output stage. Differential amplifiers have all the advantages mentioned before, and the two inputs permit both an inverting and a noninverting input signal. The equivalent circuit, shown in Figure 9-5, illustrates the key features of all operational amplifiers. The input impedance and the output impedance are practically independent of each other. Z_{in} is usually quite high and R_{out} can be any suitable value.

Figure 9-6 shows a variety of operational amplifier applications and illustrates the great versatility of this concept of analog circuit design. Frequency stabilization, phase correction, bandwidth modification, and introducing frequency-sensitive networks into the feedback path are treated in many textbooks and manufacturers' application notes. It is recommended that this highly complex subject be studied in these sources before trying actual amplifier applications.

9.3 DIGITAL LOGIC ICs

Digital circuits differ from analog circuits in that they operate either on or off, conducting or cutoff, like switches, and they are often called switching circuits. They are used in digital computers, data machines, and similar equipment where functions are performed in terms of binary (two state) logic. A variety of logic circuit families have appeared on the market, each with its special features, advantages, and limitations. They are variously labeled as TTL (transistor-transistor logic family), ECL (emitter-coupled logic family), MOS (metal-oxide semiconductor), CMOS (complementary metal-oxide semiconductor), NMOS (N-type MOS), and the Schottky TTL family. A large variety of bipolar and MOS devices now dominate the field of digital logic. Chapter 8 discusses some of these circuits, and Chapter 16 describes their applications.

Figures 9-7 through 9-10 show the basic circuits of RTL (resistor transistor logic), ECL, DTL (diode transistor logic), and TTL ICs. From the logic designer's viewpoint, a logic system can be designed using functional "black boxes," each with well-defined input and output characteristics.

Though RTL circuits are the cheapest, ECL the fastest, and DCTL the simplest to manufacture, the most widely used are the TTL and MOS circuits. All things considered, the features of the bipolar and MOS families present the best compromise among cost, speed, fan-in and fan-out capability, noise margin, power consumption, etc. Since there is considerable variation even among MOS and bipolar circuits,

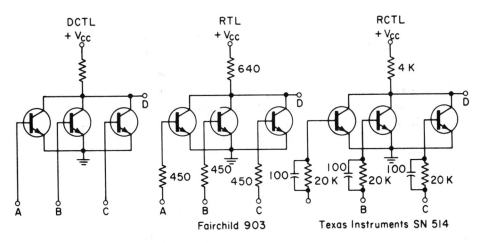

Figure 9-7.   Basic RTL circuit.

(Harry E. Thomas, *Handbook of Integrated Circuits*, © 1971. Reprinted by permission of Prentice-Hall, Inc.)

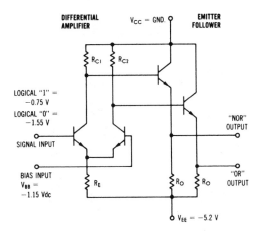

Figure 9-8.   Basic ECL circuit.

(Illustration courtesy of Motorola Semiconductor Prod.)

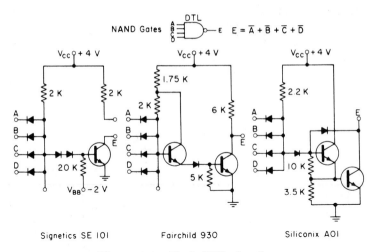

Figure 9-9. Basic DTL circuit.

(Harry E. Thomas, *Handbook of Integrated Circuits*, © 1971. Reprinted by permission of Prentice-Hall, Inc.)

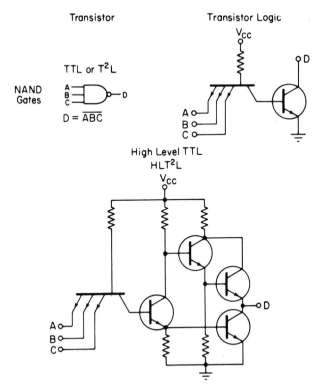

Figure 9-10. Basic TTL circuit.

(Harry E. Thomas, *Handbook of Integrated Circuits*, © 1971. Reprinted by permission of Prentice-Hall, Inc.)

the details of any one logic family are best obtained from the extensive literature, manuals, and data sheets available from the manufacturers.

The TTL logic features a multiple-emitter-base diode, each serving as an input, as illustrated in Figure 9-10. Operation of the circuit is very similar to that of the DTL circuit shown in Figure 9-9. The multiple-emitter transistor is particularly economical to fabricate in monolithic form, which accounts to a large extent for its popularity. It requires only a single isolated collector region in which a single base

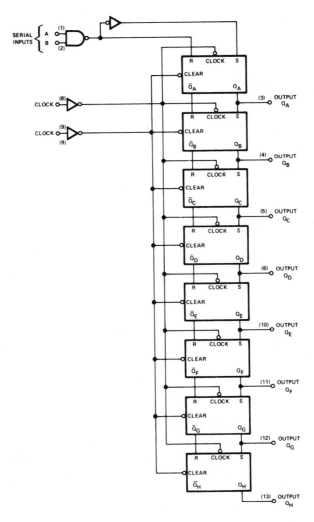

Figure 9-11. Eight-bit TTL shift register. Many ICs have a capacity of 16 and 32 bits.

region is diffused, and then several emitter regions are diffused as separate areas into the base region.

The TTL circuit is very adaptable to all forms of IC logic and produces economical circuits with high noise margins, high operating speeds, high fan-out, and relatively low power consumption.

Complex circuits can be packaged on the same silicon chip as easily as a single-gate chip, and at much lower cost per gate. Complete logic functions employing as many as several thousand interconnected gates have become standard IC function blocks. One example of such a combined logic function is illustrated in Figure 9-11, which shows the details of a commonly used shift register with a capacity of eight bits. It is packaged in a standard 14-pin dual-in-line assembly, and only those connections are brought out that are essential for interconnection to other circuits. Chapter 16 describes shift register principles.

9.4 LARGE (LSI) AND VERY LARGE (VLSI) SCALE INTEGRATED CIRCUITS

Early ICs contained enough elements to perform several complete logic operations. The shift register of Figure 9-11 is only one such example. As an arbitrary subdivision, large scale integrated circuits (LSI) refers to chips containing 100 or more equivalent gate logic elements, and very large scale integrated circuits (VLSI) refers to chips having densities as high as 10,000–25,000 gates (about 100,000 transistors). As in most other areas of electronics, CMOS is the process of choice in gate arrays. Nevertheless, there are bipolar gate arrays on the market that have densities of 8000; plus, some gate arrays are fabricated using combined bipolar-CMOS circuit techniques.

Designing with VLSI

Although the current enthusiasm for CMOS VLSI ICs is great, some designers working with signal processing, etc., are moving in four directions: N-MOS in single-chip designs; low- to medium-speed general purpose; N-MOS and CMOS special-purpose single chip; and CMOS and bipolar multichip ICs with increased speed and word size.

To increase density, gate array designers are moving to multilevel interconnection techniques and various architectures that eliminate channels. To improve speed, they are using procedures that decrease resistance and path delays as well as using other materials in place of silicon—for instance, gallium arsenide. These changes are providing gate densities of 10,000 or better with gate times as low as 1.4 ns (2.4 ns, typical), and clock speeds approaching 25 MHz.

Trends in VLSI Technology.   To attain higher speeds, there is a general shift away from the traditional methods, that is, where the operations are sequential, with the program and data sharing the same bus. Sometimes this newer method is called *pipelining*. In these designs the architecture used in the older systems is modified to provide separate data and program storage along with independent buses for data, as well as instruction transfer and control. Many of the newer 32-bit general-purpose central processing units (CPUs) and other ICs incorporate additional pipelining to allow a number of operations to be performed during the same cycle.

Special MSI and VLSI Techniques

The field effect transistor (FET), the metal oxide semiconductor (MOS), and the MOSFET operation can be understood from Chapter 8. Junction transistors are usually called bipolar devices, because two types of current carriers, the free electron and the positive hole, are involved in their operation. As indicated in Chapter 8, the MOSFET operates quite differently. The fabrication of the MOS device is based upon the planar technique described in Section 9.1. Because of its inherent construction features, the MOSFET can be fabricated in a smaller area than the bipolar transistor, allowing a higher element density and lower costs. Also, as pointed out in Chapter 8, the channels between source and drain can be used as a resistor whose value depends upon the gate potential and the transconductance of the structure. These resistors are smaller than bipolar resistors, allowing even greater density of elements. Figure 9-12 shows the relative size comparison of a MOSFET and a bipolar transistor, including a load resistor for each.

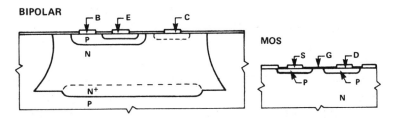

Figure 9-12.   Size comparison between MOSFET and transistor.

(Illustration courtesy of Texas Instruments, Inc.)

Some of the different semiconductor technologies used for fabricating LSI arrays are bipolar, p-channel MOS, and CMOS (complementary metal-oxide semiconductor). The bipolar and p-channel MOS techniques require load resistors through which the current passes when the transistor conducts. This wastes power and heats the chip, which limits the packing density by virtue of this "thermal barrier." The CMOS technique, however, uses both p- and n-type transistors in series, as shown in Figure 9-13. The signal that turns the n-transistor on, turns the p-transistor off—and vice versa. There is never a path to ground for the current, except through an external load. This saving in operating power is particularly advantageous in battery-operated devices.

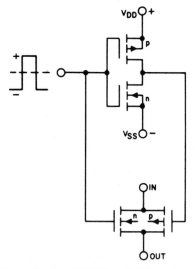

Figure 9-13.   CMOS switching circuit.

(Illustration courtesy of RCA Solid State Division.)

Packing densities can be greater and individual transistors can be smaller, and the cost per device is lower. Smaller transistor size also reduces capacitance, which provides the dual advantage of higher operating speeds and lower switching power. Integrated logic circuits using MOS techniques are very economical, especially in MSI/LSI/VLSI arrays.

Other techniques combine bipolar and CMOS during fabrication, resulting in a power dissipation at the chip level of only 650 mW at 16 MHz in gate arrays having densities up to 2500. Higher densities and lower power consumption can be achieved by combining CMOS and silicon on sapphire. There are several other processes—for example, double-layer-metal CMOS—that are used to produce 20,000 gates with delays as short as 400 ps for a two-input NAND gate.

LSI/VLSI Arrays.   Manufacturers are supplying a continuously increasing variety of "standardized" circuits, in plastic and ceramic dual-in-line packages, from 14 to 40 pins, and in a variety of TO-type transistor cans. Among them are static and dynamic shift registers featuring speeds from dc to 5 MHz and complexities up to 256 bits in the static MOS types, and speeds to 10 MHz and complexities up to 4096 bits in the dynamic MOS types. The trends are to more circuitry on the chip, including built-in clock drivers and recirculate logic, and longer bit lengths. Read-only memories (ROM) are available with capacities up to 1-megabit with access time of 120 nanoseconds. In one type (MK3901M), a control pin makes it possible to organize the IC's memory matrices as either 128 K × 8 or 64 K × 16. Of course, this means the IC can be used in 8- or 16-bit systems.

Dynamic random access memories (DRAM), with capacities of 10,000 bits and up to 1 megabit low power dissipation and access times less than 1 microsecond are competing with ferrite-core memories in many computer applications. As for all LSI circuits, the trends are to larger, faster, more economical, lower power requirements and convenient high-density packaging. For instance, the plastic zigzag dual-in-line package that has staggered leads is designed for 1 Mb DRAM chips.

Static random access memories (SRAM) are also being used in some applications, although not as much as DRAMs. CMOS technologies are ideally suited for the current generation of these fast memories because of their low power dissipation and high density. However, SRAMs are generally less dense and require more transistors per storage cell.

9.5 SPECIAL PURPOSE ICs

Analog Gates

As described in Chapter 8, the FET can be considered as a voltage-controlled resistor. When the FET is switched from full on to full off, it can be used to gate an analog signal. An array of such FETs on a single IC chip can then be used as a multiplexer for analog signals. The circuit shown in Figure 9-14 is a typical array of six analog switches in a 14-pin dual-in-line package. A signal on any of the inputs (pins 9 through 14) can be switched onto the output line, pin 7, by applying a signal to the gate of the selected channel, which then enables the MOSFET transistor.

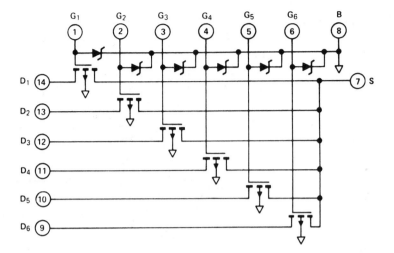

Figure 9-14. Six analog FET switches.

(Illustration courtesy of Siliconix, Inc.)

This particular array is normally operated with the substrate (pin 8) grounded and the source (pin 7) driving into a high impedance amplifier. With zero voltage on the gate (pins 1 through 6), the impedance from drain (pins 9 through 14) to source is very high, in the order of 10^{10} ohms. As the gate is made negative, the resistance between the source and drain is reduced, until, at about -30 volts, approximately 100 ohms resistance is present, and the transistor is on or closed.

Many different versions and combinations of gates are available, and detailed manufacturers' data should be consulted for a particular application.

Phase-Locked Loops (PLL)

The phase-locked loop circuit has been in use in special communications systems, servo systems, etc., but its implementation with discrete components was costly and complex. LSI-IC technology has made this circuit simple, economical, and advantageous in innumerable ways. The conventional PLL is essentially a noninductive, tunable, active filter with an adjustable band-width. The individual circuits comprising the PLL, such as the voltage-controlled oscillator (VCO), phase detector, and loop filter, are available as separate circuits on individual chips or they can be fabricated on a single chip. A basic PLL block diagram is shown in Figure 9-15.

When the phase difference between the VCO and the input signal

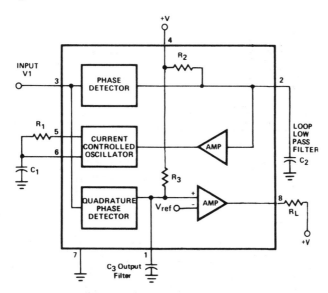

Figure 9-15.   Basic PLL block diagram.

(Illustration courtesy of Signetics Corporation.)

is constant, the phase loop is locked. If either the input or reference signal or the VCO output changes in phase, the phase detector and filter produce a dc error signal that is proportional in magnitude and polarity to the original phase change. This error voltage changes the VCO phase, so that it again locks into the reference signal. Since the error voltage is brought out, as well as the VCO frequency-controlling network, the VCO signal, and the reference input, a number of different functions can be performed by the PLL IC.

The basic PLL can serve as an FM demodulator without tuned circuits since the audio signal is simply the error signal at the output of the loop filter.

In general, PLLs can be used in FM demodulation, frequency shift keying (FKS) demodulation, frequency synthesizers, data synchronization, synchronous detection of AM signals, tone detection, FM stereo decoding, signal conditioning, tracking filters, and motor speed control.

Asynchronous Receiver/Transmitter

Another example of LSI techniques is the digital receiver/transmitter shown in Figure 9-16. The receiver section includes an input buffer data register and synchronizing circuits to prevent variations in start pulse width due to asynchronous data loading. It also includes parity

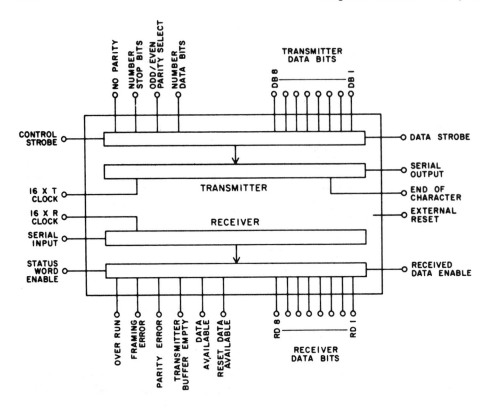

Figure 9.16. Asynchronous receiver/transmitter.

(Illustration courtesy of General Instr. Corp., Microelectronics Division.)

bit generation circuitry, end-of-transmission and buffer-empty flag out-puts, a start bit detection scheme with noise rejection features, and error-detection abilities include parity, framing, and overrun.

At the transmitter portion, all characters contain a start bit, five to eight data bits, an odd or even parity bit (or no parity), and one or two stop bits.

The digital receiver/transmitter is programmed through five con-trol inputs and a strobed or hard-wired enable input. This flexibility permits application with a wide variety of digital communications sys-tems, including teletypes 28, 32, 33, 37; the IBM 2741 communications terminal; most CRT terminals; and other computer peripherals.

Special TV IC Elements

Monochrome and color TV signal processors are available that provide video IF and final amplifiers, automatic gain control, and synchronous video detection circuits into a single IC. A second IC may house the complete audio system, while a third chip performs *all* chroma and luminance processing. Another IC that is often included in TV systems is one, or more, optocoupler. Typically the optocoupler is used to couple the horizontal drive signal to the output stages and to isolate grounds (dc and signal).

BIBLIOGRAPHY—CHAPTER 9

BUCHSBAUM, W., *Encyclopedia of Integrated Circuits: A Practical Handbook of Essential Reference Data.* Englewood Cliffs, NJ, Prentice-Hall, Inc., 1981.

GENN, R., *Illustrated Guide to Practical Solid State Circuits—With Experiments and Projects.* Englewood Cliffs, NJ, Prentice-Hall, Inc., 1983.

GREENFIELD, J., *Microprocessor Handbook.* New York, NY, John Wiley & Sons, Inc., 1985.

KAUFMAN, M., and A. SEIDMAN, *Handbook for Electronics Engineering Technicians* (2nd Ed.). New York, NY, McGraw-Hill, 1984.

Reference Data for Engineers: Radio, Electronics, Computer, and Communications (7th Ed.). Indianapolis, IN, Howard W. Sams & Co., Inc., 1985.

TILL, W., and J. LUXON, *Integrated Circuits: Materials, Devices and Fabrication.* Englewood Cliffs, NJ, Prentice-Hall, Inc., 1981.

CHAPTER 10

ANTENNAS AND TRANSMISSION LINES

10.1 FUNDAMENTALS OF ANTENNAS

The basis of most antenna work is a mathematical analysis of electromagnetic fields and waves. Because electromagnetic waves and fields exist in three-dimensional space and depend on variations in time, the mathematics required for this field involves vector analysis, differential and integral calculus, and certain special aspects, such as the *divergence* and the *curl*. Chapter 20 contains the fundamentals of this mathematical background.

Electrostatic and Magnetic Fields

Basic physics shows the interaction of electric and magnetic fields. For electrostatics, it is assumed that the electromagnetic field is zero because the electrical sources remain stationary and do not vary. The following fundamental relationships apply:

Coulomb's Law.   Deals with two small charged spheres q_1, q_2, and the force F between them as a function of distance r and the dielectric constant K.

$$F = \frac{q_1 q_2}{kr^2} \quad K = 1 \text{ in vacuum (CGS)}$$

$$F = \frac{q_1 q_2}{4\pi\epsilon r^2} \quad K = 4\pi\epsilon \text{ (MKS)}$$

$$\epsilon = 8.854 \times 10^{-12} \text{ in vacuum}$$

Electric Field Strength.   Is measured with regard to a test charge located a distance r away from the fixed charge q. The field strength E is a vector quantity, usually measured at a radial distance from the charge.

$$E = \frac{q}{4\pi\epsilon r^2} \vec{R}_1 \quad \vec{R} = \text{unit vector}$$

Electric Flux and Flux Density.   These two quantities are analogous to the magnetic flux and magnetic flux density; they are derived as follows: Electric flux $= \Psi =$ charge q (coulombs) due to Ψ. Flux density $= D = \dfrac{\Psi}{4\pi r^2} = \dfrac{q}{4\pi r^2} \vec{R} = \epsilon E$.

Gauss's Law.   The total flux through any closed surface surrounding charges equals the amount of charge enclosed. Expressed mathematically, this means:

$$\oint_{\text{surface}} D \, da = \int \rho \, dVol \qquad \begin{aligned} da &= \text{element of area} \\ \rho &= \text{charge density} \\ Vol &= \text{volume enclosed} \end{aligned}$$

Potential V.   Measured at a point P, the potential due to a charge q is given by the formula:

$$V = \frac{q}{4\pi r\epsilon} = \text{volts}$$

where the volts are equivalent to joule per coulomb. The field strength E equals the negative value of the potential gradient as in the equation: $E = -\nabla \cdot V$. The symbol ∇ (pronounced del) means a three-dimensional, partial derivative in rectangular coordinates.

The symbol $\nabla \cdot a$ means the divergence of a quantity a, where the divergence is a mathematical operation including the partial derivatives of three-dimensional rectangular coordinates.

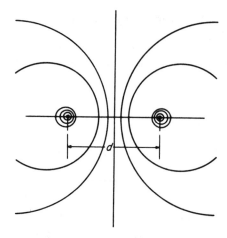

Figure 10-1. Equipotential surfaces.

(Edward C. Jordan and Keith G. Balmain, *Electromagnetic Waves and Radiating Systems*, 2nd Edition, © 1968. Reprinted by permission of Prentice-Hall, Inc.)

The expression $\nabla \times a$ (curl of a) means a matrix of partial derivatives, the original vectors, and the unit vectors in rectangular coordinates. Chapter 20 contains a more detailed mathematical derivation of these expressions.

Equipotential Surface. Refers to surfaces and space in which the potential V is everywhere the same. Figure 10-1 shows equipotential surfaces as a result of two charges.

Magnetic Terms. A detailed description of electromagnetic terminology and symbols is presented in Chapter 6. For the fundamentals of antennas, the following terms should be understood.

Magnetic flux in volt-seconds or webers.

Magnetic flux density B in webers per square meter.

Magnetic field strength H in coulombs per second-meters or amperes per meter.

Permeability μ in henries per meter.

Current density J.

Maxwell's Equations and Explanation

Maxwell's equations define electromagnetic fields and waves in terms of either the differential or integral form. It is possible to evaluate all electromagnetic field problems through one or the other of Maxwell's equations. The reader interested in learning detailed applications, theories, and proofs of this subject is referred to the references listed at the end of this chapter.

Differential   Integral

1. $\nabla \times H = \dot{D} + J$ $\oint H.ds = \int (\dot{D} + J).da$

2. $\nabla \times E = -\dot{B}$ $\oint E.ds = -\int B.da$

3. $\nabla.D = \rho$ $\oint D.da = \int \rho \, dV$

4. $\nabla.B = 0$ $\oint B.da = 0$

In most practical instances, the term *electric current* can be substituted for both *conduction currents* and the time derivative of electric displacement. The time derivative of magnetic displacement can be considered as a magnetic current, electromotive force can be considered voltage, and magnetomotive force can be considered a magnetic voltage. This permits simplification of the first two equations, by saying that the magnetic voltage around a closed path is equal to the electric current through the path, and the electric voltage around a closed path is equal to the magnetic current through the path. In a detailed interpretation of the expressions of the del operator, divergence, and the curl, it is necessary to consider the rectangular coordinates of the fields and the electric currents and voltages.

Elementary Dipole Fields

Almost all basic antenna types can be analyzed in terms of elementary dipoles. A basic dipole consists of two charges of opposite polarity or, in the case of an actual antenna, it consists of two cylindrical elements that receive signals of opposite polarity. Figure 10-2 shows the basic dipole with the voltage and current distribution, which occurs when each of the dipole elements is the length of one-quarter the wavelength at the frequency at which it is excited. In all subsequent discussions it is assumed that a sine wave signal is used to excite the antenna. Note that at the center or feed point of the antenna the voltage is zero while the current is maximum. At each of the ends of the two cylinders, the current is zero and the voltage is maximum.

Theoretical, as well as practical, considerations prove that any antenna will exhibit the same characteristics whether it is used to receive signals or to transmit signals, provided the impedance of the feed system and the surrounding radiating area is the same.

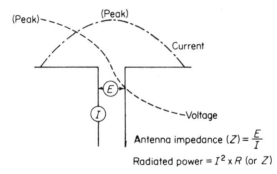

Figure 10-2. Basic dipole voltage and current.

(John D. Lenk, *Handbook of Electronic Charts, Graphs and Tables*, © 1970. Reprinted by permission of Prentice-Hall, Inc.)

The effective length of an antenna is used to indicate the effectiveness of the antenna as a radiator or as a collector of radio signals. As shown in Figure 10-3, the effective length of a dipole antenna is less than the actual length and depends on the distribution of the current at the terminals. The following formulas apply for effective length of transmitter and receiver antennas:

$$\text{Transmitter: } l_e = \frac{1}{I_o} \int_{-1/2}^{+1/2} I_z \, dz$$

$$\text{Receiver: } \quad l_e = -\frac{V_{oc}}{E}$$

In these formulas I_o is the current at the terminals of the actual antenna, I_z is the current distribution, V_{oc} is the open circuit voltage at the terminals, and E is the uniform exciting field, in volts per meter, which exists parallel to the antenna in the receiving mode.

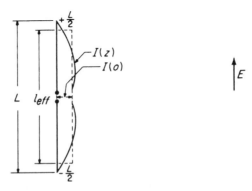

Figure 10-3. Effective length of dipole.

(Edward C. Jordan and Keith G. Balmain, *Electromagnetic Waves and Radiating Systems*, 2nd Edition, © 1968. Reprinted by permission of Prentice-Hall, Inc.)

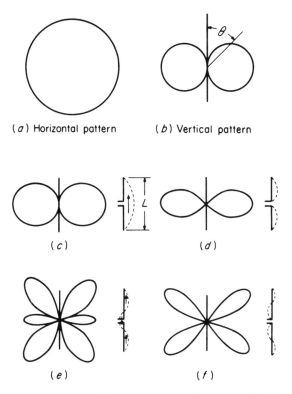

(*a*) Horizontal pattern   (*b*) Vertical pattern

(*c*)   (*d*)

(*e*)   (*f*)

Figure 10-4.   Radiation patterns of vertical dipole.

(Edward C. Jordan and Keith G. Balmain, *Electromagnetic Waves and Radiating Systems*, 2nd Edition, © 1968. Reprinted by permission of Prentice-Hall, Inc.)

The directional qualities of an antenna, that is, how it radiates or receives energy, are always of great importance. Figure 10-4 shows the horizontal and vertical patterns for dipoles of different wavelength multiples. The antenna wavelength is the sum of the two elements. Note that the horizontal pattern is the same for all vertical dipoles, that is, a circle. This means that for the half and the 1-wavelength case, the three-dimensional pattern will resemble that of a doughnut with the antenna itself at the center. Antenna patterns are generally plotted in polar coordinates with the location of the antenna itself at the center. For three-dimensional calculations, spherical coordinates are used.

Antennas can be considered as networks, and typical dipole antennas can be analyzed as the combination circuit shown in Figure 10-5. When the wavelength of the signal is the same as the actual effective length of the dipole antenna, the reactance of the inductive and the reactance of the capacitive component cancel out, and then the characteristic impedance of the antenna is simply R_a. The formula for

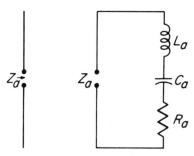

Figure 10-5.   RLC circuit equivalent.

(Edward C. Jordan and Keith G. Balmain, *Electromagnetic Waves and Radiating Systems*, 2nd Edition, © 1968. Reprinted by permission of Prentice-Hall, Inc.)

the total impedance at the antenna terminals is shown below:

$$Z_a = R_a + j\left(\omega L_a - \frac{1}{\omega C_a}\right) = \frac{E}{I}$$

Antenna impedance is affected not only by the frequency of the signal but also by the proximity of other structures and devices. Other antennas located nearby, generally less than five wavelengths away, will affect the impedance on the antenna as well as its directivity.

Polarization of Antennas

In spherical coordinates, it is apparent that the electromagnetic waves radiated from an antenna or received by it consist of two vectors, the electric E and the magnetic H vectors. Both of these vectors rotate and vary according to the sinusoidal signal. In practical antennas, however, the orientation of the transmitting antenna with respect to the earth will determine the dominant orientation of the transmitted electric field, E, signal. Simple monopole and dipole antennas that are oriented in the horizontal plane, that is, their elements are parallel to the surface of the earth, generate horizontally polarized waves. Antennas that are vertically oriented, that is, their elements point down to the earth, are considered vertically polarized. Vertically polarized signals are best received by vertically oriented antennas; horizontally polarized antennas are best suited for the reception of horizontally polarized signals. In TV broadcasting, for example, all signals are horizontally polarized, but in FM broadcasting all signals are vertically polarized.

In addition to vertical and horizontal polarization, special antenna designs (see Helical Antennas) permit circular polarization. In this method the phase relationships of signals traveling different paths become important. In circularly polarized TV signals, for example, the reflected signal that would result in ghosts, if horizontally polarized, will be canceled out. Proposals for changing TV transmission to circular polarization would permit continued reception by horizontally polarized antennas, though with some reduction in efficiency, and would eliminate or reduce ghost problems when circularly polarized antennas are used.

Spacecraft antennas often use *orthogonal polarization* (vertical/horizontal or clockwise and counterclockwise) and/or spatially separate beams to achieve more efficient use of the allocated frequency bands. The earth stations, particularly antenna systems used in deep space exploration and commercial space applications, also are designed to receive orthogonal polarizations. However, these antenna systems are parabolic reflectors using Gregorian or Cassegrain feeds (see Figure 10-15), therefore they are classified as microwave systems, that is, they are constructed using waveguides, etc. See Chapter 11.

General Antenna Parameters

While many of the general parameters are specifically derived for half-wave, center-fed, dipole antennas, they apply equally well to all types of antennas and form the basic concepts for more detailed and practical calculations.

(a) Near and Far Fields.   The electromagnetic field in the immediate vicinity of an antenna is usually called the *near field* and includes induction (magnetic) as well as electrical field and radiation field considerations. Evaluation of these effects is essential when antennas are located close to each other or to other objects and the mutual effects must be determined. Near-field effects are of primary importance in the design of antenna configurations and arrays, as described in Section 10.3. The far or radiation field is considered to start at least five wavelengths away from the antenna and is used in the reception and transmission calculations of communications systems. Unless the near field is specifically mentioned, all antenna formulas apply to the far field only.

(b) Antenna Gain.   Although the term *antenna gain* is loosely applied to both the power gain G_p and the directive gain G_d, it is often important to make the distinction. The directive gain is a function of the radiation intensity ϕ, which is defined as the power per unit solid angle in that direction and the total power radiated as indicated below.

$$G_d = \frac{\phi}{W_r}; \quad W_r = \text{power radiated, in watts}$$

Directive gain is also often called directivity D, which is defined as the maximum directive gain.

The power gain is given by the formula below:

$$G_p = \frac{4\pi\phi}{w_z}; \quad w_z = \text{total power input, in watts}$$

The ratio of power gain and directive gain provides an indication of the efficiency of the antenna, since the power loss of the antenna is included in this term. For a half-wave dipole antenna, $D = 1.64 = 2.15$ dB.

(c) Effective Area (Aperture).   Not all of the antenna area is always utilized to radiate properly polarized signals, nor is there a 100% efficient and correct impedance match between the antenna and the transmitter or receiver. The effective area of an antenna is given by the following formula:

$$A_e = \frac{\lambda^2}{4\pi} G_d; \quad \lambda = \text{wavelength}$$

(d) End Effect.   The plates of a capacitor do not provide the exact capacity indicated by the geometry because the ends of the plates are no longer completely opposed by the other plates. For similar reasons, there are end effects in antennas. The magnitude of these effects depends on the ratio of length to diameter of cylindrical antennas and to other geometric considerations, as well as to the vicinity of the antenna to other objects (in the near field). The end effect results in apparent lengthening of the antenna dimensions, meaning that they will be longer, beyond the resonance point. In practical applications, a 5% shortening of the mechanical antenna dimensions results in the most accurate electrical length.

(e) Antenna Elevation.   It is common practice to raise antennas well above the earth for a number of practical and compelling reasons. Antennas should clear the surrounding terrain, and for all frequencies above about 100 MHz, where line-of-sight transmission occurs, raising the antenna increases the range. In addition to these considerations, there is a near-field effect on the radiation resistance which is a function of antenna elevation. Figure 10-6 shows the relationship of antenna elevation as a function of wavelength, to the radiation resistance of vertical and horizontally polarized half-wave dipole antennas.

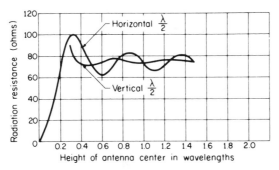

Figure 10-6. Antenna elevation vs. radiation resistance.

(Edward C. Jordan and Keith G. Balmain, *Electromagnetic Waves and Radiating Systems*, 2nd Edition, © 1968. Reprinted by permission of Prentice-Hall, Inc.)

10.2 COMMON TYPES OF ANTENNAS

Monopole Antennas

Any simple wire, connected to a source of RF, acts as a monopole antenna in that it radiates energy. One type of monopole antenna is the vertical whip antenna used for automobile radios. This type of antenna is considerably smaller than a quarter wavelength and depends on the proximity of the ground plane for its operation. The impedance and radiation characteristics of a monopole mounted on a theoretically ideal ground plane is equivalent to that of a dipole of twice the length of the monopole, if the dipole is considered to be in free space. The input impedance of such a monopole is equal to one-half of the input imped-ance of a center-driven dipole. The radiation pattern above ground is identical to the upper half of the radiation pattern of a corresponding dipole. In automobile antennas, the ground plane is effectively the chassis of the car, and the equivalents described above apply only par-tially. The radiation pattern of a vertical monopole is a circle in the horizontal plane. In the vertical plane it is an array of lobes, directed from the ground upward and depending on the wavelength fraction of the antenna size and the ground characteristics.

Monopole antennas may also be used in the horizontal direction, and Figure 10-7 shows the current distribution and radiation pattern of a monopole antenna which is two wavelengths long. Long wire anten-nas, composed of the basic element shown in Figure 10-7, are used for long-distance transmission of high-frequency signals where radiation into the Heavyside layer of the ionosphere is desirable in order to obtain reflections and long-distance communications. If a long wire antenna is terminated at the distant end with a resistance equal to its charac-teristic impedance, the radiation pattern will be considerably more di-rectional and will provide two major lobes, pointing at a small angle

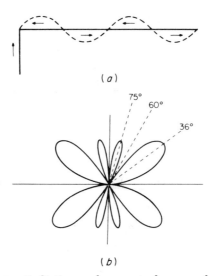

Figure 10-7. Radiation and current of monopole antenna.

(Edward C. Jordan and Keith G. Balmain, *Electromagnetic Waves and Radiating Systems*, 2nd Edition, © 1968. Reprinted by permission of Prentice-Hall, Inc.)

from the feed point of the antenna along the wire toward the terminating resistor. The principles of long wire antennas are used in designing V-type and rhombic antennas described below.

Dipole Antennas

The elementary dipole and its characteristics have been discussed previously (Section 10.1). Dipoles are widely used as basic elements in antenna arrays, and modifications of the basic dipole are used in many

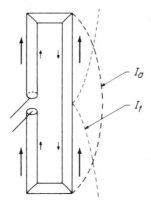

Figure 10-8. Folded dipole.

(Edward C. Jordan and Keith G. Balmain, *Electromagnetic Waves and Radiating Systems*, 2nd Edition, © 1968. Reprinted by permission of Prentice-Hall, Inc.)

applications. The simplest modification is the folded dipole shown in Figure 10-8 with its current distribution. A folded dipole has somewhat greater bandwidth than a simple dipole and its radiation resistance; that is, the characteristic impedance at resonance is approximately four times that of a simple dipole. Folded dipole antennas are widely used in television, where bandwidth and higher characteristic impedance, 300 ohms, is desirable.

Logarithmically Periodic Antennas

In order to cover an extremely wide frequency band with a constant radiation resistance, constant beam pattern, and constant gain, a special concept can be applied to the shape of the basic dipole. The principles of a log-periodic dipole array are illustrated in Figure 10-9; they show that both the spacing of dipoles d from the origin and the length of the dipoles l increases according to a constant scale factor T. Common values of the scale factor range in the order of 2 to 6:

$$\frac{l_n}{l_{n-1}} = T = \frac{d_n}{d_{n-1}}$$

The electrical equivalent circuit and the connections of the dipoles in Figure 10-10 show that, for any frequency within the range of the antenna, there is an active region in which the radiation resistance R_a is a real value. There is an unloaded and a loaded transmission line region and a reflective region. The unloaded and loaded transmission line region has little effect on the dipole action, but the reflective region provides sufficient reflection so that the beam of the antenna is directed

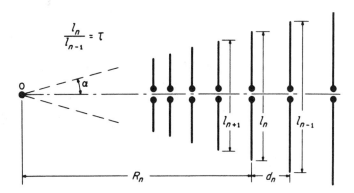

Figure 10-9.   Log-periodic dipole array.

(Edward C. Jordan and Keith G. Balmain, *Electromagnetic Waves and Radiating Systems*, 2nd Edition, © 1968. Reprinted by permission of Prentice-Hall, Inc.)

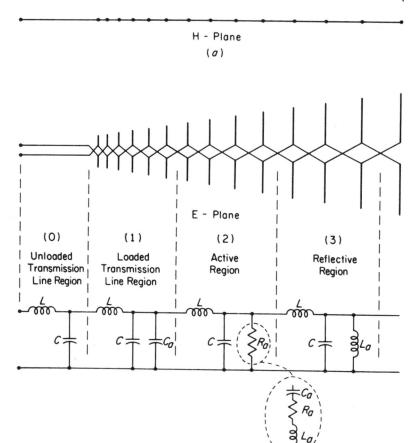

Figure 10-10. An l-p dipole and equivalent circuit.

(Edward C. Jordan and Keith G. Balmain, *Electromagnetic Waves and Radiating Systems*, 2nd Edition, © 1968. Reprinted by permission of Prentice-Hall, Inc.)

toward the transmission line end with considerable sharpness. In a typical TV antenna using the log-periodic dipole array, several adjacent dipoles will be close enough in length to resonate at a given channel, and the dipoles making up the reflective regions will provide a highly effective narrow beam. Many variations of the basic log-periodic antenna are possible. One is to convert the dipole elements into V-shaped elements, still following the log-periodic spacing and dimensions. A log-periodic V-antenna is more compact than its straight dipole equivalent and provides a slightly narrower forward beam. A second variation of the log-periodic antenna is to use two or more of them at angles to each other to obtain a beam that is narrower in the vertical plane.

Conical Antennas

When the basic monopole antenna is changed from a cylinder to a cone, the bandwidth of the antenna increases. The same thing can be done with the basic cylindrical dipole and, again, the bi-conical antenna has greater bandwidth than the cylindrical dipole antenna. The gain and radiation patterns of conical and bi-conical antennas are not very much different from a monopole and dipole.

Helical Antennas

A simple helical antenna can be constructed from a long wire monopole antenna which is wound in a helix around a cylindrical insulator. If this antenna is used close to a reflecting ground plane, such as a flat or paraboloid shaped reflector, a highly directional, broad-band antenna is obtained. The radiation from such an antenna is circularly polarized, and it is possible to obtain bandwidth ratios of 1.7 : 1 with a simple helix antenna. A typical three-turn helix can produce a 60-degree beam with 8-dB gain (referred to an isotropic radiator), and an eight-turn helical antenna can produce a 40-degree beam with 12-dB gain. Variations of the helical antenna are those in which the spacing of subsequent helical turns are arranged exponentially and those in which, instead of the cylinder, a cone is used as the form. Increased bandwidth and even narrower radiation patterns are obtainable in this fashion. It is also possible to add a field of two or more helical antennas by mounting them within a half-wavelength distance on a common ground plane.

Loop Antennas

A loop antenna is generally one in which the diameter of a circular loop or the sides of a rectangular loop are considerably smaller than one-quarter of the wavelength. The radiation pattern of such a loop antenna is essentially the same as that of a dipole, since it is possible to construct the equivalent of a loop from a folded dipole by making the distance between the two dipole elements equal to the length of the dipole. This means that maximum signal reception will occur along a line connecting the two sides of the loop and minimum signal reception will occur through the axis of the loop. When used for direction finding, loop antennas are rotated for zero signal, and then the axis of the loop points into the direction of the transmitting station. The signal voltage induced in the loop is determined by the equation below:

$$E_R = NE \frac{2\pi d}{\lambda} \cos \theta; \quad E_R = \text{induced voltage}$$

$$N = \text{number of turns}$$
$$d = \text{loop diameter}$$
$$\lambda = \text{wave length}$$
$$\theta = \text{bearing angle}$$

The angle theta (θ) in the above equation determines the azimuth or bearing of the transmitting station with regard to the axis of the loop. When theta is zero or 180 degrees, the induced voltage will also be zero.

To increase the sensitivity of a loop, it is possible to provide resonance at the received frequency by tuning the loop inductance with a capacitor. At frequencies below 3 MHz, it is common to use a ferrite core in the loop in order to increase the sensitivity and reduce the required antenna area. Such tuned ferrite-core loops are the standard built-in antennas in AM radio receivers.

Rhombic Antennas

A basic rhombic antenna is shown in Figure 10-11 and consists, essentially, of two long wire V-antennas end-to-end. A rhombus, unlike a square, has a major and a minor diagonal. In this type of antenna, the transmission line feed is connected to one end of the longer (major) diagonal; terminal resistors connect the other end of the long diagonal. Rhombic antennas are usually 3 to 10 wavelengths long and also fall

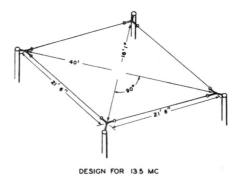

DESIGN FOR 135 MC

Figure 10-11.   Basic rhombic antenna.

(From *Television Antenna*, 2nd Edition, by D. A. Nelson, © 1951. Used with permission of Howard W. Sams & Co., Inc.)

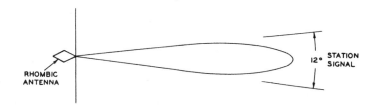

Figure 10-12.   Radiation pattern of rhombic antenna.

(From *Television Antennas*, 2nd Edition, by D. A. Nelson, © 1951. Used with permission of Howard W. Sams & Co., Inc.)

into a class called "traveling wave" antennas. Terminal resistors are necessary in order to obtain proper impedance matching and the broadband performance common to rhombic antennas. In general, terminating resistors range from 600 to 800 ohms. A rhombic antenna is always used against a ground plane, usually the earth, and must be elevated at least half a wavelength above the earth. The radiation pattern of a rhombic antenna is in a narrow beam, along the longer diagonal, facing from the feed point into the direction of the terminating resistors. When a rhombic antenna is located parallel to the earth's surface, the major lobe is directed against the Heavyside layer of the ionosphere. In order to change the direction of this beam, rhombic antennas are sometimes so constructed that their plane is at an angle with the earth's surface. Figure 10-12 shows the radiation pattern of a typical rhombic antenna used for TV reception.

Slot Antennas

At frequencies above 300 MHz, transmission lines are frequently in the form of hollow tubes or wave guides, and one way of generating an antenna is to cut a rectangular slot into the waveguide or cylindrical tube. The radiation from that slot can be compared in character to that of a dipole, provided that complementary values of electric and magnetic quantities are used. The slot antenna impedance, for example, is proportional to a dipole admittance. The slot antenna magnetic field is proportional to a dipole's electric field. Because of this relationship, it is possible to provide horizontally polarized radiation from a vertical cylinder in which a vertical, longitudinal slot has been cut. FM and TV transmitting antennas use the slot antenna principles, with multiple slots on a single vertical transmission line, to provide the desired radiation pattern for a particular reception area.

Horn Antennas

Horn antennas are usually limited to those frequencies where wave-guides are used as transmission lines, that is, above 1000 MHz. If a waveguide is suddenly terminated, the radiation will continue to move but, because of the discontinuity in the walls of the waveguide as compared to the free space at its end, a great deal of energy is lost. By flaring out the waveguide, this abrupt transmission is modified, and the horn antenna provides more directivity and somewhat better gain than would be obtained by a simple blunt ending of a wave guide. Horn antennas can be flared linearly or exponentially, with the latter arrangement providing somewhat greater bandwidth and a larger diameter tube.

The horn antennas used for microwave communications utilize a special design to provide the transition from the waveguide to that of a directional, usually parabolic or hyperbolic reflection surface, to provide a narrow beam of the radiated frequencies. In many instances, the horn antenna provides a 90-degree angle, with the angular portion specially shaped to provide maximum concentration of the resulting beam. A simple horn antenna is shown in Fig. 10-13.

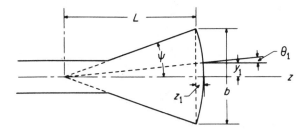

Figure 10-13.   Horn antenna.

(Edward C. Jordan and Keith G. Balmain, *Electromagnetic Waves and Radiating Systems*. 2nd Edition, © 1968. Reprinted by permission of Prentice-Hall, Inc.)

10.3 ANTENNA CONFIGURATIONS AND ARRAYS

Reflectors and Parasitic Elements

One of the basic methods of changing the radiation pattern of any type of an antenna is to place the antenna near a reflecting surface. Figure 10-14 shows the directivity pattern of a simple quarter-wavelength dipole and how this pattern is changed if the dipole is placed one-quarter wavelength away from a reflecting surface. If the reflecting surface is more or less than a quarter wavelength away at the resonant frequency, some of the reflected signal will produce side lobes. In addition to

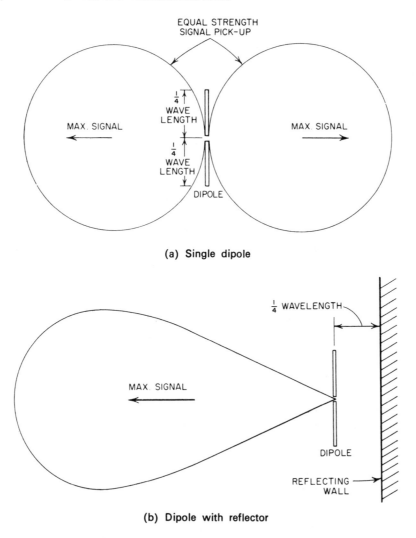

(a) Single dipole

(b) Dipole with reflector

Figure 10-14. Effect of reflector element.

(From *Fundamentals of Television*, by Walter H. Buchsbaum, © 1964. Used by permission of Hayden Book Company, Inc.)

providing directivity, the reflecting surface also provides a higher directive gain, as defined in Section 10.1.

In most practical antenna installations, a solid, infinite size reflecting surface is not available and less effective reflectors are used. If another quarter-wavelength dipole is placed one-quarter wavelength away from the first dipole, but not connected electrically, some of the current will be induced in it, due to the near field, and the other dipole will act as another radiating element. Because it does not receive energy from the transmission line, such an element is called a parasitic element. If the parasitic element is at least 5% longer than the driven dipole, it will act as a reflector, and a pattern similar to that of Figure 10-14 will be obtained. If the parasitic element is approximately 5% shorter than the driven dipole, it will act as a director, and the antenna pattern will be the same as that for a reflector but in the opposite direction.

Because a director or reflector element does not provide an infinite surface, its contribution to the radiation pattern is limited. A reflector and a director, spaced a quarter wavelength in front and behind the driven dipole, combine to produce a directional beam with relatively high gain. In practical antennas, it is customary to space parasitic elements closer than one-quarter wavelength to the driven element. As an example, two equal length dipoles, one driven, spaced 0.04 λ apart, will produce a directional gain of 3 dB. If the parasitic dipole length is increased by 5%, it acts as a reflector and the directive gain will be 6 dB. Similarly, if the parasitic element length is reduced by 5%, the directive gain will also be 6 dB. A combination of reflector and director will increase the gain of the dipole by itself by 8 dB. The theoretical maximum of 12 dB is not attained because of the interaction of the elements. The principles of a reflector and several directors are used in the yagi antenna, described below.

One method of increasing the effect of a simple rod reflector is to use a corner reflector. Such a corner can be constructed of two sheets of solid metal or close-mesh screening, with the antenna placed in a line bisecting the angle and parallel to the reflecting surfaces. This arrangement concentrates the beam pattern along the line bisecting the corner and provides considerable gain. The optimal antenna spacing from the corner varies with the angle between the two surfaces. For a 90-degree angle, for example, the best results are obtained when the antenna is 0.5 λ from the corner. For a 60-degree angle, best results are obtained when the antenna is located 0.65 λ from the corner. As the angle becomes narrower, the antenna must be spaced farther away from the corner. In order to be effective, the length of each side of the corner should be approximately twice the distance of the antenna from the corner.

Corner reflectors concentrate the beam in only one plane. When it is desired to shape the antenna pattern into a narrow, circular shape, such as a pencil beam, a parabolic reflector is most useful. As described in Chapter 13, under Radar Systems, parabolic reflectors can be used with many types of antennas. One particularly effective method is to combine the reflection from a hyperbolic surface and that of a parabolic reflector as shown in Figure 10-15. This arrangement is called a Cassegrainian antenna lens and is frequently used in precision radar systems.

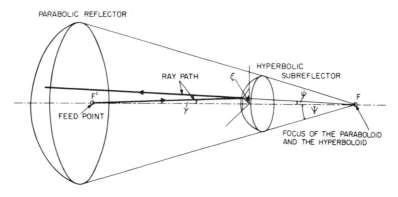

Figure 10-15. Cassegrainian antenna lens.

(R. Filipowsky and E. Muehldorf, *Space Communications Techniques,* © 1965. Reprinted by permission of Prentice-Hall, Inc.)

Yagi antennas are widely used for television reception because they are economical in design and construction, and provide a relatively narrow lobe with minimum pickup from the reflector side. A basic yagi antenna is shown in Figure 10-16, using a folded dipole as the driven element, a reflector, and two directors. Note that the directors are reduced in size by approximately 5%, as they are located farther away from the driven element. Increasing the number of parasitic elements increases the gain and directivity of the yagi antenna. Unfortunately, it also reduces the bandwidth and the frequency range over which the characteristic impedance is resistive. Specially designed

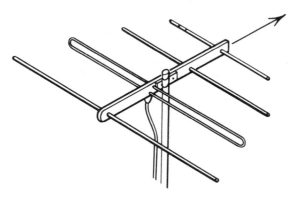

Figure 10-16. Broad-band yagi antenna.

(From *Fundamentals of Television,* by Walter H. Buchsbaum, © 1964. Used by permission of Hayden Book Company, Inc.)

broad-band yagi antennas are often used for TV reception which, in effect, are a number of multiple frequency yagis. Director elements of different lengths are interleaved with each other at carefully calculated intervals to provide the parasitic effects over a wider band of frequencies. While yagi antennas often look similar to the log-periodic types described in Section 10.2, their fundamentals of operation are entirely different.

Stacked Arrays

When a number of driven antennas are arranged to provide a specific pattern of directivity and gain, this arrangement is called an *array*. One of the two basic types of arrays is the end-fire array, where several antennas are located in the line of the main lobe and spaced so that their contributions add to the total field. The broad side array is composed of antennas arranged next to each other, so that their elements are all in a single plane. An example of a broad side array is shown in Figure 10.18. The log-periodic antenna described in Section 10.2 would be an example of an end-fire array.

In order to drive adjacent antennas, regardless of the type of array, the signals must be in proper phase and with proper impedance match, so that one antenna does not load down its neighbors. At microwave frequencies, impedance matching is frequently performed with portions of transmission line, arranged in quarter wavelength and called matching stubs, as discussed in Section 10.4 below. Figure 10-17 explains the operation and shows the formula for a quarter-wavelength transmission line impedance matching stub. The object of this imped-

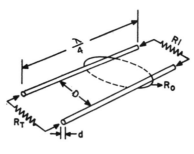

TO MAKE RI APPEAR AS R_T,

$R_0 = \sqrt{RI \times R_T}$ = STUB IMPEDANCE

$R_0 = 276 \text{ LOG } \dfrac{2D}{d}$

Figure 10-17.   Impedance matching stub.

(Illustration courtesy of *Electronics World*.)

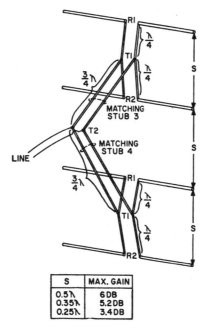

Figure 10-18.   Four dipole array.

(Illustration courtesy of *Electronics World*.)

ance matching stub is to make R1 appear the same as R_T. The characteristic impedance R_O, which is determined by the square root of the product of the two terminal impedances, can be controlled by adjusting the physical dimensions of the transmission line as shown.

In addition to proper impedance match, the physical spacing of antennas with respect to each other has a great effect on the total gain. Figure 10-18 shows a broad side array of four dipoles, with quarter-wavelength impedance matching stubs and the spacing S. A maximum gain of 6 dB is obtained when the spacing is half a wavelength. Note that, where physical dimensions make it impractical to use quarter-wavelength matching stubs, a three-quarter wavelength is used. Any wavelength in odd multiples of one-quarter wavelength is used. This will have the same effect as a quarter-wavelength transmission line.

Multiple-Beam Arrays

Multiple-beam arrays usually utilize beam-focusing microwave feed horns placed at the feed point of a Cassegrainian antenna lens (see Figure 10-15). This type antenna is ordinarily steerable and used for tracking spacecraft, satellite communications, and similar applications.

10.4 FUNDAMENTALS OF TRANSMISSION LINES

When the wavelength of ac signals transmitted over two conductors approximates the length of the conductors, transmission line theory must be considered. In practice, almost all RF signals going from the transmitter output stages to the antenna will pass over a transmission line, subject to the electromagnetic theory discussed briefly at the beginning of this chapter in Section 10.1. Maxwell's equations apply to transmission lines as well as to the antennas.

Transmission lines can be analyzed in terms of their equivalent circuit, in which the conductors themselves have both resistance and inductance, and the spacing between conductors determines the capacity. In effect, then, the equivalent circuit of a transmission line is an RLC arrangement consisting of series resistance and inductance, shunted by capacitance. Chapter 5 deals with equivalent circuits of this nature as applied to filters, and Chapter 4 deals with the basic RLC circuit.

Transmission lines, like antennas, have a characteristic impedance; in most transmission line design work, this characteristic impedance is utilized to provide impedance matching between the antenna and other parts of the system. Many different types of transmission lines, parallel wire as well as coaxial, are available commercially. Each of these transmission lines has a fixed characteristic impedance and has closely controlled and clearly specified loss characteristics at different frequencies. As outlined below, it is possible to use resonant lengths of transmission lines for improving the impedance match at UHF (and above) frequencies. Many different techniques for doing this have been developed, and an outline of some of the essential steps is presented below. The reader interested in this field should consult the references at the end of this chapter.

Voltage Standing Wave Ratio (VSWR)

If an impedance mismatch exists between the source, the transmission line, and the load, some of the transmitted energy will be reflected back into the transmission line. Where the reflected signal is out of phase with the transmitted signal, cancellation will occur, but where it is in phase, signals will be added. This means that at various points along the transmission line the voltage will be higher, while at other points it will be lower. Figure 10-19 illustrates the VSWR and its basic measurement technique. In practice, VSWR is measured by means of specially calibrated "slotted lines," a coaxial transmission line with air as

$$VSWR = \frac{1 + \text{refl. coeff.}}{1 - \text{refl. coeff.}}$$

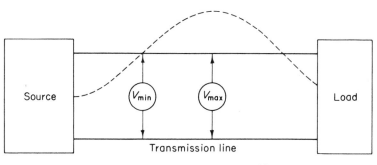

Voltage standing wave ratio $(VSWR) = \dfrac{V_{max}}{V_{min}}$

Figure 10-19. VSWR and its measurement.

(John D. Lenk, *Handbook of Electronic Charts, Graphs and Tables,* © 1970. Reprinted by permission of Prentice-Hall, Inc.)

dielectric and a probe inserted to measure the field. For two wire lines, the so-called lecher wires are used, with a probe moving between them. In both cases, the probe output is rectified and indicated on a meter. At low frequencies (below UHF) where the physical arrangement would be too large to be practical, the transmitted voltage and the reflected voltage are measured separately, and the reflection coefficient is used, which is the reflected voltage divided by the transmitted voltage:

$$VSWR = \frac{1 + \text{refl. coeff.}}{1 - \text{refl. coeff.}}$$

Smith Chart

The Smith chart shown in Figure 10-20 can be used in different ways for impedance calculations. It is particularly useful for impedance matching with transmission lines, and a number of techniques for its use are described in the references listed at the end of this chapter. The basic information available from the Smith chart is listed below and can be seen in Figure 10-20.

1. *Horizontal axis.* This logarithmic scale starts with zero at the left end, and has 1.0 at the center of the chart and 100 at the

extreme right. The horizontal axis is the equivalent of the resistance component or the conductance component, the inverse of resistance, for complex impedance notation. In transmission line work the numbers at the right of the center are used to indicate VSWR, and the axis is therefore called the VSWR axis. In ordinary impedance matching work, VSWRs will range from 1.2 to 3 or 4.0 maximum.

   2. *Curves.* The curves in the Smith chart consist of two groups. One group, the constant resistance or conductance circles, are all tangent to the main circle at the extreme right of the resistance or VSWR axis. The second group of curves represents the constant reactance of susceptance circles and are tangent to the resistance or

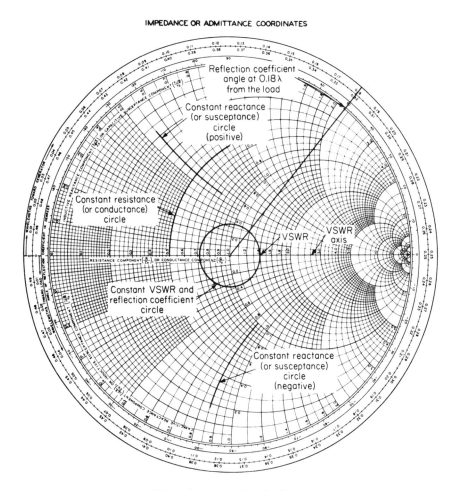

Figure 10-20.   Smith chart.

(John D. Lenk, *Handbook of Electronic Charts, Graphs and Tables,* © 1970. Reprinted by permission of Prentice-Hall, Inc.)

VSWR axis, itself at the extreme right. The actual impedances are always composed of resistance and reactance. Admittance, the inverse of impedance, is composed of conductance and susceptance. For details of the complex notation of these quantities, refer to Chapter 4.

3. *Outer scale.* The scale at the outer perimeter of the Smith chart indicates wave lengths toward generator and shows the distance in wavelengths, λ, away from the load. A particular reflection coefficient or VSWR provides the maximum and minimum, as shown in Figure 10-19, and the outer scale indicates the distance of this minimum or maximum from the load.

4. *Middle scale.* This scale, wavelengths toward load, is complementary to the outer scale and reads in the same way, wavelengths, λ.

5. *Inner scale.* The third scale along the perimeter indicates the phase angle of the reflections at particular points along the transmission line. As shown in Figure 10-20, the straight line going from the center of the Smith chart to the outer perimeter indicates phase angle of 50 degrees at 0.18 λ away from the load at a VSWR of 1.4.

Impedance of Transmission Line Sections

Figure 10-21 shows the input impedance of various lengths of transmission line sections, both open and short-circuited, which are used in impedance matching. In practical applications, the most frequently used lengths are the quarter-wave and the half-wave section. Note that a quarter-wave, short-circuited, section has exactly the same impedance as a half-wave, open circuited section. Both have the characteristics of a parallel resonant circuit. In the formulas shown in Figure 10-21 the following symbols are used:

$\beta = \omega \sqrt{LC}$, where ω is frequency in radians

l = length, for $l = \lambda/4$ or $3/4\,\lambda$, $\tan \beta l = \pm 1$, $\cot \beta l = 0$

R = resistance or loss per unit length

Resonant or tuned sections, such as the quarter-wave section described above, can be used as resonant circuits as well as step-up transformers. The actual losses in such a line are very small compared to the inductive and capacitive components, and fairly high Q circuits are possible. Quarter-wave matching stubs are used on transmission lines to provide an impedance match, if the exact location from the load is known at which to place the tuning stub. This permits tuning out standing waves.

Various lengths of transmission line sections (both coaxial cable

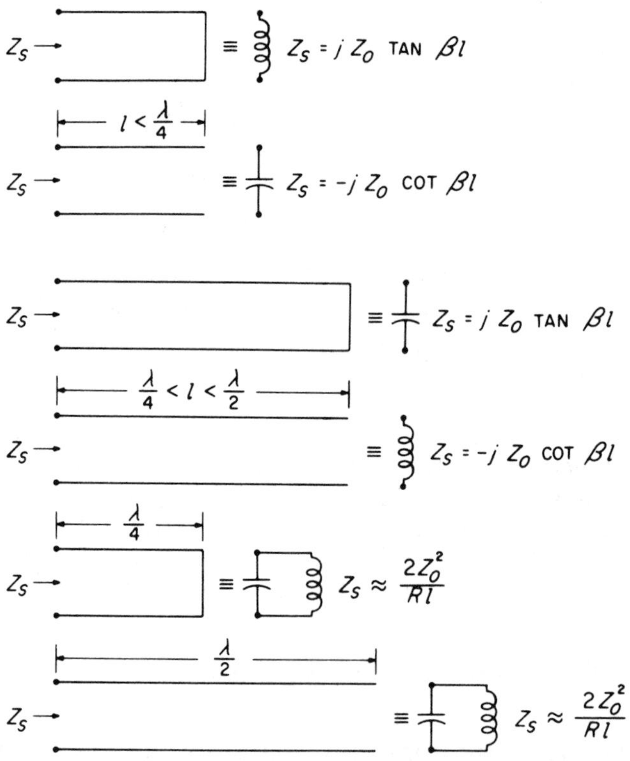

Figure 10-21. Transmission line section impedances.

(Edward C. Jordan and Keith G. Balmain, *Electromagnetic Waves and Radiating Systems*, 2nd Edition, © 1968. Reprinted by permission of Prentice-Hall, Inc.)

and waveguide) are widely used as impedance matching devices in many applications. Some examples are: satellite tracking and space communications antenna waveguide systems, microwave measurements during maintenance of such equipment, and all related applications. The Smith chart can be used to determine the length of the transmission line section needed, as well as where it should be placed for a proper impedance match.

10.5 PRACTICAL TRANSMISSION LINE TYPES

The vast majority of transmission lines in practical use are either the coaxial or the two-wire type. A large variety of coaxial cables are available, ranging from the very small, 1/8-inch-diameter flexible to the semirigid and rigid coaxial transmission line systems used in transmitters with diameters of up to 10 inches. Depending on the application, a coaxial cable may have a solid or stranded inner conductor, and

Characteristic impedance $(Z_0)=138\log\frac{B}{A}$
(where dielectric between conductors is air)

If d–c resistance is negligible

Characteristic impedance $(Z_0)=\sqrt{\frac{L}{C}}$
Where L = inductance in henrys
C = capacitance in farads

If dielectric between conductor is not air

Characteristic impedance $(Z_0)=138\log\frac{B}{A}\times\frac{1}{\sqrt{K}}$
Where K = dielectric constant of material

Figure 10-22.   Basic coaxial cable parameters.

(John D. Lenk, *Handbook of Electronic Charts, Graphs and Tables,* © 1970. Reprinted by permission of Prentice-Hall, Inc.)

the shield may consist of one or two layers of braid, a foil winding, a spiral-wound semiflexible tube or even a solid copper tube. To keep the inner conductor properly spaced from the outer conductor or shield, a large variety of different dielectric materials and constructions are employed. Solid dielectrics, such as polyethylene or Teflon, are sometimes shaped into discs, beads, or spirals to provide lower dielectric losses and allow air spacing for the majority of the coaxial cable structure. The basic design and characteristic impedance calculations for a coaxial cable of air and solid dielectric are shown in Figure 10-22. The characteristic impedance is dependent on the logarithm of the ratio of the outer conductor and the inner conductor, and, as shown in Table 10-2 at the end of this chapter, the most widely used characteristic impedance for coaxial cables are in the order of 50 and 75 ohms.

Two-Wire Transmission Line

Assuming that both wires are of equal diameter d, are separated by a distance D, and are sufficiently far away from ground with air as a dielectric, the characteristic impedance will be:

$$Z_o = 276 \log \frac{D}{d}$$

In practice, even air-dielectric, two-wire lines will contain insulating spacers at the appropriate distances. If the insulating spacer is constructed of a low-loss dielectric, such as glass or ceramic material, and if the spacer volume is no more than 0.1 of the air volume, the effect of the spacer can be neglected and the above equation will hold approximately. If such a transmission line, however, comes close to a grounded surface, within 10 times D, the effect of the ground will change the impedance. Two-wire lines are available with a solid or a foam type of dielectric between the two wires; such lines are widely used in FM and TV reception. Unfortunately, the characteristic impedance of these lines cannot be calculated readily, because part of the field is located in the dielectric while the other part is located in air. See Table 10-2 at the end of this chapter for performance data on the 300-ohm, two-wire insulated transmission lines for FM and TV.

Parallel Strip Lines

At VHF and UHF, parallel strip transmission lines are sometimes used for relatively short distances, usually with air as the dielectric. If the width W of the strips is much greater than the separation D between them, the following characteristic impedance formula applies:

$$Z_0 = 377 \, D/W$$

Numerical Data for RF Transmission Lines

In order to determine the characteristic parameters, such as inductance, capacitance, resistance, characteristic impedance, attenuation, phase shift, and phase velocity, for specific transmission line dimensions, Table 10-1 is presented, which contains this information for parallel wire lines and for coaxial transmission lines.

Commercial Transmission Lines

The list of commercially available coaxial cables is still growing, but those shown in Table 10-2 represent the most basic types. Different versions of a basic cable have different designation numbers, which only specify performance characteristics, such as the ability to operate at higher temperatures, extra waterproofing for submarine use, additional external shielding, spiral armor, and similar variations. The characteristics of the cables listed in Table 10-2 include the characteristic impedance in ohms, the capacity per foot, and the outer diameter. Attenuation per 100-foot length is given for 1, 10, 100, 400, and 1000

	Parallel Wire Lines Conductor radius = a Conductor spacing (between centers) = b	Coaxial Lines Outer radius of inner conductor = a Inner radius of outer conductor = b
Inductance L (henry/m length of line)	$\dfrac{\mu_v}{\pi}\cosh^{-1}\dfrac{b}{2a}$ or $\dfrac{\mu_v}{\pi}\ln\dfrac{b}{a}$ for $b \gg a$	$\dfrac{\mu_v}{2\pi}\ln\dfrac{b}{a}$
Capacitance C (F/m length of line)	$\dfrac{\pi\epsilon}{\cosh^{-1} b/2a}$ or $\dfrac{\pi\epsilon}{\ln b/a}$ for $b \gg a$	$\dfrac{2\pi\epsilon}{\ln b/a}$
Resistance R (ohms/unit length of line)	$\dfrac{R_s}{\pi a} = \dfrac{1}{\pi a}\sqrt{\dfrac{\omega\mu_v}{2\sigma}}$ ohms/m For copper lines $R = \dfrac{8.31\times10^{-8}f^{1/2}}{a}$ ohms/m $\approx \dfrac{\sqrt{f_{MHz}}}{a_{in}}$ ohms/1000 ft	$\dfrac{R_s}{2\pi}\left(\dfrac{1}{a}+\dfrac{1}{b}\right) = \dfrac{1}{2\pi}\sqrt{\dfrac{\omega\mu_v}{2\sigma}\left(\dfrac{1}{a}+\dfrac{1}{b}\right)}$ ohms/m For copper lines $R = 4.16\times10^{-8}f^{1/2}\left(\dfrac{1}{a}+\dfrac{1}{b}\right)$ ohms/m
Conductance G (mhos/m length of line)	ωC (dissipation factor) $\approx \omega C$ (power factor of dielectric)	

Table 10-1. Numerical data on RF transmission lines.

Characteristic impedance $Z_0 = \sqrt{\dfrac{R+j\omega L}{G+j\omega C}} \approx \sqrt{\dfrac{L}{C}}$ (air dielectric)

$$120 \cosh^{-1} \frac{b}{2a} \qquad \text{or} \qquad \frac{\eta}{2\pi} \ln \frac{b}{a} = \frac{138}{\sqrt{\epsilon_r}} \log_{10} \frac{b}{a}$$

$$276 \log_{10} \frac{b}{a} \ \text{for } b \gg a$$

Attenuation constant α (neper/m)

$$\alpha = \frac{R}{2Z_0} + \frac{GZ_0}{2}$$

Phase-shift constant β (radians/m)

$$\beta = \frac{2\pi}{\lambda} = \frac{\omega}{v_0}$$

Phase velocity v_0 (meter/sec)

$$v_0 \approx \frac{1}{\sqrt{LC}} \approx \frac{3 \times 10^8}{\sqrt{\mu_r \epsilon_r}} \ \text{for low-loss lines}$$

For air $\mu_r \approx 1$; $\epsilon_r \approx 1$;

For copper $\mu_r \approx 1$; $\dot\epsilon_r \approx 1$; $\sigma = 5.8 \times 10^7$

$\mu_v \approx 4\pi \times 10^{-7}$ $\epsilon_v \approx \dfrac{1}{36\pi \times 10^9}$

Table 10-1. Continued

[Edward C. Jordan and Keith G. Balmain, *Electromagnetic Waves and Radiating Systems*, 2nd Edition, © 1968. Reprinted by permission of Prentice-Hall, Inc.]

Type RG...U	Imp. Ohms	Cap. per ft.	Outer diam. (inches)	Attenuation 100 ft.					Remarks
				1 mc.	10 mc.	100 mc.	400 mc.	1000 mc.	
6A	76	20	.332	.21	.78	2.9	6.5	11.2	I.F. & video
9A	51	30	.420	.16	.59	2.3	5.2	8.6	Stable attenuation
11A	75	20.5	.405	.18	.62	2.2	4.7	8.2	Community TV
22AB	95	16	.420	.42	1.3	4.0	8.5	12.5	Twin conductors
23	125	12	.65 × .945		.4	1.7			Twin conductors (balanced)
25A	48	50	.565						Pulse
26A	48	50	.525						Pulse
27A	48	50	.675						Pulse
28B	48	50	.805						Pulse
34B	71	21.5	.625	.065	.29	1.3	3.3	6.0	Flexible, medium
35B	71	21.5	.945	.064	.22	.85	2.3	4.2	Low-loss video
54A	58	26.5	.250	.18	.74	3.1	6.7	11.5	Flexible, small
55B	53.5	28.5	.206	.36	1.3	4.8	10.4	17.0	Flexible, small
57A	95	17	.625	.18	.71	3.0	7.3	13.0	Twin conductors
58C	50	29	.195	.42	1.6	6.2	14.0	24.0	Test leads
59B	73	21	.242	.30	1.1	3.8	8.5	14.0	TV lead-in
62A	93	13.5	.242	.25	.83	2.7	5.6	9.0	Low capacity, small
63B	125	10	.405	.19	.61	2.0	4.0	6.3	Low capacity
64A	48	50	.495						Pulse

Table 10-2. Commercial transmission line types.

65A	950	44	.405	.25	.83	2.7	5.6	9.0	Coaxial delay line
71B	93	13.5	.250	.23	.80	2.8	5.8	9.6	Low capacity, small
83	35	44	.405	.13	.52	2.0	4.4	7.6	Semi-flexible
87A	50	29.5	.425						Teflon dielectric
88	48	50	.490						Pulse
89	125	10	.632	.19	.61	2.0	4.0	6.3	Low capacity
117	50	29	.730	.05	.20	.85	2.0	3.6	Teflon & Fiberglas
119	50	29	.465						Teflon & Fiberglas
130	95	17	.625						Twin conductors
131	95	17	.710						Twin conductors
141	50	29	.195	.35	1.12	3.8	8.0	13.8	Teflon & Fiberglas
142	50	29	.206	.35	1.12	3.8	8.0	13.8	Teflon & Fiberglas
143	50	29	.332	.24	.77	2.5	5.3	9.0	Teflon & Fiberglas
144	72	21	.405	.16	.53	1.8	3.9	7.0	Teflon & Fiberglas
174	50	30	.10				19.0		Miniature coaxial
212	52	28.5	.332	.21	.77	2.9	6.5	11.5	Small, double braid
213	52	29.5	.405	.16	.55	2.0	4.5	8.5	General purpose
216	75	20.5	.425	.18	.62	2.2	4.7	8.2	Video & I.F.
217	52	29.5	.545	.10	.38	1.5	3.5	6.0	R.F. power
218	52	29.5	.870	.06	.24	.95	2.4	4.4	R.F. Power
220	52	29.5	1.120	.04	.17	.68	1.28	3.5	Low-loss R.F.

Table 10-2. Continued

MHz. For more detailed information, the reader is referred to manufacturers' data or to the military standards which apply to many of the cables. See Chapter 11 for information pertaining to waveguides.

BIBLIOGRAPHY—CHAPTER 10

JORDAN, E., ed. *Reference Data for Engineers: Radio, Electronics, Computers, and Communications* (7th Ed.). Indianapolis, IN, Howard W. Sams & Co., Inc., 1985.

JORDAN, E., and K. BALMAIN, *Electronic Waves and Radiating Systems* (2nd Ed.). Englewood Cliffs, NJ, Prentice-Hall, Inc., 1971.

LANDSTORFER, F., and R. SACHER, *Optimization of Wire Antennas.* New York, NY, John Wiley & Sons, Inc., 1985.

LEE, K., *Principles of Antenna Theory.* New York, NY, John Wiley & Sons, Inc., 1984.

MOORE, J., and R. PIZER, *Moment Methods in Electromagnetics Techniques and Applications.* New York, NY, John Wiley & Sons, 1984.

SKOMAL, E., and A. SMITH, *Measuring the Radio Frequency Environment.* New York, NY, Van Nostrand Reinhold, 1985.

THOMASSEN, K., *Introduction to Microwave Fields and Circuits.* Englewood Cliffs, NJ, Prentice-Hall, Inc., 1971.

CHAPTER 11

RF AND MICROWAVE FUNDAMENTALS

11.1 RADIO WAVE PROPAGATION

Transmitting antennas and receiving antennas are discussed in Chapter 10, Antennas and Transmission Lines, but the propagation of radio waves between them is the topic of this section. Electromagnetic radiation from a transmitting antenna is divided into various components, according to the path these radiated waves take, as illustrated in Figure 11-1. One group of waves, called surface waves, consists of a ground wave and a guided surface wave component, propagated along the surface of the earth. Space waves, another group, also have a number of different components. In many instances, the major component is a direct signal radiated from the antenna outward from the earth. Depending on the frequency, some of these signals are reflected or refracted and return to ground. Other components, however, are first reflected from the ground and are then projected into space.

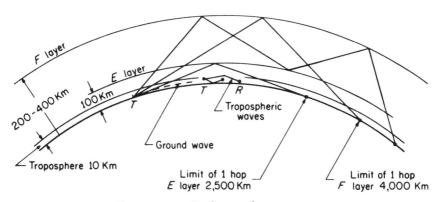

Figure 11-1.   Surface and space waves.

(Edward C. Jordan and Keith G. Balmain, *Electromagnetic Waves and Radiating Systems*, 2nd Edition, © 1968. Reprinted by permission of Prentice-Hall, Inc.)

The earth is neither a perfect dielectric nor a good conductor, but its dielectric or resistive aspects depend to a large extent on its moisture content. Seawater is an excellent conductor and acts in many cases as a reflector for ground waves up into the sky. Very dry terrain has different effects, and the roughness of the surface of the earth also has considerable effect on propagation. The United States National Bureau of Standards constantly conducts extensive studies of propagation of surface and space waves at most of the frequencies of interest. For detailed theoretical information concerning radio wave propagation, the reader is referred to the references at the end of this chapter. The following discussion is limited to practical engineering aspects of radio wave propagation.

Line of Sight.   At VHF, UHF, and higher frequencies, line-of-sight transmission and reception is the most frequently used method of communication. This propagation consists of a direct wave and a wave reflected from the ground. The roughness of the earth and the relative flatness of the terrain between transmitting and receiving antenna will determine how much of the ground wave is reflected and to what extent it is scattered. Ground reflection also is affected by the horizontal or vertical polarization of the radio waves to different extents. Figure 11-2 shows the various geometric factors involved in line-of-sight transmission. In most practical engineering applications, it is usually assumed that the transmission path goes over essentially flat terrain and that the surface wave is very small. It is also assumed that the distance d is sufficiently large compared to the antenna height so that the angles Ψ_1 and Ψ_2 are quite small. With these assumptions, it is possible to calculate the magnitude of the received signal, when the current into

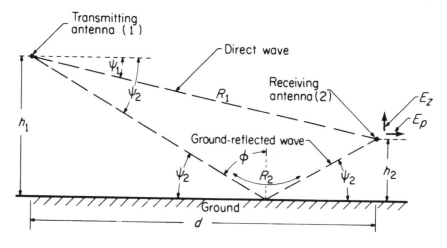

Figure 11-2. Line-of-sight propagation.

(Edward C. Jordan and Keith G. Balmain, *Electromagnetic Waves and Radiating Systems*, 2nd Edition, © 1968. Reprinted by permission of Prentice-Hall, Inc.)

the transmitter antenna and the geometrical quantities are known.

$$R_2 - R_1 \approx \frac{2h_1h_2}{d}$$

I = current in transmitter antenna
λ = wave length
$|E|$ = magnitude of received field
Other symbols as shown in Figure 11-2

$$|E| \approx \frac{60I}{d} \sin \frac{4\pi h_1 h_2}{\lambda d}; \quad \text{when } \lambda \text{ and } d \text{ are larger than 4, then}$$

$$|E| = \frac{240\pi h_1 h_2}{d^2}$$

The conclusion from the above formulas indicates that the received field can be increased in proportion to the height of the antenna, and it will decrease in proportion to the wavelength and the square of the distance. These calculations are particularly suited for FM and TV transmission and reception.

Another method of communication that is basically line-of-sight is earth station to satellite/satellite to earth station. Electromagnetic radiation in this type of communication system must pass through the atmosphere without being attenuated to any appreciable degree.

There are several "windows" (a band of frequencies that will pass through the atmosphere with minimum attenuation) that are used in commercial and government satellite communication systems. The optimum region for communication systems involving earth to space vehicles (satellite, various spacecraft, etc.) exists between 0.8 and 8 GHz. However, there are higher frequency bands in use (see Chapter 15).

Curved Paths. Returning to the subject of terrestrial communications, when the distances between transmitter and receiver are such that they are no longer within line-of-sight, transmission is possible because some curvature of the surface waves is obtained due to reflection from the troposphere and the ground wave itself. Because of the earth's curvature, there is considerable loss in the ground waves. Propagation predictions for curved paths of surface waves depend on seasonal variations caused by the earth's moisture content, by geographical variations, and by variations in the composition and height of the troposphere. The National Bureau of Standards periodically issues propagation predictions in the low-frequency region for surface wave conduction over curved paths.

Space Waves

In this category, all those waves are considered which reach the space just above the earth either directly, reflected, refracted, or reflected from the ground. Two portions of the earth's atmosphere are particularly concerned with sky waves. The layer closest to the earth's surface, ranging up to approximately 10 kilometers above the earth, is called the troposphere. Above the troposphere is the stratosphere, and above that, ranging from 50 kilometers up to as high as 1000 kilometers, is the ionosphere. The troposphere contains an appreciable amount of water, and its propagation characteristics depend largely on the water vapor pressure and temperature. The ionosphere consists of gases ionized primarily by the effects of sunlight and radiation from outer space. The ionized gases provide reflection and refraction of radio waves. Propagation phenomena of space waves have been studied extensively, and, for detailed aspects, the references listed at the end of this chapter should be consulted.

Tropospheric Propagation. The tropospheric temperature normally decreases at 6.5 °C per kilometer of altitude. Water vapor pressure decreases normally by 1 millibar per 1000 feet of altitude. When these two factors do not decrease in a normal way, conditions called temperature inversion or water content changes occur, and the refractive index of the troposphere is changed at that point. The modified refractive index is customarily plotted in so-called M-curves which show the change as a variation of height. The slope of the M-curve can indicate either substandard or superstandard propagation. When the slope of the M-curve is inverted, this indicates a phenomenon called *trapping*

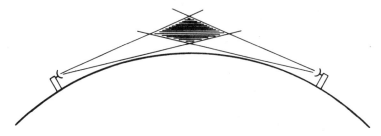

Figure 11-3. Troposcattering effects.

(Edward C. Jordan and Keith G. Balmain, *Electromagnetic Waves and Radiating Systems*, 2nd Edition, © 1968. Reprinted by permission of Prentice-Hall, Inc.)

or *duct propagation*. In this condition, signals injected into the duct by the transmitter travel within the duct with very little attenuation, somewhat as in a waveguide. Elevated ducts, similar to large waveguides parallel to the earth, are often found over Southern California and areas of the Pacific, but in order to utilize them, the signal frequency must be high enough and the receiving antenna must reach into the duct. A far more reliable effect for radio wave propagation is the so-called troposcattering effect. As illustrated in Figure 11-3, transmission well beyond the horizon is possible when areas are found where there are abrupt changes of the refractive index of a troposphere. Troposcattering usually works best between 0.1 and 10 GHz and requires large, high gain, highly directional antennas. While both slow and fast fading must be expected, it is possible to obtain reliable communications up to 350 miles. To overcome the fading effect, diversity reception, the use of two receiving sites spaced at least 100 wavelengths apart, is standard practice for troposcattering installations.

Ionospheric Propagation.   The ionosphere is also known as the Kennedy-Heavyside layer and acts as a fairly reliable reflector for frequencies between 3 and 30 MHz. At lower frequencies the signals are also reflected, but with a great deal more attenuation. The sky wave is usually a signal reflected from the ionosphere. These signals are used extensively for amateur communication and for fixed station, point-to-point radio telegraphy and telephony for commercial services. As shown in Figure 11-4, different layers in the ionosphere provide reflection at different altitudes and therefore to different points on earth. Because of their worldwide usage, extensive studies of skywaves and the reflecting properties of the ionosphere have been made, and a great deal of propagation data is constantly being generated by the National Bureau of Standards and other agencies. The charts prepared for specific areas

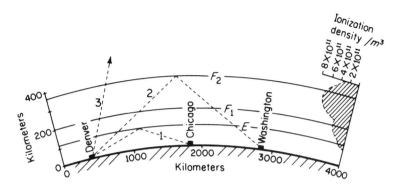

Figure 11-4.   Ionospheric reflection.

(Edward C. Jordan and Keith G. Balmain, *Electromagnetic Waves and Radiating Systems*, 2nd Edition, © 1968. Reprinted by permission of Prentice-Hall, Inc.)

refer to the maximum usable frequency (MUF) and the lowest usable frequency (LUF), as a function of time, season, and geographic location. In general, MUF rises during daytime and decreases at night. Because of the required tolerances, most radio stations do not use the MUF or the LUF but an optimum frequency which is between 50 and 80% of the MUF.

In addition to atmospheric noise of random origin, receivers monitoring sky waves often pick up a strange, brief audio signal commonly referred to as a *whistler*. These signals apparently are the result of lightning discharge pulses which bounce back and forth between the northern and southern hemispheres and are reflected by the ionosphere. Both very short and somewhat longer whistler signals have been recorded. Exact measurement of whistler signals is sometimes used to determine the condition of the ionosphere, and propagation forecasts are partially based on whistler measurements.

11.2 RF NOISE

No matter how sensitive the receiver, the useful operation is always limited by noise. When the noise level approaches the level of the signals, little can be done to improve reception. For these reasons the effects of noise have been studied extensively, and the reader interested in detailed discussions of noise is referred to the references listed at the end of this chapter. Three types of external noise interfere with radio signals arriving at the antenna, and two types of internal noise sources are considered part of the receiving system.

External Noise

Electromagnetic noise reaching the surface of the earth is generally divided into three categories: atmospheric noise, galactic noise, and man-made noise. Atmospheric noise is generally due to lightning charges in the atmosphere. This type of noise is least effective near the North and South Poles. In the temperate zones this type of noise varies largely with the weather conditions, and even in quiet spots its effects are worst below 20 MHz. Thunderstorms and lightning discharges as far as 1000 miles away may affect reception and generate atmospheric noise.

Galactic noise originates outside the earth and its atmosphere. Most of this type of noise is due to solar activities, such as sun flares and sunspots. This type of noise is usually worst between 18 and 500 MHz. Some galactic noise originates from stars millions of miles away. Many specific galactic noise sources other than the sun have been isolated by radio astronomy.

Man-made noise is due to such man-made devices as motors, neon signs, fluorescent lights, power line arcing and corona, vehicle ignition systems, medical and industrial RF equipment, etc. Generally speaking, man-made noise occurs below 20 MHz, although particular devices and particular locations exhibit man-made noise above these frequencies.

Internal Noise

One type of internal noise, so-called static noise, is due to the antenna itself. Rain, snow, hail, dust storms, etc., often cause static discharges from the receiving antenna, and these are picked up and amplified by the receiver. Antenna static is usually below 10 MHz. The second type of internal noise, thermal noise, is due to the electron motion in the molecules of resistances. For all practical purposes, thermal receiver noise is considered to be concentrated in the first stages of the receiver, including the antenna, transmission line, and input circuit. Since very little can be done about external and static noise, receiver designers have concentrated on reducing the thermal noise at the input of receivers. In theory, the lowest possible receiver input resistance or impedance will result in the lowest thermal noise voltage. Since temperature has a strong effect on the motion of electrons, one method of reducing thermal noise has been to cool the receiver inputs. For very sensitive, special purpose receivers, the input circuits are often maintained at very low temperatures, and so-called cryogenic preamplifiers are sometimes used for satellite communications. Earth station receivers cover a wide range of sensitivity and bandwidth, with noise temperatures from a few degrees Kelvin (cryogenic preamplifiers) to hundreds of degrees Kelvin in uncooled low-noise amplifiers. The following formulas apply to thermal noise voltage:

$$\text{Thermal noise voltage} = E; \quad E^2 = 4RkT\Delta f$$

$$\text{At } 17°C: E^2 = 1.6 \times 10^{-20} \, R\Delta f$$

$$k = \text{Boltzmann's constant} = 1.38 \times 10^{-23}$$

$$T = \text{absolute temperature in } °K$$

$$\Delta f = \text{bandwidth in Hz}$$

$$R = \text{Resistance in ohms}$$

Receiver Noise Factor and Noise Figure

The receiver noise factor and noise figure are used as means to evaluate the noise performance of receiver circuits. Noise factors are based on the ratio of the unmodulated RMS carrier signal and the noise RMS output of a receiver. The unmodulated RF carrier input is designated by E_s and the noise input by E_n. The square of these voltages is proportional to the ratio of the signal input power and the noise input power, as indicated below.

$$\frac{E_s^2}{E_n^2} = \frac{P_s\text{in}}{P_n\text{in}}; \quad \text{noise factor} = F = \frac{P_n\text{out}}{P_n\text{in}} \times \text{gain} = \frac{P_{1:1}}{KT\Delta f}, \text{ where}$$

$P_{1:1}$ = Power input for 1:1 carrier-to-noise ratio at the output

Noise figure = 10 log F

For microwave receivers, common noise figures range from 10 to 20 dB without the use of special input amplifiers. In the HF and VHF range, noise figures from 3 to 12 dB are common. Typical noise figures for FM and TV receivers vary from 3 to 10 dB.

Noise Factor Improvement

As mentioned in the above paragraphs, the noise factor can be improved by reducing the temperature of the input circuit and by keeping the input impedance or resistance as low as possible. With a given receiver input, that is, RF and IF section, the noise factor can be improved further, due to different types of modulation or coding schemes. Improvement in noise factor is generally evaluated by comparing the unmodulated carrier-to-noise power ratio at a convenient RF or IF point, with the output signal-to-noise ratio after demodulation. For example, considerable noise factor improvement is observed in FM receivers when the threshold of the limiter is reached. Below this threshold, the effect of the frequency modulation is very small, but once the limiter threshold is exceeded, the improvement is substantial.

The noise factor can be improved by reducing receiver bandwidth, and also by filtering the demodulated signal, since the effective bandwidth is part of the original thermal noise voltage equation. In addition to bandwidth filtering, noise factors can also be improved by increasing the complexity of decoding, such as by the use of pulse code modulation (PCM). Increasing the complexity of coding and the modulation scheme generally improves the noise factor. Where the information transmitted is of a repetitive nature, it is possible to improve the noise factor greatly by integrating successive signals and thereby cancelling the randomly occurring noise. Signal sampling techniques are another way of improving the noise factor. Most of the schemes mentioned above improve not only the internal but also the external noise characteristics of a receiver.

11.3 MICROWAVE RESONATORS AND WAVEGUIDES

Chapter 10 describes the functions and theory of transmission lines and of resonant portions of transmission lines. At microwave frequencies, transmission lines become waveguides and resonant portions of transmission lines become cavities of special designs and properties. Maxwell's equations, described in Chapter 10, are equally applicable to waveguides and microwave resonators. Because of the relationship of wavelength to physical dimensions in microwaves, however, some of the simplifications used in Chapter 10 do not apply here, while others are possible.

Principles of Guided Waves

One of the characteristics described by Maxwell's equations are boundary conditions that occur when the electric and magnetic fields travel through different materials. For the electric field equations, this means that the dielectric constant ϵ changes at the boundary. For the magnetic field, the permittivity μ changes at the boundary. The walls of the waveguide or resonator are such a boundary, and the electric field that is parallel and near to a perfect conductor must be zero. Electric lines of force, therefore, are at right angles. The magnetic field at a perfect conductor is either zero or finite, and surface currents will flow. The depth of these surface or skin currents becomes quite important because it determines the losses in the walls of the microwave resonator or waveguide. The skin depth of the current decreases at higher frequencies.

Reflection of Waves

For microwave signals, most of the same principles can be applied as for optical reflection. Chapter 19 covers the fundamentals of optics. For example, the angle of reflection from a reflecting surface is the same as the angle of incidence. The polarity of a transverse electric field, which is reflected, gets reversed. Reflective losses depend on the characteristics of the reflecting material and the depth of penetration. In many instances, the reflection principles described for parasitic antennas in Chapter 10 must be added to the optical reflection qualities of microwaves to account for some of the special effects obtained.

Transmission Line Theory

As it applies to coaxial cables, it is frequently used as the basis for explaining the operation of waveguides. Figure 11-5 shows some of the field configurations possible in a coaxial cable. In 11-5a, the so-called transverse electromagnetic (TEM) mode of propagation is shown. The solid lines with their arrows represent the electric field, which is radial from the center conductor and reaches the inner diameter of the outer conductor everywhere at right angles. The magnetic field accompanying the electric field is arranged in concentric circles around the inner conductor with polarities as indicated. The x indicates the arrow going into the paper and the dot indicates the arrow coming out of the paper. The side view of Figure 11-5a indicates reversals of polarity of both the electric and magnetic field as they would appear in a sine wave distribution. The transverse electromagnetic field arrangement can be used at any frequency on a coaxial cable.

Propagation Modes

The analysis of propagation in microwave resonators and waveguides is based essentially on the electric and magnetic fields in them. For most practical applications, the transverse electric field, TE, and the transverse magnetic field, TM, are the most important ones. These respective fields are always transverse, at right angles, to the direction of propagation of the wave. As is apparent in the illustration of Figure 11-5 and subsequent illustrations, subscripts are used with the TE and TM field. For rectangular waveguides, the subscripts indicate the num-

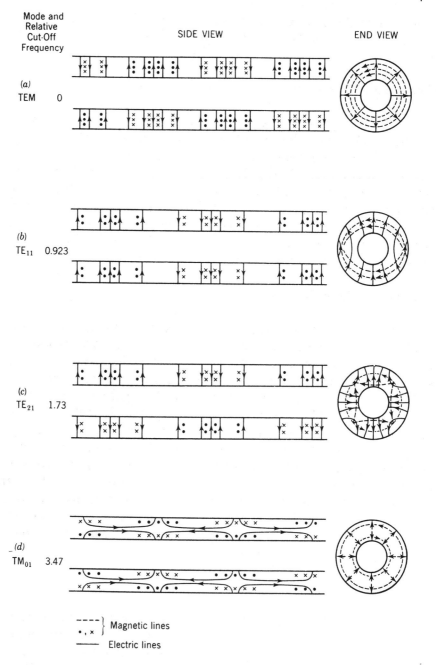

Figure 11-5.   Fields in coaxial cable.

(Illustration from *Essentials of Microwaves*, by R. B. Muchmore. Used with permission of John Wiley & Sons, Inc.)

ber of half-wave field variations in each of the two cross-section dimensions. The first number of the subscript refers to the larger dimension and the second one to the smaller dimension. In circular waveguides or coaxial cables, the first subscript number indicates the number of full-wave variations of the radial components along angular coordinates. The second subscript numeral indicates the number of half-wave variations of angular components in radial coordinates. These definitions will be apparent in the following paragraphs.

Figure 11-6 shows the basic waves that can be propagated at rectangular waveguides, with their respective electric and magnetic fields. The lowest, or dominant, mode of propagation is the TE_{10}, which has electric fields perpendicular to the sides of the waveguide as shown. The top view of the TE_{10} wave in Figure 11-6 shows the magnetic field lines as loops and the end view shows them as layers, perpendicular to the electric field. In the TE_{11} wave, the next highest order of TE waves possible in waveguides, the magnetic field loops are no longer arranged in single planes, but both the electric and magnetic field distribution is best illustrated by the end view. The lowest transverse magnetic propagation mode is the TM_{11} wave shown in side and end views in Figure 11-6. The electric field goes from the center to the edges of the waveguides, while the magnetic field loops are transverse.

Unlike coaxial cables, rectangular and circular waveguides are limited in their frequency response to the lowest frequency that will sustain the proper TE and TM fields. Because rectangular coordinates are easily used in describing the dimension and fields of rectangular waveguides, the following formulas are given for rectangular waveguides. In circular waveguides the same principles apply, but because cylindrical coordinates have to be used, Bessel functions, complex mathematical notations, result. The three essential parameters for waveguides are the propagation constant, the cutoff frequency, and the wavelength in the guide itself. The propagation constant γ is determined by the dimensions of the waveguide, the frequency ω, and by both the permittivity μ and the dielectric constant ϵ:

$$\gamma = \sqrt{\left(\frac{m\pi}{a}\right)^2 + \left(\frac{n\pi}{b}\right)^2 - \omega^2 \mu\epsilon} \quad m, n = \text{any integer}$$

$$a = \text{width (larger)}$$

$$b = \text{height}$$

If the propagation constant is a real number, there can be no propagation. This means that the negative portion under the square root above

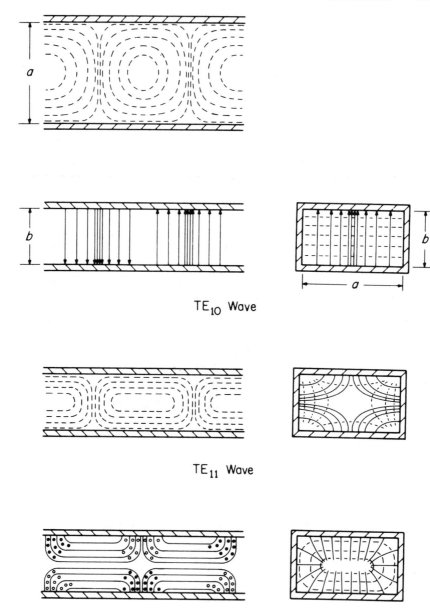

TE$_{10}$ Wave

TE$_{11}$ Wave

TM$_{11}$ Wave

Figure 11-6. Fields in rectangular waveguides.

(Edward C. Jordan and Keith G. Balmain, *Electromagnetic Waves and Radiating Systems*, 2nd Edition, © 1968. Reprinted by permission of Prentice-Hall, Inc.)

must be larger than the sum of the two positive square terms. From this it is possible to arrive at the cutoff frequency:

$$f \text{ cutoff} = \frac{1}{2\pi\sqrt{\mu\epsilon}}\sqrt{\left(\frac{m\pi}{a}\right)^2 + \left(\frac{n\pi}{b}\right)^2}$$

The wavelength of a signal in a waveguide is not the same as that in free space, and its exact dimensions are given by:

$$\lambda = \frac{2\pi}{\sqrt{\omega^2\mu\epsilon - \left(\frac{m\pi}{a}\right) - \left(\frac{n\pi}{b}\right)^2}}$$

In circular waveguides the same principles apply, but the lowest mode for both TE and TM is the 01 mode as in Figure 11-7. The TE_{11} mode in circular waveguides is best understood by observing the end view of Figure 11-7 and then the separate side and top views. The TE_{11} wave is quite different between circular and rectangular waveguides, as a comparison between Figures 11-6 and 11-7 will show.

Cavity Resonators

In the paragraphs on transmission lines in Chapter 10, the principles of standing waves and VSWR were discussed. In microwave circuits, these principles are utilized to provide resonant circuits. If a waveguide which is one-half of a wavelength long is closed at both ends, a signal at the resonant frequency can be injected into it and a standing wave pattern generated, which operates in the same manner as a high-Q parallel resonant circuit. Figure 11-8a shows the simplest of these resonators, a cylinder in which a transverse electric field is generated at a half wavelength. In this example, the dimensions were chosen so that one half-wave standing wave pattern is obtained in all three directions. For that reason this mode of excitation is described as TE_{111}, the subscript denoting the number of half-waves in each of the three directions. In the example of Figure 11-8b, the subscript indicates that two half-wave patterns are obtained in two directions and a single half-wave pattern in one direction. Figure 11-8 shows the electric field lines for different modes of excitation and different types of resonators. The magnetic field lines are not shown, but in every case they are perpendicular to the electric lines. As is apparent from the illustration, a huge variety of standing wave patterns, with resonant modes for each, are possible

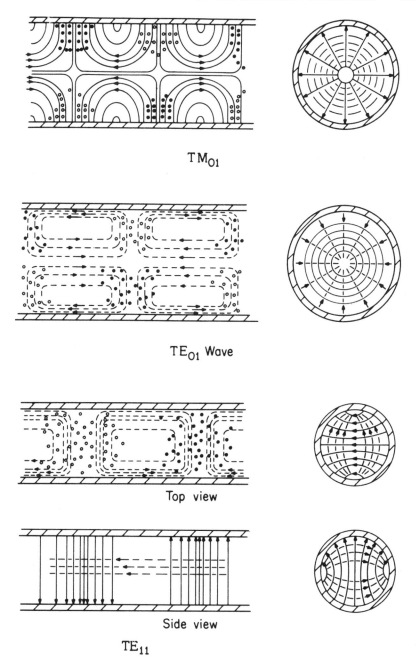

TM_{01}

TE_{01} Wave

Top view

Side view

TE_{11}

Figure 11-7.   Fields in circular waveguides.

(Edward C. Jordan and Keith G. Balmain, *Electromagnetic Waves and Radiating Systems*, 2nd Edition, © 1968. Reprinted by permission of Prentice-Hall, Inc.)

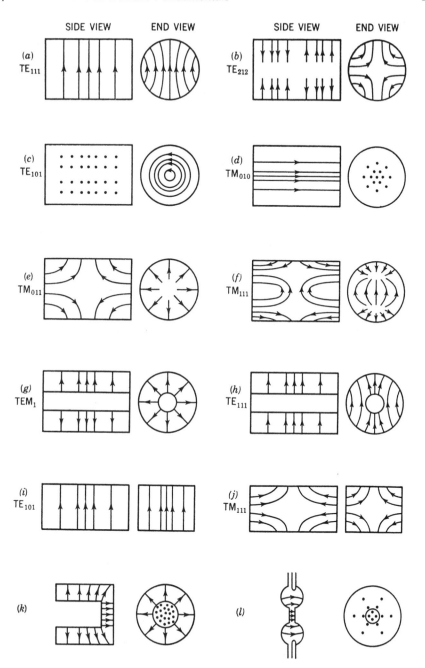

Figure 11-8.   Fields in cavity resonators.

(Illustration from *Essentials of Microwaves*, by R. B. Muchmore. Used with permission of John Wiley & Sons, Inc.)

in cavity resonators. The exact pattern obtained depends not only on the geometry and dimensions of the cavity but also on the method of injecting the signal.

Excitation Methods

Waveguides as well as cavity resonators must obtain the RF energy from a generating source. This generating source may be an antenna or an amplifier, but in every case there must be some means of exciting the waveguide; that is, a method of setting up the proper field in the cavity. The principle employed in excitation circuits is that the probe should produce lines of electric and magnetic fields that are approximately parallel to the lines of those fields expected in the desired mode. Figure 11-9 shows four basic methods of exciting rectangular waveguides, together with the propagation modes that are generated in this way. To

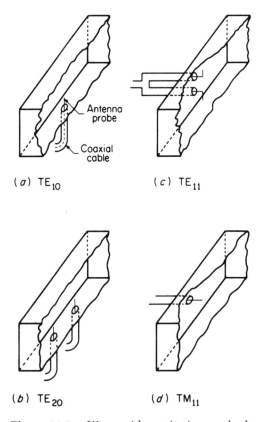

Figure 11-9.   Waveguide excitation methods.

(Edward C. Jordan and Keith G. Balmain, *Electromagnetic Waves and Radiating Systems*, 2nd Edition, © 1968. Reprinted by permission of Prentice-Hall, Inc.)

generate a TE_{10} mode, a radiating element is inserted at the end of the waveguide such that it is parallel to the Y-axis and generates electric field lines in the Y direction and magnetic field lines which lie in the X-Z plane. The section of a waveguide in which the excitation circuitry is located is frequently called a *launcher*. Commercially available launchers feature a coaxial input and special flange connections to mate with the waveguide end of the system.

Using the proper excitation method for a particular mode does not insure that other modes will not simultaneously exist in the waveguide, if they are well above the cutoff frequency for this particular mode. To eliminate unwanted modes, special waveguide sections are sometimes used in which the unwanted modes are suppressed by the geometry of the waveguide, and the dominant mode only is permitted to propagate.

It is also possible (and often done) to select and/or adjust the geometry of a metallic or dielectric piece of material to be placed inside a waveguide so that it will produce specific area concentration of either electric or magnetic energy. The result is essentially either a lumped inductance or capacitance very similar to that shown in Chapter 10, Figure 10-21. This permits tuning out standing waves, thus lowering line loss. The inserted elements are usually metal screws or thin diaphragms, and a Smith chart can be used to determine the correct placement in the waveguide system.

Standard Rectangular Waveguide Table

Rectangular waveguides, like the coaxial cables described in Chapter 10, are available in a number of standard sizes and characteristics. Table 11-1 lists these by the Electronic Industry Association's (EIA) waveguide designation, which starts with WR followed by the number. The number designation is arranged according to the frequency band covered; the largest number, WR-2300, covers the lowest frequency band.

The table shows the operating range for the TE_{10} mode and the cutoff for the TE_{10} mode KMHz as frequency and in centimeters as wavelength. The theoretical continuous wave (CW) power rating, from the lowest to the highest frequency, is shown in megawatts (1 million watts), and the theoretical attenuation in dB per 100-foot length is shown for the lowest to the highest frequencies. The material of which the waveguide is made is also shown. Joint Army-Navy (JAN) waveguide designations start with RG/U, just as for coaxial cable described in Chapter 10. Two types of flanges are available for JAN waveguide types: those using a choke flange and those using a cover flange. The inside dimensions and their tolerances, as well as nominal wall thicknesses, are shown in inches.

EIA WG Designation WR ( )	Recommended Operating Range for TE₁₀ Mode		Cut-off for TE₁₀ Mode		Theoretical cw power rating lowest to highest frequency megawatts	Theoretical attenuation lowest to highest frequency (db/100 ft.)	Material Alloy	JAN WG Designation RG( )/U	JAN FLANGE DESIGNATION		DIMENSIONS (inches)				Wall Thickness Nominal
	Frequency KHz	Wavelength (cm)	Frequency KHz	Wavelength (cm)					Choke UG( )/U	Cover UG( )/U	Inside	Tol.	Outside	Tol.	
2300	0.32-0.49	93.68-61.18	0.256	116.84	153.0-212.0	.051-.031	Alum.				23.000-11.500	±.020	23.250-11.750	±.020	0.125
2100	0.35-0.53	85.65-56.56	0.281	106.68	120.0-173.0	.054-.034	Alum.				21.000-10.500	±.020	21.250-10.750	±.020	0.125
1800	0.41-0.625	73.11-47.96	0.328	91.44	93.4-131.9	.056-.038	Alum.	201			18.000-9.000	±.020	18.250-9.250	±.020	0.125
1500	0.49-0.75	61.18-39.97	0.393	76.20	67.6-93.3	.069-.050	Alum.	202			15.000-7.500	±.015	15.250-7.750	±.015	0.125
1150	0.64-0.96	46.84-31.23	0.513	58.42	35.0-53.8	.128-.075	Alum.	203			11.500-5.750	±.015	11.750-6.000	±.015	0.125
975	0.75-1.12	39.95-26.76	0.605	49.53	27.0-38.5	.137-.095	Alum.	204			9.750-4.875	±.010	10.000-5.125	±.010	0.125
770	0.96-1.45	31.23-20.67	0.766	39.12	17.2-24.1	.201-.136	Alum.	205			7.700-3.850	±.005	7.950-4.100	±.005	0.125
650	1.12-1.70	26.76-17.63	0.908	33.02	11.9-17.2	.317-.212	Brass	69		417A	6.500-3.250	±.005	6.660-3.410	±.005	0.080
						.269-.178	Alum.	103		418A					
510	1.45-2.20	20.67-13.62	1.157	25.91	7.5-10.7						5.100-2.550	±.005	5.260-2.710	±.005	0.080
430	1.70-2.60	17.63-11.53	1.372	21.84	5.2-7.5	.588-.385	Brass	104		435A	4.300-2.150	±.005	4.460-2.310	±.005	0.080
						.501-.330	Alum.	105		437A					
340	2.20-3.30	13.63-9.08	1.736	17.27	3.1-4.5	.877-.572	Brass	112		553	3.400-1.700	±.005	3.560-1.860	±.005	0.080
						.751-.492	Alum.	113		554					
284	2.60-3.95	11.53-7.99	2.078	14.43	2.2-3.2	1.102-.752	Brass	48	54A	53	2.840-1.340	±.005	3.000-1.500	±.005	0.080
						.940-.641	Alum.	75	585	584					
229	3.30-4.90	9.08-6.12	2.577	11.63	1.6-2.2						2.290-1.145	±.005	2.418-1.273	±.005	0.064
187	3.95-5.85	7.59-5.12	3.152	9.510	1.4-2.0	2.08-1.44	Brass	49	148B	149A	1.872-0.872	±.005	2.000-1.000	±.005	0.064
						1.77-1.12	Alum.	95	406A	407					
159	4.90-7.05	6.12-4.25	3.711	8.078	0.79-1.0						1.590-0.795	±.004	1.718-0.923	±.004	0.064
137	5.85-8.20	5.12-3.66	4.301	6.970	0.56-0.71	2.87-2.30	Brass	50	343A	344	1.372-0.622	±.004	1.500-0.750	±.004	0.064
						2.45-1.94	Alum.	106	440A	441					
112	7.05-10.00	4.25-2.99	5.259	5.700	0.35-0.46	4.12-3.21	Brass	51	52A	51	1.122-0.497	±.004	1.250-0.625	±.004	0.064
						3.50-2.74	Alum.	68	137A	138					
90	8.20-12.40	3.66-2.42	6.557	4.572	0.20-0.29	6.45-4.48	Brass	52	40A	39	0.900-0.400	±.003	1.000-0.500	±.003	0.050
						5.49-3.83	Alum.	67	136A	135					
75	10.00-15.00	2.99-2.00	7.868	3.810	0.17-0.23						0.750-0.375	±.003	0.850-0.475	±.003	0.050
62	12.4-18.00	2.42-1.66	9.486	3.160	0.12-0.16	9.51-8.31	Brass	91	541	419	0.622-0.311	±.0025	0.702-0.391	±.003	0.040
						6.14-5.36	Alum. Silver	107	—	—					
51	15.00-22.00	2.00-1.36	11.574	2.590	0.080-0.107						0.510-0.255	±.0025	0.590-0.335	±.003	0.040
42	18.00-26.50	1.66-1.13	14.047	2.134	0.043-0.058	20.7-14.8	Brass	53	596	595	0.420-0.170	±.0020	0.500-0.250	±.003	0.040
						17.6-12.6	Alum.	121	598	597					
						13.3-9.5	Silver	66							
34	22.00-33.00	1.36-0.91	17.328	1.730	0.034-0.048	— —					0.340-0.170	±.0020	0.420-0.250	±.003	0.040
28	26.50-40.00	1.13-0.75	21.081	1.422	0.022-0.031	Brass Alum.		—	600	599	0.280-0.140	±.0015	0.360-0.220	±.002	0.040
22	33.00-50.00	0.91-0.60	26.342	1.138	0.014-0.020	21.9-15.0 Silver		96	—	—	0.224-0.112	±.0010	0.304-0.192	±.002	0.040
						— — Brass									
19	40.00-60.00	0.75-0.50	31.357	0.956	0.011-0.015	31.0-20.9 Silver		97		383	0.188-0.094	±.0010	0.268-0.174	±.002	0.040
15	50.00-75.00	0.60-0.40	39.863	0.752	0.0063-0.0090	— — Brass		—		385	0.148-0.074	±.0010	0.228-0.154	±.002	0.040
						52.9-39.1 Silver		98							
12	60.00-90.00	0.50-0.33	48.350	0.620	0.0042-0.0060	— — Brass		—		387	0.122-0.061	±.0005	0.202-0.141	±.002	0.040
						93.3-52.2 Silver		99							
10	75.00-110.00	0.40-0.27	59.010	0.508	0.0030-0.0041						0.100-0.050	±.0005	0.180-0.130	±.002	0.040
8	90.00-140.00	0.333-0.214	73.840	.406	0.0018-0.0026	152-99	Silver	138	—	—	0.080-0.040	±0.0003	0.156 DIA	±.001	—
7	110.00-170.00	0.272-0.176	90.840	.330	0.0012-0.0017	163-137	Silver	136	—	—	0.065-0.0325	±0.00025	0.156 DIA	±.001	—
5	140.00-220.00	0.214-0.136	115.750	.259	0.00071-0.00107	308-193	Silver	135	—	—	0.051-0.0255	±0.00025	0.156 DIA	±.001	—
4	170.00-260.00	0.176-0.115	137.520	.218	0.00052-0.00075	384-254	Silver	137	—	—	0.043-0.0215	±0.00020	0.156 DIA	±.001	—
3	220.00-325.00	0.136-0.092	173.280	.173	0.00035-0.00047	512-348	Silver	139	—	—	0.034-0.0170	+0.00020	0.156 DIA	±.001	—

Table 11-1. Commercial rigid waveguide types.

(Illustration courtesy of Microwave Development Laboratories, Inc.)

11.4 RF AND MICROWAVE TUBES

Experience has proven that at very high frequencies the utility of electron tubes is limited by the capacity between the tube elements, the cathode, grid, and anode, and by the transit time, the time it takes electrons to travel between the elements. Special designs such as planar triodes are one solution for frequencies up to the UHF range. Planar construction means that the cathode, the grid, and the anode are parallel planes, unlike the cylindrical construction of ordinary electron tubes. Planar construction reduces capacity considerably and, if the elements are spaced close enough, it also reduces the transit time. However, specially designed transistors that are interfaced to stripline and microstrip are usually used in today's high-frequency receivers.

For power amplifier tubes, planar construction is not as useful because of the possibility of breakdown and arcing between elements. For these applications, special screen grid tubes, tetrodes, and klystron tubes are used. These tubes operate at high voltages and are capable of producing high power outputs up to 2000 MHz and above. These special tubes are often found in the transmitters of VHF and UHF TV stations and in communications transmitters.

For operations in the microwave range, generally above 1000 MHz, entirely different electron tubes have been developed; the three basic types are described below.

Klystron Tubes

The simplest explanation of klystron operation would be to compare them to RF amplifiers, which have a parallel resonant circuit at the input and at the output. Because klystrons operate in the microwave band, these resonant circuits are cavity resonators, located inside the tube (some are located outside the tube, but the principle of operation is the same), so that the electron stream, traveling from cathode to anode, passes through these cavities. As illustrated in Figure 11-10, each of these cavities contains a fine grid through which the electron stream can travel. The signal injected through a coaxial cable and coupler into the first cavity, closest to the cathode and control grid structure

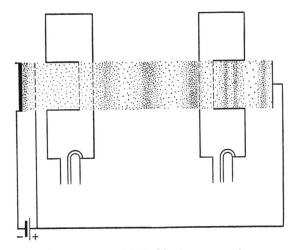

Figure 11-10.   Basic klystron operation.

(Illustration from *Essentials of Microwaves*, by R. B. Muchmore. Used with permission of John Wiley & Sons, Inc.)

in Figure 11-10, modulates the stream of electrons flowing through its wire grid. The stream of electrons is accelerated due to the high voltage at the anode, and the weak input signal is amplified by the electron stream until it is intercepted at the second resonator, where the output is obtained. A more rigorous explanation includes the fact that the electron stream travels much slower than the frequency of the input signal, and there is therefore some delay in signal between the input and output due to the electron stream. As shown in Figure 11-10, the klystron would be used as an amplifier, and in order to use it as an oscillator, all that is necessary is to connect the input and output together, since a 180° phase shift, with the proper spacing, is already available from the tube construction.

A different version of the klystron is in wide use under the name "reflex klystron"; it uses only a single cavity with a reflector or repeller which causes the electrons to enter this same cavity again. Figure 11-11 shows a simplified diagram of such a device. The voltage on the reflector or repeller can be varied to determine the point at which the electron stream is bent back. The location of this reflection point greatly influences the frequency and phase of the signal, and the reflector voltage is used to control the frequency of oscillation, since most of these tubes are used as oscillators. Two separate terminals are connected to the cavity, one for input and one for output. Reflex klystrons are particularly suited for frequency modulation as the modulation signal can be simply applied to the reflector, in addition to the high dc voltage.

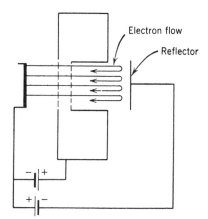

Figure 11-11.   Reflex klystron operation.

(Illustration from *Essentials of Microwaves*, by R. B. Muchmore. Used with permission of John Wiley & Sons, Inc.)

A more detailed discussion of klystron operation is contained in the references listed at the end of this chapter. For practical applications of klystrons, manufacturers' data should be carefully reviewed.

Traveling Wave Tubes (TWT)

Traveling wave tubes, often used in space communications applications, are a great improvement over klystrons because they provide considerably greater bandwidth and much more gain. Like the klystron, TWTs depend on the interaction between an electron stream and the microwave signal. Whereas the klystron utilizes an electron beam only at the points where it passes through cavity resonators, the TWT consists of a carefully designed waveguide, including resonators, with an electron stream moving in the center of the waveguide. In the TWT, therefore, the electron stream interacts with the microwave signal over the length of the tube.

Because the signal injected into a TWT is confined by a guide structure, many different modes of propagation are possible, and TWTs are therefore able to operate in many different ways. One basic type of TWT is illustrated in Figure 11-12 and consists of a coaxial structure with the center conductor arranged in the form of a helix. Input and output cavities provide the necessary impedance matching. A stream of electrons flows through the center of the helix and, with proper design, the speed of the electron stream approximates that of the propagated wave. For this reason, the interaction between the electron stream and the microwave signal is much greater than in a klystron,

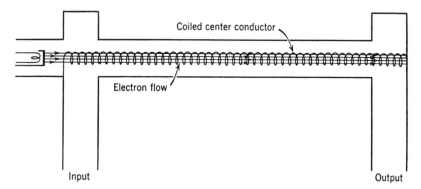

Figure 11-12.   Traveling wave tube principles.

(Illustration from *Essentials of Microwaves*, by R. B. Muchmore. Used with permission of John Wiley & Sons, Inc.)

and substantially more gain can be obtained. In some TWT designs, cylindrical waveguides with partitions are used to provide a "loaded line" effect to slow down the traveling wave. Other microwave structures can also be used, but the principle of matching the propagation speed of the electron stream to that of the microwave itself remains the same for all TWTs.

Because of their essentially "waveguide" construction, TWTs have a number of limitations. One of these is the possibility that some of the propagated wave is reflected backward. The backward wave may also be amplified and, if the parameters are right, oscillation may be set up. So-called backward wave oscillators provide highly stable and efficient microwave sources.

While the microwave signal is confined by the waveguide itself, the electron stream traveling down through the center will have a tendency to spread out. This means that an external force is required to maintain the path of all of the electrons in a straight line. Most TWT amplifiers use an external magnetic, longitudinal field to focus the electrons in the stream.

Magnetrons

The third type of microwave electron tube depends on the interaction of an external magnetic field with the electric field generated between the cathode and the anode, in a cylindrical configuration. Figure

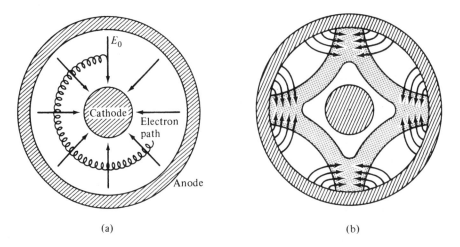

(a)   (b)

Figure 11-13.   Magnetron principles.

(K. Thomassen, *Introduction to Microwave Fields and Circuits,* © 1971. Reprinted by permission of Prentice-Hall, Inc.)

11-13a shows the basic structure of a magnetron with the cylindrical cathode emitting electrons which are attracted by the surrounding cylindrical anode. The direction of the external magnetic field is perpendicular to the electric field, going into the page. In the klystron and TWT, a straight-line stream of electrons was modulated by the microwave signal, but in the magnetron the electrons, because of the interaction of the electric and magnetic field, rotate in a spokelike pattern as shown in Figure 11-13b. This rotating field takes the place of the stream of electrons, and its modulation by the microwave signal depends on the geometry of cathode and anode, a circular waveguide structure, and the design of the anode itself. Although shown as a smooth surface in Figure 11-13, a practical magnetron anode consists of an even number of segments, similar to the stretched-out, flat version shown in Figure 11-14. A total of six segments are shown here, which

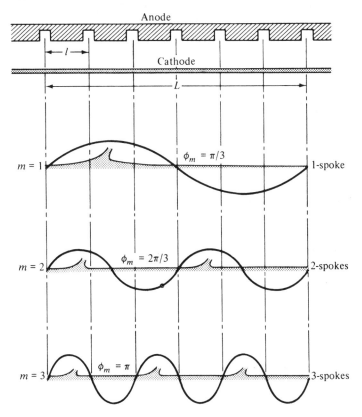

Figure 11-14.   Practical magnetron modes.

(K. Thomassen, *Introduction to Micowave Fields and Circuits.* © 1971. Reprinted by permission of Prentice-Hall, Inc.)

give rise to three possible modes of oscillation. Which of the three modes is in operation is determined by the voltage applied between cathode and anode. In most practical systems, the π mode is the one used.

A variety of different magnetron designs are in use, with many of them containing resonating cavities between anode segments. These cavities make the magnetron an essentially single frequency device. Voltage-tunable magnetrons avoid the use of resonant cavities to permit some frequency variation by adjustment of the anode voltage.

Because of their high efficiency in generating microwave signals, magnetrons are widely used in radars and in mircowave ovens. A conversion efficiency, dc into microwaves, of 80% is typical of many magnetron tubes. Magnetrons are available for small 10- or 20-watt sizes, up to the megawatt range where watercooling is necessary. For a detailed application and specifications of individual magnetrons, the reader is referred to the manufacturers' data.

11.5 RF AND MICROWAVE SEMICONDUCTORS

As in the case of electron tubes, the inherent limitations to the application of semiconductors at higher frequencies are the transit time, the capacity, and the geometry of the device. Diodes and transistors are available which operate in the microwave band, generally up to 2, 2.3, and 4 GHz, and in some cases even higher. At these higher frequencies, special designs for the terminations of diodes and transistors are required to minimize the inductance and capacitance due to the pin connecting arrangements. In the case of diodes, two different approaches are widely in use. In one method the diode leads are flat conducting strips emerging from the diode in opposite directions. The second technique provides a coaxial arrangement in which one end of the diode connects to the center and the other end to the outer conductor. These diodes are usually mounted in specially designed coaxial diode mounts and are used for mixing, detecting, and the special functions described below. Transistor mountings are usually in the form of flat strips emerging at different points of the transistor package. For UHF and microwave applications, the terminal strips of a transistor are incorporated into a stripline configuration (see Figure 11-15). Most UHF and microwave transistors are state-of-the-art devices that do not comply with any established standards. Detailed manufacturers' data and application notes are essential when dealing with such devices.

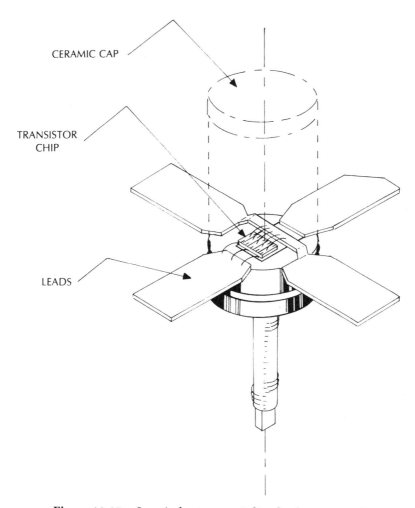

CERAMIC CAP

TRANSISTOR
CHIP

LEADS

Figure 11-15. Low-inductance stripline leads connected to a transistor chip.

Microwave Diodes

A number of specialized mirowave diodes are currently on the market, and a brief description of each type is given below.

- *Voltage variable capacitance diodes*—using a specially doped *pn* junction, this type of diode uses the capacity variation with reverse bias to provide frequency multiplication. Available in silicon as well as gallium arsenide for the upper frequency band, varactors can produce as much as 15 watts at 1 GHz and 1 watt at 5 GHz, with appropriately higher power inputs at lower frequencies.

- *Step-recovery diodes*—specially heavy doping near the junction enhances the charge-storage principle of the step-recovery or charge-storage diode. It can be used for frequency multiplication up to 10 GHz at very low power levels, usually below 100 milliwatts.

- *Tunnel diodes*—depend on a voltage-controlled negative resistance effect for amplification of microwave frequencies. Tunnel diodes are capable of low signal-level amplification but cannot be used for large power handling. Because of technical problems in mass production, the tunnel diode has found only limited application in the microwave field.

- *Gunn diodes*—named after their inventor, they consist of a single piece of semiconductor material, such as N-type gallium arsenide. At a certain voltage, this material develops a negative resistance region and spontaneously generates microwave signals.

- *Hot-carrier diodes (Schottky barrier diode)*—these devices are designed primarily for ultra-fast switching circuits and other ultra-high-frequency applications.

- *PIN diodes*—when used in the microwave part of the RF spectrum, a PIN diode behaves like a variable resistor with its value controlled by a dc bias current. There are several commercial microwave attenuators that are designed using PIN diodes. These are usually called *PIN-diode attenuators*.

Transistors used in microwave applications are usually packages in *stripline opposed emitter* (SOE) transistor packages. Figure 11-15 shows how the transistor chip leads are connected to the flat rectangular strips used as connecting leads. This illustration is shown in a submounted package (SOE package) with the top cap removed so that the inside connections are visible. It should be pointed out that microwave technology is constantly improving. For example, Hewlett-Packard Co. currently has a signal generator on the market that will produce a signal up to 18 GHz.

BIBLIOGRAPHY—CHAPTER 11

HAYT, JR., WM., *Engineering Electromagnetics* (3rd Ed.). New York, NY, McGraw-Hill, 1974.

Hewlett-Packard Co., *Microwave Theory and Application*. Englewood Cliffs, NJ, Prentice-Hall, Inc., 1969.

JORDON, E., and K. BALMAIN, *Electromagnetic Waves and Radiating Systems*. Englewood Cliffs, NJ, Prentice-Hall, Inc., 1971.

LEWIN, L., *Theory of Waveguides*. New York, NY, John Wiley & Sons, Inc., 1975.

LIVINGSTON, D., *The Physics of Microwave Propagation*. Englewood Cliffs, NJ, Prentice-Hall, Inc., 1970.

Reference Data for Engineers: Radio, Electronics, Computers, and Communications (7th Ed.). Indianapolis, IN, Howard W. Sams & Co., Inc. 1985.

THOMASSEN, K., *Introduction to Microwave Fields and Circuits*. Englewood Cliffs, NJ, Prentice-Hall, Inc., 1971.

CHAPTER 12

RADIO BROADCASTING AND RECORDING

12.1 FUNDAMENTALS OF RADIO AND BROADCASTING

In the past few years the technical aspect of the broadcasting industry has grown into a highly advanced and very broad field that involves ultrasophisticated equipment and operating techniques. During the decade of the 1980s, direct broadcasting to the home from satellites (television receive only and radio programming, especially National Public Radio programming) has become commonplace. Throughout the country, AM, FM, and TV stations have stereo transmission; several other changes such as Subsidiary Communications Authorization (SCA) and FM and TV translators have been authorized.

Network distribution of broadcast program signals was usually accomplished by the use of terrestrial microwave stations. However, satellite program distribution is being used extensively by the broadcast industry today. Also, direct broadcast satellite (DBS) transmission has been approved by the FCC.

RF Frequencies and Allocations

RF energy emanating from a radiator, such as an antenna, is usually considered in terms of the near field, usually within a distance of less than five wavelengths, and the far field. Where radiation of the far field is appreciable, regulation by both international and national agencies is imperative. To avoid the chaos that would occur if radio transmission were not closely regulated, a number of international and national bodies are set up to allocate frequencies for different applications and to govern their use.

International Regulation.   Worldwide frequency assignments and control of transmissions is in the hands of the International Telecommunications Union (ITU) in Geneva, Switzerland. Through its different committees, this international body controls and allocates all frequency assignments in conjunction with the national agencies of different countries. For purposes of frequency assignment, the world is divided into three major regions. Region 1 comprises all of Europe, Africa, the Soviet Union, and the Republic of Mongolia. Region 2 consists of North and South America, and includes Hawaii. Region 3 consists of Australia, New Zealand, Oceania, and all of Asia, excluding the Soviet Union and Mongolia. In addition to these three major regions, special assignments of frequencies are made to broadcast stations in the tropical zone because of the high noise levels in the medium frequency band in the tropical zone.

The ITU periodically updates the various volumes, generally known as the *Green Books*, concerning frequency assignments, radiated power assignments, and various other technical aspects of radio transmission. In addition, the ITU has different committees which concern themselves with standards in telephony, telegraphy, and other forms of communication.

The CCIR (International Radio Consultive Committee) is a branch of the ITU and is the committee that deals with broadcasting radio signal transmissions.

National Regulation.   Most countries exercise central control over all radio transmission through their Post, Telegraph, and Telephone Ministries, which also operate the various local broadcast systems. In the United States, a somewhat different system is in use. The Federal

Communications Commission (FCC) controls all those services not owned by the government. Government-owned services are controlled by the Director of Telecommunications Management, who is responsible to the President and who is advised by the Interdepartmental Radio Advisory Committee. A series of coordinating subcommittees and groups exists between the FCC and the Director of Telecommunications Management. Seven separate volumes of FCC rules and regulations are available from the commission which contain, in detail, all regulations concerning frequency assignments and the use of these frequencies for nongovernment services.

Authorized Radio Services in the United States. Radio services can be classified into two major groups. The first group is aimed at communications between individual stations such as those used in air traffic control, police radio, radio telephone operations, etc. The second group consists of broadcasting facilities where no response is expected from the receiving station. Communications services between individual stations fall into the following categories: fixed stations, land mobile (automobiles, trucks, etc.), maritime mobile (ship to shore, etc.), aeronautical (aircraft radio), citizens band, amateur radio, meteorological aids, radio navigation and radio location aids, space and radio astronomy. A host of different frequency bands from the low frequencies (LF) to the super-high frequencies (SHF) are assigned for these various services.

Recent broadcast radio services that are authorized by the FCC include: AM stereo transmission, subsidiary communications authorization for FM stations, FM translators, and satellite broadcasting. *FCC Rules and Regulations*, Part 73A, covers standard broadcasting; Part 73, Subparts B and C, cover FM broadcasting; and Part 74 covers the various auxiliary broadcast services (remote pickup, studio transmitter links, and the like). Broadcasting in the U.S. includes the following major services:

AM broadcast stations—from 535 to 1605 kHz.

International AM broadcasting—from 5.95 MHz to 26.1 MHz, in seven bands.

FM broadcast stations—from 88.1 MHz to 107.9 MHz.

Television broadcast stations—from 55.0 MHz to 88.0 MHz.

174.0 MHz to 216.0 MHz, 470.0 MHz to 890.0 MHz.

Direct broadcast satellite transmission—12.2 to 12.7 GHz for down link and 17.3 to 17.8 for up link has been approved by the FCC. The use of these frequencies has the disadvantage of signal degradation during snow and rain periods, plus cost of equipment. One of the main advantages is reduction in antenna size—less than one meter for practical earth station reception.

Detailed operations of broadcasting systems are described in Sections 12.4, 12.5, and 12.6, and in Chapter 15 under TV Broadcasting Systems.

A third and very special type of broadcasting is performed by the National Bureau of Standards, which operates WWV in Ft. Collins, Colorado. WWV transmits, at scheduled times, frequencies of 2.5, 5, 10, and 15 MHz, modulated by alternating tones of 440 and 680 Hz, transmitted in one-second pulses. These transmissions are used by many amateur and professional radio operators for comparison and adjustment of the frequencies of their own equipment.

Modulation

In order to transmit information by radio, it is necessary that the radio frequency carrier be modulated with the information. A large variety of different modulation techniques exists, the most basic of which is simply to interrupt the continuous wave (CW) signal at the pulse rate. This type of system was used in the original radiotelegraphy. More sophisticated systems, however, are now in widespread use. Chapter 14 contains more detail on modulation techniques.

Continuous Modulation.   Amplitude modulation (AM) is the simple technique of varying the amplitude of the CW signal in accordance with the information to be transmitted. The broadcast band from 535 to 1605 kHz and the international broadcast band both use amplitude modulation. Many amateur and citizens band operations also use amplitude modulation. In this method, both sidebands are transmitted. This means that a single frequency RF carrier, modulated by a band of frequencies, such as audio at 15 kHz, will result in two sidebands, one 15 kHz above and one 15 kHz below the carrier frequency.

An alternate method of double sideband AM is to suppress the carrier by using a balanced modulator. Probably the most widely used AM for certain types of communications is the single sideband, suppressed carrier (SSB), in which a balanced modulator is used, followed by a filter which permits only one of the two sidebands to be transmitted. This method has the advantage of conserving spectrum space. In a compromise method, only a portion of one of the two sidebands is suppressed; this is called vestigial sideband suppression and is used in TV transmission.

Frequency modulation (FM) involves varying the frequency of the carrier at a rate analogous to the information to be transmitted. FM is widely used in UHF communications and for high-fidelity broadcasting because it is relatively independent of atmospheric noise.

Phase modulation (PM) involves a variation of phase of a single frequency carrier in order to convey information. PM is used in color TV to transmit the chroma or color information. Details concerning this technique are found in Chapter 15, Television Systems.

Pulse Modulation. A more detailed description of the different types of pulse modulation will be found in Chapter 14, Communications Systems. The most basic type of pulse modulation is to interrupt, or key, the CW signal at the pulse rate, the method used in early radio telegraphy. Pulse amplitude modulation (PAM) is a method of modulating the RF carrier by transmitting it at various amplitude levels corresponding to the pulse data. Pulse duration or width (PDM), pulse position (PPM), and pulse frequency (PFM) modulation are all different ways of varying the characteristics of the RF carrier with pulse data. Probably the most widely used method is pulse code modulation (PCM), in which fixed numbers of pulses occurring at fixed repetition rates transmit the desired information by means of binary or other code schemes.

When the transmitted information consists of only two states, on-off or mark-space, the modulation is generally called *keying*; three basic types of such keying exist. Amplitude shift keying (ASK) is essentially the same as PAM. Frequency shift keying (FSK) is a form of frequency modulation, except that in this type of modulation only two frequencies exist, one for the "on" state and one for the "off" state. The third keying method is called phase shift keying (PSK) and uses a CW signal on which the phase is instantaneously changed either by 90 or 180 degrees, in accordance with the "on" or "off" state of the keying signal.

RF Propagation

The propagation of RF signals in the vicinity of the earth, at the ionosphere, the troposphere, and interspace, depends on the frequency and, in some instances, the polarization of the signals. Chapter 11 contains more information on the type and range of different propagation modes. In general, the effects of humidity are greater at higher frequencies. Losses due to reflections and refraction by mountains, absorption by vegetation, etc., also increase with frequency. Conduction along the ground and the resulting influence of the character and nature of the ground is greater at the lower frequencies.

VLF, up to 30 kHz.   At these frequencies RF transmission is dependent on ground or surface waves which travel great distances, well beyond the curvature of the earth. Because of the large wavelength at these frequencies, the reflections from the ionosphere may be quite strong. Considerable differences in reflected or sky waves are found between daylight and night-time. The ground waves will vary due to the conductivity of the ground. Seawater provides the best conductivity and therefore the strongest ground wave. Dry desert sand has the opposite effect.

LF and MF, 30 kHz to 3 MHz.   In this frequency range the ground waves do not reach as far nor are they as strong as at VLF, but the sky waves are generally very strong. At medium distances the sky waves frequently fade in the LF and MF range, but they get stronger at long distances, particularly at night and at the higher frequencies. Daytime sky waves favor the lower frequencies. The effects of different ground conductivity, as well as variations due to seasons, high ionospheric disturbances, sunspots, etc., are particularly strong at the MF range.

HF, 3 to 30 MHz.   Transmission in this frequency range depends to some extent on ground waves up to about 100 miles. Beyond this range transmission depends only on ionospheric reflections, and the effects of changes in the various layers of the ionosphere are particularly pronounced in the HF region. The reflections vary greatly, causing skip signal effects which are often a function of frequency. Different frequencies will be reflected at different points, and it is therefore important to be able to select the proper frequency in order to provide the best reflection of the sky wave for a particular type of communication. Propagation forecasts for this band are issued monthly, for a three-month period, by the Institute for Telecommunication Sciences, a part of the Environmental Sciences Services Administration (ESSA).

VHF, UHF, and SF, above 30 MHz.   At these frequencies transmission is essentially line-of-sight, although some bending around the earth due to atmospheric refraction is observed, particularly below 100 MHz. This refraction effectively changes the earth's horizon to a "radio horizon" which is greater than the earth's horizon. Above 30 MHz the reflection by the ionosphere rapidly disappears and the signals travel, with some absorption, directly into outer space. For this reason, space communication always uses signals well in the UHF or SHF band. Some scattering type of reflection occurs in the troposphere, and this is used for tropospheric scatter propagation, generally effective a few hundred miles beyond the horizon. Chapter 11 presents more details. UHF and SHF signals generally are beamed directly at particular points in the troposphere with carefully calculated angles, so that the weak refracted field is received at the desired location.

12.2 SATELLITE AND SPACE COMMUNICATIONS SIGNAL PROPAGATION

As has been explained, electromagnetic waves being propagated between earth and space encounter various regions: the troposphere, stratosphere, ionosphere, and outer space. The outer free space is lossless and has a refractive index of unity (1). The troposphere and ionosphere that were discussed in Chapter 11 have refractive indices other than unity, thus refraction and absorption are a necessary consideration.

In addition to all the ionosphere phenomena described in Chapter 11—that is, the ionosphere consists of gases ionized by the effects of sunlight and outer-space radiation—this layer also induces Faraday rotation. All the phenomena previously described in this chapter and Chapter 11 affect space and satellite broadcast and communications systems.

Faraday Effect

Faraday effect can be regarded as the rotation of the plane of polarization of electromagnetic waves as they pass through the ionosphere in the earth's magnetic field. This effect may produce a signal path loss between two linearly polarized antennas.

Typically, this problem is circumvented by the use of circular polarization. Modern antenna waveguide feed systems usually can produce left-hand, right-hand, and circular polarization, if needed. Faraday rotation effects are negligible below 1 GHz and above 10 GHz.

Ionospheric Effects on Satellite Communications

The variation in the state of the ionization of the ionosphere, strong short-term (1 to 15 s) variations that are beyond the normal observed random day-to-day variations of amplitude, phase, polarization angle, and angle of arrival of electromagnetic waves reaching earth from outer space are known as *ionospheric scintillations*. Figure 11-1, in Chapter 11, shows the approximate locations of the E layer and F layer. The fluctuations of electron density in the sporadic E layer and the spreading of the F layer are the principle causes of the ionosphere scintillations.

Depending on the geographical location of the earth station, season of the year, time of day, and the solar activity, the ionospheric scintillations can be anywhere from 1 to 15 s and have a large magnitude. Earth station location and time of day are important. For example, near the earth's magnetic equator (between the earth's magnetic north and south poles) large-magnitude scintillations between 6 and 12 can be observed. These scintillations are affected by sunspots and are strongest in the fall and spring seasons of the year.

Due to the fact that scintillations have large magnitudes near the equator and in the Northern and Southern Hemispheres, the in-between latitudes are better choices for earth stations operating in the lower microwave bands. Generally speaking, communications systems operating at gigahertz frequencies are not troubled with strong scintillations when located in the intermediate latitudes.

Tropospheric Effects on Space Communications

In Chapter 11, under Tropospheric Propagation, the effects for terrestrial communications such as direct propagation and troposcattering were discussed. Other effects that can be important in certain space communication systems are: ray bending, scintillation, attenuation, and increased sky noise temperature, all of which are major considerations when choosing earth station antenna systems that must operate at very low angles of elevation (military and deep space exploration applications). On the other hand, the tropospheric effects are not significant to the designer.

12.3 MICROWAVE RELAY SYSTEMS

There are several frequency bands allocated for remote pickups and studio-to-transmitter links. The 7 GHz region is the most popularly employed band of frequencies for these applications.

In FM and television relay practice, the emitted waves are concentrated into a narrow beam by the use of a very directional antenna such as a parabolic reflector. These units, antenna, transmitter, and mounting tripod are portable and allow exact control while aiming the emitted signal toward the receiver and adjusting for maximum signal strength at the receiver. The trade-offs are:

1. The greater the area of a parabolic reflector in terms of wavelength, the greater is the gain.
2. The larger the diameter of the transmitter antenna, the less portability.

The basic formula for the power gain of a communication system using a parabolic reflector is:

$$P_G = 4\pi a/\lambda^2$$

where:

P_G = power gain.

a = the effective area of the reflector.

λ = the wavelength of the frequency being utilized.

The effective area of a reflector is greatly affected by the type feed system being used. For example, a Cassegrainian antenna (see Chapter 10, Figure 10-15) with a "shaped" hyperbolic subreflector can illuminate 80 percent or more of the reflector with minimal spillover. However, it is easier to ascertain the projected area of a basic parabolic reflector than the effective area, due to the type feed system used, etc., and since the effective area is approximately 0.65 times the projected area, the following formula applies.

$$P_G = (0.65) \, 4\pi A/\lambda^2$$

where:

P_G = power gain.

A = projected area.

λ = wavelength of the frequency being utilized.

This formula results in only a slight error, providing the diameter of the reflector is large compared to the wavelength of the frequency, which is usually the case in practical applications. In practice, an antenna operating in the 6.8 to 7.05 GHz range obtains a gain of 37 dB (a relative power gain of 5000) with 4-foot reflector, and a gain of 40 dB (a relative power gain of 11,500) with a 6-foot reflector.

The power gains are only valid when the antenna is aligned precisely with the incoming beam so that maximum power is available. For instance, a 6-foot receiving antenna only 1.7 degrees off the central axis, operating in the 7 GHz range, will provide only one-half the maximum power available, assuming a 20-mile signal path.

The practicality of relaying a signal from a remote location to another specific location depends on antenna height and the distance to the optical horizon. The basic formula to determine distance to the optical horizon is

$$D = 1.23\sqrt{H}$$

where:

D = distance to horizon (miles).

H = height of transmitting antenna (feet)

For example, if the antenna is placed on top of a 100-foot-high building, the theoretical distance to the optical horizon (line-of-sight) is 12.3 miles. But, if the receiving antenna is also situated on top of a 100-foot-high building (assuming both buildings are at the same elevation in respect to sea level), the effective line-of-sight path becomes:

$$D = 123(\sqrt{H_T} + \sqrt{H_R})$$

where:

D = line-of-sight distance (miles).

H_T = height of transmitting antenna above sea level (feet).

H_R = height of receiving antenna above sea level (feet).

If both antennas are at a height of 100 feet, as in this example, the theoretical line-of-sight path length would be 24.6 miles.

For practical applications, the engineer usually must have a map of the area showing elevation contours (height above sea level). Then the height of the structure where the antenna is to be placed is added to the overall height (H_R). Normally, the height of the studio antenna has already been calculated and is in the station engineering records. Figure 12-1 shows a typical microwave relay antenna arrangement.

Practical RF Path Calculations

The microwave signal beamed between two relay antennas actually takes various paths depending on atmospheric conditions and the earth's surface, etc. The energy in the direct line-of-sight is said to be in the first *Fresnel zone*. Actually, many Fresnel zones exist, but the engineer is primarily interested in the first.

The radius of the first Fresnel zone at the point of a major obstruction in the signal path (such as a large building, etc.) may be calculated

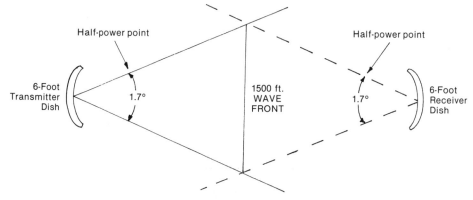

Figure 12-1. Microwave antenna arrangement, assuming a 20-mile separation that will produce a 1500-foot wave front at the halfway point to the receiving antenna.

using the equation:

$$R = 72\sqrt{AB/PF}$$ **(12-1)**

where:

   R = radius (feet).

   A = distance between first antenna and obstruction (miles).

   B = distance between second antenna and obstruction (miles).

   P = total signal path length (miles).

   F = frequency in GHz.

A value of 0.6 times the radius of the first Fresnel zone is usually taken as the absolute minimum clearance the signal must have to pass over an obstruction.

Path length (miles)	⅛ & ⅞ Distance	¼ & ¾ Distance	½ Distance
5	13	16	19
10	21	27	32
15	29	38	45
20	38	49	57
25	46	59	69
30	54	71	82

Table 12-1. Minimum microwave signal path clearance above the earth's surface, in feet. Calculated using the curvature of the earth on a radius of 1.333 times its true value.

The signal path profile, and obstruction on this path, may be plotted using topographic maps. The heights of the obstructions are added to the terrain sea level altitudes. The clearance over the earth curvature should be at least that shown in Table 12-1.

To find the correction for the earth curvature, use the equation

$$h = 0.5 \ d_1 \ d_2 \qquad \text{(12-2)}$$

where:

h = highest point of earth's curvature at point of major obstruction.

d_1 = distance from the earth's obstruction at near end of the signal path.

d_2 = distance from the earth's obstruction at far end of the signal path.

Using Equation 12-2 and plugging in example values, the correction factor for the earth's curvature is:

$$h = 0.5 \ d_1 \ d_2$$

Assuming equidistance,

$$h = 0.5 \ d^2 = 0.5(10)^2$$
$$= 0.5(100) = 50 \text{ feet}$$

Using Equation 12.1 to find the radius of the first Fresnel zone for 7 GHz,

$$R = 72 \sqrt{(10) \ (10)/(20) \ (7)} = 60.85$$

The minimum clearance is approximately

$$0.6(61) = 36.6 \text{ feet}$$

Assuming there is a building located in the signal path and it is 100 feet tall, our example has shown that the earth curvature correction factor is 50 feet and the clearance should be approximately 61 feet. Therefore, $100 + 50 + 61 = 211$ feet. Of course, further corrections must be made if one antenna is located at a higher elevation than the other. Also, since microwave beams are sometimes affected by atmospheric conditions, not only adequate clearance, but also an adequate signal-to-noise ratio must be established.

In general, the signal-to-noise ratio requirement depends on the distance between the microwave stations (studio transmitter link, etc.). For example, a path length of 20 miles should have a signal-to-noise ratio of about 40 dB to assure continuous communication. This would be in the vicinity of 45 dB for 25 miles, and 48 dB for 30 miles.

It is possible to estimate the signal-to-noise ratio by use of the antenna system gains and free-space attenuation. Free-space attenuation can be calculated using the formula

$$\text{free-space loss in dB} = 37 + 20 \log f + 20 \log D$$

where:

f = operating frequency in MHz.

D = distance in miles.

12.4 STANDARD (AM) BROADCASTING

The standard AM broadcast band extends from 535 to 1605 kHz, consisting of 10-kHz-wide channels each. Broadcast stations are licensed by the FCC and must be operated by licensed operators. The following basic standards apply:

Modulation: 85% to 95% at authorized operating power without exceeding required audio-frequency distortion limits.

Audio frequency distortion: Transmission characteristics between 100 and 5000 Hz within 2 dB, referenced to 1000 Hz, from microphone to antenna output. Harmonics less than 5% (voltage measurement of arithmetic sum or root sum squared) from 0 to 84%; not over 7.5% when modulating 85% to 95% from 50 to 7500 Hz.

System noise: At least 45 dB, unweighted root sum squared, below 400 Hz 100% modulation for frequencies from 30 to 20,000 Hz.

Carrier amplitude: Carrier shift less than 5% at any level of modulation percentage.

Out-of-band radiation: Referenced to unmodulated carrier level, attenuation of emissions removed from the carrier 15 kHz to 30 kHz greater than 25 dB, 30 kHz to 75 kHz greater than 35 dB, and more than 75 kHz 43 + 10 $\log_{10}$ (power in watts) or 80 dB whichever is less.

Broadcast stations fall into a number of different classifications within the three major types shown below.

Clear Channel Stations.   No interference is permitted within its primary and secondary regions, which comprise city, business, factory, residential, and surrounding rural areas. Within the primary service area, the ground wave signal is expected to be from 0.1 to 50 millivolts per meter, depending on the type of area. The secondary service area is defined by the presence of more than 500 microvolts per meter of sky wave signal, at least 50 percent of the time. Within the clear channel designation, the following classes of transmitters are established:

Class IA —50 kW maximum

Class IB —10 to 50 kW maximum

Class IIA—0.25 to 50 kW maximum during daytime and 10 to 50 kW maximum at night

Class IIB and IID—0.25 to 50 kW maximum

Regional Channel.   This channel has a primary service area consisting of a city and its surrounding rural areas, but its primary service area may be limited by interference from other licensed broadcast stations. The following classes are established for regional channels:

Class IIIA—1 to 5 kW maximum

Class IIIB—0.5 to 5 kW during daytime and 0.5 to 1 kW at night

Local Channel.   This channel has a primary service area comprising a city, town, and its surroundings, which may be limited by interference from other broadcast stations. Local channels have only class IV stations with 0.1 to 1 kW maximum during the daytime and 0.1 to 25 kW minimum at night.

AM Stereo Transmission

Standard broadcast stations are authorized to broadcast programs in stereo. Three methods to accomplish this are:

Mixed Mode: The carrier is amplitude modulated with the ($L + R$) signal, and phase or frequency modulated with the ($L - R$) signal.

Independent Sideband: The upper and lower sidebands of the carrier are modulated with the L and R signals, respectively.

Quadrature: Two phase-locked carriers are amplitude modulated and combined at a fixed phase angle.

Detailed rules for operations of standard AM broadcasting stations are issued by the FCC and should be consulted for further information.

12.5 INTERNATIONAL BROADCASTING

International AM broadcasting stations are specifically licensed for transmission of broadcast programs to foreign countries. While the transmitter standards are essentially the same as those described in Section 12.4 for standard AM broadcasts, the transmitter output must be a minimum of 50 kW, and the antenna power gain toward the target area must be at least 10. Antenna power gains and directivity are discussed in Chapter 10. International broadcasting uses sky waves only and the FCC assigns specific frequencies for daily specified hours and for specified target areas. Because of the sky wave propagation dependency, international broadcast stations will operate on one frequency for a given time and then shift as propagation forecasts require it. The following seven bands are assigned in the United States for international broadcasting:

A— 5.95–6.20 MHz	E—17.70–17.90 MHz
B— 9.50–9.775 MHz	F—21.45–21.75 MHz
C—11.70–11.975 MHz	G—25.60–26.10 MHz
D—15.10–15.45 MHz	

The band 7100 to 7300 kHz is also allocated for broadcasting, except in the Western Hemisphere. Nongovernment international broadcasting in the United States is licensed by the FCC in accordance with Part 73, Subpart F of the *FCC Rules and Regulations*. The last band (G) listed is expected to be reduced to 1560–1600 kHz when seven new bands that were authorized by the World Administrative Radio conference of 1979 become effective. They are (in kHz):

9,775–9,900
11,650–11,700
11,975–12,050
13,600–13,800
15,450–15,600
17,550–17,700
21,750–21,850

12.6 FM BROADCASTING

Frequency-modulation (FM) broadcasting stations are authorized for operation on 101 allocated channels, each 200 kHz wide, extending consecutively from channel 200 on 87.9 MHz to channel 300 on 107.9

MHz. Commercial broadcasting is authorized on channels 221 (92.1 MHz) through 300. Noncommercial educational broadcasting is permitted on any channel, but channels 200 through 220 are reserved for such use (see Part 73 of the *FCC Rules and Regulations*). The following basic standards apply to all FM stations:

Carrier frequency stability: ±2000 Hz.

Modulation: Frequency modulation capability of 100%, corresponding to ±75 kHz of center frequency.

Audio frequency response: 50 to 15,000 Hz, with preemphasis according to the standard preemphasis curve for FM and TV audio.

Audio frequency distortion: The maximum combined harmonic RMS voltage in the system output must be less than the following:
50 to 100 Hz—3.5%
100 to 7500 Hz—2.5%
7500 to 15,000 Hz—3%

Noise: FM noise must be at least 60 dB below the 100% modulation over the band from 50 to 15,000 Hz at 400 Hz modulating frequency. AM noise in the band from 50 to 15,000 Hz will be at least 50 dB below 100% levels.

Out-of-band radiation: At least 25 dB below the level of the unmodulated carrier at 120 to 240 kHz away from the carrier frequency, 35 dB below between 240 and 600 kHz and 80 dB below beyond 600 kHz.

FM Station Classification

Four classes of stations, with varied assignments according to locality, region, antenna height, power radiated, etc., have evolved into a complex pattern of allocations in order to provide maximum service without interference over the entire United States. For allocation details of the overall plan or for any specific location, the appropriate office of the Federal Communications Commission (FCC) should be consulted. See Part 73, Subparts B and C.

Class A Stations.   With a minimum effective radiated power of 100 watts and a maximum of approximately 300 watts at an antenna height above average terrain of 300 feet, class A stations are intended primarily for relatively small communities and their surroundings. Twenty different channels are reserved for class A stations. A proposal to permit Class A operation on any channel was submitted to the FCC in 1982. See *FCC Rules and Regulations* for current status.

Class B Stations. With a minimum effective radiated power of 5 kW and maximum of 50 kW at an antenna height of 500 feet above average terrain, class B stations are intended for service in large communities, and are usually assigned to the northeastern part of the United States (zone 1) and Puerto Rico, the Virgin Islands, and California, south of 40 degrees latitude (zone 1A).

Class C Stations. With a minimum effective radiated power of 25 kW and a maximum effective radiated power of 100 kW at an antenna height of 2000 feet, class C stations are intended for service in large communities and are usually assigned in Alaska, Hawaii, and those parts of the United States not in zone 1 or 1A.

Class D Stations. Limited to transmitter output power of 10 watts, class D stations are intended for noncommercial educational channels such as FM stations operated by colleges, community service organizations, etc.

FM Stereo Operation

For transmission of stereophonic programs, FM stations use the main channel and a subchannel. The major FCC requirement is that monophonic and stereophonic programs must be received without degradation and full compatibility. While the main FM channel modulates the carrier from 50 to 15,000 Hz, the stereo subchannel is a band of frequencies from 23 to 53 kHz. A pilot subcarrier frequency of 19 kHz, ± 2 Hz, is used to modulate the main carrier between 8% and 10%. The second harmonic of this pilot subcarrier frequency, 38 kHz, contains the stereo subcarrier and its various sidebands, extending from 23 to 53 kHz. Figure 12-2 shows the frequency spectrum. The stereo-audio signals consist of a left-hand and a right-hand signal, which are added

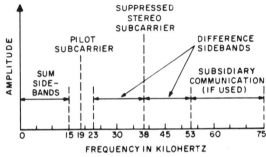

Figure 12-2. FM stereo frequency spectrum.

(From *Reference Data for Radio Engineers*, 5th Edition, by H. P. Westman, © 1968. Used with permission of Howard W. Sams & Co., Inc.)

together to modulate the main channel and thereby produce the signal received by monophonic receivers. The difference between the left-hand and right-hand signals is used to modulate the stereo subcarrier at 38 kHz. In reception, the 38 kHz subcarrier must be regenerated and synchronized with the 19 kHz pilot in order to produce both the sum and the difference signals of the original audio pickup.

Other Services

In addition to stereo signals, FM transmitters are also permitted to broadcast simplex facsimile during periods not devoted to audio broadcasting, provided these periods do not exceed one hour between 7 A.M. and midnight. Facsimile may be multiplexed at other times, together with the regular programs, provided there is no degradation. To transmit facsimile either by AM or FM, a subcarrier between 22 and 28 kHz is used.

Other services, such as background music and "store casting," can be multiplexed on the main channel by FM subcarriers in the range of 20 to 75 kHz for monaural stations, and in the range of 53 to 75 kHz for stereophonic stations. This enables commercial establishments to receive uninterrupted programs of music, special announcements, weather forecasts, etc., while the FM station is transmitting its regular program. In each case the subcarrier has to be restored, detected, etc.

Translators (FM)

Subpart L, Part 74 of *FCC Rules and Regulations* permits the use of very low power (1 watt east of the Mississippi, 10 watts west) translators to rebroadcast the signals of FM stations in areas where the station signals are below minimum standards. The translator receives the main station signal and rebroadcasts it on a different channel.

12.7 AUDIO RECORDING STANDARDS

AM and FM broadcasting frequently involves the use of recorded audio material, and for this purpose standard parameters have been adopted for disc and tape recordings. The standards used for disc recordings are also used for the prerecorded discs available to the general public. Only a portion of the tape recording standards adopted for broadcasting are used for the prerecorded magnetic tapes available to the public. In particular, the cassettes used for broadcasting are quite different from the variety of prerecorded magnetic tape cassettes accessible to the general public.

Disc Recording Standards

Speed—33-1/3 or 45 ± 0.1% rpm, clockwise, starting from the outside.

Size—11-7/8 ± 1/32 inches, 9-7/8 ± 1/32 inches or 6-7/8 ± 1/32 inches.

Deviation in pitch—0.04% of the mean speed from 0.5 to 200 Hz.

Flutter—Pitch deviation below 5 Hz, typical around 10 Hz.

Wow—Pitch deviation below 5 Hz, typical around 1 rpm.

Frequency—A specific frequency response curve with 0 dB at 1000 Hz has been standardized. Typical values are: +13 dB at 100 Hz, +18 dB at 38 Hz, −10 dB at 7000 Hz, and −17 dB at 15,000 Hz.

Grooves—For monophonic recording, a 90-degree ±5-degree angle, with a top width of minimum 2.2 mils (0.0022 inches) and a bottom radius with a maximum of 0.25 mils. For reproduction, a stylus should be used with a tip radius of 1 mil and an angle of 40 to 55 degrees.

Stereo—Grooves have a 90-degree angle, ±2 degrees, with a top width of at least 1 mil and a bottom radius of 0.2 mil maximum. For reproduction, the stylus tip radius should be from 0.5 to 0.10 mils with an angle of 40 to 55 degrees. Stereo modulation is accomplished on both sides of the grooves, and at a 45-degree angle with the disc surface. The outer wall of the groove contains the right-hand channel, and the inner wall, toward the center of the disc, contains the left-hand channel.

Tape Recording Standards for Broadcasting (Analog)

The major difference between the reel-to-reel and the cartridge method of tape recording, as far as broadcasting is concerned, is that in the cartridge operation only the 7-1/2-inch-per-second ±0.4% speed is available, and the thickness of the 1/4-inch tape that is used in cartridge standards is a maximum of 1.6 mils (0.0016 inches). The National Association of Broadcasters (NAB) has standardized on three sizes of cartridges, type A, type B, and type C. All three types are 0.9375 inches high and range from 4 × 5-1/4 for the type A, to 7-5/8 × 8-1/2 inches for the type C. Cartridges are generally available for playing in one direction only. Monophonic tape has a program track and a cue track, on which three different tones, 115, 1000, and 8000 Hz, are recorded for control of the tape program. In the stereo-tape arrangement, two program tracks and one cue track are available. The electrical performance of cartridge tape is essentially the same as that of reel-to-reel described below.

Speed—7-1/2 inch per second, ± 0.2%, is the preferred speed but 15 ips and 3-3/4 ips are also used.

Size—1/4-inch tape is 0.246, ±0.0002 inches wide, with a maximum thickness of 2.2 mils and typical thicknesses of 1.5 and 1.0 mils.

Length—Depending on reel size, 3 inches to 14-inch diameter reels; thickness, anywhere from 125 to 7200 feet of tape can be accommodated.

Tracks—When tape moves from left to right, plastic side to observer, track 1 is on top, track 2 is next lower. Track width for a single track is 0.238 inches, for two tracks the width is 0.082 inches each. In stereo recording, track 1 is always the left-hand track. For two tracks, the individual track width is 0.043 inches, with left-hand tracks being used as tracks 1 and 4.

Cross talk—For monophonic operation, cross talk must be −60 dB minimum from 200 to 10,000 Hz between adjacent tracks, while for stereo only −40 dB minimum is required from 100 to 10,000 Hz.

Distortion—Less than 3% rms at 400 Hz, when recorded 6 dB above the standard reference level. (400 Hz and 7-1/2 ips, 8 dB below the 3% third harmonic distortion point.)

Frequency—Including the amplifier response, a specific curve is prescribed for 7-1/2 ips tape. Typical points on this curve are 0 dB at 200 to 1000 Hz, −5 dB at 30 Hz, +5 dB at 4000, and +10 dB at 10,000 Hz.

Digital Audio Recording

Digital audio recording has several advantages when compared to audio analog recording using conventional disk and tape. For example, the dynamic range of the recording can be greater; when tapes are copied there is virtually no deterioration; and defects in the recording tape can be easily concealed (eliminated). Digital sound recording specifications are:

Sampling bits per digital word—14 to 16 bits.

Digital sampling rate—48 kilobits per second (48 kb/s).

Digital specifications for home equipment are slightly lower values, usually 44.3 kb/s.

BIBLIOGRAPHY—CHAPTER 12

FCC Rules and Regulations, Part 73, Subpart A. "Standard Broadcasting."

FCC Rules and Regulations, Part 73, Subparts B and C. "Frequency-Modulation Broadcasting."

FCC Rules and Regulations, Part 73, Subpart F. "International Broadcasting Standards."

FREEMAN, R., *Reference Manual for Telecommunications Engineering*. New York, NY, John Wiley & Sons, Inc., 1984.

International Telecommunications Union, Geneva, Switzerland.

Reference Data for Engineers: Radio, Electronics, Computers, and Communications (7th Ed.). Indianapolis, IN, Howard W. Sams & Co., Inc., 1985.

SLATER, J., and L. TRINOGGA, *Satellite Broadcasting Systems, Planning and Design*. New York, NY, John Wiley & Sons, Inc., 1985.

CHAPTER 13

RADAR AND NAVIGATION FUNDAMENTALS

13.1 BASIC RADAR PRINCIPLES

The word *radar* is derived from the original description of *radio detection and ranging*. Basic radar systems consist of a transmitter, which sends out radio waves, and a receiver, which receives reflections or echoes from objects such as planes, ships, buildings, etc. The time period from the transmission of the original signal to the reception of the echo or return signal depends on the range or distance of the reflecting object or target from the radar set. If the propagation velocity of radio signals is 300,000 kilometers per second, it takes approximately 12 microseconds (12×10^{-6} seconds) for a signal to travel one nautical mile and back. While some radars operate in a continuous wave (CW) mode, the majority transmit pulses, with a sufficiently long period between pulses to allow the return or echo signal to appear. While there

311

is a large variety of different types of radar systems, described in Section 13.2, the fundamental considerations are the same for all.

The Radar Range Equation

The range of any radar system is limited by many factors, such as the power transmitted, the efficiency of the antenna, the sensitivity of the receiver, and, of course, the size of the target and the amount of energy it reflects. With a given set of equipment parameters and target parameters, however, the maximum range can be determined by the radar range equation presented below.

$$R_{max} = \left[\frac{P_T A^2_e \sigma}{4\pi\lambda^2 S_{min}} \right]^{1/4}$$

where

P_T = peak transmitted power in watts.

A_e = effective area of the antenna in square inches.

σ = ratio of the transmitted to reflected energy due to the target.

λ = free space wavelength of the transmitted signal.

S_{min} = minimum detectable signal. This minimum detectable signal will depend, to a great extent, on the noise and interference surrounding the return signal at maximum receiver sensitivity.

This equation shows that the range varies as the fourth root of the transmitted power. This means that, in order to double the range of a given radar unit, it is necessary to increase the transmitted power 16 times. Two parameters that can often be changed to increase range are an increase in the effective antenna area (A_e) and a decrease in the wavelength (λ) of the transmitted frequency.

A Basic Radar System

Figure 13-1 shows the functional blocks of a basic radar system. The control circuit provides the timing for the indicator and controls the actual pulses to the transmitter, which consists of a conventional driver, a modulator, and the RF oscillator. While the functions of all radar transmitters are essentially the same, their design and construction will vary greatly according to the frequency range and power levels. In most cases, the block labelled "oscillator" is a power amplifier, driven by a

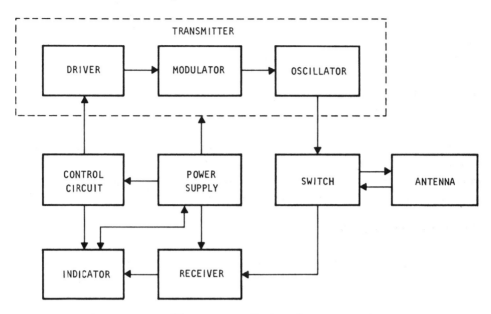

Figure 13-1. Basic radar system.

(Gershon J. Wheeler, *Radar Fundamentals*, © 1967. Reprinted by permission of Prentice-Hall, Inc.)

crystal-controlled frequency multiplier chain to provide exact RF frequencies. The RF pulse generated at the transmitter goes to the antenna, but during this same period, a switching device shuts off the input to the receiver section so that the high-level RF transmitter pulse does not burn out or overload the low-level, sensitive receiver input. In addition, in most radar systems, the control circuit sends out a keying signal to all of the receiver amplifiers to turn them off during the period of the radar pulse transmission. After the transmission of the pulse, the switch device changes again so that the antenna is now connected to the receiver, which has been "enabled" for maximum sensitivity by the control circuit. The receiver then receives the echo pulse signal, amplifies it, and displays it on the indicator. This process is repeated at the pulse repetition frequency (prf) assigned to the particular radar system.

Most radars operate at microwave frequencies, and many use a magnetron oscillator as a transmitter. Also, pulse-type radar systems are very common. However, in some systems, high average power or controlled modulation is required. In these designs the transmitter is frequently a power amplifier such as a klystron, traveling-wave tube, or a crossed-field amplifier. For example, klystron power amplifiers are used in most radio astronomy earth stations where high average power is desirable.

Low-noise receivers (some using cryogenic technologies, some not) using a parametric amplifier or low-noise transistor are often used in modern radar systems. Digital processing (computer) has made it possible to track many targets with accuracies superior to any equipment used in the past. For example, range resolution can be of the order of a few inches, and even more precision can be achieved by using laser radars that operate in the infrared and optical regions of the spectrum. Nevertheless, sophisticated systems such as described have been restricted in application due to cost and complexity.

Transmitter Parameters

The main parameters that must be defined for a particular radar system are as follows:

1. *Peak power*—this is the maximum power, in RF peak amplitude, which the transmitter can deliver to the antenna during the pulse.

2. *Average power*—the average power is dependent on the peak power, the duration of the pulse (τ), and the repetition frequency (fr). Figure 13-2 shows the relationship between these characteristics.

$$P_{Av} = P_{Peak} \times \tau \times fr$$

3. *The duty cycle*—this is the product of pulse width and repetition frequency ($\tau \times fr$). It is usually expressed in percent and obtained as follows: A pulse lasting 1 microsecond ($\tau = 10^{-6}$) with

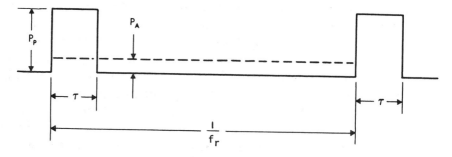

Figure 13-2.   Peak and average pulse power.

(Gershon J. Wheeler, *Radar Fundamentals*, © 1967. Reprinted by permission of Prentice-Hall, Inc.)

a pulse repetition frequency of 800 times/second ($fr = 800$) will produce a duty cycle of 0.0008, which is equivalent to 0.08%. If the peak transmitted power is 1 kW, and the duty cycle 0.008%, the average power will be 0.8 watts.

4. *Pulse rise time*—in order to be able to measure and observe the echo signal accurately, it is important that the transmitted signal have a sharp rise time; that is, the modulator and power amplifiers of a transmitter must start up very quickly.

5. *Fall time*—the end of a transmitted pulse should be as sharp as possible for similar reasons, as described above.

Receiver Parameters

1. *Sensitivity*—should be such that very weak signals, appearing almost submerged in noise, should be readily detectable.

2. *Bandwidth*—must be sufficient to pass all the components of the reflected pulse. The receiver bandwidth, however, should also be narrow enough to reject unwanted and spurious signals.

3. *Dynamic range*—because the amplitude of the returned echo signal depends on the range from the radar antenna to the target, the radar receiver must have sufficient dynamic range to receive very weak as well as relatively strong echoes.

4. *Keying response*—as explained above, the receiver is effectively turned off during the transmission of the pulse and must recover sufficiently quickly to gain full sensitivity before the return echo is received. It also has to be able to be keyed off just before the radar pulse is transmitted.

Antenna Parameters

The basic radar system depends on knowing the azimuth and the angle of elevation of the antenna (in some applications, such as radio astronomy, the hour angle and declination), in order to be able to indicate the location as well as the range of a target. If the antenna were omnidirectional, the only information obtainable would be the range, or the distance, of the target from the radar transmitter. This would mean, in the case of an aircraft, that it is located on a circle surrounding the radar site. Radar antennas are usually specified according to the following parameters:

1. *Horizontal beam width*—this is the width of the antenna beam in the horizontal plane in angular degrees.

2. *Vertical beam width*—this is the width, in angular degrees, of the antenna beam in the vertical plane. In some instances, where altitude is not measured, the antenna radiation pattern may be fanshaped in the vertical plane. Ordinarily, however, in order to radiate maximum energy, the beam is limited to a fixed angle in the vertical plane as well as in the horizontal plane.

3. *Antenna gain*—this describes the amount of increased radiating power for the transmitted pulse, or increased sensitivity for the received echo, of the particular antenna as compared to an isotropic radiator.

4. *Rotation speed and accuracy*—refers to antennas that are rotated in the horizontal and/or vertical plane. To search or scan for targets, it is essential that antennas be capable of reporting their instantaneous position, rate of rotation or scanning, etc., accurately to the indicator for appropriate displays. (This parameter depends on the electromechanical arrangements for mounting and rotating the antenna itself.)

Indicator Parameters

Two basic types of indicators are used. The first is a simple oscilloscope, which gives a presentation as shown in Figure 13-3. The hori-

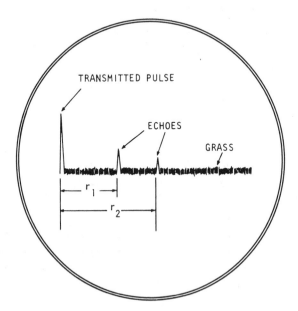

Figure 13-3.   A-scope presentation.

(Gershon J. Wheeler, *Radar Fundamentals*, © 1967. Reprinted by permission of Prentice-Hall, Inc.)

zontal axis represents time or the distance to the target, while the vertical axis represents the amplitude of a received signal. In this presentation, only the range is indicated. The second, more frequently used type of indicator is called PPI (plan-position-indication) and is shown in Figure 13-4. It involves a cathode-ray tube (CRT) in which the deflection yoke is effectively rotated in synchronism with the rotation of the antenna. Each time the transmitter sends out a pulse, the electron beam is deflected from the center to the outer rim. Whenever an echo is received, the electron beam is intensified. In this manner,

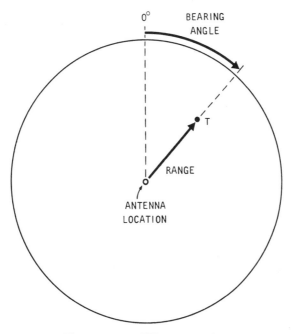

Figure 13-4. PPI presentation.

(Gershon J. Wheeler, *Radar Fundamentals*, © 1967. Reprinted by permission of Prentice-Hall, Inc.)

targets are indicated as bright dots on the CRT and both range and bearing, or azimuth, are shown. By overlaying geographic maps, it is possible to pinpoint targets accurately and locate them geographically.

Stationary radar systems using phased-array antenna systems to produce 3-D displays (such as used in air surveillance radars for aircraft landing, etc.) display an entire map or image of the area under surveillance. An example of a mobile system is called *synthetic aperture radar* (SAR). The received echos are usually stored in an electronic memory before being displayed on the viewing screen in the form of an image of the screen being scanned.

Stationary radars using Doppler shift measurements also image a moving target; this method is called *inverse synthetic aperture radar*. The Doppler shift is proportionate to the velocity of the moving object as it approaches or recedes (see CW Doppler radar systems).

Target Area Effects

One of the essential parameters of the radar equation presented at the beginning of this chapter is the quantity sigma (σ), which represents that portion of the original radar signal that is reflected and which determines the actual echo. Sigma is dependent almost entirely on the area, shape, and material of the target itself. Extensive studies have been made of radar cross-sections of various aircraft and missiles, and materials have been developed with which such targets can be coated to minimize the reflection. In fundamental radar theory, however, it is assumed that a given radar target can be represented by a metallic sphere. The basic equation for sigma is:

$$\sigma = \frac{4\pi P_r}{\alpha P_i}$$

in which (P_r) is the amount of power reflected toward the transmitting antenna and in which the incident power (P_i) is that amount of power that falls on the target. The angle alpha (α) represents the unit solid angle.

     The radar cross-section of a given target depends on the relationship between the diameter of the target and the wavelength of the transmitted signal. For a sphere of radius a, the following relationships hold:

$$\text{for } \frac{2\pi a}{\lambda} \gg 1; \quad \sigma = \pi a^2$$

$$\text{for } \frac{2\pi a}{\lambda} \approx 1; \quad \sigma = +5.7 db\pi a^2 \text{ or } -4 db\pi a^2$$

$$\text{for } \frac{2\pi a}{\lambda} \ll 1; \quad \sigma \ll 2\pi a$$

13.2 RADAR SYSTEMS

Simple Pulse Systems

Probably one of the most elementary radar systems is used as an altimeter in aircraft. The radar antenna is located underneath the aircraft and beams the transmitted signal downward toward the ground. The time interval from the transmitted pulse to the received echo is interpreted as distance and displayed to the pilot as altitude. In this system, the target is the ground underneath, and there is no question as to the target cross-section, nor is there any need for the antenna to indicate a specific position. Only distance or range is measured.

Other pulse radar systems, however, are more complex in that they depend on the shape and location of the beam radiated by the antenna for the location of targets. Search radar beams are either pencil- or fan-shaped, with the antenna rotating through a sector, or an entire hemi-sphere, to scan the skies. In order to obtain adequate resolution, it is usually necessary to send out at least five to ten pulses for each angular degree of rotation. This means that the rotation speed must be related to the pulse repetition frequency. Targets are displayed on a PPI indicator in terms of range and angular displacement from reference north.

In addition to the search capability, most radar systems can also be used for tracking targets. Basically, it is only necessary to note the range and azimuth of a target during successive sweeps, and, as this range and azimuth change, a track can be computed. A somewhat different approach uses a very narrow beam to illuminate the target. As the target appears to move away from the center of the beam, this is sensed as an error signal, and the antenna is then repositioned so that it maintains the target essentially, always in the center of the beam. A more detailed description of tracking methods is contained in Section 13.3.

CW Doppler Radar System

A continuous wave (CW) radar system can be used when a target moves radially, toward or away from the transmitting antenna, because of a special phenomenon called the Doppler effect. This effect, applicable to visible light (and sound) as well as to radio waves, predicts a change in the transmitted frequency, dependent on the radial velocity of a target and the transmitted frequency. The Doppler frequency, the change in the transmitted frequency, is given by the formula shown below:

$$f_D = \text{Doppler frequency in Hz}$$

$$f_D = \frac{2V_r}{\lambda}$$

$$V_r = \text{radial velocity in m/sec}$$

$$\lambda = \text{wave length in m}$$

At a radar frequency of 10 GHz, the Doppler frequency will be 894 Hz if a target is moving at 30 miles per hour (44 feet per second) toward the transmitter or away from the transmitter.

In order to separate the transmitted signal from the received signal in CW radars, special devices known as isolators, circulators, duplexes, or "magic T's" are used. These devices provide different attenuations for the transmitted signal going to the antenna and for the received signal going from the antenna to the receiver. These devices take the place of the electronic switch described in Figure 13-1.

Figure 13-5 shows the block diagram of a basic CW radar, such as

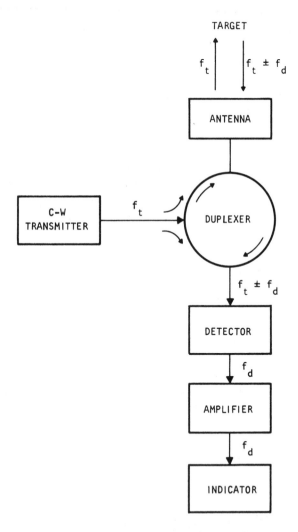

Figure 13-5. Block diagram of CW radar.

(Gershon J. Wheeler, *Radar Fundamentals*, © 1967. Reprinted by permission of Prentice-Hall, Inc.)

may be used in a motor vehicle speed-detecting system. In this system, the duplexer acts as a mixer as well, because a small portion of the transmitted frequency is mixed in with the echo signal which contains the Doppler frequency. Note that the Doppler frequency can be either above or below the transmitted frequency, because a target approaching the antenna causes an increase in frequency, while a target moving away from the antenna causes a reduction in the transmitted frequency. The Doppler frequency, f_D, is amplified and can be indicated either on a meter, a digital display, or in an earphone or loudspeaker. By application of the formula presented above, the velocity, converted into miles per hour, can be displayed or printed out directly.

FM Doppler Radar

The CW Doppler radar provides no indication of distance or range and only indicates the speed of a target. One simple approach to provide range information as well is to modulate the frequency of the transmitter in a predetermined pattern, such as in the triangular FM pattern shown in Figure 13-6. The solid line indicates the rate at which the transmitted frequency is modulated or changed. The dotted line, displaced by the time period T, represents the echo from a stationary target.

$$T = \frac{2R}{c}$$

where

$$R = \text{range in meters}$$

$$c = 300,000 \text{ km/sec}$$

If the target is approaching, the dotted line in Figure 13-6 will effectively be raised by the Doppler frequency but will not be changed as far as the horizontal, or time, axis is concerned. If the target moves away from the radar antenna, the portion of the dotted line of Figure 13-6, corresponding in time to the target echo, will be lowered, but again, the displacement T will not change. FM Doppler radars frequently use PPI indicators with a separate readout, usually a counter or meter, to indicate the velocity of the target.

Pulsed Doppler Radar System

The Doppler effect can also be used in a pulsed radar system, and then provides the capability for distinguishing moving targets from station-

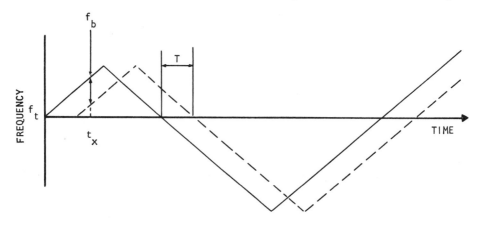

Figure 13-6.   Triangular FM Doppler signal.

(Gershon J. Wheeler, *Radar Fundamentals*, © 1967. Reprinted by permission of Prentice-Hall, Inc.)

ary targets. In many radar systems, the moving target indication (MTI) is a feature that is selected at certain times and neglected at other times. Further refinements of the principles described below permit selection of targets moving at specific minimum or maximum velocities, and this permits radar systems to distinguish between such slow-moving targets as flocks of birds and such fast-moving targets as jet aircraft. MTI operation also permits the elimination of fixed terrain features from the PPI display so that moving targets—such as vehicles, or a man, or aircraft—are more easily detected. The basic principles of pulsed Doppler radar systems and the MTI feature are described below.

A simple target MTI-pulsed Doppler radar is shown in block diagram form in Figure 13-7. The generation of pulses, modulation, duplexing, and receiver are basically the same as in a pulsed radar system. The key to the MTI feature lies in the two blocks labelled STALO and COHO. The former abbreviation stands for *stable local oscillator*, and the abbreviation COHO stands for *coherent oscillator*. The STALO provides a very stable frequency reference to the receiver mixer and the locking mixer, which in turn correlates the output of the high-frequency oscillator with the phase of the STALO. The coherent oscillator provides a continuous signal which contains the phase reference of the transmitted signal, based on the STALO and the actual transmitted signal. This coherent signal is necessary because of the pulsed operation and the fact that there would be no reference signal between pulses.

The echo signal is passed through a mixer and IF amplifier and is then supplied to a phase detector. Here its phase and frequency are

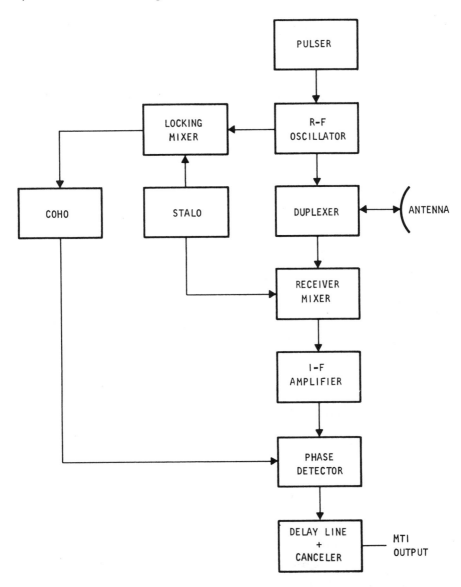

Figure 13-7. Block diagram of pulsed Doppler system.

(Gershon J. Wheeler, *Radar Fundamentals*, © 1967. Reprinted by permission of Prentice-Hall, Inc.)

compared to the coherent signal and an output is provided to the delay line and canceller section, which then generates the MTI indication as well as the actual range indication. In a typical pulsed Doppler radar, there are a number of additional feedback loops, timing control networks, and logic functions not shown in the simplified diagram of Fig-

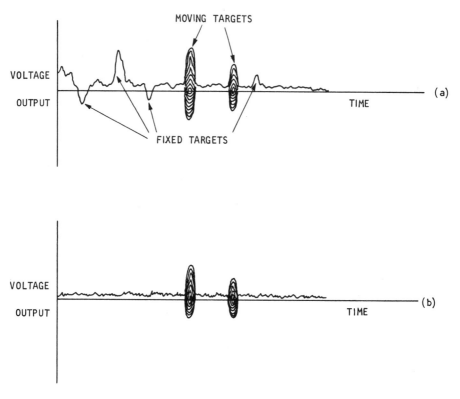

Figure 13-8. Pulsed Doppler output signals.

(Gershon J. Wheeler, *Radar Fundamentals*, © 1967. Reprinted by permission of Prentice-Hall, Inc.)

ure 13-7. Figure 13-8 illustrates two of the typical output signals of a pulsed Doppler radar system as they would appear on an A-scope presentation. When the signals are put through the delay line and canceller section, repetitive signals, such as the fixed targets, are delayed one pulse repetition period, and if they appear the same in the next pulse repetition period, they are cancelled out. This results in the presentation shown in Figure 13-8b, where the fixed targets, which always appear at the same amplitude and phase, are cancelled out while only the moving targets are shown.

A version of the pulsed Doppler radar which operates on amplitude variations only, instead of phase, does not need a COHO element and is therefore called noncoherent. In this system, fixed targets are separated from the moving targets by comparison of the frequency content of each echo, and a noncoherent system therefore depends on fixed or stationary targets for its cancellation.

Pulse Compression Systems

The velocity of propagation of radar signals is roughly 1000 feet per microsecond. This means that when two targets are close enough to each other so that their distance corresponds closely to the pulse width of the transmitted radar pulse, they will not be distinguishable. Pulse width, therefore, determines the degree of fine resolution of targets close to each other. In the case of aircraft flying in formation, this can present a quite a problem. It is therefore desirable to have radar pulses as short as possible to distinguish closely spaced targets, while having them long enough so that they are easily separated from the noise. One method of improving range resolution is to compress the transmitted pulse and then expand the received pulse electronically in the receiver.

Pulse compression can be accomplished by frequency modulation of the transmitted pulse. If, for example, the transmitted pulse is 3 microseconds in duration, it is possible to change frequency linearly from f_i at the start to f_z at the end, 3 microseconds later. The return pulse will have the same characteristic and can be filtered from the noise by means of a matched filter in which the velocity of propagation depends on the frequency. The lower frequencies at the start of the pulse are slowed down with respect to the higher frequencies at the end of the pulse, so that instantaneous received energy can be added up, and the output of the pulse compression filter is a shorter pulse with increased amplitude. Compression ratio is usually given in terms of the frequency change times the pulse period. Compression ratios of $100:1$ or greater are frequently possible, and this would mean that the 3-microsecond pulse of the above example could be compressed to 30 nanoseconds, corresponding to a range resolution slightly greater than 15 feet.

Pulse compression systems are usually used with long distance, high-power radars and can be provided with moving target indication (MTI) features as well. For MTI operation, however, the difference between the frequency modulated pulse and the resulting Doppler signal must be carefully filtered, and this implies complex and very sophisticated filtering circuits.

Interrogator-Transponder Systems (Beacons)

Essentially consisting of two separate radar systems, the interrogator and the transponder, or beacon, each contains its own receiver and transmitter. The interrogator transmits a coded pulse signal which is received and decoded by the transponder. When the transponder receives the correct address or interrogation code, it replies by transmitting another set of pulse coded signals. The interrogation frequency is usually somewhat different from the response frequency, although it is possible to have the both the interrogation and the reply on the same frequency.

One of the major advantages of this system is that the transmitter power and receiver sensitivity can be less than in other radars, since the return is not an echo, but the signal generated by the transponder. A second advantage is that it permits identification both of the interrogator and the transponder, and this system is therefore used for radio navigation, identification—friend or foe (IFF), and altitude and identity transponders in civilian aircraft.

In the navigation system a moving aircraft sends out interrogation signals, and when a fixed beacon on the ground receives them, it responds with its identity and geographic location. This permits the navigator to determine the range and exact location of his aircraft. In IFF operations, elaborate coding schemes are used to protect the password so that only friendly aircraft can respond with the proper coded reply. All major airports handling passenger aircraft use the transponder system to provide altitude information, as well as the aircraft's identification, which is then displayed to the ground controller. Commercial aircraft in the United States must carry transponders for this purpose.

13.3 RADAR ANTENNAS

Chapter 10 is devoted entirely to antennas and transmission lines and contains the fundamentals which apply to the specialized antennas required in radar systems. Chapter 11 deals with microwave transmission lines, antennas, and circuits; since most radars operate in the microwave region, the reader is directed to review the principles presented in Chapter 11.

Almost all radar antennas are intended to provide clearly defined radiation patterns. Except when operating in an interrogator-transponder system, where one of the two stations will be omnidirectional, almost all other radar functions require a relatively narrow beam, either in the vertical or horizontal plane or, in many instances, in both planes.

A theoretical isotropic radiator (dipole) is considered to have a basic gain of 1. Practical antennas radiate in specific directions, and therefore can be expected to have a gain greater than 1. Because the entire area of an antenna is generally not completely coupled to the transmission line, the concept of the effective area, or the efficiency of illumination, is introduced. The gain of an antenna can be calculated from the following formulas, in which η is the efficiency of illumination as defined below.

$$\eta = \frac{A_e}{A}; \quad A_e = \text{effective area of antenna}$$

$$\text{Gain} = G = \frac{4\pi A_e}{\lambda^2} = \frac{4\pi\eta A}{\lambda^2}; \quad \lambda = \text{wavelength of transmitted signal}$$

Waveguide Antennas

At frequencies above 1 GHz, the wavelength becomes small enough so that the signals can be transmitted over rectangular or circular waveguides as well as coaxial cables. Waveguides are described in detail in Chapter 11. When a waveguide suddenly stops, the transmitted signal will travel from the confined space of the waveguide into free space and will suffer some loss due to the transition.

When a specific window is prepared to improve the impedance of this transmission, the basic waveguide antenna, called the *slot antenna*, is obtained. Slot antennas are not very efficient and are rarely used in practical radar systems. A modification of the slot antenna is the waveguide horn antenna, which provides a more gradual transition from the waveguide dimensions to the electric and magnetic field configuration in free space. While horn antennas provide a great improvement over slot antennas, they usually do not furnish the beam characteristics desired in radar and are therefore used, usually, in conjunction with other devices, such as the parabolic reflector or various types of lenses discussed below.

A typical antenna beam is shown in Figure 13-9, complete with the main lobe and the side and rear lobes. The beam width of the antenna is measured at the points where the intensity is 3 dB down from the maximum, and, in the example of Figure 13-9, the beam width is approximately 24 degrees. The beam pattern also indicates that the two major side lobes are 15 dB down from the main lobe, with the two back lobes being approximately 22 dB down.

Parabolic Reflector Antennas

Parabolic reflectors are used in optical systems to provide a parallel beam from a point source. The basic parabolic reflectors shown in Figure 13-10 would have a slot antenna or a waveguide horn located at point F. Actual radar systems contain a large variety of parabolic re-

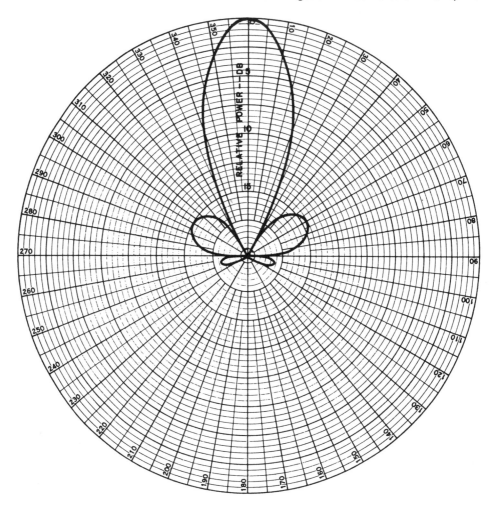

Figure 13-9.   Radar beam pattern.

(Gershon J. Wheeler, *Radar Fundamentals*, © 1967. Reprinted by permission of Prentice-Hall, Inc.)

flectors, some with only portions of a parabola, some with modified parabolic characteristics, and some with the parabolic characteristics available only in either the horizontal or vertical plane.

Lens Antennas

Many radars using microwave frequencies are able to employ optical principles, since microwaves, just like light waves, are electromagnetic radiation. The principle of the lens lies in the fact that the dielectric material of which it is made has an index of refraction (μ) which equals

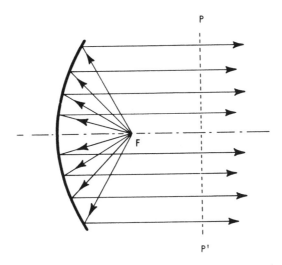

Figure 13-10.   Parabolic reflector principles.

(Gershon J. Wheeler, *Radar Fundamentals*, © 1967. Reprinted by permission of Prentice-Hall, Inc.)

the square root of the dielectric constant (ϵ). μ represents the difference in velocity of propagation of electromagnetic waves between free space and the dielectric material. For microwave antennas, typical values of μ range around 1.5.

The wavelengths used in radar generally require lenses that are quite bulky and heavy; for this reason, a method called *zoning* or *stepping* is used to reduce the bulk of the lenses. A typical lens is reduced in thickness by using only steps so arranged that the maximum depth of each step is one wavelength or an even fraction thereof. This process of stepping reduces the bandwidth of the lens and also affects the general pattern of the resultant beam.

In addition to natural dielectrics, such as polystyrene and similar plastics, it is also possible to produce an artificial dielectric by either one of two methods. An artificial dielectric antenna is obtained by locating small metal objects, spheres, cylinders, etc., much smaller than one wavelength, at controlled locations, forming an array which is then the effective dielectric lens. The other approach uses a plastic dielectric that contains empty spaces, each of which is considerably smaller than one wavelength.

Typical of practical radar lens antennas is the so-called Luneberg lens, essentially a dielectric sphere with a point source, usually a waveguide horn antenna, at the perimeter. This type of lens is capable of providing a narrow pencil beam, comparable in many respects to that obtained from a parabolic reflector assembly.

Other Radar Antennas

Chapter 10, Antennas and Transmission Lines, describes a number of basic antennas capable of producing selected beam shapes. Almost all of these antennas are found in one type of radar system or another. Low-frequency radars, intended to go through the Heavyside layer in the ionosphere, generally use large, fixed arrays, in which the antenna elements are switched electronically to form beams and move them as required. Helical antennas, as well as the yagi antennas, are used not only in television, but also in many radar applications, usually in conjunction with some reflecting or focusing elements. Even rhombic and so-called long wire antennas are used for low-frequency radars, particularly in the navigation field.

Antenna Beam Motion

The basic search antenna beam moves in a circle because the basic radar antenna simply turns on its pedestal at a regular rate, providing a PPI display of a 360-degree horizon. A basic height finder radar moves in a nodding motion up and down to scan a sector of the sky and pick up targets at different angles of elevation. Some radars combine vertical and horizontal motion. Other radars combine a circular scan with a conical rotation to perform a track-while-scan operation. In the conical scan mode, the antenna beam is shifted, either electronically by shifting the frequencies slightly or mechanically by shifting the horn antenna with respect to the focal point of a parabolic reflector, in such a way that the antenna beam swings back and forth, or rotates in a narrow solid angle, while the entire assembly is rotated in azimuth. The conical motion is used when tracking a target because it provides different levels of echo amplitude as the target is first illuminated by the center of the lobe and then each side of the lobe. This permits clear indication of accurate target location. By adding and subtracting the signals received at either side of the antenna lobe and dividing them, the center of the antenna beam is accurately determined.

Many long-range radars use large arrays of antennas and reflectors to provide so-called *electronic scanning*, in which the motion of the electron beam is generated by switching the transmitter output signal among different antenna elements. This method also permits changing the shape of the electron beam electronically.

One of the most widely known characteristics of radar systems is the *radome*, generally the dielectric enclosure of the moving radar antenna. Radomes are designed to provide minimum loss to the radar signal and to give protection against wind, rain, and other elements for the moving antenna. This is particularly important in aircraft, where the radar antenna would otherwise be subjected directly to very strong airflow. Aside from the small amount of attenuation, radomes do not generally play an important part in the electronic aspects of the radar systems they house.

13.4 NAVIGATION SYSTEMS

Navigation systems can be divided broadly into three major types. In one type, a moving vehicle navigates by direction finding. In another type, navigation is performed by measuring distances to fixed beacons. In the third type, the difference in distance between fixed stations is measured. Combinations of these three systems, including the use of space satellites as beacons, make up most of the presently implemented navigation systems. A fourth type, in very limited use in military aircraft, uses a pulsed radar to map the ground below and compares this radar map with a cartographic map.

Direction-Finding Systems

Direction finding (DF) is based on the directivity of either a transmitted or received antenna pattern and uses azimuth (direction) as its main parameter. A simple loop antenna, aimed at a known transmitter location, will result in obtaining a bearing or azimuth, as described in Chapter 10. When two or three different bearings can be obtained, the intersection of these bearings reveals the location of the receiver. The VHF Omnidirectional Radio Range (VOR), used in the continental United States, provides bearing information for aircraft of all classifications. VOR stations operate on twenty different channels in the 108- to 118-MHz range. The transmitter antenna pattern is rotated at 30 Hz. A separate antenna, operating at the carrier frequency, is modulated at 9960 Hz with AM and a 30-Hz superimposed FM signal. The receiver, mounted in the aircraft, contains circuitry to resolve the AM and FM information, together with the varying intensity due to the rotating antenna pattern, and produces accurate bearing information from a single VOR station.

Distance-Measuring Equipment

The basic distance-measuring equipment (DME) consists of fixed beacons located on the ground, which respond to interrogations by an aircraft radar. When illuminated by radar signals, the beacon transponder returns a set of pulses that permit the aircraft radio to measure the time, and therefore the distance, between the interrogation and the response. The returned pulse pattern identifies the particular beacon as to its location.

DME beacons in the United States receive on 126 channels ranging from 1025 to 1150 MHz, and respond on 126 channels in the band from 962 to 1024 MHz and 1151 to 1213 MHz. These beacons have omni-directional antennas.

A special-purpose DME system is the Shoran system, which operates in the 300-MHz region and offers accuracies to 75 feet.

The Tacan system, originally used on aircraft carriers, combines the features of the DME with an advanced type of direction-finding system. A rotating cardiod antenna pattern, superimposed on another antenna pattern which contains nine equally spaced lobes, provides the accurate bearing information.

A combination of VOR (see above) and Tacan is called VORTAC; it is in limited use, mostly for military applications. Both Tacan and VORTAC provide range and bearing information with great accuracy from a single fixed station.

To provide accurate position in the standard DME system, it is necessary to measure range to two separate beacons. The resulting over-lapping circles are then plotted, and possible ambiguities are resolved by knowing the original path of the aircraft.

Microwave Landing System

The International Civil Aviation Organization (ICASO) has approved a microwave landing system (MLS) that is currently being developed and is expected to be operational worldwide by the end of this century. The new system will replace the current instrument landing system (ILS) now in use throughout most of the world. The MLS is based on time-referenced scanning beams referenced to the airport runway that enable the aircraft unit to accurately determine azimuth and elevation angles. The azimuth and elevation angle functions are provided by 200 chan-nels in the 5-to-5.25-GHz band. Range information is provided by dis-tance-measuring equipment (see above) operating in the 960-to-1215-MHz band.

Satellite Navigation Systems

Two satellite navigation systems (one in use, the other being developed) are the TRANSIT system operated by the U.S. Navy and the NAVSTAR/

Global Positioning system being developed by the U.S. Department of Defense (DOD).

The Transit satellite navigation system broadcasts computed positions continuously at 150 and 400 MHz. A receiver measures sequential Doppler shifts of the signal as the satellite approaches or passes the vessel. The geographic position of the vessel is calculated using the satellite position data (received from the satellite every two minutes) and the Doppler shift measurement. Generally, only the 400-MHz frequency is used for navigation purposes.

A far more accurate system (depending on the capability of the user's equipment) is the NAVSTAR/Global Positioning system being developed by the DOD. This system is expected to become operational in the late 1980s (about 1989). The projected positioning accuracy that will be available to civilian users is 152.4 ft (500 m). The satellite signals are transmitted at two frequencies, 1575.42 and 1227.6 MHz. The two transmissions are for the purpose of correcting for ionospheric delays in propagation (as are the two signals in the Transit system). In this system the two signals also are continuously modulated with a navigational data-bit stream (50 bits per second) plus additional signals that require sophisticated equipment and will provide greatly improved positioning accuracy—approximately 6.069 ft (20 m) horizontally and 9.144 ft (30 m) vertically.

Long-Range Navigation

Long-range navigation (LORAN) depends on measuring the time difference of arrival between two signals originating from two different fixed stations. When all possible points are plotted at which the difference in distance between two fixed points is equal, a hyperbola results. In the LORAN system, the two fixed points are made up of a master and slave pulse transmitter. The master transmitter sends a pulse signal along the ground wave, direct path to the slave station to synchronize it to transmit a pulse at the same frequency. An aircraft or a ship receiving both signals will first receive the pulse from that station to which it is closest and then will receive the second pulse. The difference in the time of reception of the two pulses is equivalent to the difference in distance the aircraft is located from the two fixed stations. Detailed charts for all possible time differences within the transmitting range of any pair of LORAN stations have been plotted and published.

To locate itself, the aircraft or ship must receive a second pair of transmitters, one of which may belong to the first pair of stations, so that two hyperbolas are obtained. The point where the two hyperbolas intersect indicates the instantaneous position of the aircraft or ship. In practice, LORAN receivers provide a direct digital readout of the time difference of the first and second pair of LORAN stations; the navigator merely looks up these two numbers on a table and can then plot his position directly on a chart.

In the LORAN-A system, the transmitted pulses are 45 microseconds in duration and occur at a rate of 20 to 34 per second. Fixed stations operate at 1850, 1900, and 1950 kHz. Because of the ground wave transmission characteristics at these frequencies, master and slave stations are usually located approximately 300 miles apart. In the LORAN-C system, the same principles are used, but the transmitted frequencies are in the 100-kHz region, permitting master and slave stations to be much farther apart, and therefore allowing a smaller number of LORAN stations to cover a larger area. Currently, there are 17 chains consisting of 50 transmitting stations.

The Omega navigation system uses the LORAN principles, but instead of pulses, it uses continuous wave navigation (CW) signals and depends on phase differentials rather than pulse-time delays. The Omega system operates in the 10-kHz band; this makes it possible to locate master and slave stations as much as 5000 miles apart.

The British Decca system has the basic LORAN geometry and uses continuous wave (CW) signals, with differential phase detection in place of pulse time delays. In the Decca system, the master and slave stations operate at different frequencies which, however, are harmonically related to each other.

BIBILOGRAPHY—CHAPTER 13

BARTON, D., and H. WARD, *Handbook of Radar Measurements*. Englewood Cliffs, NJ, Prentice-Hall, Inc., 1969.

BECK, G. E., *Navigation Systems—A Survey of Modern Electronic Aids*. London, Van Nostrand Reinhold Co., 1971.

BIRD, G. J. A., *Radar Precision and Resolution*. New York, NY, John Wiley & Sons, Inc., 1974.

Reference Data for Engineers: Radio, Electronics, Computer, and Communications (7th Ed.). Indianapolis, IN, Howard W. Sams & Co., Inc., 1985.

SKOLNIK, M., *Introduction to Radar Systems*, (2nd Ed.). New York, NY, McGraw-Hill Book Co., 1980.

CHAPTER 14

COMMUNICATIONS SYSTEMS

14.1 BASIC COMMUNICATIONS PRINCIPLES

Communication has been defined as any form of transmission of information between two or more sources and destinations which implies a two-way action. In the broadcast systems described in Chapter 12, information is usually transmitted in one direction without necessarily a return of information. Chapter 14 deals exclusively with communications intended for a specific recipient, and some means of confirming the receipt of this information is always included.

Fundamentals of Information Theory

The specialized field of information theory (and its various aspects as they apply to communications) is the subject of an extensive literature. The reader interested in details is referred to the references appearing

335

at the end of this chapter. To understand the basic principles of communications, only the fundamentals of information theory are required.

Information is defined as the reduction of uncertainty. When uncertainty does not exist, no information can be transmitted. If the exact time of day is known, no information is obtained from looking at a clock. Based on this definition of information, a unit of information is defined as the selection between two equally probable choices, and this unit of information is called a "bit." The pilot light of a power supply supplies one bit of information because it is illuminated when the supply is on and is not illuminated when power is shut off. This selection between two equally probable choices results in a binary system, a system based on two equally probable events. If there are N equally probable choices, where N is any power of two, then the total amount of information equals $H = \log_2 N$. In the more general case, the amount of information H is expressed as follows:

$$H = -\sum_{i=1}^{k} p(i) \cdot \log_2 p(i)$$

In this expression, $p(i)$ is the probability distribution of equally probable choices and i and k are its boundaries. Note that $\log_2$ is used, which is different from $\log_{10}$ and from ln (natural log). When messages of length n are transmitted, made up of k choices, the overall probability and the resulting information content of the messages is obtained as shown below.

Overall probability $P = p_1^{np1} \times p_2^{np2} \dots p_n^{npn}$;

$$H_n = \log_2 \frac{1}{P}$$

H is also called the entropy of the source, an expression derived from thermodynamics, indicating the uncertainty or randomness of a system. An explanation of the concept of entropy is contained in the references listed at the end of this chapter. H is zero when only one choice can occur, that is, when the probability equals 1. H will be maximum when all choices have equal probability. In communications, where the English alphabet is used, H has been computed for the 26 letters and one space, and $H = \log 27 = 4.76$ bits per letter or symbol. Similar calculations can be made for any other array of symbols.

Elements of Information Transfer Systems

Having defined communications as the transfer of information between two or more sources, an information transfer system must consist of the following basic functions:

1. *Transmitter*—uses a source-to-signal encoder, a transducer such as a microphone, teletypewriter, TV camera, etc., to change the information into an electrical signal. Next is a signal-to-signal encoder, a modulator of some kind which changes the electrical signal into a format suitable for transmission. The third element of a transmitter is the transmitter-to-medium encoder, such as an antenna which converts the electrical RF signal into electromagnetic radiation traveling through space.

2. *Medium*—the pathway of the information going from the transmitter to the receiver; it may consist of a variety of different phenomena. In voice communications between people, the medium is acoustic variations in the pressure of air. In electromagnetic radiation, the medium may be the atmosphere, various layers of it, or space itself. Optical communications also use electromagnetic radiation, but at a much higher frequency. Most electronic communication uses either electromagnetic radiation or wires, coaxial cables, waveguides, etc.

3. *Receiver*—uses a medium-to-receiver decoder, such as the receiving antenna or line circuit. Next there is a signal-to-signal decoder, a demodulator to remove the information signal from the signal used for transmission, the carrier. The third element of the receiver is the signal-to-destination decoder, another transducer such as a loudspeaker, a cathode-ray tube, a teletypewriter, etc.

In considering the operation of the three elements of information transfer systems—transmitter, medium, and receiver—it is apparent that the information will change somewhat between its source and destination. Each of the three functional elements will contribute some loss and some additional noise. The transmitter and the receiver are usually designed to compensate for the loss incurred in each of them and in the medium by means of suitable amplifiers. Similarly, the noise introduced by the transmitter and receiver can be carefully controlled by proper design. The noise contributed by the medium, however, is usually beyond the control of the communications systems designers. Chapter 11 describes the different types of external noise as well as the fundamentals of internal receiver noise. In designing information transfer systems or communications channels, the effects of noise form one of the most important parameters.

Information Transducers

Communications systems are usually designed to convey information from one human being to another or to a group of others. Even in the instances of communications systems between computers, the ultimate source of information and destination is human beings. Human beings receive and transmit information primarily through the senses of sight and hearing, although tactile (touch) sensations have been used for communications between blind people. A variety of transducers is used to change the acoustic and visual information into electrical signals. Chapter 17 deals with transducers and control elements, primarily those used in industrial electronics, but the fundamentals of these transducers are also used in communications. For converting acoustic information into electrical signals, microphones of a variety of different types are used. To convert electrical signals into acoustic information, loudspeakers and earphones have been in use for many years. Visual information is converted into electrical signals by means of TV cameras. Some systems use an indirect process in which photographic cameras convert a picture into a photographic image which can then be scanned and converted into electrical signals. To convert electrical signals back into pictorial information, cathode-ray tubes, such as TV picture tubes and liquid crystal displays (LCDs) can be used. A modulated light beam can scan a photosensitive or heat sensitive surface, and subsequently a photographic or thermographic image of the original visual information can be obtained.

An intermediate step is to convert the electrical signal obtained from the transducer into a magnetic pattern of thin plastic, magnetically coated tape or into some other kind of recording medium such as a video cassette recorder (VCR). This is simply a means of storing the information between the time it emanates from its human source to the time when it is transmitted.

All transducers, no matter how excellent in their performance, introduce some limitations and some noise to the original information. When a microphone limits the frequency band over which it efficiently converts sound into electrical signals, it automatically limits the information content of the original source. Almost all communication systems impose some limitation on the total information content, but if the system is properly designed, the amount of information transferred is sufficient for the human recipient to understand fully the original information.

The transfer of written information is somewhat different from that of acoustic or visual information because it is inherently limited to the symbols used in the writing or printing process. Little information is lost, for example, between a hand-written telegram and the printed-out version delivered to the recipient. As long as all the letters and numerals are properly transmitted, the message to the recipient is exactly the same, whether he receives it in handwritten or typewritten form. The computer, most keyboard devices, and computer peripheral equipment such as line printers usually operate on the principle of digital rather than analog information. By using a limited set of agreed-upon symbols, such as the English alphabet and Arabic numerals, it is possible to obtain great economy in information transfer systems. In analog communications it is important to transfer the information as closely as possible to the form in which it is originated at the source. In digital communication systems it is only necessary that the correct individual symbols be transmitted, regardless of their form. The size or form of letters and numerals does not materially contribute or detract from an understanding of a printed message.

Fundamentals of Modulation and Coding

Another functional element in a transmitter and receiver is the signal-to-signal encoder and decoder, the device that converts the electrical signal into a form suitable for transmission and back again into the electrical signals. If the medium is electromagnetic radiation through the atmosphere or space, this automatically means that the original information signal has to be encoded onto an RF signal. Most long distance wire transmission systems also require some kind of carrier signal, and modulation and demodulation are therefore essential functions in almost all communication systems.

It is theoretically possible to modulate a carrier signal in only three basic ways. As illustrated in Figure 14-1, a carrier encoder can use amplitude modulation, frequency modulation, or phase modulation. Although audio and video information could be used as well, for simplicity of illustration Figure 14-1 shows only a pulse-type binary signal as the original information. Each of the three basic modulation schemes requires different types of circuitry, produces different sidebands, acts differently under different noise conditions, and is applied in a number of different variations. The simple amplitude modulation (AM), for example, can be used as a single sideband, suppressed carrier; the frequency modulation (FM) can be used in a double FM system or as frequency shift keying (FSK); and the phase modulation (PM) can be used as phase shift key (PSK), with a variety of different phase angle combinations. Detailed discussions of these techniques and their uses in a wide variety of applications will be found in the references at the end of this chapter. The fundamental characteristics of each, however, are presented here briefly.

Amplitude Modulation (AM). Amplitude modulation results in a signal containing the original carrier signal, plus one sideband higher

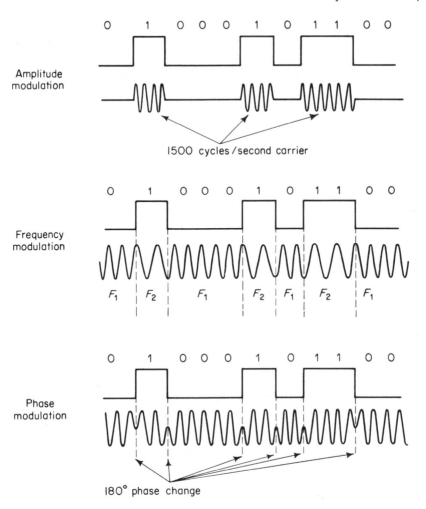

Figure 14-1. Three basic modulation methods.

(James Martin, *Telecommunications and the Computer,* © 1969. Reprinted by permission of Prentice-Hall, Inc.)

in frequency and one lower in frequency than the carrier. The same information is carried in both sidebands. The amplitude relation of the information signal to the carrier is expressed as the modulation factor or modulation index. Figure 14-2 shows the appearance of an AM carrier when the modulation index is 1.0 or when it is greater or less than 1.0. For most efficient operation, a modulation index close to but not quite 1.0 is generally selected. Since the AM carrier contains no information, it is suppressed in many systems just to save transmission power. The upper and the lower sidebands both contain the same information and, where bandwidth is a problem, one or the other is usu-

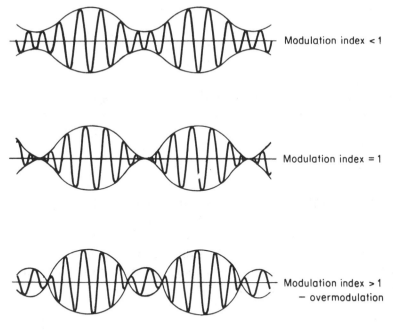

Figure 14-2. AM modulation levels.

(James Martin, *Telecommunications and the Computer*, © 1969. Reprinted by permission of Prentice-Hall, Inc.)

ally suppressed. Single sideband, suppressed carrier AM provides a noise factor improvement of 4 dB over double sideband AM. Because the carrier is suppressed, however, the detection method cannot be a simple envelope detector, but a synchronous or coherent detector circuit is required, and this usually means that the suppressed carrier must be reconstituted or locally generated. In simple AM detection, the received signal is rectified and the carrier is filtered out.

Frequency Modulation (FM).   Frequency modulation is much less susceptible to noise in the medium than AM, because amplitude variations are disregarded at the receiver, and only the variations in frequency originated at the transmitter encoder contain information. The modulation index of an FM signal is the ratio of the maximum frequency deviation that can occur to the frequency of the information signal. For transmission of binary data, two versions of FM are used. In one version two different tones are used, one to indicate a zero and one to indicate a one, and the carrier is alternately modulated with either of these tones. The second method is to shift the frequency of the carrier itself between two points, according to the zero or one condition.

Phase Modulation (PM). Similar to frequency modulation, phase modulation is able to provide a somewhat better performance with respect to noise than FM. Phase and amplitude modulation together are used to carry the color information in commercial color TV systems, as described in detail in Chapter 15. In communication systems, the most frequently used type of phase modulation is called phase shift keying (PSK), in which a 90-degree or 180-degree change of phase of the carrier signal indicates the beginning of a zero or one. PSK provides a 3-dB noise factor improvement over FSK.

In addition to the above three basic ways of modulating a carrier, it is also possible to transmit information in the form of a number of pulses. In radar systems, as described in Chapter 13, the RF signal is pulsed in order to obtain the range information. In communications systems, pulse information can be modulated on a carrier in either the AM, FM, or phase modulation techniques. Regardless of the modulation methods, four different basic methods of converting information into the form of pulses are available.

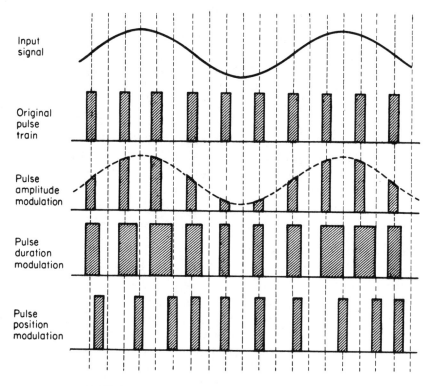

Figure 14-3. Pulse modulation methods.

(James Martin, *Telecommunications and the Computer,* © 1969. Reprinted by permission of Prentice-Hall, Inc.)

Figure 14-3 shows how an input sine wave signal and a pulse train can be used to provide pulse amplitude modulation (PAM), pulse duration modulation (PDM), and pulse position modulation (PPM). Mathematical analysis of these three techniques shows that there is great similarity between amplitude modulation, frequency modulation, and phase modulation, as discussed above. Each of these three techniques has different advantages and disadvantages for different types of systems.

The fourth technique, pulse code modulation (PCM), is illustrated in Figure 14-4 and involves coding the amplitudes of a sampled signal. In the example shown in Figure 14-4, it has been assumed that only

① The signal is first "quantized" or
  made to occupy a discrete set of values

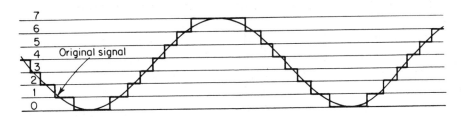

② It is then sampled at specific points. The
  PAM signal that results can be coded
  for pulse code transmission

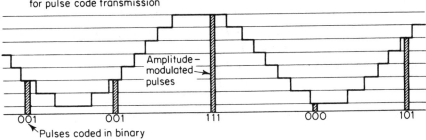

③ The coded pulse may be transmitted
  in a binary form

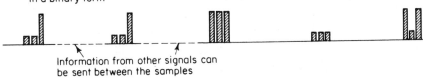

Information from other signals can
be sent between the samples

Figure 14-4.   Pulse code modulation.

(James Martin, *Telecommunications and the Computer,* © 1969. Reprinted by permission of Prentice-Hall, Inc.)

seven different amplitude levels will be required to specify a particular information signal. Each level of amplitude corresponds to a binary number which will then be transmitted in binary form. As shown in the illustration, on line 3 the pulse corresponding to a zero is shown at a low level and the pulse corresponding to a one at a high level. The frequency of sampling the original signal will depend on many factors, but for simplicity's sake only a few samples have been shown in Figure 14-4.

The basic PCM technique described here can be greatly improved by adding complexity, redundancy, parity bits, and other features to reduce the likelihood of errors.

Communication Channels

Communication channels are generally considered the band of frequencies over which messages, in analog or digital form, are transmitted. The basic problem of how much information can be sent, in how short a time, over a given bandwidth, has occupied research scientists and engineers for considerable time; most of the modulation and coding methods are designed to provide the maximum channel capacity, consistent with particular transmission characteristics. One way of looking at channel capacity is to realize that, with a bandwidth Bw, it is possible to carry two times Bw the number of separate voltage values per second. In simple teletype, where the voltage has one of two possible values, one can send two Bw bits per second. By using four possible voltage levels (L) at any one instance, it is possible to use the two Bw voltage values per second to code four Bw bits per second. The channel capacity, neglecting noise, is then:

$$C = 2 \text{ Bw } \log_2 L$$

When noise is considered, channel capacity changes and

$$C = \text{Bw } \log_2 \left(1 + \frac{S}{N} \right)$$

To determine the number of data bits which can be sent over a channel in a given time, simply multiply the channel capacity by the time. For example, a 2600-Hz bandwidth and a 30-dB signal-to-noise ratio result in a channel capacity of 25,900 bits per second. This assumes that the sequence of symbols or bits is unpredictable, that is, each subsequent choice is equally likely. Many different coding techniques have been devised to increase channel capacity in the presence of noise by making the appearance of some information dependent on the past appearance of others, by introducing cyclical variations and other means to improve

noise characteristics. Figure 14-5 shows the relationship between channel capacity and bandwidth for different levels of Gaussian noise. Gaussian or white noise is the type of random noise due to thermal effects.

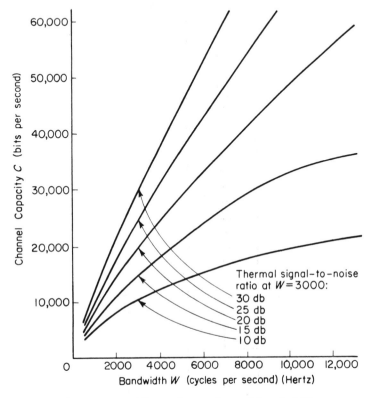

Figure 14-5.   Channel capacity and bandwidth.

(James Martin, *Telecommunications and the Computer*, © 1969. Reprinted by permission of Prentice-Hall, Inc.)

14.2 DIGITAL COMMUNICATIONS

A basic block diagram for a digital communications system is illustrated
in Figure 14-6. The transmitter consists of analog source, analog/digital
converter, and modulator. The receiver consists of demodulator, digi-
tal/analog converter, and the analog termination.

Typically, a *null modem adapter* allows two digital devices such
as microprocessor units (MPUs) to communicate directly or by use of
a telephone. In this case, the modem (modulator/demodulator) changes
a digital signal to a pulsed audio tone. Frequently, this two-level signal

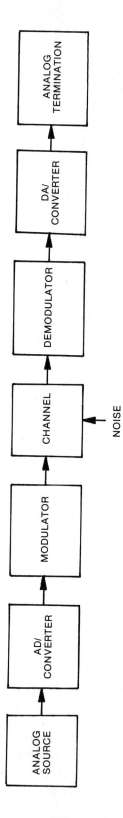

Figure 14-6. Digital communication system block diagram.

is called *space and mark* and, in many modems, the pulses are used to modulate an audio tone (or tones) that can be transmitted to some type of receiving device.

Digital Modulation Techniques

Amplitude-Shift Techniques.   A sequence of pulses whose amplitudes are modulated to represent the messages, commonly referred to as digital AM or ASK. A basic example is binary ASK modulation, that is, on and off, called *on-off* keying.

A modification of ASK is known as quadrature ASK, also referred to as quadrature AM—in other words, the modulation of two AM carrier components 90 degrees apart in phase by separate modulating functions. However, in some applications the signal consists of sequences of binary pulses derived from a single binary input. In this case, odd-numbered bits go to the in-phase channel and the even-numbered bits go to the out-of-phase channel (the quadrature channel).

Binary Phase-Shift Keying.   It is possible (and often done) to transmit a binary coded message by shifting the phase of an RF carrier by ±90 degrees, depending on whether the data bit is a zero or a one. This technique may be called *binary phase modulation* or *binary amplitude modulation*. The only requirement is that in reference to the two sinusoidal signals, they must differ in phase by a total of 180 degrees (±90 degrees in reference to the carrier).

Quadriphase Shift Keying.   Basically, this method of modulation is the same as binary phase-shift keying except it allows the phase waveform to be modulated with more than one set of off and on signals—that is, in sets of four. In other words, quaternary modulation is used instead of binary modulation.

There is one very important advantage to using quadriphase shift keying rather than binary phase-shift keying if we restrict the discussion to only the bandwidth required. Binary phase-shift keying requires twice the bandwidth of quadriphase shift keying. Furthermore, it has been proven that if the two signals have equal power and the same baud rate, they also have the same energy per data bit.

Binary Frequency-Shift Keying.   In the beginning of this section, a null modem adapter was discussed. The modem changes a digital signal to pulsed audio tones for transmission over a telephone, for example. A somewhat similar scheme is used to produce binary frequency-shift keying. In this case, the signals may be generated by switching between two oscillators or by applying a binary pulse train to a device such as a voltage-controlled oscillator, or to any other type of frequency modulation circuit. As was mentioned in the preceding pages discussing the modem, the signals are often referred to as tones, and are called *mark* and *space* (mark = binary digit 1, space = binary digit 0).

Although there are a large number of different digital modulation techniques, in general, even the most sophisticated—such as *multiple frequency shift keying*—are designed using binary frequency-shift keying to modulate with frequency tones. However, the commonly used binary phase-shift keying only requires two sinusoidal signals; M-ary systems utilize more than two.

Data Communications Systems

Communications systems between computers and their remote terminals usually operate much faster than the keyboard type of teletype systems, although both use PCM. Individual characters, consisting of a number of bits, can be transmitted either in parallel or in series. For parallel transmission over short distances, individual wires—one per bit—are sometimes used. For parallel transmission over long distances, frequencies are assigned to each bit. At the receiver, the individual frequencies are detected and their presence, or absence, indicates the one or zero condition of the particular bit.

Series transmission requires five, seven, or eight serial bits to transmit a single character. By speeding up the rate at which these bits are transmitted, it is possible to achieve the same transmission speed as in the parallel, separate frequency method.

Series and parallel transmission both require a modulator to encode the original information onto the carrier, and a demodulator at the receiver to convert the carrier information into the true digital signal. Since each station can receive and transmit, these devices are incorporated into a single unit, generally known as a modem, as was explained in the beginning of Section 14.2. Standard modem speeds for transmission over telephone circuits are 600, 1200, and 2400 baud. For higher speeds, the Bell System Data Set 301B is available, which operates at 40.8 kilobaud. This is accomplished by using a 30.6-kHz carrier which is four-phase modulated and covers the band from 10.2 to 51 kHz. So-called wide-band channels of this type are available in all major cities, and usually offer the following wide-band facilities for transmission:

Series 800 channel: 48 kHz (12 voice channels)
Type 5700 channel: 240 kHz (60 voice channels)
Type 5800 channel: 1 MHz (240 voice channels)

In transmitting digital data in series, it is possible to transmit them asynchronously, as described above for teletype signals; that is, each character is preceded by a start and followed by a stop signal. The requirement for start and stop signals for each character reduces the maximum transmission speed, and synchronous transmission is, therefore, used for high-speed transmission. In synchronous transmission, the receiving station relies on exact synchronization with the transmitting station for its ability to decode the message properly. Most synchronous systems use precise oscillators, controlled by phase-locked loops and similar circuits, to maintain absolute synchronism. In addition, framing signals are sent out at certain intervals which denote the end of one group of characters and the beginning of another.

14.3 VOICE COMMUNICATIONS (TELEPHONY)

The telephone is the most widely used means of electronic voice communications, and its technology has gradually evolved in the twentieth century. The present telephone system is a conglomeration of older and newer techniques and is constantly changing. Telephone engineering is supported mostly by the Bell System, other large telephone companies, and to some extent, by their equipment suppliers. Technical information, design details, and operating theory are generally available only from these sources. At the end of this chapter, we present a number of readily available references which contain some, however limited, technical details on telephony. Perhaps *Reference for Engineers: Radio, Electronics, Computer, and Communications* (7th ed.) presents the most technical information (see Bibliography).

Basic Telephone Operation

The functional circuit of a typical rotary dial telephone is shown in Figure 14-7. Connected to the central office by two wires and sometimes a ground, only the bell or "ringer" is always across the circuit. The

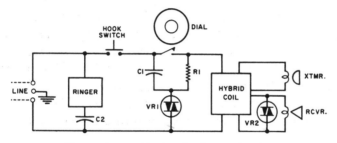

Figure 14-7.   Basic telephone circuit.

(Illustration courtesy of *Electronics World*.)

subscriber ringing signal is a 20-Hz sine wave ranging from 70 to 90 volts rms. When the hook switch is closed, the hybrid coil converts the two-wire line into an appropriate four-wire connection for the transmitter (microphone) and the receiver (earpiece). The dial contacts are normally closed, but when the dial rotates back, a cam actuates the contacts, which open, once for each fingerhole, the dc line. This results in dc pulses going to the central office, and these pulses determine the called number the calling subscriber wants to speak to.

Part of the hybrid coil assembly is a special network to limit the coupling between receiver and transmitter to a desired level, the so-called side tone, which indicates to the user that the instrument is "alive." VR1 and VR2 in Figure 14-7 are limiting diodes to prevent excessive amplitude signals from going through. In other versions the dial mechanism is replaced by a push-button assembly. The telephone operates in the same basic manner except that instead of interrupting the dc line, the depression of a push button causes an oscillator to generate two tones which are transmitted to the central office. This system, called dual-tone multi-frequency (DTMF) signaling, is described in more detail below.

An individual telephone is connected to the telephone company, or to the local private branch exchange (PBX) through a subscriber loop. This is a dc path, energized by −48 volts from the telephone company equipment and with a dc resistance which depends on the length of the loop, but which usually does not exceed 5000 ohms. The general specifications for the subscriber loop characteristics are given in Table 14-1.

Two types of signaling are required for telephone service. The so-called supervisory signaling consists of on-hook, off-hook, dial tone, busy tone, and the ringing signal. The dialing information from the subscriber set is either in the form of dc pulses or DTMF. Dial pulses occur at a nominal 10-Hz rate, with the break period nominally 61% or approximately 60 milliseconds and the closed circuit period approximately 40 milliseconds long. A digit is indicated by the number of break periods. The digit 3, for example, consists of three openings of the dial contacts. The digit 0, however, consists of ten breaks. The time interval between digits depends, of course, on the speed with which the human operator can dial, but a minimum interdigital period of 600 milliseconds is built into the dial mechanism.

The DTMF frequency assignment is shown in Table 14-2. The minimum duration of each tone and the minimum spacing between digits is 40 milliseconds.

In communications between switching centers, two other methods

VOICE TRANSMISSION

Transmission loss:	0 dB
Frequency range:	300–3000 Hz
Impedance:	600 ohms (nominal)
Speech transmission level:	–2 dB to –12 dB over a 3-second interval

OUT-OF-BAND LIMITS

3995–4005 Hz	At least 18 dB below in-band
4000–10,000 Hz	Less than –16 dBm
10,000–25,000 Hz	Less than –24 dBm
25,000–40,000 Hz	Less than –36 dBm
Above 40,000 Hz	Less than –50 dBm

INTERNAL IMPEDANCE (subscriber's equipment)
600 ohms in voice band (resistive)

SIGNALING CONSIDERATIONS
Because of tone-signaling devices used for network-control functions, customer devices should have no signals with energy solely in the 2450–2750 Hz band. If such signals are present, at least equal energy in the 800- to 2450-Hz band must be present.

D.C. SIGNALS AND POWER
Any d.c. must be less than 10 mA into the voice transformer of the coupler.

Max. between any conductor and ground:	135 volts d.c.
	50 volts d.c.

Table 14-1.   Subscriber loop standards.

(Illustration courtesy of *Electronics World*.)

of signaling are in common use. One is the multi-frequency (MF) signaling method, which permits extremely rapid transmission of dialing information. Table 14-3 shows the frequencies assigned to the ten digits and the two control characters. The key pulse (KP) and the stop pulse (SP) are also used in confirmation signaling systems. MF signaling speed is nominally seven digits per second, with an interdigit time of 73 milliseconds nominally.

Digit	Pushbutton Designation	Tone pair
Digit 1	1	697 and 1209
Digit 2	2	697 and 1336
Digit 3	3	697 and 1477
Digit 4	4	770 and 1209
Digit 5	5	770 and 1336
Digit 6	6	770 and 1477
Digit 7	7	852 and 1209
Digit 8	8	852 and 1336
Digit 9	9	852 and 1477
Digit 0	0	941 and 1336

Table 14-2. DTMF assignments.

Single frequency signaling (SF) is used on certain trunks, and substitutes a 2600-Hz sine wave for the dc, where it is interrupted in dc signaling. SF signaling, then, consists of a continuous 2600-Hz tone, interrupted at a 10-Hz dial pulse rate.

Telephone Network

Private subscribers are generally connected to a local central office (CO), while organizations having a number of phone extensions usually operate from a private branch exchange (PBX) or a private automatic branch exchange (PABX). In a PBX the subscribers can dial each other, usually by a 2- or 3-digit address, but to dial an outside number they must call the operator, who then establishes the outside connection. All incoming calls reach the operator, who will ring and then connect the desired extension.

Digit	Group Tone Pair (Hz)
1	700 and 900
2	700 and 1100
3	900 and 1100
4	700 and 1300
5	900 and 1300
6	1100 and 1300
7	700 and 1500
8	900 and 1500
9	1100 and 1500
0	1300 and 1500
KP	1100 and 1700
SP	1500 and 1700

Table 14-3. MF assignments.

In PABX operation, the function of the operator is greatly reduced because most subscribers can reach outside lines directly by dialing a special code, usually the numeral 9. Incoming calls still reach the operator. In PABX operation, it is possible to connect certain subscriber loops so that they do not have access to the outside lines. Conference calls and other special features can also be implemented with modern PABX equipment.

A slightly different version of the PABX is the Centrex operation, in which a block of numbers is assigned to a particular organization and incoming calls are directly routed to the called subscriber. Usually, the first three digits of the seven-digit telephone number are common to all subscribers within a Centrex, and the last four digits are assigned on an individual subscriber basis. The Centrex function can be located either at the central office or at the customer's premises, depending on the particular installation size and available equipment.

The central office is usually located within a few miles of the subscriber's phone, although, particularly in rural areas, longer subscriber loops are possible. Calls which can be completed by going only through one central office are usually classified as local calls. In the present direct distance dialing (DDD) system, the hierarchy of telephone centers is as follows:

Class 5   end office
Class 4   toll center (TC) or toll point (TP)
Class 3   primary center
Class 2   sectional center
Class 1   regional center

A typical sequence of routing a long distance call is shown in Figure 14-8. There are nine regional centers in the United States; they are interconnected by direct trunk facilities.

Telephone Transmission

Telephone connections from the individual subscriber, PBX, or PABX to the CO are usually in the form of individual wire pairs, one pair per subscriber. From Class 5 to Class 4 offices and further up, individual wire pairs would be highly uneconomical; therefore, different techniques are used. Frequency division multiplexing (FDM) is the most widely used method of carrying many different telephone conversations over a single cable. Figure 14-9 illustrates the fundamental methods of multiplexing 12 different signals into a channel group from 60 to 108 kHz. Five channel groups are further multiplexed to form a super group,

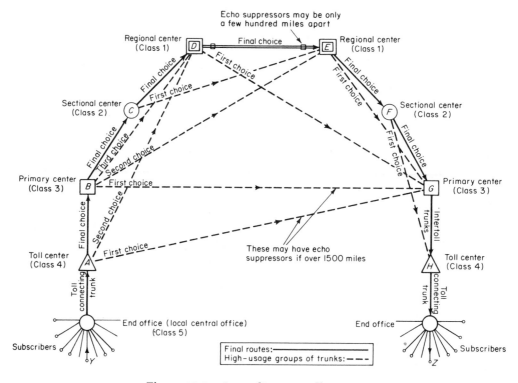

Figure 14-8. Long distance call routing.

(James Martin, *Telecommunications and the Computer,* © 1969. Reprinted by permission of Prentice-Hall, Inc.)

and ten super groups are combined into a basic master group. The coaxial cable used in the Bell L-3 system has a bandwidth of 8 MHz and can carry a total of 1860 voice channels or 600 voice channels and a television channel, as illustrated in Figure 14-10. Similar arrangements are used for microwave transmission, where an even greater number of master groups can be carried.

Time division multiplexing (TDM) is also used, but requires first the encoding of the voice signal into pulse code modulation (PCM). Figure 14-11 illustrates the principle of TDM, and Figure 14-12 shows the bit pattern used to multiplex 24 voice channels on the Bell T-1 system. Synchronization between transmitting and receiving modems is performed by means of the framing bits illustrated.

The actual transmission of telephone channels may involve overhead wires, underground or submarine cable, microwave relay links, or even a communications satellite relay link. This complex technology is generally of interest only to the technical personnel of the communications companies, since the customer simply pays for the channels

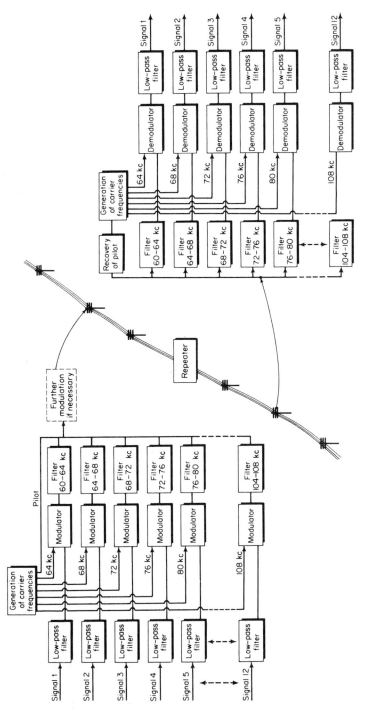

Figure 14-9. Frequency division multiplexing.

(James Martin, *Telecommunications and the Computer*, © 1969. Reprinted by permission of Prentice-Hall, Inc.)

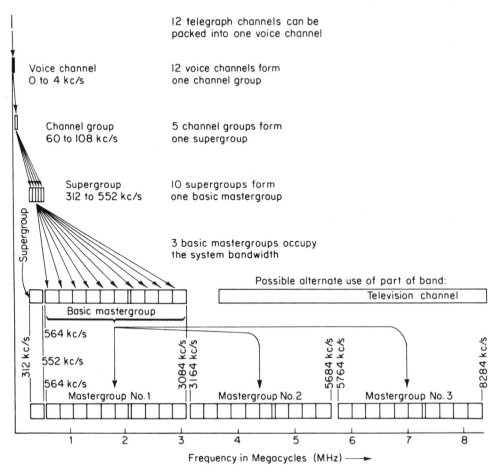

Figure 14-10.   FDM in the Bell L-3 system.

(James Martin, *Telecommunications and the Computer,* © 1969. Reprinted by permission of Prentice-Hall, Inc.)

used, regardless of routing. Chapters 10, 11, and 12 cover the principles of antennas, RF and microwaves, and broadcasting, and the references at the end of this chapter contain further material relevant to the transmission of electrical signals.

Telephone Switching Techniques

In the early manual switching centers, individual telephone operators provided the interconnection by plugging in the wire coming from the calling subscriber to certain links, which were then connected to the called subscribers' lines. Early automatic telephone exchanges used a step-by-step switch which had a 10 × 10 switching field to connect

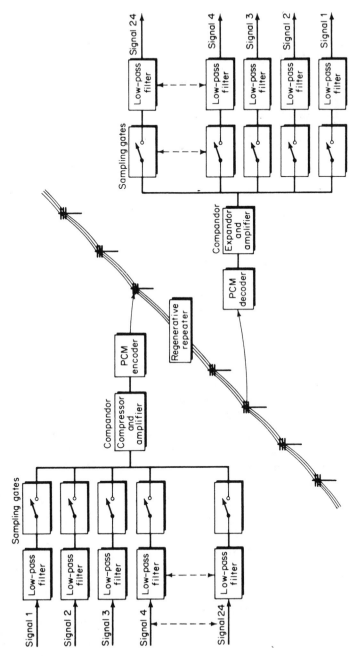

Figure 14-11. Time division multiplexing.

(James Martin, *Telecommunications and the Computer,* © 1969. Reprinted by permission of Prentice-Hall, Inc.)

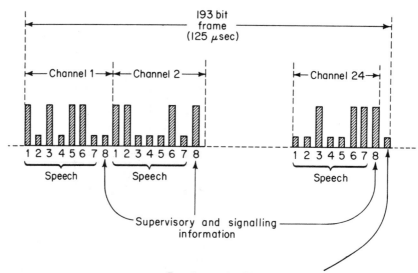

Figure 14-12. TDM in the Bell T-1 system.

(James Martin, *Telecommunications and the Computer*, © 1969. Reprinted by permission of Prentice-Hall, Inc.)

two digits' worth of dial information at one time. The more modern automatic exchanges usually use crossbar switches, a rectangular matrix of electromagnets which activate mechanical linkages to close a set of contacts. The crossbar switches are controlled by relay logic, which is arranged to receive and store the incoming dial pulses. A typical automatic exchange serving 5000 to 10,000 subscribers contains a large number of crossbar switches arranged in a series of primary, secondary, and tertiary matrices. The relay control logic sets up connections, based on the availability of connecting links, between these switching matrices to provide a direct path from the calling line circuit to either a distant trunk or else to another local line subscriber circuit. At the end of this chapter the reader will find some references for detailed information concerning switching techniques.

Many telephone companies are now in the process of installing a new, all-electronic, automatic exchange, the Bell Type No. 1 electronic switching system (ESS). This system, capable of serving from 5000 to 65,000 lines, is controlled by computers and contains memory elements to store both directory information and the "calls in process" status.

This advanced system can provide a number of special features such as call forwarding (forwarding calls to a new number when a subscriber has dialed in this information), call waiting (sending a special tone to a busy subscriber that another call is waiting), conference calls, and other special services.

Regardless of the type of switching center, however, each subscriber loop at the local exchange is first connected to the protector frame (PF), which contains the protection circuits against lightning, overvoltage, excessive current, etc. Every subscriber's wires also appear at the main distribution frame (MDF) and then go to the individual subscriber's line circuit.

The hold relay, which establishes a connection when the subscriber takes his handset off the hook, is located at the line circuit. When the hold relay is actuated, this indicates a service request to the line scanner or line finder section, which continuously scans all of the subscriber lines for either a service request or a disconnect signal. When a service request is detected, the subscriber line is connected to a register, sometimes called a register-sender, which signals the subscriber by sending a dial tone that it is ready to accept dial information. After dialing is complete, the control logic attempts to establish a path through the various switching matrices between the calling and the called line. If the called line is busy, or if a path through the available links cannot be established, the register sender will send a busy tone to the caller. If the called line can be connected, the line circuit of the called line will receive a 20-Hz ringing signal and the calling line will receive a ring-back tone. When the called subscriber answers, the voice path is finally connected.

In locations where subscribers are charged on a per call basis, a unit call counter is attached to the subscriber's circuit at the local exchange. Automatic toll recording equipment is usually installed in toll points and sometimes in toll centers.

Connections to the Telephone System

Since the 1966 Carter phone decision, many individuals and organizations have developed devices that can be connected directly or indirectly to the public telephone network. Both the technical and legal implications of interconnections are beyond the scope of this text, and the reader is referred to current information available from the FCC, the Electronics Industries Association (EIA), and local telephone companies, for further information.

Common Carrier Satellite Communications

An example of a nationwide digital network is known as *skyline*. This network consists of twenty-five switching centers in major U.S. cities linked by Satellite Business Systems Inc., a time-division multiple-access satellite system. Additional cities are attached to the network by terrestrial facilities.

Data encryption has become mandatory for many businesses, especially for government contractors who must pass sensitive information from point to point.

Security requirements pertaining to encryption are contained in "Federal Standard 1027: General Security Requirements for Equipment Using Data Encryption Standard."

In reference to telephony's long-awaited integrated services digital network, several companies—Southern Bell and Pacific Bell are two— are conducting field trials. However, at this time (1986) industry standards committees are still deciding the eventual standards. Companies and associations involved are: AT&T Network systems, U.S. Telephone Association, U.S. Telecommunications Suppliers Association, and Bell operating companies.

14.4 OPTICAL COMMUNICATIONS

As in the common carrier satellite communications field, the past few years have produced significant advances in fiber optic transmission technology. Fiber optic systems are widely used in telephone and data communications systems. Detailed information on device and optical transmission systems can be found in Chapter 19, Opto-Electronic Devices.

BIBLIOGRAPHY—CHAPTER 14

FEHER, K., *Digital Communications: Microwave Applications*. Englewood Cliffs, NJ, Prentice-Hall, Inc., 1981.

FREEMAN, R. L., *Telecommunication Transmission Handbook* (2nd Ed.). New York, NY, John Wiley & Sons, Inc., 1981.

GAGLIARDI, R. M., and S. KARP, *Optical Communications*. New York, NY, John Wiley & Sons, Inc., 1976.

HAMMING, R., *Coding and Information Theory*. Englewood Cliffs, NJ, Prentice-Hall, Inc., 1980.

HOLMES, J. K., *Coherent Spread Spectrum Systems*. New York, NY, John Wiley & Sons, Inc., 1982.

LINDSEY, W., and M. SIMON, *Telecommunications Systems Engineering*. Englewood Cliffs, NJ, Prentice-Hall, Inc., 1975.

Reference Data for Engineers: Radio, Electronics, Computers and Communications (7th Ed.). Indianapolis, IN, Howard W. Sams & Co., Inc., 1985.

RODEN, M., *Analog and Digital Communications Systems*. Englewood Cliffs, NJ, Prentice-Hall, Inc., 1979.

SEIPPEL, R., *Optoelectronics*. Reston, VA, Reston Publishing Co., Inc., 1980.

SPILKER, J., *Digital Communications by Satellite*. Englewood Cliffs, NJ, Prentice-Hall, Inc., 1976.

CHAPTER 15

TELEVISION SYSTEMS

15.1 BASIC TV PRINCIPLES

Introduction

This chapter presents first the basic principles of human vision, optics, and colorimetry, followed by the electronic principles and standards which form the basis of all television systems. Then TV broadcasting systems and TV receiver principles are described. The fundamentals of different video recording systems are followed by a brief outline of the most frequently used nonbroadcast television systems.

Characteristics of Human Vision

For the purposes of understanding television, the basic mechanisms of the eye and three specific properties of human vision are important.

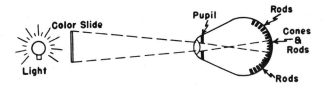

Figure 15-1.   Basic mechanism of the human eye.

(Walter H. Buchsbaum, *Color TV Servicing*, 2nd Edition, © 1968. Reprinted by permission of Prentice-Hall, Inc.)

Mechanisms of the Eye.   As shown in Figure 15-1, the basic mechanism of the human eye consists of the lens, which focuses the external image onto the retina, the retina itself, and the pupil, which controls the amount of light reaching the retina. At low levels of illumination, such as in moonlight, the pupil opens wide, allowing the entire retina to "see" the projected image. In the center of the retina are most of the cones, photosensitive cells that can detect colors. The rods, distributed over the entire retina, are also photosensitive cells; however, they can only detect various shades of gray. At low light levels, when the image is focused over the entire retina, color vision is lost and we only see various shades of gray. Only when sufficient brightness is available to contract the pupil and have the image focused onto the fovea, a spot near the center of the retina where cones predominate, can we see colors.

Spatial Integration.   This refers to the human eye's ability to integrate small areas into a larger one. A typical example of that is our ability to recognize a newspaper photograph which actually consists of many small picture elements. Spatial integration depends on the viewing distance. In most systems, a television picture consists of individual lines created by a moving dot of light. At the proper viewing distance from the screen, individual lines are not distinguishable and the whole picture is seen.

Color Integration.   This characteristic of the human eye permits the use of certain primary colors to be combined to reproduce a gamut of others. As an example, when a field of small blue dots is intermeshed with a field of small red dots, the human eye, at the correct viewing distance, will see purple. Color printing as well as color photography depends on this effect. Color pictures are usually printed by combining several separate single-color (monochrome) pictures so as to obtain the overall multi-color (polychrome) effect. The color integration mechanism of the human eye makes it possible to see a wide range of different colors even though only three primary colors are produced on the color picture tube screen.

Temporal Integration. This characteristic of the human eye permits us to see a series of separate events occurring in sequence as a single smooth action. A typical example of temporal integration is motion picture film where 24 individual pictures are projected each second, each interrupted once by a shutter to produce 48 separate pictures per second on the screen. Temporal integration in television allows the viewer to see a steady picture and continuous motion where, in actuality, a small dot of light is scanned across the TV screen in endless sequence.

Spatial and color integration depend on viewing distance, but temporal integration is very largely dependent on illumination levels. As the speed of individual picture presentation is increased, the flickering effect reaches the point where the flicker disappears. This point is called the *critical fusion frequency*. Critical fusion frequency and light intensity are related logarithmically, and an increase in illumination will also bring an increase in the critical fusion frequency. Because television screens are generally much brighter than moving picture screens, the frame frequency for television is also higher. While forty-eight pictures per second are sufficient for movies, television systems in the United States use sixty pictures per second.

Photoelectricity and Optical Principles

Photoelectric Properties. Photoelectricity concerns the interaction between light and the flow of electric current. For television systems, these effects are classified either as *photovoltaic* or *photoresistive*. In photovoltaic devices, a light-sensitive material changes its electric charge according to the intensity of the light striking it. Photoresistive materials change their electrical resistance according to the intensity of the light striking them. Photovoltaic effects are used in such highly sensitive television cameras as the imageorthicon, while photoresistive effects are used in the widely used vidicon type of cameras.

Television cameras are covered in Section 15.2. For TV cameras, a photoelectric device must have a fairly uniform frequency response over the visible spectrum. To show fine detail, the photoelectric material must consist of very small cells, comparable to the photosensitive emulsion on film. Photoelectric materials for television must also be able to change their electric state very rapidly, as required by the scanning principles described in more detail below. For TV cameras, photoelectric materials should be efficient in converting variations of light into electrical signals.

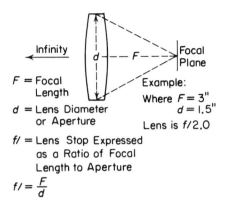

Figure 15-2. Lens characteristics.

(Gerald L. Hansen, *Introduction to Solid-State Television Systems: Color and Black and White,* © 1969. Reprinted by permission of Prentice-Hall, Inc.)

Optical Quantities. The following optical quantities are frequently used in television work (see Chapter 19).

$$\text{Luminous flux} = F = \text{lumens}$$

$$\text{Intensity} = I = \frac{\text{lumens}}{\text{steradians}} = \frac{F}{W}; \quad I = \text{candlepower}$$

$$\text{Illumination} = E = \frac{\text{lumens}}{\text{m}^2} = \text{meter-candle}$$

$$\text{Brightness or luminescence} = B = \frac{\text{candle}}{\text{m}^2} = \frac{\text{lumens}}{W \times \text{m}^2}$$

$$= \text{meter-lamberts}$$

$$\text{Luminous emittance} = L = \frac{\text{lumens}}{\text{m}^2}$$

Lenses. In television systems, lenses are used to focus a scene onto the photoelectric surface inside the camera tube. Basically, lenses are either concave or convex. Convex lenses are thicker at the center than at the edges, while a concave lens is thicker at the edges than at the center (see Chapter 19). The following characteristics of lenses are important in television systems:

- Transmission coefficient = T, varies from 0.4 to 0.7 and indicates the amount of light losses due to the lens material.
- Diameter = d, the useful diameter of a lens.
- Focal length = F, the distance from the center of the lens to the point at the focal plane onto which an object at a great or infinite distance can be focused.

- Object distance = D, the distance from the center of the lens to the object to be projected through the lens onto a focal plane.
- Aperture—the actual area of the lens through which light can reach the focal plane, often equal to the lens diameter.
- Lens speed or stop = $f = \dfrac{F}{d}$, see Figure 15-2.
- Depth of field—the distance between the two points closest and farthest from the center of the lens over which the scene appears in focus, dependent on object distance, focal length, and aperture. Greater focal length, greater aperture, and greater object distance will increase the apparent depth of field.

Television cameras make use of an array of concave and convex lenses to produce compound lenses which contain color correction, spherical aberration correction, extended focal point, and many other features. Compound lenses are available with wide and narrow viewing angles, shallow and deep field depth, telescopic enlarging qualities, and the so-called zoom lenses, in which these characteristics can be mechanically varied. Lens manufacturers supply detailed specifications, including all of the parameters and characteristics given above.

Principles of Colorimetry

The human eye can perceive colors through an as yet unexplained process involving the cones in the retina. Color vision varies somewhat among individuals, and for this reason all the data in colorimetry are based on the average of a great many observations. The subject of human color perception and the physics of colors is quite complex, and the concepts outlined below are limited in their application to color television systems.

Subjective Color Aspects. Light reaching the human eye can be classified, according to its source, as reflected or transmitted light. Reflected light produces colors that depend on the original color of the light and on the color absorption and reflection qualities of the reflecting surface (i.e., green velvet will appear quite differently when viewed in daylight, under fluorescent lighting, or under the yellow light of a sodium vapor lamp). Transmitted light reaches the eye directly from its source, such as an incandescent lamp, a cathode-ray tube screen, etc. The rules concerning the mixing and combination of colors are different for reflected and transmitted light. The type of color mixing that occurs in painting applies to reflected light only.

In describing colors, we can distinguish three different characteristics: saturation or purity as against the amount of pastel effect. Red is a pure color, but pink is its pastel counterpart. Hue, or tint, describes the color as such and determines whether a particular color is more closely related to one or the other of the saturated colors. A particular material may have a bluish or a greenish tint, meaning that the dominant saturated color would be blue or green. Color brightness, or luminance, means the intensity of illumination of the particular color. On a sunny day colors appear brighter than on a rainy day. To sum up, a color is usually described by three characteristics: saturation or purity, hue or tint, and brightness or luminance.

ICI Standard Chromaticity Diagram.   Figure 15-3 shows a chromaticity diagram, part of the ICI (International Committee on Illumination) standards designed to permit specifications of particular colors in terms of X and Y coordinates. Pure or saturated colors are located along the horseshoe-shaped rim. White lies in the center and is a mixture of at least two other colors. Its saturation is zero. The straight line

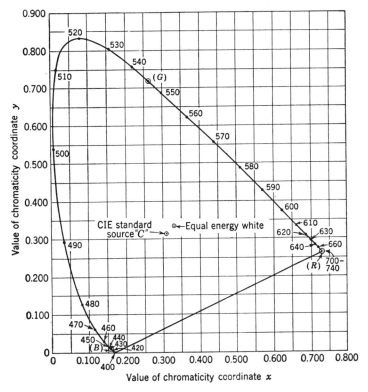

Figure 15-3. Chromaticity diagram.

(From *Fundamentals of TV Engineering*, by Glenn M. Glasford. Copyright 1955 by McGraw-Hill Book Company. Used with permission of McGraw-Hill Book Company.)

between the red and the blue presents those colors that do not appear in the natural color spectrum, such as those generated by a prism or a rainbow.

Figure 15-3 applies only to transmitted light and indicates that color mixing is possible by straight-line addition or subtraction.

Example: Saturated yellow at X = 0.5, Y = 0.5, can be mixed with a bluish color at X = 0.05, Y = 0.3. If equal light intensities of these two colors are mixed on a reflecting surface, the resulting color will be represented by the pastel green, which is determined by drawing a straight line between the chosen yellow and blue and picking the center of that straight line.

In similar fashion, subtraction is also possible. Optically, subtraction is performed by means of colored filters.

TV Color Standards.   For color television, three primary colors— red, green, and blue—are chosen; all other colors are derived from that. Referring to the ICI standard chromaticity diagram of Figure 15-3, the primary colors for color TV are specified as follows:

> Red: X = 0.67, Y = 0.33
> Green: X = 0.21, Y = 0.71
> Blue: X = 0.14, Y = 0.08

The colors reproduced by these three primaries lie within the triangle described by connecting the three points on the chromaticity diagram.

White is obtained by mixing the following relative light intensities:

> White: 0.30 red + 0.59 green + 0.11 blue

The ICI standard chromaticity diagram of Figure 15-3 represents only one fixed level of brightness or luminance. The real world contains many such levels, but since color perception varies as a function of illumination, the number of colors distinguishable is greatly reduced as the intensity of the light, brightness or luminance, is reduced. As a result, it is useful to think of a color pyramid with its apex at black, its center a shaft of white, and its three sides given by the hues of the primary colors. Brightness is then given as a function of the distance of the chromaticity diagram from the apex of the color pyramid. In this method, brightness is unrelated to a particular hue and can be related directly to the black-and-white TV signal.

Scanning and Synchronization Principles

Television pictures are generated both at the camera and at the picture tube by the motion of their respective electron beams. This motion consists of scanning from left to right, with a quick return to the left again, and from the top downward in subsequent lines. The number of lines and the timing has a relationship to the shape of the scanned image. In commercial television broadcasting this shape is rectangular, with an aspect ratio, width over height, of 4:3. A commercial TV picture is therefore four units wide and three units high.

Figure 15-4 shows the horizontal scanning sequence for commercial television. Each complete picture, or frame, consists of 525 horizontal lines, and 30 frames are scanned per second. A complete frame, however, consists of two fields, each of which has 262-1/2 lines, arranged in alternate odd-even sequence as illustrated in Figure 15-4. This method is called the 2:1 interlaced standard scanning ratio.

Not all of the 525 horizontal lines contain picture information. Between 7 and 8% are used up during the vertical retrace action, the return from bottom to top, and some of the top and bottom lines are blanked out. Only 483 to 490 active scanning lines are used for video information. Closed-circuit and special-purpose TV systems often use different aspect ratios and different scanning line standards.

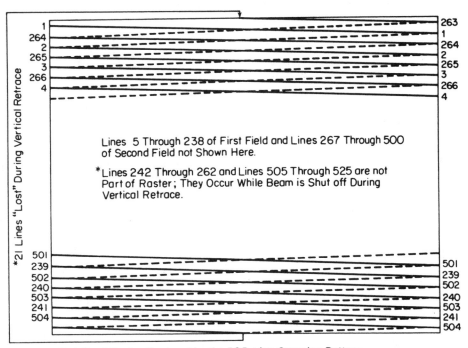

Details of Raster Produced by the 525-Line Scanning Pattern

Figure 15-4.   Interlaced scanning raster.

(Gerald L. Hansen, *Introduction to Solid-State Television Systems: Color and Black and White*, © 1969. Reprinted by permission of Prentice-Hall, Inc.)

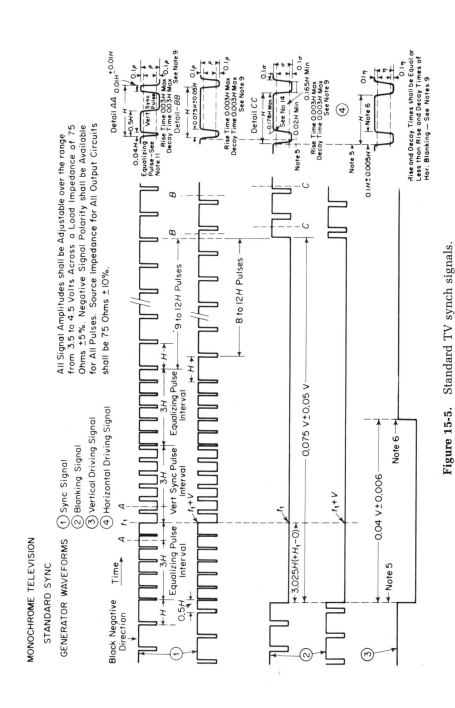

Figure 15-5. Standard TV synch signals.

(Gerald L. Hansen, *Introduction to Solid-State Television Systems: Color and Black and White,*
© 1969. Reprinted by permission of Prentice-Hall, Inc. (Figure originally courtesy of EIA.))

Synchronization of the received picture with the transmitted picture is accomplished by synchronizing pulses which are sent out with the video signal to indicate the beginning of each horizontal line and the beginning of each vertical field. To generate 525 lines, the horizontal synchronizing and line frequency is 15,750 Hz. To maintain horizontal synchronization during the much longer vertical synchronizing and retrace period, the vertical synchronizing signal is interrupted by horizontal pulses. Figure 15-5 shows the standard transmitted synchronizing signals during two successive fields and their vertical synch-pulse interval. The difference in the width of the horizontal pulses during the equalizing pulse interval, the vertical synch-pulse interval, and the ordinary line-by-line synchronizing period, are used in an integrating network in the vertical synchronizing section of the TV receiver to regenerate the vertical synchronizing pulse itself.

Bandwidth and Resolution Principles

The video signal bandwidth required to reproduce the maximum resolution of a particular image is usually expressed in terms of the number of horizontal lines. This is based on the assumption that the number of picture elements contained in a line is limited by the thickness of the line; that is, each picture element is a square with a thickness of the line as controlling dimension.

The following formula applies:

$$\text{Bandwidth (frequency)} = 1/2\ KAN^2F$$

where:

K = aperture misalignment (approximately 0.64 for commercial TV quality).
A = aspect ratio (width over height).
N = number of horizontal lines making up a complete picture.
F = number of frames per second.

For commercial TV the bandwidth is approximately 4 MHz and the channel width is 6 MHz.

U.S. and Foreign TV Broadcast Standards

Table 15-1 shows a comparison between the United States and the major foreign TV broadcast scanning standards.

TV Channels in the U.S.:

Channel 1	not used	v.h.f. band----line of sight
Channels 2–4	54 to 72 MHz	
Channels 5–6	Gap at 72–76 MHz	reception with some fringing around earth's curvature. No Heavyside reflection
	76 to 88 mc	
	Gap at 88–174 MHz	
Channels 7–13	174 to 216 mc	
	Gap at 216–470 MHz	
Channels 14–83	470 to 890 MHz ⟶	u.h.f. band----same characteristics as v.h.f., but greater losses

MONOCHROME SCANNING STANDARDS

U.S. and Japan	British Only	French Only	European Standard
525 Lines	405 Lines	819 Lines	625 Lines
60 Fields	50 Fields	50 Fields	50 Fields
15,750 cps horiz.	10,125 cps horiz.	20,475 cps horiz.	15,625 cps horiz.
2:1 interlace	2:1 interlace	2:1 interlace	2:1 interlace

Table 15-1.   Commercial TV transmission standards.

15.2 TV BROADCASTING SYSTEMS

Camera Characteristics

A wide variety of different television camera tubes is available, each suitable for particular applications. The following essential characteristics apply, regardless of the type of camera tube:

Resolution—usually specified in number of lines; resolution also depends on the speed with which charge and discharge occurs.

Lag or persistence—another indication of the speed with which the photosensitive surface reacts.

Sensitivity—usually stated in footcandles per volt output.

Gamma characteristics—describe the conversion linearity of light input versus voltage output. Ideally, a completely linear relationship would exist between them. However, practically, the gamma characteristics of camera tubes are not quite linear, requiring gamma correction circuits.

Spectral response—refers to the relative sensitivity of the photosensitive surfaces to visible light from 3800 to 7000 Angstroms.

Dynamic range—the maximum useful range of illumination; that is, the difference between the minimum usable light levels and the maximum light that does not cause overloading or "blooming."

Transmitter Parameters (Commercial Broadcast Service)

The following parameters generally apply to television broadcast transmitters:

Frequency assignment—determined by the FCC under the overall TV station allocation plan for a particular station. Table 15-2 shows the frequency allocation of TV stations in the United States.

Video modulation—the video signal is amplitude modulated (AM) on the carrier, and the lower sideband is suppressed. FCC standards require negative AM; that is, the black level is reached at 75% of maximum amplitude and the white level occurs at 12.5% of amplitude. The synchronizing signal reaches into the "blacker than black" region from 75% to 100%.

Audio modulation—is performed by means of frequency modulation (FM) with ± 25 kHz deviation about the audio carrier which is 4.5 MHz above the video carrier, as indicated in Figure 15-6.

Power output—authorized by the FCC for a particular station and area coverage, ranging from 100 watts to 5000 kw maximum.

Color Matrixing and Modulation

The color information is transmitted within the same bandwidth and over the same channel as the monochrome information, which permits full compatibility with monochrome receivers. The original color scene is converted into electric signals by three separate camera tubes, one each for the red, green, and blue primary color components (refer to Principles of Colorimetry above). These three color components are related to the monochrome or brightness signal according to the equation:

$$E_Y = 0.3 \, R_{Red} + 0.59 \, E_{Green} + 0.11 \, E_{Blue}$$

For convenience in transmission, two signals, E_Q and E_I, were chosen, and they are related to the three primary color signals as follows:

$$E_Q = 0.21 \, E_{Red} - 0.52 \, E_{Green} + 0.31 \, E_{Blue}$$
$$E_I = 0.6 \, E_{Red} - 0.28 \, E_{Green} - 0.32 \, E_{Blue}$$

E_Q and E_I are modulated on a separate color subcarrier by means of amplitude and phase modulation as shown in Figure 15-7. The E_I signal is in the orange and the E_Q in the magenta hue range. The total video signal obtained after matrixing is described by the equation:

$$E_M = E_Y + \left[E_Q \sin (\omega t + 33°) + E_I \cos (\omega t + 33°) \right]$$

Chan. No.	Frequency Band (MC)	Video Carrier	Audio Carrier	Chan. No.	Frequency Band (MC)	Video Carrier	Audio Carrier
2	54-60	55.25	59.75	43	644-650	645.25	649.75
3	60-66	61.25	65.75	44	650-656	651.25	655.75
4	66-72	67.25	71.75	45	656-662	657.25	661.75
5	76-82	77.25	81.75	46	662-668	663.25	667.75
6	82-88	83.25	87.75	47	668-674	669.25	673.75
7	174-180	175.25	179.75	48	674-680	675.25	679.75
8	180-186	181.25	185.75	49	680-686	681.25	685.75
9	186-192	187.25	191.75	50	686-692	687.25	691.75
10	192-198	193.25	197.75	51	692-698	693.25	697.75
11	198-204	199.25	203.75	52	698-704	699.25	703.75
12	204-210	205.25	209.75	53	704-710	705.25	709.75
13	210-216	211.25	215.75	54	710-716	711.25	715.75
14	470-476	471.25	475.75	55	716-722	717.25	721.75
15	476-482	477.25	481.75	56	722-728	723.25	727.75
16	482-488	483.25	487.75	57	728-734	729.25	733.75
17	488-494	489.25	493.75	58	734-740	735.25	739.75
18	494-500	495.25	499.75	59	740-746	741.25	745.75
19	500-506	501.25	505.75	60	746-752	747.25	751.75
20	506-512	507.25	511.75	61	752-758	753.25	757.75
21	512-518	513.25	517.75	62	758-764	759.25	763.75
22	518-524	519.25	523.75	63	764-770	765.25	769.75
23	524-530	525.25	529.75	64	770-776	771.25	775.75
24	530-536	531.25	535.75	65	776-782	777.25	781.75
25	536-542	537.25	541.75	66	782-788	783.25	787.75
26	542-548	543.25	547.75	67	788-794	789.25	793.75
27	548-554	549.25	553.75	68	794-800	795.25	799.75
28	554-560	555.25	559.75	69	800-806	801.25	805.75
29	560-566	561.25	565.75	70	806-812	807.25	811.75
30	566-572	567.25	571.75	71	812-818	813.25	817.75
31	572-578	573.25	577.75	72	818-824	819.25	823.75
32	578-584	579.25	583.75	73	824-830	825.25	829.75
33	584-590	585.25	589.75	74	830-836	831.25	835.75
34	590-596	591.25	595.75	75	836-842	837.25	841.75
35	596-602	597.25	601.75	76	842-848	843.25	847.75
36	602-608	603.25	607.75	77	848-854	849.25	853.75
37	608-614	609.25	613.75	78	854-860	855.25	859.75
38	614-620	615.25	619.75	79	860-866	861.25	865.75
39	620-626	621.25	625.75	80	866-872	867.25	871.75
40	626-632	627.25	631.75	81	872-878	873.25	877.75
41	632-638	633.25	637.75	82	878-884	879.25	883.75
42	638-644	639.25	643.75	83	884-890	885.25	889.75

Table 15-2. TV channel assignments.

(From Walter H. Buchsbaum, *Fundamentals of Television*, © 1964. Used by permission of Hayden Book Company, Inc.)

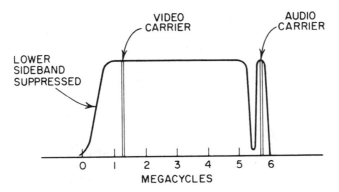

Figure 15-6.　TV channel spectrum.

(Walter H. Buchsbaum, *Fundamentals of Television*, © 1964. Used by permission of Hayden Book Company, Inc.)

Matrixing of the original RGB signals into the Y, I, and Q signals is accomplished by means of amplifiers, voltage dividers, and phase shifters.

　　The color subcarrier is chosen as the 445th multiple of the modified horizontal line frequency of 15,734.264 Hz and equals 3.579545 MHz. This requires a change in the field rate from 60 Hz to 59.94 Hz. These frequencies were chosen to deviate approximately 0.1% from the frequencies used in monochrome television and provide an improvement in the color pictures as concerns noise and various types of interference. Because the color subcarrier is an odd multiple of half the line frequency, its visual effect on both monochrome and color pictures tends to cancel out on successive fields. It also permits transmission

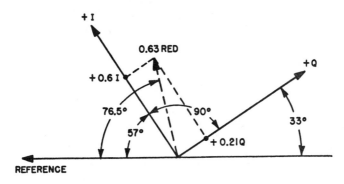

Figure 15-7.　Color subcarrier vectors.

(Walter H. Buchsbaum, *Color TV Servicing*, 2nd Edition, © 1968. Reprinted by permission of Prentice-Hall, Inc.)

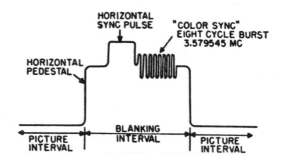

Figure 15-8.   Color synch burst.

(Walter H. Buchsbaum, *Color TV Servicing*, 2nd Edition, © 1968. Reprinted by permission of Prentice-Hall, Inc.)

along with the monochrome or brightness signal in an interleaved fashion, since the video information is carried in harmonics of the horizontal scanning frequency, and the color subcarrier frequency spectrum bursts fall halfway between the frequency spectrum bursts of the Y or brightness signal (frequency interleaving).

In order to detect the color subcarrier phase modulation, the receiver must have an exact phase reference signal from the transmitter. A reference burst of eight cycles of 3.579545 MHz is transmitted on the pedestal of the horizontal blanking pulses as shown in Figure 15-8.

The E_I signal lies in the region of orange colors, and the E_Q signal is in the magenta region. Color detail resolution in the magenta area is much less than that in the orange area, and for that reason, to conserve spectrum space, the bandwidths of the two signals have been arranged as shown in Figure 15-9.

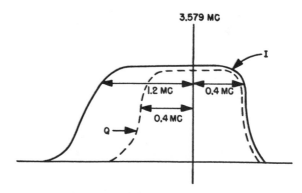

Figure 15-9.   I and Q channel spectrum.

(Walter H. Buchsbaum, *Color TV Servicing*, 2nd Edition, © 1968. Reprinted by permission of Prentice-Hall, Inc.)

Color Camera Characteristics

Actually, a television camera may fall within one of several categories: studio, portable, or telecine (also called a *film island*). The telecine is a nonportable device that is used in studio applications. It includes a 16-mm projector and slide projector that are focused into a TV camera for use of motion picture film for TV broadcasting. Portable cameras are widely used for live TV news coverage and in conjunction with consumer products such as video cassette recorders.

Modern TV cameras are entirely solid state, except for the actual camera tube. Even this tube may be a solid state device called a *charge-coupled device* (CCD).

Studio color cameras consist basically of three separate camera tubes, one for each of the three primary colors, red, green, and blue. A special system is used to split the incoming light into the components of the three primary colors, which are then focused on the respective cameras. In some color TV cameras a fourth camera tube is used, which receives the complete signal just like a monochrome camera tube would. The characteristics of studio color TV camera tubes are the same as those of monochrome camera tubes except for the following:

1. Gamma characteristics must be matched between all three or four camera tubes.
2. Spectral response must take into account the effects of the color filters.
3. Convergence, the degree to which all three images are optically focused on the camera tube photosensitive surfaces, is very critical. This implies that the horizontal and vertical size controls and the scanning circuits must produce identical rasters in each camera tube.

Broadcast Antenna Types and Area Coverage

All television transmission uses horizontally polarized antennas. Broadcast stations aim for maximum coverage of the area they serve, generally using omnidirectional antennas (circular radiation). Where a particular area includes uninhabited portions, such as the ocean, the antenna configuration is designed to provide maximum radiation over the inhabited area, with minimum radiation toward the ocean. Similarly, radiation patterns are arranged for maximum horizontal radiation with minimum vertical power losses.

All TV radiators used in broadcasting evolved from the dipole. (See Chapter 10 for details on antennas and transmission lines.)

Most TV antennas, especially for the VHF bands, are built to provide circular radiation. Circular radiation is achieved by crossing the dipoles in a turnstile arrangement and feeding them in quadrature; that is, the currents are fed to the radiating elements 90 degrees out of phase.

The *bat-wing*, *butterfly*, or *superturnstile* (various names but basically all the same antenna), is another popular type used for TV broadcasting. Antennas used in the UHF bands are of several different types. *Zee panel*, *polygon*, and *Vee-Zee* are three well-known designs approved for public broadcasting.

15.3 TV RECEIVER PRINCIPLES

Monochrome Picture Tube Parameters

Monochrome picture tubes are cathode-ray tubes (CRT) with a phosphor deposited on the screen which has a short persistence and glows in a bluish-white light. CRT fundamentals are covered in Chapter 7. The electron gun consists of filament, cathode, control grid, first anode, and focusing element. The second anode receives between 10 and 18 kV. Deflection is magnetic. Monochrome picture tubes are either constructed of all glass, with a graphite coating on the inside to act as second anode, or with a metal cone connecting the glass neck to the glass screen. In the latter construction, the metal cone acts as second anode and is insulated by a plastic cover to protect the high voltage area. The most recently produced picture tubes are rectangular, but round picture tubes are still on the market.

Magnetic deflection is accomplished by the deflection yoke mounted on the neck of the tube, close to the flare which goes to the screen. Short-necked tubes have a wider deflection angle. Maximum deflection angles available are approximately 120 degrees. The basic elements of a monochrome picture tube are shown in Figure 15-10.

Picture Tube Designation

The first two digits indicate screen size—diameter for round tubes and diagonal for rectangular tubes. The second group of letters indicates the type of tube. The designation P4 indicates the phosphor persistency and color. The letters following indicate the particular version or model of this type of picture tube.

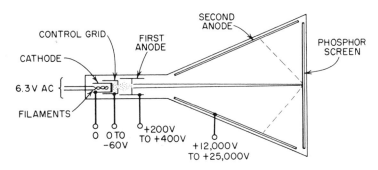

Figure 15-10.   CRT voltages and elements.

(Walter H. Buchsbaum, *Fundamentals of Television*, © 1964. Used by permission of Hayden Book Company, Inc.)

Color Picture Tube Parameters

Shadow-Mask Picture Tube.   Three separate electron guns provide three electron beams, each modulated with a video signal, representing the primary color with which this beam illuminates the screen. The three electron beams, passing through an aperture mask, strike three adjacent dots of differently colored phosphor. The electron beam modulated with the red video signal will strike all those dots which emit a red glow; the electron beam modulated with the green video signals will strike all those dots which produce a green light, etc. Figure 15–11 illustrates this basic principle.

In order to produce a sharp and clear color picture, it is necessary that each respective electron beam strike only those dots assigned to it. A magnetic field, external and common to the three electron guns, is used to align all three electron beams to their respective colors. This field is generated by the "purity" magnet assembly.

To assure that the three electron beams arrive at a particular hole in the shadow mask at the proper angle to strike the correct phosphor dots, three separate electromagnetic fields are generated, each located to control the electron beam producing the respective color. These electromagnets receive a steady state component, which is adjustable by means of a permanent magnet so that the three electron beams can be converged at the center of the screen. This is called *static convergence*. To correct for the arc described by the motion of the electron beam in the vertical and horizontal planes, as opposed to the flat color dot pattern on the screen and the equally flat shadow mask, two additional components of the electromagnetic field must be generated in each con-

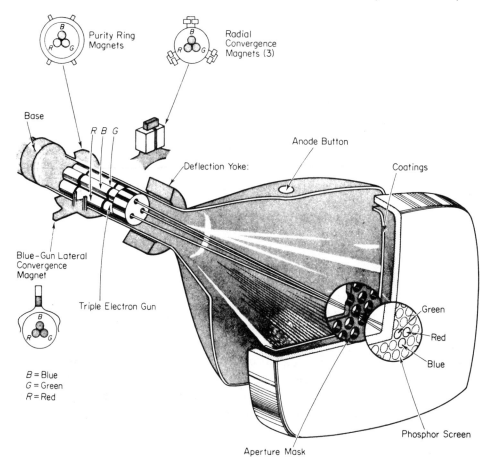

Figure 15-11. Shadow-mask color picture tube.

(Gerald L. Hansen, *Introduction to Solid-State Television Systems: Color and Black and White,* © 1969. Reprinted by permission of Prentice-Hall, Inc. (Figure originally courtesy of Sylvania Electric Products, Inc.))

vergence magnet. One is the vertical dynamic signal and the other the horizontal dynamic signal, occurring at the respective vertical and horizontal sweep frequencies and having essentially a paraboloid wave shape. Convergence adjustments therefore consist of the following:

R, G, B static convergence

R, G, B vertical dynamic convergence

R, G, B horizontal dynamic convergence

To provide a reference alignment of one of the three beams, a separate, adjustable, permanent magnet is located over the blue electron gun. This magnet controls the lateral movement of the blue electron beam and is necessary to assure correct static convergence.

Color picture tubes generally require an anode voltage of 10 to 30 kV, depending on the size of screen.

To overcome the effects of stray magnetic fields, a special degaussing coil is used around the screens of most modern color TV picture tubes.

Designation of color picture tubes is the same as that for monochrome tubes, except that the phosphor type is P22 or P24.

Single-Gun Color Picture Tubes.   A single electron gun is used to drive a screen made up of thin lines of the three primary colors. A wire mesh, located a short distance behind the phosphor line screen, is driven by the 3.579-MHz color subcarrier reference signal and deflects the single electron beam back and forth between the three lines. This, in effect, results in excitation of the correct red, green, and blue portions of each of the lines. In actuality, each color is reproduced in sequence and not, as in the shadow-mask tube, simultaneously. The eye of the observer, however, provides sufficient integration, so that a complete color TV picture is visible.

In order to deflect the electron beam electrostatically between the three adjacent primary color lines, a fair amount of power at 3.579 MHz must be available to drive the deflection grid. To prevent unauthorized radiation at this frequency, adequate shielding must be provided.

The parameters of deflection and electronic operation of the picture tube are the same as those for monochrome picture tubes. Unlike the shadow-mask tube, the single-gun color TV picture tube requires no convergence or purity adjustments.

Monochrome Receiver Functions

As in AM radio receivers, all the active portions of the typical monochrome TV receiver are contained on one or more ICs. Only the resonant networks, and possibly a few others, have to be provided externally. Typically, all the blocks shown in Figure 15-12 (except the picture tube, which may be CRT or crystal display) are contained within the integrated circuits.

The functions of each of the blocks contained in an IC or other solid state circuit are described below:

RF tuner—tunes to the selected TV channel and amplifies the RF signal. A local oscillator, also tuned to the proper frequency, heterodynes with the RF signal to produce the IF signal.

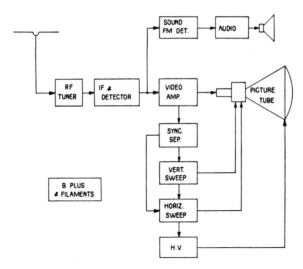

Figure 15-12. Monochrome receiver block diagram.

(Walter H. Buchsbaum, *Color TV Servicing*, 2nd Edition, © 1968. Reprinted by permission of Prentice Hall, Inc.)

IF amplifier and detector—provides amplification for the audio and video IF signals. At the last stage of the IF amplifier chain, a diode detector and IF filter remove the video signal and the 4.5-MHz intercarrier, frequency modulated sound signal.

Sound, FM detector—audio and video carriers are separated by 4.5 MHz. As a result, the detector following the IF amplifier will produce a 4.5 MHz signal, which contains the frequency modulated sound information. This signal is amplified, limited, and then detected by an FM detector.

Audio—several stages of wideband amplification provide a video signal of sufficient amplitude to drive the picture tube. The video amplifier also serves as source for the composite video signal, from which the synchronizing information is obtained.

Synchronizing separator—using clipping and limiting, followed by integrating and differentiating filters, the synchronizing separator circuit separates the horizontal and vertical synchronizing pulses and supplies them to their respective sweep sections.

Vertical sweep—contains a 60-Hz oscillator synchronized with the vertical synchronizing pulses, which, after suitable amplification, provides a vertical sawtooth deflection current to the deflection yoke.

Horizontal sweep—contains a 15,750-Hz oscillator synchronized by the horizontal synchronizing pulses, which, after suitable amplification, provides the horizontal sawtooth current to the deflection yoke. The horizontal output amplifier also drives the high-voltage flyback circuit.

High voltage supply—uses the "stepped-up" horizontal flyback pulses for rectification and filtering to generate the high voltage required for the anode of the picture tube.

B plus and filament power supply—converts the ac line voltage into the required dc voltages for vacuum tubes and transistors and provides ac filament voltage for the picture tube and any other vacuum tubes in the receiver.

Color Receiver Functions

The color TV receiver diagram of Figure 15-13 differs only in a few respects (usually it has a few more ICs) from the monochrome receiver diagram of Figure 15-12. The following receiver functions are unique for color receivers:

"Y" amplifier—essentially the same as the video amplifier in monochrome; it includes a small time delay, so that the Y or brightness information arrives at the color picture tube at the same time as the color information, which has to go through a decoder, etc.

Decoder, three-color amplifier—the color decoder consists of the

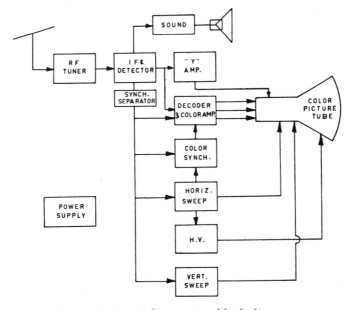

Figure 15-13.   Color receiver block diagram.

color subcarrier band-pass amplifier and the synchronous demodulator. The band-pass amplifier amplifies the color subcarrier and its sidebands. In the synchronous detector, the amplitude and phase modulation of the color subcarrier is compared against the reference signal, obtained from the color synchronizing section. The resultant output of two synchronous detectors is then matrixed into three color difference signals, corresponding to the red, green, blue components of the original image.

Color synchronizing section—contains a local 3.579-MHz oscillator, which is synchronized with the color reference burst, removed from the horizontal synchronizing and blanking pulse. The resultant reference signal is supplied to the decoder.

Color picture tube control circuits—provide convergence, focus, and blanking signals to the color picture tube (not shown in Figure 15-13). They utilize signals from the horizontal and vertical sweep sections and from the high-voltage supply. (See Color Picture Tube Parameters, discussed previously.)

15.4 NONBROADCAST TV SYSTEMS

When television is used to connect a camera with a receiver by coaxial cable, many of the requirements imposed on the commercial broadcast system can be eliminated. The simplest closed circuit TV system is one in which a single camera supplies information to a single receiver or monitor at a remote location. In this system, the vertical and horizontal scanning information can be transmitted over separate wires, directly linking the camera tube and the picture tube and thereby eliminating the need for the elaborate synchronizing circuits required in commercial broadcasting. Depending on the length of transmission line, however, suitable amplifiers must be provided. In general, where more than 100 feet of cable are involved, and where more than a single receiver is used, the synchronizing information is transmitted together with the video information, similar to the composite video signal of commercial broadcasting. In those cases, synchronizing circuits are again required.

CATV Principles

Community antenna TV systems generally utilize receive-only satellite TV earth stations to receive signals and retransmit them, via coaxial cable, to the individual subscribers' homes. Often the original broadcast frequencies are changed, for convenience in transmission over cable.

To enable transmission of a large number of TV channels, within the VHF band, the frequency spectrum between channels 6 and 7 (88 to 174 MHz) which is assigned to other services is used within the cable system to carry up to 14 additional channels.

CATV systems employ a large variety of different amplifiers, equalizers, line splitters, and terminations in order to distribute and amplify the RF video signals properly. In addition, some CATV systems provide locally originated programs, modulated on an available commercial TV channel, which are then transmitted over the cable. Some CATV systems offer two-way communication, either by using the low-frequency portion for the return or by using two separate cables.

In addition to providing free commercial TV, many hotels also offer special entertainment for pay on some of the unused channels. In this system, the basic CATV RF distribution system is used to bring the output of a master antenna and its amplifiers to each room. A special program channel carries video-recorded advertisements for the pay-TV channels. The viewer can then either call the hotel operator or put quarters into a special switch in the room and select one of the TV channels advertised on the program channel. The hotel phone operator can switch the desired channel onto that room's cable, or else some automatic coding and decoding scheme permits reception. In some systems the pay-TV channels are always on all cables, but are transmitted without the horizontal synch signals. When the hotel operator connects the room to the desired program, only the horizontal synch signal is really added.

A number of schemes for over-the-air pay TV have been developed and are under evaluation. In most of them, special decoding and synchronizing circuits are the key to letting the customer pay for what he wants to see.

15.5 TELEVISION RECEIVE-ONLY SATELLITE SYSTEMS (TVRO)

Satellite TVRO systems receive satellite signals from 3700 to 4200 MHz (called the C-band). At these frequencies transmission line loss is substantial; therefore the first stage of the satellite receiver (a down-converter) is located at the antenna, which is usually a parabolic reflector and low-noise amplifier designed using the Cassegrainian antenna lens principle (see Chapter 10, Figure 10-15).

In the simplest system, the output frequency from the antenna system is much lower (70 MHz) and can be transmitted over coaxial cable without appreciable loss.

In a basic system the single IF is transmitted over the coax to the receiver for further processing and is usually viewed on Channel 3 or 4. However, because there is only a single frequency transmitted over the coax, program selection must be done at the first IF of the satellite receiver (ordinarily called *master location*).

The TVRO system just described is obviously not desirable for motels, hotels, and other similar establishments. The satellite TVRO system used in this type of application is called *block down-conversion*. In this design the first IF stage passes the entire band of signals being transmitted by the geostationary satellite, placed at 22,300 miles above the earth. This permits the viewer to select any one of the programs being transmitted by the satellite since the selection can be done at the TV receiver, which has access to all satellite frequencies being beamed down. Now each TV receiver can be used to select the signal desired. A modern satellite can carry as many as 24 separate TV channels simultaneously, and there are numerous operating satellites transmitting TV signals to the continental United States. Assuming the receiving dish is capable of being moved from one satellite to another, a TVRO can receive as many as eight to ten (depending on geographical location, equipment design and quality, etc.). The general rule is the larger the dish, the more sensitivity the system will have.

Another factor that has considerable influence on the quality of TVRO received signals is the sensitivity of the low-noise amplifier (LNA). The lower the temperature in degrees Kelvin, the better the amplifier. In general, solid state LNA noise temperatures are somewhere between 50°K and 120°K. In practice, TVRO LNAs are usually 85° to 100°K.

Several applications use other arrangements rather than individual low-noise amplifiers; for instance, low-noise block down-converters in one package with low-noise converters, a combination of a LNA and a standard down-converter circuit.

High-performance low-noise amplifiers such as used in radio astronomy and deep space exploration generally require liquid helium (4.2°K or colder); some satellite systems use liquid nitrogen (77°K). In practice the entire LNA is immersed in the liquid coolant and placed at the focal point of the reflector of the earth station antenna system.

TV Network Distribution

Satellite and earth stations are being used extensively for distribution of television programming. For example, the Public Broadcasting Service (PBS) distributes its TV programming almost exclusively by use of satellite systems. Cable television companies are also using satellite TV channels to receive a great amount of their programming.

The present domestic satellite system uses mostly C-band transponders, as was explained in the preceding section (3.7–4.2 GHz down link and 5.925–6.425 up link). Currently under investigation are satellite systems using the Ku-band (12/14 GHz). However, there is no program distribution of television programming in the Ku-band being used on a day-to-day basis in the United States. It is predicted that such systems will be operational by 1990.

15.6 DIGITAL TELEVISION

There are three basic steps required in digital television: (1) sampling, (2) quantitizing, and (3) coding of an analog video (picture) signal. Typically, an analog-to-digital converter (A/D converter IC) receives a standard analog video signal. The analog-to-digital converter changes the analog signal to a digital signal, then this signal is processed before changing it back to analog by use of a digital-to-analog converter (D/A converter IC).

Much digital equipment used in the United States is designed with a sampling rate of three to four times the television color subcarrier frequency (10.7 or 14.3 MHz). This corresponds to 2.5 or 3.4 times the picture carrier bandwidth of 4.2 MHz.

The next step in digitizing the analog TV signal is to *quantitize* it. The NTSC television system can be properly quantitized using 256 levels (2 to the eighth power).

Finally, coding is required. Coding is simply assigning a binary code to designate each of the 256 levels. In general, 8-bit binary code groups are used to designate each of the 256 levels (see Chapter 16 for basic digital concepts).

15.7 MULTICHANNEL TELEVISION SOUND (STEREO TV)

The system chosen by the Broadcast Television Systems Committee as the current standard for stereo TV is one developed by Zenith; the noise reduction system was developed by dbx Corporation. This system has been approved by the Federal Communications Committee. Together, the two are referred to as the MTS system; they are the ones approved by the FCC for use in the United States.

FM stereo operation (see Chapter 12, Section 12.6) is somewhat similar in respect to the encoding methods shown in Figure 12-1. In the MTS system the stereo output contains L + R, L − R and pilot signals. The major difference in this system is the dbx compressor. The output is AM double-sideband suppressed carrier, modulated by a compressed L − R signal and summed with preemphasized (75 μs, same as radio and mono TV) L + R signal. The output from the stereo generator used in the system contains the stereo sum (mono) signal, the stereo difference signal, and the pilot tone.

Finally, one other component of the system is called a SAP generator. The outputs of the stereo generator and SAP generator are added together to form the composite signal that frequency modulates the TV audio carrier.

BIBLIOGRAPHY—CHAPTER 15

BLEAZARD, G.B., *Introducing Satellite Communications.* New York, NY, John Wiley & Sons, Inc., 1985.

"Cable Television." *FCC Rules*, Part 76.

FEHER, K. *Digital Communications, Satellite/Earth Stations Engineering.* Englewood Cliffs, NJ, Prentice-Hall, Inc., 1977.

HAMMING, R., *Coding and Information Theory.* Englewood Cliffs, NJ, Prentice-Hall, Inc., 1980.

"Instructional Television Fixed Service." *FCC Rules*, Subpart 1, Part 74.

International Radio Consulative Committee (CCIR), *Study Group 11, Broadcasting Service (Television).* Geneva, Switzerland, 1982.

"Remote Pickup Broadcast Stations." *FCC Rules*, Subpart D, Part 74.

SPILKER, J. J., *Digital Communications by Satellite.* Englewood Cliffs, NJ, Prentice-Hall, Inc., 1977.

"Television Auxiliary Stations." *FCC Rules*, Subpart F, Part 74, and Subpart H, Part 74.

"Television Broadcast Translators." *FCC Rules*, Subpart G, Part 74.

CHAPTER 16

DIGITAL LOGIC

16.1 FUNDAMENTALS OF DIGITAL LOGIC

The word *digital* is always associated with discrete steps, units of a given size or number, as opposed to *analog*, which refers to continuous variations. Digital signals are usually pulses, while analog signals may represent the human voice (audio), light variations (video), or other physical phenomena. In digital logic, discrete signals, such as pulses or voltage levels, are manipulated to produce results in terms of discrete quantities, such as numbers. Relays and switches are the original digital logic devices, and much of that terminology persists in today's solid state digital technology. The terms *switches* and *gates* are used interchangeably to refer to the logic elements. These devices can have only two states—they can have open or closed contacts. Two relays with contacts in series must both be energized to allow current to flow through their contacts. If the contacts are connected in parallel, either relay can control current flow. These are the concepts used in the two basic digital logic elements—the AND and OR circuits.

A simple diode AND circuit is shown, together with its logic symbol, in Figure 16-1. The various possible combinations of input and output signals are summarized in the truth table, in terms of 0 and 1. For convenience, we designate a positive voltage as 1 and a ground or negative voltage as 0. This convention is called *positive logic*, and the opposite, a positive voltage meaning 0 and a ground meaning 1, is called, appropriately enough, *negative logic*. Looking at the truth table of Figure 16-1, it is apparent that a 1 output can only occur when both A *and* B input are 1. This corresponds to the case where both relays must be energized, because their contacts are in series.

The second function is shown by the circuit, logic symbol, and truth table of Figure 16-2. Here a 1 output will occur if either the A *or* B input, or both, are a 1. This case corresponds to the two relays having contacts in parallel. Energizing either relay, or both, will allow current to flow.

In the circuits shown so far, the input and output signals are of the same polarity, but if transistors are used instead of diodes, it is

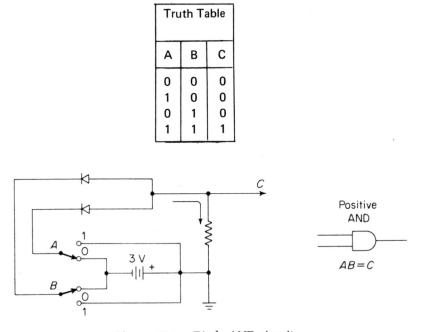

Truth Table		
A	B	C
0	0	0
1	0	0
0	1	0
1	1	1

Positive
AND

$AB = C$

Figure 16-1. Diode AND circuit.

(M. Mandl, *Electronic Switching Circuits: Boolean Algebra and Mapping*, © 1969. Reprinted by permission of Prentice-Hall, Inc.)

Truth Table		
A	B	C
0	0	0
1	0	1
0	1	1
1	1	1

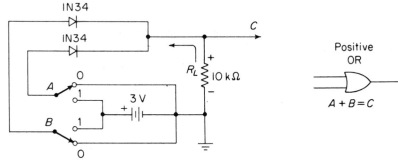

Figure 16-2.   Diode OR circuit.

(M. Mandl, *Electronic Switching Circuits: Boolean Algebra and Mapping,* © 1969. Reprinted by permission of Prentice-Hall, Inc.)

Truth Table		
A	B	C
0	0	1
1	0	1
0	1	1
1	1	0

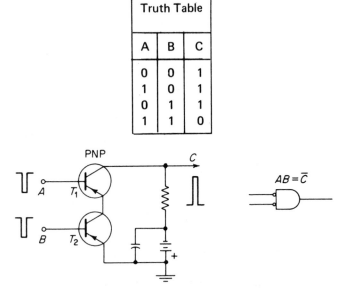

Figure 16-3.   Transistor NAND circuit.

(M. Mandl, *Electronic Switching Circuits: Boolean Algebra and Mapping,* © 1969. Reprinted by permission of Prentice-Hall, Inc.)

possible to invert the polarity and perform the logic function in a single stage. It is therefore simple to change the AND circuit into a circuit in which there is a 1 output only when both inputs are *not* 1. Since this circuit, shown in Figure 16-3 with its truth table and symbol, has an output only in the NOT AND condition, it is called the *NAND circuit.*

Applying the same reasoning to the OR circuit, it is possible to obtain the NOT OR or *NOR circuit,* shown with its truth table and symbol in Figure 16-4.

By combining the basic AND and OR circuits with inverting functions, it is possible to obtain the NAND and the NOR logic elements. By further combining these logic elements, it is possible to make up a very large variety of logic functions which can ultimately perform extremely complex computations, data processing, and digital control tasks.

Digital logic elements usually operate in rapid sequence, with pulse

Truth Table		
A	B	C
0	0	1
1	0	0
0	1	0
1	1	0

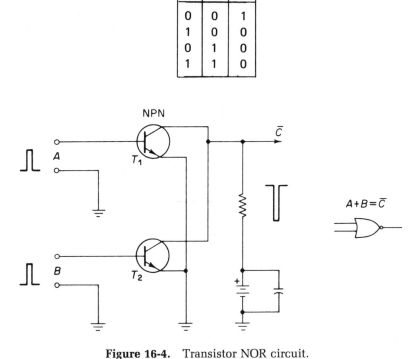

Figure 16-4. Transistor NOR circuit.

(M. Mandl, *Electronic Switching Circuits: Boolean Algebra and Mapping,* © 1969. Reprinted by permission of Prentice-Hall, Inc.)

trains as input and output signals. In order to provide some means of storing information, or delaying portions of the signal, it is possible to connect two gates (either NOR or NAND) in the circuit shown in Figure 16-5 to produce a flip-flop or multivibrator. For digital logic applications, the circuit of Figure 16-5 acts as a toggle switch which can be set to two different positions by the input signal. If the input to R (Reset) is unchanged when the input to S (Set) changes, the polarity at both output leads will change. If both inputs change in the same way (both positive or both negative) at exactly the same time, the output will not change. In logic applications, there are a number of combinations of flip-flops and logic gates to provide different functions. Flip-flops are represented symbolically by a rectangle with the input and output leads labeled.

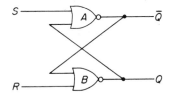

Figure 16-5.   RS flip-flop circuit.

(T. Sifferlen and V. Vartanian, *Digital Electronics with Engineering Applications*, © 1970. Reprinted by permission of Prentice-Hall, Inc.)

A typical combination of flip-flop and gates is shown in Figure 16-6, in which a pulse train or clock changes the flip-flop according to the 0 or 1 condition of the S and R inputs. When a set of such flip-flops is connected as shown in Figure 16-7, a binary counter is obtained. As successive pulses enter into the first trigger or clock input, the 0 or 1 condition of the flip-flops change. Binary arithmetic is described in a later section.

The binary counter shown in Figure 16-7 can be used in a variety of ways. Each flip-flop (FF) stage can act to divide the number of pulses by two, because only the rising part of the pulse can change its state. That means two successive pulses are needed to produce one complete output pulse. When used in this manner, the binary counter becomes a binary divider. Two stages divide by four, three stages divide by eight, and so on.

With the addition of AND gates a chain of FFs can become a shift register, as illustrated in Figure 16-8. A data signal is applied at the set input of the first FF, together with the clock signal. At the same time the inverse data signal together with the clock signal is applied to the

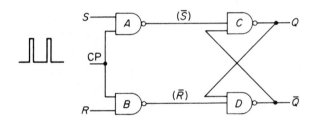

Figure 16-6. Clocked flip-flop circuit.

(T. Sifferlen and V. Vartanian, *Digital Electronics with Engineering Application*, © 1970. Reprinted by permission of Prentice-Hall, Inc.)

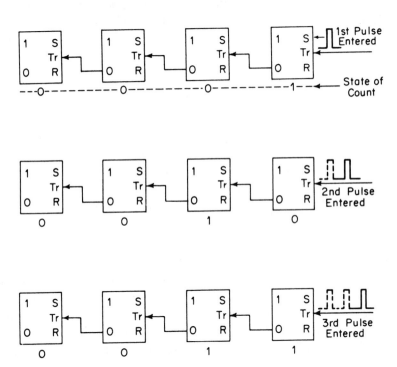

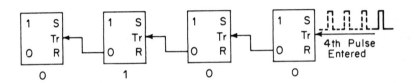

Figure 16-7. Binary counter operation.

(M. Mandl, *Electronic Switching Circuits: Boolean Algebra and Mapping*, © 1969. Reprinted by permission of Prentice-Hall, Inc.)

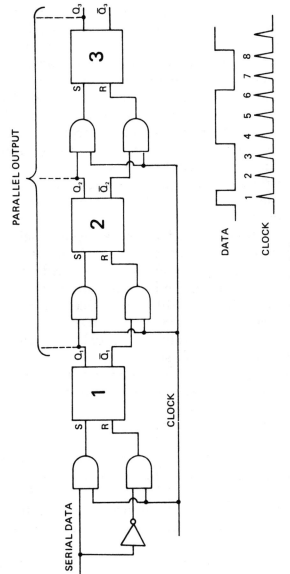

Figure 16-8. Simple shift register.

(Walter H. Buchsbaum, *Complete Guide to Digital Test Equipment,* © 1977. Reprinted by permission of Prentice-Hall, Inc.)

reset input. As the clock signal pulses progress, the data signal is shifted, step by step, into the FFS. At the end of the desired number of clock signals, eight in this example, the data is contained in eight FFs (only three are shown in Figure 16-8). It is possible to stop and restart the shifting process at a later time, using the shift register as temporary storage for the data signal. It is also possible to obtain a parallel output of the data by connecting to the Q terminals of the FFs.

Shift registers can be obtained which can enter data in the parallel mode and read them out serially. By proper gating it is possible to shift in both directions, obtain the complement of a data train (connecting to the Q terminals), and recirculate the data repeatedly through the shift register. In real life applications a variety of shift registers are obtainable with large scale integrated (LSI) devices that perform essential functions in computers and other digital systems.

The basic shift register, in a special arrangement, is also used in Random Access Memories (RAM) and Read Only Memories (ROM), described in Chapter 17.

LSI devices are available which combine specific logic functions, starting from binary-to-decimal converters up to complete microcomputers. Figure 16-9 shows a widely used Binary-Coded-Decimal (BCD) to decimal converter, with all of the logic contained on a single chip. Another LSI is shown in Figure 16-10. This device converts BCD numbers to the seven-segment format necessary to drive numerical displays. Note that in this device a relatively large number of gates are used, followed by seven transistor driver stages. Provision for lamp test (LT) and multiplexed driving power is included in this chip. Chapter 19 covers the display applications served by such devices.

Standard Logic Symbols and Functions

MIL-STD 806B "Graphic Symbols for Logic Diagrams" is probably the most widely accepted standard because the Department of Defense has been the largest customer of digital devices. The MIL-STD logic symbols define the logic functions to be performed rather than the electronic circuit and are therefore equally applicable for positive and negative logic.

In positive logic, the positive voltage level at the inputs is defined as a 1 and the negative voltage level is defined as a 0. In negative logic, the negative voltage level at the inputs is defined as a 1 while the positive voltage level is defined as a 0. The logic functions along with the symbols are shown in Table 16-1.

The lines entering or leaving the symbols are the signal flow lines. A line entering or leaving a circle indicates a "low" signal level; a

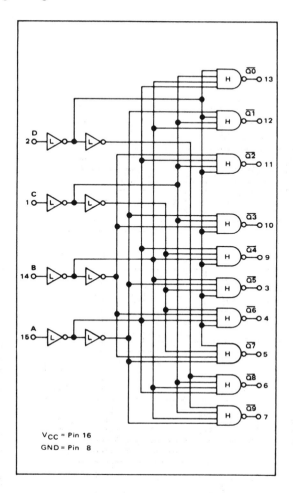

V_CC = Pin 16
GND = Pin 8

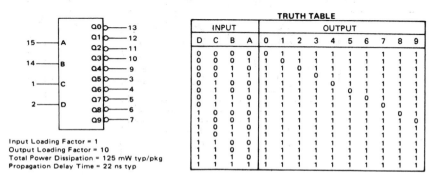

Q0	—13							

Input Loading Factor = 1
Output Loading Factor = 10
Total Power Dissipation = 125 mW typ/pkg
Propagation Delay Time = 22 ns typ

				TRUTH TABLE									
INPUT				OUTPUT									
D	C	B	A	0	1	2	3	4	5	6	7	8	9
0	0	0	0	0	1	1	1	1	1	1	1	1	1
0	0	0	1	1	0	1	1	1	1	1	1	1	1
0	0	1	0	1	1	0	1	1	1	1	1	1	1
0	0	1	1	1	1	1	0	1	1	1	1	1	1
0	1	0	0	1	1	1	1	0	1	1	1	1	1
0	1	0	1	1	1	1	1	1	0	1	1	1	1
0	1	1	0	1	1	1	1	1	1	0	1	1	1
0	1	1	1	1	1	1	1	1	1	1	0	1	1
1	0	0	0	1	1	1	1	1	1	1	1	0	1
1	0	0	1	1	1	1	1	1	1	1	1	1	0
1	0	1	0	1	1	1	1	1	1	1	1	1	1
1	0	1	1	1	1	1	1	1	1	1	1	1	1
1	1	0	0	1	1	1	1	1	1	1	1	1	1
1	1	0	1	1	1	1	1	1	1	1	1	1	1
1	1	1	0	1	1	1	1	1	1	1	1	1	1
1	1	1	1	1	1	1	1	1	1	1	1	1	1

Figure 16-9.   BCD-to-decimal decoder.

(Walter H. Buchsbaum, *Complete Guide to Digital Test Equipment*, © 1977. Reprinted by permission of Prentice-Hall, Inc. Illustration courtesy of Motorola Semiconductor Products Div.)

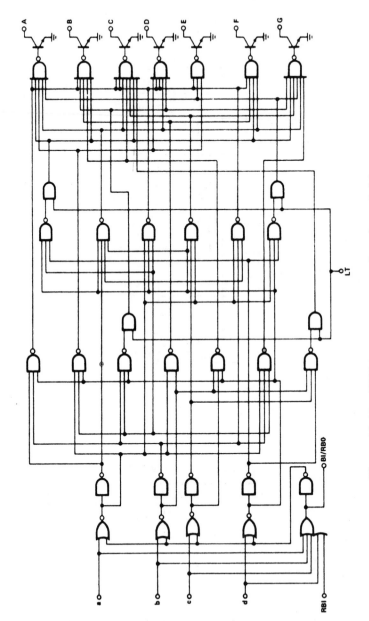

Figure 16-10. BCD-to-seven segment decoder and driver.

(Walter H. Buchsbaum, *Complete Guide to Digital Test Equipment*, © 1977. Reprinted by permission of Prentice-Hall, Inc. Illustration courtesy of Signetics.)

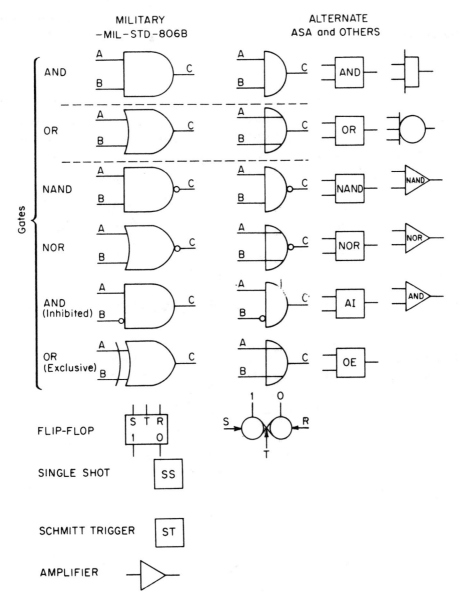

Table 16-1.   Logic functions and symbols.

(Harry E. Thomas, *Handbook of Transistors, Semiconductors, Instruments, and Microelectronics,* © 1968. Reprinted by permission of Prentice-Hall, Inc.)

plain line indicates a "high" signal level. The symbols are defined for the active state of the device.

Numbering Systems

All numbering systems are mathematically constructed in identical fashion in that they have a base number where powers are used to provide increments of magnitude. For example, in the 10 or decimal system, the number $32,684_{10}$ is made up of:

$$3 \times 10^4 + 2 \times 10^3 + 6 \times 10^2 + 8 \times 10 + 4 \times 1$$

$$\text{or } 4 \times 10^0 = 4 \times 1 \qquad = \qquad 4$$

$$8 \times 10^1 = 8 \times 10 \qquad = + \qquad 80$$

$$6 \times 10^2 = 6 \times 100 \qquad = + \qquad 600$$

$$2 \times 10^3 = 2 \times 1000 \qquad = + \qquad 2000$$

$$3 \times 10^4 = 3 \times 10,000 = + \underline{\quad 30,000}$$

$$\text{Total} \qquad\qquad\qquad 32,684_{10}$$

In the binary system, the base $n = 2$ is used to form any number by assigning fixed positions to each power of 2. For example:

2^4	2^3	2^2	2^1	2^0
16	8	4	2	1
1	1	0	0	1

Where is summed up as follows

$$1 \times 2^0 = 1$$

$$0 \times 2^1 = 0$$

$$0 \times 2^2 = 0$$

$$1 \times 2^3 = 8$$

$$1 \times 2^4 = \underline{16}$$

$$\text{Total} \qquad\qquad 25_2$$

Table 16-2 lists the binary equivalents of decimal numbers for converting numbers from decimal to binary.

Binary Arithmetic

Binary arithmetic follows the same rules as those used for decimal arithmetic. The basic functions of addition, subtraction, multiplication, and

Decimal to Binary Conversion Table

Decimal number	Binary number	Number of "1's" in binary number
0	0	0
1	1	1
2	10	1
3	11	2
4	100	1
5	101	2
6	110	2
7	111	3
8	1000	1
9	1001	2
10	1010	2
11	1011	3
12	1100	2
13	1101	3
14	1110	3
15	1111	4
16	10000	1
17	10001	2
18	10010	2
19	10011	3
20	10100	2
21	10101	3
22	10110	3
23	10111	4
24	11000	2
25	11001	3
26	11010	3
27	11011	4
28	11100	3
29	11101	4
30	11110	4
31	11111	5
32	100000	1
33	100001	2
34	100010	2
35	100011	3

Table 16-2.   Binary equivalent numbers.

(Gerald A. Maley, *Manual of Logic Circuits*, © 1970. Reprinted by permission of Prentice-Hall, Inc.)

Decimal number	Binary number	Number of "1's" in binary number
36	100100	2
37	100101	3
38	100110	3
39	100111	4
40	101000	2
41	101001	3
42	101010	3
43	101011	4
44	101100	3
45	101101	4
46	101110	4
47	101111	5
48	110000	2
49	110001	3
50	110010	3
51	110011	4
52	110100	3
53	110101	4
54	110110	4
55	110111	5
56	111000	3
57	111001	4
58	111010	4
59	111011	5
60	111100	4
61	111101	5
62	111110	5
63	111111	6
64	1000000	1
65	1000001	2
66	1000010	2
67	1000011	3
68	1000100	2
69	1000101	3
70	1000110	3
71	1000111	4
72	1001000	2
73	1001001	3
74	1001010	3
75	1001011	4
76	1001100	3

Table 16-2. Continued

Decimal number	Binary number	Number of "1's" in binary number
77	1001101	4
78	1001110	4
79	1001111	5
80	1010000	2
81	1010001	3
82	1010010	3
83	1010011	4
84	1010100	3
85	1010101	4
86	1010110	4
87	1010111	5
88	1011000	3
89	1011001	4
90	1011010	4
91	1011011	5
92	1011100	4
93	1011101	5
94	1011110	5
95	1011111	6
96	1100000	2
97	1100001	3
98	1100010	3
99	1100011	4
100	1100100	3

Table 16-2. Continued

division are still performed, but digital circuits perform these arithmetic operations in binary forms instead of using the paper and pencil method people have learned for the decimal system.

The basic function in any arithmetic is addition, with subtraction the inverse of addition. Multiplication can be considered a series of successive additions. The product of 5 times 4 really means "What is the sum of 5 added to itself 4 times?" Division is the inverse of multiplication and can be considered as "How many times can one number be subtracted from another?" If we look at $12/3 = (12 - 3) - (3) - (3) - (3)$, we see that four successive subtractions are performed, and the answer is 4.

In binary arithmetic, the subtraction function can be performed by a form of addition. There are two techniques for performing this operation; both require "complementing" the number to be subtracted.

The complement of a binary number is obtained by changing all 1's to 0's and all 0's to 1's. The complement of 11001 (25) is 00110 (6). Note that 11111 equals 31, which is also the sum of 25 + 6.

In the 2's-complement method, 1 is added to the complement of the subtrahend (number being subtracted), and the minuend (number being subtracted from) is then added to it.
Example: 15 − 9 = + 6;

01111 (15)	01001 = 9
−10111 (2's complement of 9)	10110 = complement of 9
100110 sum	10110 + 1 = 10111
	= 2's complement of 9

Since the sixth digit to the left above is a 1, the difference is positive and the resulting difference is (+00110) = + 6. If the $n + 1$ digit (n = number of digits in the minuend) had been a zero, then the answer is negative and equal to the 2's complement of the sum.

The 1's-complement method works in a similar manner. The subtrahend is complemented and the minuend is added to the subtrahend's complement. If the $n − 1$ position is 1, the difference is positive and 1 is added to the sum to get the correct difference. This operation is termed an *end-around carry*.
Example:

$$\text{to subtract } 8 − 6 = 2 \quad 0110 = 6$$

$$
\begin{array}{r}
1000 = 8 \\
+ \quad 1001 \quad \text{(1's complement of 6)} \\
\hline
10001 \\
+1 \\
\hline
10010 \rightarrow 0010 = +2
\end{array}
$$

If the sum has a 0 in the $n + 1$ digit position, the difference is negative and the correct answer is found by complementing the sum.

The 1's complement with "end-around carry" is more commonly used than the 2's complement.

Multiplication and division operations are performed as a continuous series of addition and subtraction operations. Detailed explanations of binary arithmetic are found in the references at the end of this chapter.

Fundamentals of Boolean Algebra

Boolean algebra, originated by George Boole (1815–1864), set symbolic logic into mathematical form. It is currently taught as the algebra of sets, as compared to the more common algebra of fields, and is based on the following postulates:

$$AND \qquad OR \qquad NOT \text{ (complement)}$$

$$0 \cdot 0 = 0 \quad 0 + 0 = 0 \quad \overline{1} = 0$$
$$0 \cdot 1 = 0 \quad 0 + 1 = 1 \quad \overline{0} = 1$$
$$1 \cdot 0 = 0 \quad 1 + 0 = 1$$
$$1 \cdot 1 = 1 \quad 1 + 1 = 1$$

Axioms

A, B, C, etc. represent variables, each of which is a literal symbol. The complement of a variable is represented by $\overline{A}$, $\overline{B}$, $\overline{C}$. Variables can only have values of 1 or 0; therefore, proofs become simple.

The symbols + (OR) imply ADDITION and · (AND) imply MULTIPLICATION.

Identity Laws (1 and 0 laws)
$$0 \cdot A = 0; \quad 0 + A = A$$
$$1 \cdot A = A; \quad 1 + A = 1$$

Commutative Law $\quad A + B = B + A; \quad A \cdot B = B \cdot A$

Associative Law $\quad A + (B + C) = (A + B) + C;$
$$A(BC) = (AB)C$$

The order of the numbers does not change the result. These laws are the same as in algebra (see Chapter 23).

Distributive Law (1) $\quad A(B + C) = A \cdot B + A \cdot C$
Same as in algebra. Multiplying the sum of two variables by another variable is the same as multiplying each by the third and adding the sum.

Distributive Law (2) $\quad A + (B \cdot C) = (A + B) \cdot (A + C)$
New for Boolean algebra. Adding a variable to the product of two variables is the same as multiplying the sum of the first and second by the sum of the first and third.

Absorption Law $\quad A + AB = A \quad A(A + B) = A$
New for Boolean algebra. $\quad A + \overline{A}B = (A + B)$
Since all variables can have only two values, 1 or 0, the above relationships hold true.

Complementary Law $\quad A + \overline{A} = 1 \quad A \cdot \overline{A} = 0.$
New for Boolean algebra. A variable can have only two values, 1 or 0, and only one state can exist at a time; therefore, the above law holds true from the first postulates.

Indempotence Law $\quad A + A = A \quad A \cdot A = A$
New for Boolean algebra. In Boolean algebra there are no expo-

nents (where $A \cdot A$ results in A^2), and the values are confined to 1 and 0. For this condition, the law is true. This law can be demonstrated using switching logic. If $A \cdot A$ are two switches in series, each has the same circuit; when one is closed, the other is closed. The two can be replaced with one. If $A + A$ are two switches in parallel, each having the same circuit, when one is closed the other is also closed. The two can be replaced by one.

De Morgan's Theorem $(\overline{A + B}) = \overline{A} \cdot \overline{B}$ $\overline{A \cdot B} = \overline{A} + \overline{B}$

New for Boolean algebra. The complement of a sum of N variables is equal to the product of the separate variables' complement. The complement of a product of n variables is equal to the sum of the separate variables' complement.

Principles of Karnaugh Maps

A multivariable function can be reduced to minimum form by the use of the Boolean algebra laws stated above; however, for more than three variables, it is simpler to visualize and manipulate the logic by the use of a Karnaugh Map. A Karnaugh Map allocates a position for all the truth table possibilities.

Figure 16-11 shows the buildup of the Karnaugh Map. A single variable has two states, 1 and 0. This is shown in Figure 16-11a. The 0 state is labeled $\overline{A}$ and the 1 state is labeled A. The map contains a cell (square) for each of the states of the variable. A two-variable map is generated by folding down the one-variable map as shown in Figure

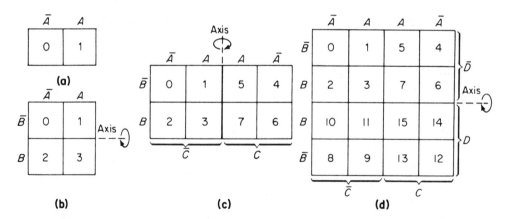

Figure 16-11.   Karnaugh Map construction.

(T. Sifferlen and V. Vartanian, *Digital Electronics with Engineering Applications,* © 1970. Reprinted by permission of Prentice-Hall, Inc.)

16-11b. The folding is indicated by the axis, and the second variable becomes a mirror image of the first. A three-variable map is generated by folding the two-variable map along the axis shown in Figure 16-11c. The increase of the map is always twice its size, so that any map has 2^n cells where n is equal to the number of variables. The three-variable map, for example, has 8 cells; $2^3 = 8$. The original cells are represented by $\overline{C}$ and the new cells by C. The four-variable map is obtained by folding down the three-variable map along the axis as shown in Figure 16-11. An examination of the maps developed shows the following relationships:

1. Going from one cell to an adjacent cell, only one variable changes. This is true going horizontally or vertically. It also holds true for a movement from one end cell to the other end cell in the same row or column.

2. The vertical outside columns of the map can be placed next to each other without affecting the map.

3. In like manner, the horizontal outside rows of the map can be placed next to each other without affecting the map.

4. Each time the map is enlarged, the original cells are the NOT portion of the new variable and the new cells are the TRUE portion of the new variable.

5. Adjacent cells form a couple, which eliminates one variable in the function.

6. In enlarging the map, the numerical representation of the new cells is obtained by adding 2^{n-1}, where n = number of variables in the map to the corresponding cells. In the four-variable map, the new cells are numerically represented by adding 8 ($2^{4-1} = 2^3 = 8$) to the corresponding cells of the three variable map.

7. The true values of the variables from a truth table are entered in the proper squares of the map as 1. The false values are entered as a 0.

8. All adjacent 1's are circled in groups of powers of 2; that is, 2, 4, 8, etc.

Using the Karnaugh Map

Figure 16-12 is the map of

$$F = A\overline{B}C + \overline{A}\,\overline{B}C + AB\overline{D} + \overline{A}B\overline{C} + BC\overline{D} + BCD$$

Three quadruples (groups of four adjacent 1's) can be formed. (Note:

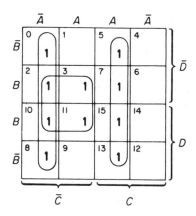

Figure 16-12.   Use of Karnaugh Map.

(T. Sifferlen and V. Vartanian, *Digital Electronics with Engineering Applications,* © 1970. Reprinted by permission of Prentice-Hall, Inc.)

The cells of numerals 7 and 15 could have been joined with 3 and 11 instead of 6 and 14 for solution.) The selection is left to the designer, based upon which signals are available. The function is reduced to

$$F = A\overline{C} + \overline{A}C + BC = C(\overline{A} + B) + A\overline{C}$$

The map also represents those functions which have a 0 marked in them. The NOT function could be determined for these:

$$F = \overline{\overline{A}\,\overline{C} + A\overline{B}C}$$

The designer, in some cases, may find this easier to implement, particularly if using NOR logic gates.

Logic network functions can be categorized as one of the two types, either a product of sums or a sum of products.

A Karnaugh Map generates a sum of products logic network, that is, a group of ANDed functions being summed or ORed to give a single function output.

16.2 FUNDAMENTALS OF LOGIC DESIGN

Basic Flow Charting

Flow charting is a technique used in various phases of digital logic and computer design, and is generally used at the following levels:

1. Circuit design analysis

2. System signal flow

3. Computer program design (same level as system signal flow)

Circuit Flow Charting

Logic systems are made up of large numbers of sequentially operating circuits, and the outputs of these circuits are functions of both present and past input signals. These input signals may be external signals or internal feedback or loop signals. Timing diagrams are used to show the sequence of operation, but timing diagrams do not represent the circuit operations for different input sequences. Just like individual circuits, whole networks have stable and unstable states. The stable state is shown in the truth tables, as described in the previous pages. The unstable state occurs normally during the transient or transitional period between two stable states as a result of an input signal. The unstable states can cause race conditions and false inputs to other circuits.

Circuit flow charting utilizes the Karnaugh Map to show diagrammatically all the possible output states for all the input signals. There are seven basic steps to network analysis by flow chart.

1. Label all the gates (AND, OR, NOR, etc.) that make up the network in some appropriate manner (by decimal number, letter, etc.).

2. Write the Boolean expression for the output of each of these gates. This is done from the immediate inputs of the gate.

3. Draw the flow chart, which looks the same as a Karnaugh Map, using one column for each external input and one for each internal input combination.

4. Enter the Boolean expression for the output of gate 1 to the left of each column, using the Karnaugh Map as a chart.

5. Enter the Boolean expression for the output of the rest of the gates to the right of gate 1, in proper sequence, the highest gate number farthest to the right.

6. Identify the stable and nonstable states. When the binary number within the square is equal to the binary number labeling the row, the state is stable; all others are unstable.

7. To go from one stable state to another requires going through all unstable states. The binary number within the square indicates the row to which the operating point will move. The movement stays

within the same column since it is assumed that the external inputs have not changed during this time. The movement of the operating point is depicted by arrows going from one state to another.

System Signal Flow Charting

System signal flow charting uses some of the same symbols used in computer system flow charting, as shown in Figure 16-13. Flow charting lists the sequence of events to be performed, step by step, to complete a function.

For example: Multiply two 4-bit numbers ($5 \times 7 = 35$).

$$
\begin{array}{r}
0101 = 5 \\
\times \quad 0111 = 7 \\
\hline
0100011 = 35
\end{array}
$$

To perform the multiplication requires the use of four registers: Register A—Multiplicand (number being multiplied) Register, Register B—Adding Register, Register C—Multiplier Register, and an Accumulator Register for the answer. Multiplication is performed by successive addition and shifting. Writing out the sequence of events:

1. Put Multiplicand into Register A.
2. Put Multiplier into Register C.
3. Clear Register B.
4. Clear Accumulator Register.
5. Look at Least Significant Bit (LSB) of C.
6. If zero, shift LSB of B to Accumulator.
7. If 1, Add A to C and put sum in B.
8. Shift LSB of B to Accumulator.
9. Shift Register C 1 bit.
10. Look at LSB of C.
11. If zero, shift LSB of B to Accumulator C.
12. If one, add A to C and put sum in B.
13. Shift LSB of B to Accumulator.
14. Repeat 5.

The flow chart of Figure 16-13 explains the previous steps in a pictorial form. Note that only three different ''actions'' symbols (Input/Output, Decision, and Sequential Flow) are required to describe all steps.

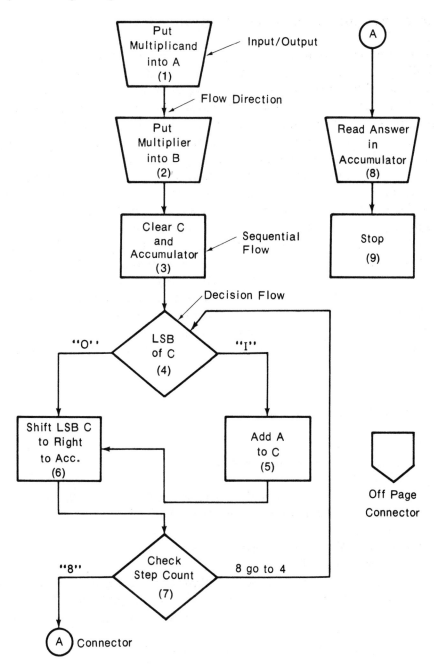

Figure 16-13.   Flow chart of a multiplication.

Synchronous and Asynchronous Logic Systems

A synchronous logic system operates from a fixed time source (clock), and all operations are related (or synchronized) to some submultiple of the source. A synchronous system runs continuously after being started initially as the result of some external action. The resulting sequence of events is in synchronism with the starting event. Synchronous systems are easier to design and implement since all operation occurrences are assigned a unique time period within the sequence.

Asynchronous systems have operations occurring in apparently random fashion, since the controlling factor is the function to be performed. Asynchronous operation allows for greater flexibility and a savings in hardware and system cycle time, though such systems are more difficult to design. Asynchronous systems are the result of a large number of random timed inputs of varying length and frequency.

Synchronous systems with multiple inputs poll or multiplex the inputs. In these systems, each input is polled in its proper turn; if information is available, it is acted upon before going on to the next input line. Asynchronous systems utilize a technique of "First in, first out" (FIFO), whereby the input advises the system that it has information.

BIBLIOGRAPHY—CHAPTER 16

BUCHSBAUM, W. H., *Encyclopedia of Integrated Circuits: A Practical Handbook of Essential Reference Data.* Englewood Cliffs, NJ, Prentice-Hall, Inc., 1986.

GENN, R. C., *Digital Electronics: A Workbench Guide to Circuits, Experiments and Applications.* Englewood Cliffs, NJ, Prentice-Hall, Inc., 1982.

McCLUSKEY, E. J., *Logic Design Principles.* Englewood Cliffs, NJ, Prentice-Hall, Inc., 1986.

MICHELSON, A., and A. LEVESQUE, *Error-Control Techniques for Digital Communication.* New York, NY, John Wiley & Sons, Inc., 1985.

RHYNE, T. V., *Fundamentals of Digital System Design.* Englewood Cliffs, NJ, Prentice-Hall, Inc., 1973.

TROY, H. NAGLE, Jr., D. IRWIN, and C. IRWIN, *An Introduction to Computer Logic.* Englewood Cliffs, NJ, Prentice-Hall, Inc., 1975.

CHAPTER 17

COMPUTER PRINCIPLES

Modern computers have their origins in a differencing machine built over a century ago by Babbage. The first large-scale computers were developed after World War II in university laboratories: the ENIAC by the University of Pennsylvania and the EDSAC by Cambridge University in England. These were the start of the so-called "first generation" of modern computers; they were characterized by vacuum tube elements, extremely large size, frequent failure rate, and slow operation (by today's standards). The second generation of computers uses discrete solid state elements, such as diodes and transistors, is considerably smaller in size than the first generation, and is more reliable and faster. The third or current generation of computers is characterized by integrated circuit elements, high speed, excellent reliability, and high capacity. Possibly the biggest advantage of third generation machines over their predecessors is a sophisticated internal logic which enables a number of functions to take place almost simultaneously (called multiprocessing).

17.1 ANALOG COMPUTERS

When using the word *computer* today, generally we mean *digital computer*. There is, however, another class of computers–*analog computers*–still in use. The basic difference between analog and digital computers is that analog machines deal with voltage or current as analogous to such physical phenomena as light, heat, and gravity, while digital computers deal with numbers directly. To calculate the trajectory of a missile, for example, the muzzle velocity is equated to a starting voltage in an analog computer, while it is entered as the number of meters per second into a digital computer. The analog computer output will be a curve tracing, but the digital computer will print out a set of numbers. Analog computers operate in an environment which is time-based, either real time as in a manufacturing process, or some fictitous time base which is 10 to 100 times faster than real time. Analog computers use elements that add or intergrate voltages or their time rates of change. The programming complexity and accuracy (approximately slide-rule equivalency) of analog computers has made them slowly give way to digital computers.

17.2 DIGITAL COMPUTERS

Digital computers range in size and complexity from the so-called "computer on a chip" to a system being developed by Princeton University researchers. The researchers are attempting speeds of about 60 gigaflops using a 128-node system. Each node in the Princeton computer has twenty-four 32-bit emitter-coupled-logic microprocessors and up to 2 gigabytes of memory.

Other specialized computers are called *supercomputers* and *minisupercomputers*; in these applications, most use 64-bit rather than the 32-bit hardware devices found in the newer business and personal computers.

Due to the complexity and variety of today's computers, the following discussion of computer hardware and software (programming) must of necessity be quite general and limited to fit into the scope of this work. Those interested in a more detailed study of this field are referred to the bibliography at the end of this chapter.

17.3 BASIC STRUCTURE

A digital computer, no matter what the size, is composed of the following elements: a control unit, an arithmetic unit, a memory, and an input/output unit. These units interact as shown in Figure 17-1. The

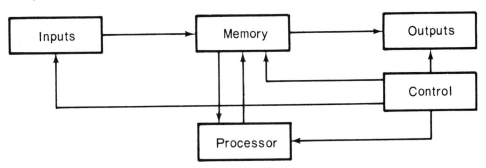

Figure 17-1. Basic computer structure.

size and complexity of these elements vary from computer to computer. The boxes labeled Inputs and Outputs may in actuality consist of many peripheral devices such as disk drives, line printers, and even several other computers. The basic function, however, is that of interfacing the computer to some other device so that there can be a transfer of digital data.

The different elements are connected by data buses, groups of wires or optoelectronic devices which carry the signals between the elements. Moving numbers between elements consists of moving one or more words of data (which could be instructions) in either a serial or parallel form. If parallel movement is used, the bus will consist of one wire for each bit of the word. It is quite possible to have parallel operation between certain sets of elements and serial operation between others. Most computers with their high-speed operational requirements would use parallel operation, and are also designed to handle serial operation. Naturally, each element must contain the required gating logic to ensure appropriate operation.

Before discussing each of the functional computer elements, some general aspects must be considered. A digital computer is basically a device that can follow a logical set of instructions (program) sequentially. The most basic form of these instructions is either to move data or to compare data. These instructions can consist of any of the following:

1. Moving a number between memory and the processing unit.

2. Adding one number to another.

3. Moving a number between an input/output device and the processing unit.

4. Comparing a number to zero and, on the basis of the comparison, either continuing with the instruction sequence or jumping to an instruction further on in the sequence to begin execution there (known as branching).

This last capability allows the computer to go through what is known as a looping operation. The computer can execute the same set of instructions more than once, applying them to different data. For example, we can ask the computer to set the value of an index to 1, then instruct it to perform the following:

1. Take the value of the index and output it.
2. Multiply the value of the index by itself (done by successive additions).
3. Output the squared value.
4. Add 1 to the index value.
5. Subtract 1200 from the index value and test whether the result is greater than 0. If so, stop execution; if not, branch to the first instruction.

By using the comparing and branching capability of the computer, the operations of sorting and merging can be performed. Modern computers operate with great speed—for example, gigaflops (on-off)—and the industry is talking femto flops (10^{-15}) for supercomputers.

17.4 CENTRAL PROCESSING UNIT (CPU)

The control and arithmetic units of a computer are the heart of the device. The two are sometimes lumped together into what is called a central processing unit (CPU). The basic microprocessors usually consist of only the control and arithmetic units, but most manufacturers are now including memories with their microprocessors. Figure 17-2 shows the basic CPU block diagram.

The control unit determines the address of the next instruction to be processed, determines the source of the next data, fetches it, processes it, and returns the result to a predetermined location. The control unit also includes a timing unit and activates or deactivates all the gating and buffering devices in the computer. It determines when data is to be passed into memory or when it goes into an input/output device, by opening or closing the respective logic gates. If it takes three timing pulses, for instance, to retrieve a word of data from memory, there must be a pause of that long a time to ensure proper operation. If the data were acted on before it was fully retrieved, wrong answers would result.

The timing or clock pulse frequency must be compatible with the frequency capabilities of the operating circuits. Usually, the faster the clock pulse, the faster the speed of operation of the computer.

Some systems have what is known as decentralized control. In this

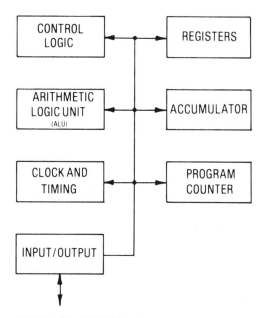

DATA, PROGRAM, etc.

Figure 17-2.   Block diagram of central processing unit.

case, there would be a main control unit which controls the peripheral
units. This sort of control is particularly applicable to input/output
buffers, where it is desired to have the buffers load or unload themselves independent of what the computer is doing. Each buffer indicates
when it is ready to have its data transferred to the computer in a read
operation or to stop having the computer transfer data to it in a write
operation.

The arithmetic unit performs the arithmetic and logical (comparison) operations. Although all sorts of schemes have evolved for arithmetic operations, the basic computer need only have the capability to
add two numbers and to shift the digits in one number to the next lower
or higher order position. That is, shifting the digits of the number 123
to the next higher order position would result in the number 1230.
Subtraction is achieved by adding the complement of one number to
the other. Multiplication (or division) involves adding two numbers,
then shifting one number to a higher order position (or lower for division), adding again, and so on.

In order to perform these operations, the arithmetic unit contains
a number of subunits called *registers*. Registers and their operation are
described in Chapter 16. A typical simple computer would have at least
three arithmetic registers, while some machines have many more. Three

registers are required to perform all necessary arithmetic operations, but a special purpose computer could have less. A greater number of registers enables a good programmer to get more powerful use out of the computer. In the case of only three arithmetic registers, one is for multiplication, one for division, and one to cumulate the result of all arithmetic operations, called the *accumulator*. The accumulator is, therefore, the register from which all results are moved to memory or to an output unit. The particular instruction being worked on is normally also in a register called the *instruction register*. In addition, there is also a register called a *program counter* which simply keeps track of which step of the program is to be executed. Thus, at the beginning of a program execution, the program counter is set to the value of the memory location in which the first program instruction is stored. This instruction is fetched into the instruction register, acted on, and the program counter incremented by one. The cycle then repeats with the instruction in the next memory location being moved into the instruction register, the instructions residing in memory in sequential order. In the case of a branching instruction, the program counter is incremented by some number greater than one (or decremented) so as to point to a memory location not within the original sequence.

In a typical CPU many additional logic functions are included, such as the logic which senses when the product of a multiplication or

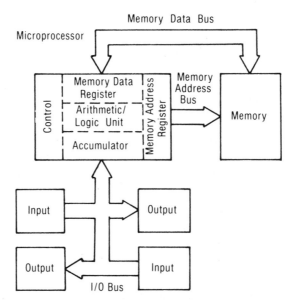

Figure 17-3.   Microcomputer with multiple bus.

(Hilburn, J. and Julich, P., *Microcomputers/Microprocessors: Hardware, Software and Applications*, © 1968. Reprinted by permission of Prentice-Hall, Inc.)

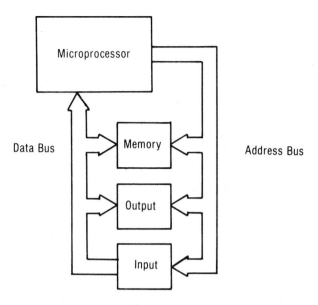

Figure 17-4. Microcomputer with single data bus.

(Hilburn, J. and Julich, P., *Microcomputers/Microprocessors: Hardware, Software and Applications*, © 1976. Reprinted by permission of Prentice-Hall, Inc.)

division will exceed the capacity of the machine (as when dividing by zero). Sometimes such logic is under partial or full programmer control.

For large computers, all sorts of configurations exist, with the control unit interacting directly with all other units. For small computers, the interaction could be indirect. For example, Figure 17-3 shows a microcomputer with a data bus for memory and for input/output, while Figure 17-4 shows a microcomputer with a single data bus.

17.5 INPUT/OUTPUT UNITS (I/O)

The input/output (I/O) element (for example the RS-232C, a serial interface) consists of one or more devices which bridge the gap between the computer and the outside world. Typical peripheral devices are tape recorders, disk drives, teletypewriters, printers, CRT terminals (Figure 17-5), card readers and punches, drum memories, optical scanners, etc. Most computers have a buffering device attached to some or all of the input/output elements. This buffering device is simply a separate memory, which temporarily holds a block of data. This is useful in a multiprocessing environment where the computer is processing several programs almost simultaneously. When a particular program calls for data to be read off a disk, the computer simply sends an in-

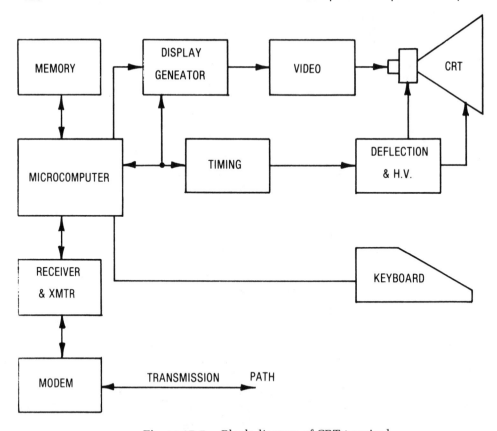

Figure 17-5. Block diagram of CRT terminal.

struction to an appropriate buffer to get the data, then proceeds to process the next program, checking back every so often to see whether the buffer has the data. If not, processing of the next program commences again. If the data is ready, it is moved into memory from the buffer, and the particular program which required it continues its processing where it left off. Thus, the high-speed computer does not have to wait while the buffer fills up from the relatively low-speed tape or disk drive or other element. If data is to be transferred from memory to an input/output element, the same type of operation is used. (In actuality, for input the buffers fill up as soon as program execution begins, and refill with new data just after the original data is transferred to memory.) On some pocket computers, I/O buffers are not required.

There are a variety of input/output devices. For a simple, special purpose computer the output device might be just a light on a control panel, or perhaps a voltage applied to or removed from a microwave oven. Commonly used peripheral devices, however, are:

1. Terminals (CRT and/or hard copy)
2. Line printers
3. Paper tape readers and punches
4. Magnetic tape recorders
5. Magnetic disks
6. Magnetic drums
7. Data cells
8. Punched cards
9. Optical or magnetic character readers
10. Mark sensors
11. Plotters

Almost all the above devices can be connected directly to a computer through a serial or parallel I/O port. A communications link such as microwaves, satellite, or telephone can connect the peripheral to the computer by use of the proper modem and baud rate. Appropriate devices are required at each end of the link for encoding and decoding the signal.

Devices that transmit in one direction at a time are said to be operating in a half-duplex mode. Devices that operate in both directions simultaneously are said to be operating in a full-duplex mode. For example, a person operating a terminal in full-duplex mode can send an interrupt or break signal to the computer while receiving data to tell the computer to stop its operation. In the half-duplex mode, the terminal operator would have to wait for a break in the computer's transmission before the interrupt signal gets to the computer. Full duplex mode is preferable, but more costly, because two separate channels are required. All the devices have operator controls, and most of them work with some form of stored data. A type-laser or line printer output can be considered stored because there is a permanent impression on paper. Simple CRT terminals such as TV sets, etc., usually have no storage ability, but most modern computers have hard copy capability.

Terminals and Line Printers

Terminals are usually in the form of QWERTY keyboards, liquid crystal, or CRTs, and usually the keyboards are for inputting data or instructions. Terminals exist which only output and thus have no keyboards, only a few control keys. Often a telephone is used as a terminal for data entry to a computer. CRT terminals attached directly to a computer or its buffer can operate at very high speeds, but CRT terminals or keyboards connected via a communications link or even keyboards attached to a computer operate much slower. The slow speed of remote control terminals is dictated by the bandwidth of the communications link and by the speed limitations of the electromechanical keyboard.

Magnetic Tape and Disks

Magnetic tape and disks represent magnetic storage or memory devices and are the ones that carry the heaviest load in terms of words of data. Speeds of a million characters per second are not uncommon. In addition, magnetic devices allow for high data density, such as 1600 characters per inch on tape. Of the two devices, magnetic tape is the one that is sequentially accessed; the other is randomly accessed.

Disks are magnetically recorded devices that are known by several names, such as *hard disks* and *floppy disks*, and are available in different sizes required by individual computer disk drive systems. Also, they are listed as double-sided, double-density, preformatted, and include information pertaining to digital word capacity. Examples of hard disk capacities are 10 meg, 20 meg, and 100 meg. However, existing computers are limited by the type and size of disks they are designed to use. Some of the various sizes are: 3½ inches, 5¼ inches, and 8 inches. Some floppy disk drive systems can read 1.2 megabytes and 360 k formats, which are needed with many formatted diskettes.

Magnetic tape used for recording data in home and business applications is usually cassette tape. When larger amounts of data storage are required, reel-to-reel recorders are often employed.

Most small computers are capable of being used with an ordinary cassette tape recorder. But unless it is built in, a cassette interface is required (usually a RS-232C) in addition to the cassette recorder. Although tape cassette systems that will read or store data are adequate for storing data, they are slower than disk systems.

Another piece of equipment often needed to transfer data from a computer to a cassette recorder is called a *null modem adapter*. The word *modem* is an acronym for *modulator/demodulator*. This device changes a digital signal to a pulsed audio tone.

The binary signal must be transmitted at a specified rate, called *baud rate*. Baud is a unit of transmission speed derived from the duration of the shortest signal element. Speed in bauds is the number of code elements transmitted per second; normally, one element is one data bit interval. Typically, a low baud rate is 300 and a high is 1500. Although much higher rates are possible, the error rate usually becomes significant.

Character Readers and Mark Sensors

Character readers are perhaps the most interesting of the peripheral devices, since they can read printed or written characters and are able to distinguish one character from another. Some readers can only read numbers and a few symbols specially formatted for the purpose (see, for example, the lower edge of a bank check), while other readers can read letters of the alphabet or numbers printed by hand. The symbols and numbers on the bottom of checks are not only optically readable, but are also magnetically coded. There are readers that can sense different denominations of dollar bills and other graphic material.

Supermarket packages have a small area on them with the pertinent product code marked in the form of parallel lines. Where supermarkets and other stores have adopted the system, the salesperson merely passes the optical sensor over the lines and the information is passed to the computerized cash register, which displays the item code, sometimes the item description, and the price. Not only does this information go on the receipt, but it is also stored internally to build up sales records for the store. Typically, the sensors (at the cash register) consist of a source light (usually a laser) and a photocell.

Plotters

Plotters are devices that can draw graphs, charts, etc. Basically, they consist of a plotting surface (which can be flat or in a drum shape, utilizing a roll of paper), with one or more styli or pens positioned by computer. Most modern plotters have several styli, to plot multicolored graphs. The styli are lowered onto the plotting surface, raised, or moved to another position by computer command. Some styli move only in a straight line across the surface; motion in the other direction is achieved by moving the surface itself (as in the case of drum-shaped plotters) or by moving an arm which moves in a direction perpendicular to the styli (as in the case of planar plotters).

Computers and programs are available that can plot any graph, symbol, or complete design as they appear on the computer screen. A laser printer is usually used in these work station systems (computer-aided engineering, CAE, and computer-aided design, CAD).

17.6 MEMORY UNIT

The memory element is where program instructions and data are stored. Memories can be classified as either internal or external. An external memory could be a disk, drum, tape, etc., while an internal memory could be made of relays (early computers), magnetic cores, semiconductor elements, or any device that can exhibit a two-state behavior, one state to represent binary 0, the other to represent a binary 1. All computers today have internal memory. Most also have what is known as a *disk storage*. This is not really an internal memory, but only acts like one. The disk storage contains data that resides external to the computer, say, on disks. When called for by a program, part or all of that data is transferred to internal memory prior to use, giving an appearance to the user as if all the memory were internal.

Memory can be classified by how it may be accessed. Access can be either sequential or random. Data on tape can only be accessed sequentially; that is, if a particular data item is in the middle of a reel of tape, all the data before it must be read before getting to the required data. Data residing on a disk is considered to be randomly accessible, since the disk is something like a phonograph record in shape and has a movable read/write arm. For data in the middle of a disk, the arm moves to the middle of the disk before starting to read, then it has to read at most the data for one revolution of the disk before reaching the required data. Disk memory is of such a nature that any word of it may be accessed without having to read any other words.

Memory is also classified by whether it can be altered by the programmer or whether it is fixed permanently. Fixed memory is termed "read only memory" or ROM. Random access core memory is sometimes called RAM. An example of ROM would be the bootstrap program used on a computer start-up. This is not something the user programs every time he needs it. Different manufacturers have different types of ROMs. Some are alterable under special electronic or optic conditions, though the difficulty involved in changing the ROM dictates that they would not be programmed often. Some ROMs are programmable only by the manufacturer. Others, such as field-programmable ROMs and EEPROMs, are programmable by the user.

A subclassification of random access memory is whether it is static or dynamic. A static memory retains its setting as long as the circuit is powered. A dynamic RAM needs periodic refreshing. Dynamic RAMs are useful because they provide faster access time and involve less hardware than static memories.

The term *bit*, used by computer people, simply stands for binary digit. Another common term is *byte*, which means a group of 8 bits. The term *word* is also used. A computer word is usually the smaller number of bits on which the computer operates (either arithmetically or by moving it in or out of the central processor). Different computers have different-length words, although most manufacturers today have oriented their word lengths to either 8, 16, or 32 bits. The significance of the term *byte* is that any letter, numerical digit, symbol, or control character in common use may be uniquely represented by 8 or less bits of information. In fact, using the standard ASCII code, only 7 bits are required, leaving the eighth bit for checking purposes. Some computers will add the number of 1 bits in the 7-bit character, and will put a 1 bit in the eighth position if the sum is odd or put a 0 bit in the eighth position if the sum is even (known as *even parity*), so as to keep the total sum of 1 bits even. Computers with odd parity will put a 1 bit in the eighth position only if the sum of 1 bits is even, so as to always maintain an odd number of bits. By counting the number of 1 bits after the movement of a number, the computer can insure itself to some extent that no information has been lost. (There are many higher-order checks performed, but most of them are based on the same principal of parity checking.)

17.7 ARITHMETIC OPERATIONS

For numbers to be used arithmetically, variations in storage exist too. For example, with a 20-bit word, an integer number anywhere between 0 and slightly over 1,000,000 can be stored. Numbers that are too small or too large can be scaled before storage and rescaled afterward. If the result of an arithmetic operation is expected to exceed capacity (be larger than 1,000,000) or be too small, further scaling can take place. Negative numbers can have 1,000,000 added to make them positive. Before output the 1,000,000 is then subtracted.

Obviously, this scaling becomes very tedious, error-prone, and involves a great deal of planning by the programmer. By adding one bit to the word length, negative numbers can be accommodated simply by letting that added bit have the value of 1 if the number is negative and 0 if the number is positive. In practice, negative numbers are stored in a complementary form which facilitates arithmetic operations (see Chapter 16).

Without adding any more bits, the computer can be designed to do its own scaling by converting the number into two pieces, a decimal number and an exponent. If 5 bits are used for the exponent (1 bit for the sign, 4 for the number), the exponent can range in size from -15 to $+15$, while the 15 remaining bits representing the decimal could store a number as great as 32,767. Thus, the computer would scale all numbers by dividing (or multiplying) them by powers of 10 until the number is just under 32,767. The power of 10 used then becomes the exponent. This scaled number is now stored and arithmetic operations can be done with this type of number with far greater capacity than before.

This division of a number by the machine is quite common, and numbers converted and stored this way are known as *floating point* decimal numbers. In essence, all such numbers are converted to numbers in the form of 0.XXXXXXXX multiplied by 10 to some power or other. This is expressed in unformatted computer output as $+0.XXXXXXXXE+XX$ where the X's represent digits and the E represents exponent. The "+" signs could, of course, also be "$-$" signs. Thus, any number greater than or equal to 1 would be divided by 10 repeatedly until the number is less than 1 (i.e., the decimal point would be shifted to the left, or floated), and the resultant number then expressed as above, with the value of the exponent being the number of divisions by 10. Similarly, any number less than .1 would be multiplied repeatedly until that number was just greater than or equal to .1. The number of digits carried is a function of the word length of the machine. For example, the number 12345.678 would be converted to $+0.12345678E+05$ and the number .000012345678 would be converted to $+0.12345678E-04$. In practice, the latter storage techniques are used whenever possible.

Systems known as *superminicomputers* can perform millions of instructions per second (mips), and about 6.15, 6.4, and 6.5 mips are typical. In general, the system contains a floating point extended arithmetic unit and a coprocessor that handles mathematically complex portions of programs. The coprocessor supports all floating operations, including prescaling, exponentiation, trigonometric functions, and conversion between formats. A high-capacity 32-bit memory can handle up to 24 megabytes needed for running large applications programs.

17.8 MODES OF OPERATION

Computers can be operated in various modes, depending on the nature of the operation, the capabilities of the hardware, and the requirements of the users.

On-Line and Off-Line Operation

The most common mode of operation is that used in hand or desk calculators, most desktop computers, and most computers constructed from microprocessors. This mode of operation is called *on-line operation*; it means that the user is always connected directly to the computer. There may be a small buffer in between which would transfer its data to the computer on the push of a button, such as the add key on a hand-held device. At any rate, as soon as a command is given, the computer acts on it immediately and the results are output as soon as they are ready. *Off-line operation* means operation with the computer turned off, or with the operator not connected to the computer. For a desktop computer with an attached tape cassette drive, the operator could load data onto the tape from the keyboard with the computer off, then later turn on the computer and read the tape data into it at a much higher speed. This type of operation makes sense when there is a cost constraint on computer time.

Batch and In-Line (Continuous) Processing

Processing where data is accumulated in batches is known as *batch* or *periodic processing*. An example would be a store where data is accumulated throughout the week and processed over the weekend to generate statistics for the week's sales. Continuous processing would mean that the data is processed as soon as it is input to the computer, perhaps daily or hourly. Presumably, batch processing requires a storage medium such as tape or disk which can accumulate the data. Many large installations actually use a combination of the two types of processing, doing some editing on the data as soon as it is input, and doing the bulk of the processing only on a periodic basis.

Real-Time Processing

The term *real-time* refers to the computer responding almost instantly to any inquiry or command. In order to achieve this, an operator or a peripheral must be on-line with the computer. An example would be the commands of a desktop computer which is dedicated to the use of the person at the keyboard or, possibly, a radar antenna system feeding real-time target location coordinates to a computer. Operation in a time-sharing mode (described below) is not in real-time, although it is on-line.

Time Sharing

Time sharing is the simultaneous use of one computer by several users. The users each have a terminal, usually called a work station (which can be at a remote location serviced by a communications link), and all users can apparently input, or output, to the computer simultaneously. In point of fact, each user accesses the computer for only a fraction of a second (termed a *time slice*) and then another user gets it. This is made possible by multiprocessing techniques and appropriate hardware in the communications links.

Present-day time-sharing systems sell their services worldwide, and can have thousands of "simultaneous" users connecting to the computer via telephone lines, satellites, etc. Each of these users gets to use the computer as if he were the only one on it. When usage becomes heavy it is evidenced by longer-than-usual delays to user command. Each user can write and store programs and/or data into a catalog assigned to him. He must identify himself to the computer with appropriate passwords and can run his own programs or those of a library made available by the time-sharing service. He can process his own data, access certain data in the service's library (such as stock quotations, etc.), and, in fact, do almost anything he could do with his own desktop computer. Of course the time-sharing service provides a much larger file storage capability (see beginning of this chapter "Section 17.2") and a more sophisticated use of several programming languages.

Time-sharing users are said to be on-line with the computer. Most commercially available time-sharing services are accessible via a terminal attached to a local telephone, but there are also systems which are hard-wired into the same computer, as, for example, the terminals in a large manufacturing plant. Also, several satellite systems provide such services.

Parallel and Multiprocessing

Multiprocessing refers to the computer processing several programs together. Although it seems to an observer that the programs are processed simultaneously, the computer actually processes one program first up to some point (for example, where that program must wait for an I/O buffer, or simply when that program has used up enough processing time), then the computer proceeds to process the next program, eventually returning to the first program and repeating the cycle. When one program is finished, another program is read into the computer so it can become part of the processing cycle, in this manner keeping the computer perpetually busy and keeping a high turnover rate for programs.

Sophisticated logic is necessary to minimize delays between the processing of one program and the next. It is possible for the programs to be temporarily stored on disks while waiting to be processed. It is also possible for a large computer to have more than one disk drive, but only being able to access one at a time. Multiprocessing was probably the greatest boon to data processing since the introduction of semiconductors. It has tremendously speeded up the processing capabilities of computers.

Parallel processing refers to the actual simultaneous processing of more than one portion of the same program at the same time. For example, if the square and cube roots of a number are required, a parallel processor can do both calculations simultaneously, so both results are available at almost the same time for the next step of the operation. In actuality, parallel processing should be used for performing lengthy iterative routines which are capable of being processed simultaneously. To achieve this, there must be two or more central processors.

17.9 OPTICAL COMPUTERS

Perhaps optical disk drives are the most available components of optical computing systems. These devices are designed to work as personal computer peripherals that can access more than 550 megabytes of, for all practical purposes, indestructible data. The recording is made by laser on a single 20,000-track, 5-inch disk. For the personal computer, these devices are usually "compact-disk read-only memory" (CDROM). The other type disk in this field is a write-once version.

Bistable devices that can be used in optical computer systems are currently being developed in many research laboratories. These include AT&T Bell Laboratories, Optical Society of America, and several others throughout the world. See Chapter 19, Opto-Electronic Devices, for additional information.

BIBLIOGRAPHY—CHAPTER 17

ANGELL, I. O., *Advanced Graphics with IBM Personal Computers*. New York, NY, John Wiley & Sons, Inc., 1985

CHATTERGY, R., and V. POOCH, *Introduction to Microcomputers*. Cambridge, MA, Winthrop Publishers, Inc., 1981.

GENN, R. C., Jr., *Illustrated Guide to Practical Solid State Circuits—with Experiments and Projects*. Englewood Cliffs, NJ, Prentice-Hall, Inc., 1983.

ILIFFE, J., *Advanced Computer Design*. Englewood Cliffs, NJ, Prentice-Hall, Inc., 1981.

MARTIN, J., *Computer Networks and Distributed Processing: Software Techniques and Architecture*. Englewood Cliffs, NJ, Prentice-Hall, Inc., 1981.

MAYNE, A. J., *Linked Local Area Networks*, (2nd Ed.). New York, NY, John Wiley & Sons, Inc., 1985.

WAKERLY, J. F., *Microcomputer Architecture and Programming*. New York, NY, John Wiley & Sons, Inc., 1986.

CHAPTER 18

PROGRAMMING SYSTEMS AND LANGUAGES

18.1 LANGUAGE DEVELOPMENT

Digital computers are programmed by a sequence of instructions, each in the form of a binary number that represents a specific computer operation (see Chapter 17). These binary number sequences are called *machine language*; different computer models use different machine languages.

Since it is extremely cumbersome to set up an involved program in machine language, symbolic language is used to program most computers. With symbolic language, a programmer uses mnemonic equivalents of machine instructions in writing a program. Instead of the binary number 10110101, representing the instruction to move the word of data in memory location 56 to register 1, the instruction could simply read "MOVE 1,56" or even "M 1,56." The computer must of course be programmed to translate these mnemonics into binary words. This translation progam would have to be in machine language, the only language the computer understands.

Originally, languages were simply mnemonic translations of machine commands. Then, higher-order languages came into being, where one command in the symbolic language (called a *macro instruction* by some manufacturers) corresponds to several instructions in the machine language.

Symbolic languages are often classified as either machine-oriented or problem-oriented. Machine-oriented languages are called *assembler languages*, and consist of mnemonic symbols which represent either one or a series of machine language instructions. The translation of an assembler language program into machine language is called a *compiler*, which must obviously be loaded into the computer first, before the computer can read the assembler language program. Unfortunately, the assembler language for one computer often bears little resemblance to the assembler language of another computer. The programmer would have to learn a host of different languages to be able to work with different computers. To avoid this, several programming languages have been developed which are used (with minor variations) by most computers.

These languages are called *problem-oriented* because they were designed not for a particular computer, but rather to solve certain types of problems. Problem-oriented languages require an additional translator to translate the problem-oriented language into the assembler language of the machine. Different computers would have different compilers since their assembler languages are different.

Problem-oriented languages are higher-order languages, where one statement can correspond to many assembler language statements which, in turn, correspond to even more machine language statements. The most popular of these languages are COBOL, FORTRAN, and BASIC. There are other problem-oriented languages, but their use is either limited to one manufacturer or the military (Ada, for example, which is supported primarily by the military), or are special purpose languages such as those used to control machine tool operations and Pascal for business applications. COBOL, FORTRAN, and BASIC are the only three languages in truly universal use and these will be discussed in more detail in this chapter.

BASIC was developed as a simplified language to be used on time-sharing terminals by busy executives who do not have time to learn a more complex language. As the name implies, it is (or at least was originally) a relatively simple language intended to be used for business applications.

FORTRAN (formula translation) was designed primarily for scientific as opposed to business purposes.

COBOL (common business-oriented language) was specifically designed for business programming, accounting, inventory control, etc. COBOL was originally almost readable by a nonprogrammer, with the idea being that after a program is written by a programmer, it could be understood by anyone with very little programming background.

All three languages have grown in complexity over the years to the point where only a programmer can really understand them. Their versatility has also increased over the years to such a point that a scientific or business programmer can use any of the three languages to achieve desired results. However, ease of use and efficiency considerations usually restrict the choice of languages, depending on the function.

Most business institutions today use COBOL in their data-processing centers, while FORTRAN is still used in scientific applications as well as in time sharing. BASIC is primarily used in time sharing today.

Scientific applications were once thought to require a lot of processing of small amounts of data, while business applications were thought to require little processing of large amounts of data. While no longer true, this influenced the development of COBOL and FORTRAN. COBOL was designed not only to be an easily readable language, but also to provide the programmer with maximum control of the input/output facilities. By knowing in advance the types and magnitudes of the data and by expeditious programming, the COBOL programmer can optimize data handling, an important feature when large quantities of data are involved. Although similar optimization can be achieved in FORTRAN, it requires many extra programming steps.

A few pages of a book are not enough to enable one to learn a programming language, and the description will be more general to give the reader some idea of what is involved in working with each language. Even though these three languages are supposed to be standardized, differences will be found, sometimes in the form of options, extra capabilities, or even less capabilities, due to the nature of the hardware involved.

To learn a language, consult one of the books mentioned in the bibliography, as well as the programming manuals supplied by the computer manufacturer. Prior to working on a particular computer, it is essential to consult the manufacturer's programming manuals as well as systems description manuals. These will enable the programmer to understand what variations are to be found in the equipment.

18.2 FEATURES COMMON TO ALL LANGUAGES

All languages have to follow certain grammatical rules (called *syntax*) in order for them to be understood by a compiler. The full syntax of a language can only be learned from a language manual, either one supplied by the computer manufacturer or a general language manual such as the ones listed in the Appendix. An addition and a subtraction symbol, for example, can never be found adjacent to each other as $- +$ or $+ -$, but can be found separated by parentheses: $+ (-$ or $- (+$.

Listed below are the more commonly used definitions. The list, and the definitions, are limited by the space available in this book.

Program

A program consists of a sequence of statements. These statements are of two varieties—executable and nonexecutable. The first causes an action to occur; the second only gives information to the computer. There is another type of statement called a comment, or remark, which is ignored by compilers. Its only purpose is to provide the programmer with little notes to himself.

Data

Data can be in the form of a constant or a variable. A constant is data that has the same value throughout the program, while the value of a variable may change. Each constant and variable is assigned a memory location upon program execution. This memory location remains the same regardless of the value of data in it. An exception is in a multiprocessing environment, where only relative memory locations are assigned, such as the location relative to the starting point of the program. In this case, the relative location remains fixed, while the starting location changes each time that particular program gets its turn at the central processor.

Constants

Constants can be numeric or character in nature.

Numerical Constants

Numerical constants consist of numbers such as: 0, 1, 584.32, -77.8, 0.005, 3.14159, etc. These are constants that would be used in arithmetic or comparison operations.

Character Constants

Character constants consist of symbols, numerals, or characters, usually enclosed by quotations marks or apostrophes. Examples are "able," "A24," "3%4," "X Y," "ANDERSON, C." These constants could not be used arithmetically, but could be used comparatively, such as in sorting operations.

Variables

A variable is simply a symbolic representation for a memory location. Just as there are numerical and character constants, there are numerical and character variables. Either type can be used for comparison, but only the numeric type can be used arithmetically. The variable x, for example, can be assigned a particular memory location (either permanent or relative to the starting location of a program), no matter what its value. Since it is a variable, its value can change. One would not expect, however, that the constant 3.14159 could alter its value in a program. This is not strictly true, because there are ways to do it by poor programming in certain types of compilers.

Dimensioned Variables or Arrays

Variables can have either single values or can have several values, i.e., they can occupy several locations in memory as opposed to just one. The number of locations occupied is the dimension of the variable. An array is a dimensioned variable. The variable may have more than one dimension. It may be a 3×5 array consisting of 15 memory locations, for example, but these locations would be referred to by two subscripts, one for the first dimension and one for the second. The computer has an algorithm which changes these two subscripts into one particular memory location. Variables are declared to be arrays by a DIMENSION statement in BASIC and FORTRAN, and by an OCCURS clause in COBOL.

Subscript

A subscript is a number that refers to a particular element (or memory location) of a dimensioned variable. A dimensioned variable can only have its elements referred to by subscripts. One subscript is required for each dimension of the variable. For example, if a variable is designated to describe two features of an employee, say the salary and length of employment, and if there are from 100 to 170 employees, an array variable having dimensions of 2 × 200 could be defined. The value of the variable DESC with the subscript 1, 1 would refer to the salary of employee number 1; the subscript 2, 1 to the length of service of employee number 1; the subscript 1, 110 to the salary of employee number 110; etc.

18.3 EXPRESSIONS

An expression consists of one or more variables (or constants) connected by operators. These operators can be arithmetic, relational, or logical. For example, the expression Y + 5 combines the variable Y and the constant 5 with an addition operator. The following table lists the operators for the three languages. Most present-day compilers allow FORTRAN arithmetic operators to be used optionally for BASIC or COBOL, and FORTRAN relational operators to be used optionally for BASIC. Therefore, many of the later examples will use the FORTRAN symbology:

	FORTRAN	BASIC	COBOL
Arithmetic			
	+	+	ADD
	−	−	SUBTRACT
	*	*	MULTIPLY
	/	/	DIVIDE
	**	Δ	EXPONENTIATE
Relational			
	.EQ.	=	EQUAL TO
	.NE.	< >	NOT EQUAL TO
	.LT.	<	LESS THAN
	.GT.	>	GREATER THAN
	.LE.	< =	LESS THAN OR EQUAL TO
	.GE.	> =	GREATER THAN OR EQUAL TO
Logical			
	.AND.		AND
	.OR.		OR
	.NOT.		NOT

Although expressions can contain several operators of the same type and even operators of different types, care must be exercised to insure that the expression represents its intended meaning. Making subexpressions out of expressions with the use of parentheses can clarify most expressions and is highly recommended. For example, A/B*C is equal to (A/B)*C but is not equal to A/(B*C).

Arithmetic expressions are evaluated by first exponentiating (if there is such an operation), then dividing or multiplying, finally adding or subtracting. If more than one exponentiating operator is present, the leftmost one is performed first. The same applies if more than one of the other operators is present. For example, in the expression A**2 + B**3, first A is squared; then B is cubed; finally, the results are added together. When subexpressions exist, the innermost ones are evaluated first. The only exception to the above rules occurs when exponentiating operators are adjacent to each other without another operator between them, as in A**B**C. Obviously, if the operators were addition or multiplication, the order of operation is immaterial. In this case, B is raised to the power of C first, then A is raised to the resultant power. This is perhaps clearer by rewriting the expression with parentheses: A**(B**C). The use of parentheses can also reverse the order of operation: (A**B)**C. If A = 2, B = 3, and C = 2, the expression A**(B**C) yields 512, whereas (A**B)**C yields 64.

Relational operators are used in expressions such as A.NE.B. Logical operators are used either with logical variables (not discussed here) or in conjunction with relational operators such as in the expression A.EQ.B.OR.C.EQ.D. An expression such as A.EQ.B.OR.C. is meaningless; it must be written A.EQ.B.OR.A.EQ.C. Note that relational operators must be used between like constants or variables. Thus, A and B above must both be either arithmetic or both be character variables, but it is possible for A and B to be character variables in the expression A.EQ.B.AND.C.EQ.D, while C and D are arithmetic.

18.4 ASSIGNMENT OR DATA MOVEMENT STATEMENTS

An assignment statement is one that assigns a value to a variable. That value can be a constant, another variable, or an expression. A character value cannot be assigned to an arithmetic variable and an arithmetic value cannot be assigned to a character variable. For example, A = B (FORTRAN), LET A = B (BASIC), MOVE B TO A (COBOL). Again, many current compilers optionally allow BASIC and COBOL to be in the FORTRAN syntax. All three statements mean the same thing. There is a memory location designated by the variable A which has its value altered to whatever is in memory location B. The value of memory location B does not change. In FORTRAN and BASIC, and some COB-

OLs, B can be a constant or an expression, for example A = X + 1. If A were a character variable, one could have an expression like A = "CHARLES," but one could not have A = "CHARLES" + "ED-WARD."

Branching and Transfer of Control

Branching or transfer of control is accomplished by changing the sequence of execution of a program's statements. The statements that accomplish this are called *control statements*. Two types will be discussed—transfer statements and loops.

18.5 TRANSFER STATEMENTS

There are three types of transfer statements—unconditional transfer, conditional transfer, and computed transfer.

Unconditional Transfer

The statement "GO TO A" directs the sequence of execution to continue at A, where A is either a statement number in FORTRAN, a line number in BASIC, or the name of a paragraph in COBOL (in COBOL, the term NEXT SENTENCE is also used). Thus, examples would appear as: GO TO 100 (FORTRAN or BASIC), GO TO PARAGRAPHA (COBOL). Some FORTRANS allow names as well as numbers. It must be pointed out that since each BASIC statement now has a number associated with it, any FORTRAN statement may have a number associated with it, and that COBOL statements are written in sentences and these are lumped together in paragraphs, each of which has a name.

Conditional Transfer

Transfers that take place only under certain conditions are called conditional transfers. These statements are called "IF" statements. They appear as: IF( ) GO TO A (parentheses not necessary in BASIC or COBOL), where the parentheses denote a logical (as opposed to arithmetical) expression. If this expression is true, control is transferred to statement or paragraph A, otherwise execution continues with the next statement. For example, the expression could be: X.GT.Y (X Y for BASIC or X GREATER THAN Y in COBOL). The word THEN is optionally used in BASIC and COBOL to replace "GO TO." Also, the words ELSE and OTHERWISE can be used in COBOL. Examples are:

```
FORTRAN:   IF (X.GT.Y) GO TO 100
BASIC:     IF (X > Y) GO TO 100
           IF X > Y GO TO 100
           IF X > Y THEN 100
COBOL:     IF X GREATER THAN Y THEN PARAGRAPHA
           IF X GREATER THAN Y THEN PARAGRAPHA ELSE
           NEXT SENTENCE
           IF X GREATER THAN Y GO TO PARAGRAPHA
           IF X IS GREATER THAN Y GO TO PARAGRAPHA
```

The word "IS" is optional in COBOL. The equivalent of the next sentence in COBOL is the next statement in FORTRAN or BASIC. That is, if the argument is false, if X is less than or equal to Y, the flow of control continues as before and no branching occurs. Note that all the above examples produce the same effect.

There is another form of the "IF" statement in FORTRAN, called an arithmetic IF. It appears as: IF( ) A,B,C where the parentheses denote an arithmetic expression and A, B, and C denote three statement numbers. If the arithmetic expression is less than 0, control is transferred to statement A, if equal to 0 to statement B, if greater than 0 to statement C. Thus, the equivalent of the previous FORTRAN example using a so-called logical IF would be: IF(X-Y) 80,80,100 where 80 would now be the number of the next statement. If X is greater than Y, control is transferred to statement 100. The expression in parentheses could be only a single variable such as: IF(X) A,B,C. Note that A,B, and C do not have to be three different statement numbers.

It should be pointed out that the incorrect use of transfer statements can put the program into an infinite loop. Take for example the FORTRAN statements:

```
    X = 8.6
100 Y = 3.7
    IF(X.GT.Y) GO TO 100
    X = Y + 1
```

The program loops forever between the second and third statements and never gets to the fourth. Although obvious in the example, even experienced programmers have accidentally built infinite loops into their programs in a far less obvious manner.

Computed Transfer

A computed transfer is a branch to a choice of two or more locations, dependent on the value of a particular variable. Thus, the statements:

FORTRAN: GO TO (100, 200, 300), X
BASIC: ON X GO TO 100, 200, 300
COBOL: GO TO PARAGRAPHA, PARAGRAPHB, PARAGRAPHC,
                   DEPENDING ON X

are all equivalent and transfer control to statement 100, 200, or 300 (PARAGRAPHA, B, or C for COBOL) respectively, depending on whether X equals 1, 2, or 3. Any other value for X is not allowed, and either produces a program abort or an unpredictable transfer.

18.6 LOOPS

The infinite loop discussed above is an accidental example of a loop. There are, however, deliberately constructed loops. For example, if one wanted to compute the squares of the numbers 1 through 50, it could be achieved as follows:

```
FORTRAN:   X = 1
      100   Y = X*X
            OUTPUT X AND Y
            X = X + 1
            IF(X.LE.50) GO TO 100
               ⋮
```

```
BASIC:   90 LET X = 1
        100 LET Y = X*X
        110 OUTPUT X AND Y
        120 LET X = X + 1
        130 IF X < = 50 THEN 100
               ⋮
```

```
COBOL:         ⋮
            MOVE 1 TO X.
PARAGRAPHA. COMPUTER Y EQUALS X TIMES X. OUTPUT X
AND Y.
ADD 1 TO X. IF X GREATER THAN 50 NEXT SENTENCE ELSE
PARAGRAPHA.
               ⋮
```

Instead of the sentence "COMPUTE Y EQUALS X TIMES X.", for the COBOL example, any of the following could also be used: "Y EQUALS X TIMES X.", "Y = X * X.", "COMPUTE Y = X * X." or "MULTIPLY X BY X GIVING Y.". PARAGRAPHA is the name of the paragraph that starts with the sentence "COMPUTE Y EQUALS X TIMES X."

However, there is a way to define loops which has fewer programming steps. In FORTRAN this is done with a "DO" statement. It appears as follows: "DO 200 X = 1, 50, 12." This means that X is set to the value 1, and all statements up to and including statement number 200 are executed with X equaling 1. Then X is incremented by 12 and the loop starts over with the statement immediately following the DO statement, with X having the value 13. This continues until the loop is satisfied. This occurs when X is greater than or equal to 50 after statement 200 is executed. Statement 200 does not have to be executable; it can be a dummy statement called a CONTINUE statement. Such a statement would be used, for example, if the last executable statement were an arithmetic IF statement. When the increment is 1, it may be left out of the DO statement. Thus, "DO 200 I = 1, 50" is equivalent to "DO 200 I = 1, 50,1." The previous looping example would now appear as:

$$
\begin{array}{l}
\text{DO 200 X = 1,50} \\
\text{Y = X*X} \\
\text{200 OUTPUT X AND Y} \\
\qquad \vdots \\
\text{OR} \\
\text{DO 200 X = 1,50} \\
\text{Y = X*X} \\
\text{OUTPUT X AND Y} \\
\text{200 CONTINUE} \\
\qquad \vdots
\end{array}
$$

In BASIC, the equivalent of the DO and CONTINUE statements are the FOR and NEXT statements:

```
100 FOR X = 1 to 50
110 LET Y = X*X
120 PRINT X; Y
130 NEXT X
```

X starts at 1 and keeps that value until the 'NEXT' statement where it is incremented by 1. Then the loop begins again at statement 110 and continues until X equals 50 at statement 130. Execution then continues with the statement following 130. If an increment other than 1 is desired, line 100 above would appear as:

100 FOR X = 1 TO 50 STEP 12

for an increment of 12.

In COBOL, looping is accomplished with a PERFORM statement. There are many loops available with the use of the PERFORM statement, and it is hoped that the few options shown below are self-explanatory, since they all accomplish the same thing. COBOL sentences may be written one after the other, or one below the other, as is done here for clarity:

:

PERFORM PARAGRAPHA 50 TIMES.

PARAGRAPHA.   MOVE 1 to X.
Y = X*X.
OUTPUT X AND Y.
ADD 1 TO X.

PARAGRAPHC.   ....
OR
:

PERFORM PARAGRAPHA VARYING X FROM 1 BY 1 UNTIL X = 50.

PARAGRAPHA.   Y = X * X.
OUTPUT X AND Y.

PARAGRAPHC.   ....
OR
:

PERFORM PARAGRAPHA THRU PARAGRAPHB VARYING X FROM 1 BY 1 UNTIL X = 50.

PARAGRAPHA.   Y = X * X.

PARAGRAPHB.   OUTPUT X AND Y.

PARAGRAPHC.   ....

18.7 INTEGERS AND REAL NUMBERS

Although not mentioned before, a distinction is made in storage whether a numeric variable is an integer or a noninteger (termed real number). The distinction is that an integer has no decimal portion; therefore, an integer variable that is set equal to (assigned) the result of a computation is always truncated at the decimal point prior to storage. FORTRAN and COBOL allow variables to be classified as integers. The result of the computation "X/Y", if X were equal to 3 and Y to 2, would be 1.5. If stored as an integer, that is, if I were an integer, then I = X/Y would result in I being stored in memory location I. Some FORTRAN compilers only allow integers to be used as indices to "DO" loops or computed "GO TO" statements.

The subscript of an array must also be an integer in certain FORTRAN compilers; others allow real variables or expressions, but truncate them to integers before use as a subscript. An important point occurs when comparing real numbers which are themselves the results of computations rounded to whatever number of places of accuracy the computer has. If X equals 6.4 and Y equals 3.2, the quotient Z of X divided by Y might appear in memory as 1.9999997 (to eight-place accuracy), while the value of W in: W = V**.5, where V is 4, might have the value 2.0000001 in memory. When comparing Z to W in a statement such as: IF(W.EQ.Z) GO TO 100, the program would not branch as desired. If, however, integers are compared, this type of error would not occur. For example, if I and J are integers, one can convert W and Z by adding a trifle for round-off, say, .000001:

I = W + .000001
J = Z + .000001

The expressions on the right evaluate to 2.0000011 and 2.0000007, respectively, and upon truncation both I and, J have the value 2. Now, if the comparison is done on I and J: if(I.EQ.J) GO TO 100, branching will result.

18.8 SUBPROGRAMS

A *subprogram* or *subroutine* is a set of executable statements which appears in form just as the statements of a program appear. A subprogram is branched to at some point in a program, and when execution of all the statements of the subprograms are completed, control branches back to the next statement in the program after the branching statement. Usually, the statements of a subroutine are placed at the end of those of the main program, but it is possible to use subprograms that exist in a library accessible to all users. Presumably, values of variables used or called for in the subprogram must be passed between the subprogram and the main program.

In FORTRAN, subprograms are entities separate from the main program; that is, they are compiled separately (even though they may physically appear after the end of the main program). In COBOL and BASIC, subprograms are compiled with the main program and are, in fact, included with the statements of the main program. In COBOL, a subprogram is simply a paragraph consisting of one or more statements called for in another paragraph by means of a PERFORM statement. A library subprogram is called for in COBOL by a COPY statement which copies the routine from a library into the specified location in the main program.

In BASIC and FORTRAN (and some COBOLs too) there are two types of subprograms—functions and subroutines. Function subprograms are called in BASIC with a CALL statement, but in FORTRAN they are simply used like a variable, and may therefore appear in expressions. Subroutines in BASIC (except library routines) are always called with a GOSUB call statement. Examples are:

FORTRAN SUBROUTINE CALL:	CALL S CALL S(X) CALL S(X, Y, Z, ...)
FORTRAN FUNCTION CALL:	W = F(X) Z = A + F(X) Q = A + B + F(X, Y, Z, ...) + C V = 2**F(X)
BASIC SUBROUTINE CALL:	GOSUB 100
BASIC FUNCTION CALL:	CALL(F)
COBOL SUBPROGRAM CALL:	PERFORM PARAGRAPHA. PERFORM PARAGRAPH A 3 TIMES. PERFORM PARAGRAPHA VARYING 1 FROM 20 BY 3 UNTIL 1 EQUALS 57. PERFORM PARAGRAPHA UNTIL X EQUALS 25.

100 is the statement number of the BASIC statement where the subprogram starts. X, Y, and Z are arguments of the subprograms. That is, they are variables used in the main program whose values are either passed to the main program from the subprogram or passed from the main program to the subroutine. The variables in the argument list of the main program's calling sequence do not have to have the same names as the variables in the argument list of the actual subroutine. What they must have, however, is the same relative position in the argument list and be of the same type (i.e., integer, character, etc.). If the argument list in the main program has the variables X, Y, and Z in that order, and if X is an integer, Y real, and Z a character variable, the subprogram could, for example, call the variables with the names VAR1, VAR2, and VAR3, but must use VAR1 as an integer, VAR2 as a real, and VAR3 as a character variable.

FORTRAN subroutines do not need an argument list. BASIC subprograms (except library routines) do not have an argument list in their calling statements. All variables used in the main program may be used in the subprogram, and vice versa, and retain their meaning. Of course, subroutine S, function F, statement 100, and PARAGRAPHA exist somewhere. A peculiarity of function subprograms is that the name of the function, since it is used as a variable, must also be assigned a value somewhere within the statements comprising the function.

18.9 INPUT/OUTPUT ROUTINES

Although in previous examples the term 'OUTPUT X and Y' has been used, it must be realized that this is great oversimplification, just used for illustrating other points. Input/output routines are many and varied, and peculiar to each particular language. In general, input can be done from (or output to) a terminal, a line printer, magnetic tape, or a variety of their media. The formats of such data vary. Certainly, one would not expect to output the same way onto a 160-character-wide printer terminal as onto an 8-line, 40-character device. Furthermore, it may be desirable to preface magnetic tape with a so-called header record (and end it with a trailer record) or label which tells what the tape is, in case a physical label stuck onto the tape reel falls off. It also serves the purpose of identifying the tape reel to the program to make sure it is the correct tape, when using it for input. For printer output, it may be desirable to start various sections of the output on a new page.

It might be desirable to output onto (or input from) several media simultaneously. To simplify matters slightly, one could consider each input/output medium as a file (including printers). Then, by designating each file with a number or name, one can simply refer to them in a program by their names or numbers. Of course, there must be a way for the compiler to determine what medium the number or name refers to, and also if it is an existing file (such as a floppy disk) or a new file (such as a blank tape that needs mounting in a tape drive when called for).

On output, there must be a way to determine when the input medium has run out of data. Since certain compilers have a limit to the number of files that can be written on or read from, there must exist a way of opening and closing files at will. When outputting onto a tape that had previously been written on (most tapes are reused very often), there must exist a way of designating the end of the new data, so that, when used as input to another program, the old data is not read in after the new data is exhausted. For input/output using random access files (disks, drums, data cells), there must be a way to define the size and nature of the file space required and a way of defining locations on the file so that these locations may be accessed at will. As memory consists of binary numbers, when converting it to typewritten or printed characters, the program must determine whether it is an arithmetic number or a character, and for either case just how it will be output. The number 1.23456 could be output simply as 1, or as 1.234 (or rounded to 1.235), or as 001.23456, or as 1.23456000, etc.

If data is being input from punched cards, the computer must be told how the data is arranged on each card in order to make sense out of the input. Thus, each read or write command carries with it an actual or implied format to be followed. In addition, each read command carries with it a test to see if the input file is exhausted, and an action to take if it is (a branching instruction). To go into further detail at this point would be far beyond the scope of this work. It is hoped, however, that the reader has been made aware of some of the complexities involved in input/output operations. The highest degree of input/output control is achieved by COBOL, which stands to reason, as COBOL is the language originally designed to do few operations with much data.

Ending Execution

Execution is halted when a STOP statement is reached. The STOP statement is the last executable statement in a program. However, it need not be the last statement in the program. Due to branching capabilities, it can be anywhere in the program. For example, to output the squares of the numbers 1 to 50 in a FORTRAN program (without using a DO loop), one could incorporate a STOP statement as follows:

```
        X=1
100     Y=X*X
        OUTPUT X AND Y
        X=X+1
        IF (X.GT.50) STOP
        GO TO 100
```

18.10 LANGUAGE DETAILS

Fortran

A FORTRAN program consists of a series of statements which may have statement numbers. The last statement in the program is an END statement which tells the compiler it can start compiling once it has read the prior statements in. At the start of the program are definitional statements defining names and sizes of arrays; file characteristics; which variables are integers, which are real, which are characters; and, if desired, a DATA statement which gives an initial value to the variables identified by the DATA statement. All variables are assumed to be arithmetic unless defined otherwise; that is, if used arithmetically they need not be defined as such. By convention, any variables beginning with the letters I through N are assumed to be integers, while all other variables are assumed to be real. The maximum capacity of each variable is fixed, and all variables of the same type have the same capacity. That is, if the manual for a particular computer states that integer variables cannot be larger than six digits, then this refers to all integer variables whether defined implicitly (starting with the letter I, J, K, L, M, or N) or explicitly by a type declaration statement at the beginning of the program. Variable names always start with a letter followed (if desired) by other letters and/or numbers. Some compilers also allow a few symbols to be used. The maximum length of the name varies from computer to computer, but most allow between five to eight characters.

Subprograms appear after the END statement. Each subprogram starts with the name SUBROUTINE XXX or FUNCTION XXX and ends with an END statement. Each subprogram has a statement called a RETURN statement which returns control to the main program when encountered. A variety of mathematical functions exists for the FORTRAN program, such as trigonometric functions, random number generators, etc. Communication with subprograms is either by means of an argument list (such as the argument Y in: X=SIN(Y)) and/or by means of a statement called a COMMON statement, which lists all variables that share common memory locations. The name of the variable in a subprogram need not be the same as that in the main program, since the subprogram is compiled separately; it need only be of the same type (real, integer, character) and occupy the same relative position in the common statement or argument list.

Basic

A BASIC program consists of a series of statements all of which require a statement number. The last statement in the program is the END statement. At the start of the program are definitional statements similar to FORTRAN, except that variables are all considered to be real. Variable names are either a letter of the alphabet or a letter followed by a single digit (0 through 9). Reading may be done from a file or from one or more DATA statements placed anywhere in the program. Subprograms are normally placed at the end of the main program, however, prior to the END statement. Function subprograms have an argument list in their definition statement, but not in their calling statement. The variables in the argument list are variables used only within the function, and have no value outside the function. Subroutines do not have an argument list.

Any variable may be used in the main program or a function or subroutine subprogram interchangeably. Functions are defined with a DEF statement which consists of the word DEF followed by the name of the function in parentheses, followed by an argument list. After the DEF statement are the executable statements which the function consists of, and which are ended by a statement which simply reads FNEND. Subroutine subprograms are simply a series of statements, indistinguishable from other statements, except that the last statement of the series is a RETURN statement. The subroutine is called by the statement "GOSUB 100" where 100 is the line number where the subroutine begins. Communication with subprograms is achieved by using the same variables within as without the subroutine (with the exception of those variables named in the argument list of a function definition). Those variables have the same values wherever they are used, and retain those values until changed by a LET statement (e.g., LET Y = X).

Cobol

COBOL is by far the most complex of the three languages discussed, due partially to all sorts of options available with most statements.

A COBOL program is written like an English composition. It consists of four parts called divisions: the IDENTIFICATION division, which is used to identify the title of the program, name of the programmer, and other commentary type of data as is felt necessary, which has no effect on the compiler; the ENVIRONMENT division, which describes the nature of the particular hardware for the compiler and assigns names to the various equipments; the DATA division wherein all variables and constants are defined, as well as the names and nature of all input/output files; and the PROCEDURE division, which contains the actual operations or executable instruction of the program.

Each division consists of one or more parts called sections; each section consists of one or more statements. In the PROCEDURE division, these statements take the form of sentences complete with nouns, verbs, and connectives. The variables are the nouns, the verbs are the names of the various executable procedures (such as the MOVE verb in: MOVE A TO B). There are actually four types of nouns or names: data names given to all variables, condition names used to indicate values for data names, procedure names used to name paragraphs or sections (groups of one or more sentences) in the PROCEDURE division, such as PARAGRAPHA, and special names assigned in the environment division and used to name the hardware required. There are also reserved words (such as the MOVE or PERFORM verbs) which cannot be used as variables.

Names may be up to 30 characters in length, may consist of letters and/or numbers, and hyphens, but all names must have at least one alphabetic character. Spaces are not allowed. Thus, PARAGRAPHA, or PARAGRAPH-A, or PARAGRAPH-A-SQUARING-ROUTINE are acceptable names.

The ENVIRONMENT division consists of two sections, the CONFIGURATION section, which describes the computer configuration and any special names (for example, telling the compiler to use a comma instead of a decimal point and decimal point instead of a comma when outputting numbers to a printer, so as to be in European format), and the INPUT-OUTPUT section, which associates files with actual hardware (e.g., tells the compiler that a certain file is on tape while another is a printer, etc.), and defines special input/output techniques.

The DATA division consists of two sections, a FILE section, which gives specifics about each file used (header and trailer labels, how data is arranged on the file, and the description of data variables), and the WORKING STORAGE section, which describes all variables used in the procedure division, but not described as part of the data for any particular file in the file section. One of the more important aspects of COBOL is that variables can be described which are really a group of variables, but can be moved as if they were a single variable. As part of the description of variables, there is a hierarchy, defined by the programmer, which defines certain variables as being subsets of other variables. For example, one may define on a so-called first level the variable named MASTER. On a second level—that is, as a subset of MASTER—one can define the variables that make up MASTER. If the program is concerned with an inventory situation, the second level could for example consist of the variables PART-DESCRIPTION and PRICE-INFORMATION. Under PART-DESCRIPTION, as a third level, could be the variables PART-NUMBER, MINIMUM-ORDER-QUANTITY, NAME-OF-SUPPLIER, etc. The lowest-level variable needs to have a detailed description of how much storage space is occupied by that variable and what type of variable it is (integer, etc.). Thus, the compiler knows how much memory to allocate to each subvariable, hence how much to allocate to each variable. For example, if the two third variables named above each took 3 words of storage, then the second-level variable PART-DESCRIPTION requires 6 words. If the other second-level variable PRICE-INFORMA-TION requires 18 words, then the variable MASTER would require a total of 24 words. In the FILE section, the variable called MASTER belongs to (is described as part of) a particular file. The WORKING STORAGE section has only lowest-level variables; subvariables or levels do not exist in this section. Each variable is at the same level, and its description determines the storage requirement for it. COBOL allows for variables to take up fractions of memory words, as well as several words per variable. BASIC and FORTRAN variables always take up the same amount of memory per variable of each type (some compilers use different amounts of memory for an integer as for a real variable).

COBOL also has special techniques available for table handling and sorting, two of the main functions of many COBOL programs.

18.11 PROBLEM-SOLVING TECHNIQUES

There are four major steps to solving any problem: defining the problem, analyzing the problem and deciding on procedures necessary for a solution, implementing the solution, and documenting the solution.

Defining the Problem

A problem is defined by means of a statement indicating what the problem is and what is required in the way of a solution. The problem statement must contain sufficient information to enable analysts to solve it.

Analyzing the Problem

A problem is analyzed by determining a set of procedures necessary for a solution. This should include a flow diagram showing the sequence of steps to be followed, decisions to be taken, data needed to formulate a solution, data generated as part of the solution, and the form of the data. Timing considerations must also be included, i.e., when certain data is available, when various procedures should be carried out, etc. The media used for problem solving is also necessary for adequate analysis. Is a computer to be used or can certain steps be carried out manually? Should punched cards be sorted off-line or should their data be read into a computer and sorted by the computer, then repunched onto new cards? Are cards necessary at all or could they be discarded after being entered into a computer and sorted onto tape perhaps? Obviously, the problem analyst must know what resources are available (manpower, computer hardware, etc.) and the cost of using these resources (time and money).

Implementation

After a set of procedures has been thoroughly defined, these procedures are implemented using whatever resources are required. After initial implementation, a testing procedure should be applied to ascertain not only that the implementation adheres to the analysis, but that the analysis actually does the required job. As most often happens, especially in complex problems, the analyses are neither thorough, nor accurate, nor do they present the optimum solution, and are therefore usually revised several times after initial implementation.

Documentation

When implementation is deemed correct and complete, the problem and its solution are documented. This serves the purpose of informing anyone who reads the documentation what was required, what was done about it, and how it was done. In case circumstances require a change of procedure for any reason—say, for example, that the problem definition changes slightly—then the documentation provides a convenient takeoff point. Without it, the whole problem might have to be solved over from the beginning, especially if different personnel are working on it. Normally, complex problems are documented as analysis proceeds, allowing for a change of personnel in midstream.

It should be noted that certain problems are complex enough to require years before they are completely analyzed and implemented. What often happens in these cases is that the definition of the problem changes in that time, or the cost of certain resources changes (perhaps manpower becomes very expensive, while computer power decreases in cost), indicating a change in the analysis.

Sample Problem

The solution of a rather simple problem will be shown below using FORTRAN language. FORTRAN is chosen because it is the language most likely to be used by readers of this handbook.

A company has less than 100 personnel who are paid on a monthly basis. The employee information is kept on a magnetic tape which has, for each employee, an employee number and his monthly salary. Data for new employees, as well as changes in data for old employees (i.e., changes in salary, retirements, etc.) are put onto punched cards which update the tape file once a month, just before the payroll is printed out onto checks. The punched cards simply have on them an employee number (one card for each change), and the new salary. Thus, if the employee number does not exist in the tape, that is a new employee. If the salary is zero, that employee has left the company, and if the employee number exists (assume that no employee number is assigned twice) and the salary is not zero, then that indicates a salary change. There are never more than 50 punched cards.

To solve this, once a month the tape is read into the computer memory, then the cards are read in, one at a time, and as each card is read in, the tape data now in memory is modified. Then, when all cards are exhausted, the tape data in memory is used to output the payroll, the tape is rewound and rewritten with the new data. This is neither the best nor the only solution available, but just one of many which was chosen here simply for illustrative purposes.

A flowchart of the procedure to be followed appears in Figure 18-1. Although many symbols have been devised for flowcharting, only two are used here so as not to confuse issues—a rectangular block and a diamond-shaped decision block.

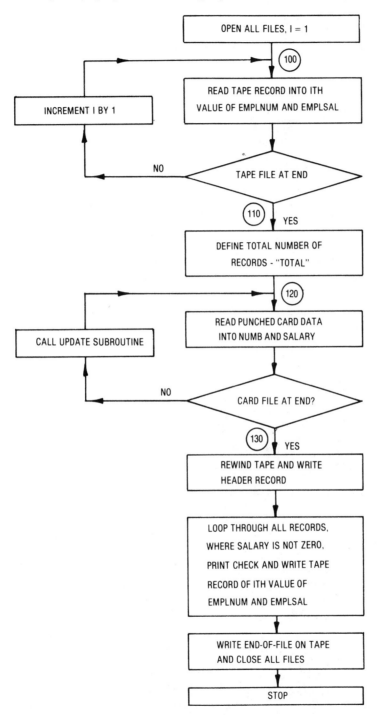

Figure 18-1a.   Main program.

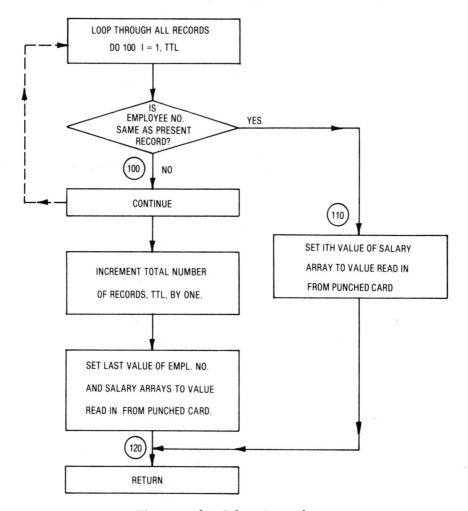

Figure 18-1b.   Subroutine update.

The following is a listing of the program. The numbers on the left are simply reference numbers to be used for commentary. The file handling statements are not according to the program language, but written in English for illustration:

```
          11:04EDT 10/21/78
1         DIMENSION EMPLNUM(150), EMPLSAL(150)
2         INTEGER EMPLNUM,TOTAL
3         OPEN TAPE FILE, CARD FILE, AND PRINT FILE
4         I=1
```

5	100	READ A TAPE RECORD INTO I'TH VALUE OF EMPLNUM AND EMPLSAL, IF TAPE IS AT END, GO TO STATEMENT 110
6		I=I+1
7		GO TO 100
8	110	TOTAL=I−1
9	120	READ THE DATA FROM A PUNCHED CARD INTO THE VARIABLES NUMB AND SALARY, IF NO MORE CARDS, GO TO STATEMENT 130
10		CALL UPDATE (EMPLNUM,EMPLSAL,NUMB, SALARY,TOTAL)
11		TO TO 120
12	130	REWIND TAPE REEL AND WRITE APPROPRIATE HEADER ONTO TAPE
13		DO 140 I=1,TOTAL
14		IF(EMPLSAL(I).LT..01) GO TO 140
15		ADVANCE PRINT FILE TO NEXT CHECK
16		PRINT EMPLOYEE NUMBER AND SALARY ONTO CHECK, FROM THE I'TH ELEMENTS OF THE ARRAYS EMPLNUM AND EMPLSAL
17		WRITE ONTO THE NEXT TAPE RECORD THE I'TH ELEMENTS OF EMPLNUM AND EMPLSAL
18	140	CONTINUE
19		WRITE END OF FILE DESIGNATION ONTO TAPE SO THAT WHEN IT IS NEXT READ, THE READ ROUTINE STOPS AT THAT POINT
20		CLOSE ALL FILES
21		STOP
22		END
23		SUBROUTINE UPDATE (NUMARRAY,SALARRAY, N,S,TTL)
24		DIMENSION NUMARRAY(150),SALARRY(150)
25		INTEGER TTL
26		DO 100 I=1,TTL
27		IF(NUMARRAY (I).EQ.N) GO TO 110
28	100	CONTINUE
29		TTL=TTL+1
30		NUMARRAY(TTL)=N

```
31              SALARRAY(TTL)=S
32              GO TO 120
33 110          SALARRAY(I)=S
34 120          RETURN
35              END
```

Commentary

At line 1, the two variables EMPLNUM and EMPLSAL are declared to be arrays each with dimension 150, the maximum necessary. At line 2, the variables EMPLNUM and TOTAL are declared to be integers. Line 4 sets an index, I, to 1. Line 5 reads a tape record into the Ith elements of the arrays. Line 6 increments I and line 7 branches back to the READ statement. Assuming there are 70 records, I will reach the value 70 and the 70th elements of the two arrays will have their values assigned by the 70th tape record. Next, I takes on the value 71 and another attempt to read is made. Since there are no more tape records, the execution branches to statement 110, where the total number of records, TOTAL, is defined to be I-1, or 70. Line 9 begins the punched-card reading routine. A card is read into the variables NUMB and SALARY, and then the subroutine UPDATE is called. Next, an attempt is made to read another card. When there are no more cards, the execution branches to statement 130, line 12, where the tape file is rewound (so writing can start at the beginning), and a header record is written onto the tape (if desired). This header record can contain such information as the date the tape was last modified. Next a DO loop is started at line 13, with the index I varying from 1 to TOTAL. The value of TOTAL at this point may not be 70, since the subroutine may have changed it. The Ith value of the array EMPLSAL is tested at line 14. If the salary is less than one cent, a branch is made to statement 140, where the index I gets incremented and the loop is started again with line 14. The point of the test is to ignore all employee numbers where the salary has become zero, i.e., for employees who have left the firm. If a branch is not made in line 14, a check is printed out with the employee's number and salary on it (the employee's name has been left out of this discussion so as not to confuse it further), and a tape record is written with the same information.

When the loop is exhausted, execution continues with line 19, where an end-of-file designation is written onto the tape. At line 20, all files are closed; next, the program comes to a stop. The END statement tells the compiler that it can start compiling the above statements.

The subroutine UPDATE begins at line 23 with the name of the routine and the argument list. Note that the names used in the subroutine are not the same as those used in the main program. The subroutine is compiled separately; it could even be compiled at a different time and be stored in a library. The only essential point is the order of the variables in the argument list. The array EMPLNUM is the first variable in the argument list of the main program, and corresponds to the array NUMARRAY in the subroutine. The nonarray variables NUMB, SALARY, and TOTAL in the main program correspond respectively with the nonarray variables N, S, and TTL in the subroutine. Note that the integer array EMPLNUM needs to be declared an integer in the main program because it does not begin with the letters I through N, but needs not be so declared in the subroutine because the array NUMARRAY does begin with an N, hence is implicitly an integer.

Line 26 starts a DO loop, using the index I which varies from 1 until the value TTL. The first time through, the value of TTL is 70. Line 27 checks if the employee number as read from a punched card is part of the existing records of employee numbers. If it is, then this is an indication that a salary change has taken place. Either the salary was modified or the employee left the company, in which case the salary went to zero. Either way, a branch is made out of the loop to statement 110, where the Ith value (the same value of the index where a match was found in the loop) of SALARRAY is assigned the new value of the salary, S. That is, if the 14th employee number in the array NUMARRAY matched the number on the punched card, N, a branch would be made out of the DO loop with I retaining the value 14. Then, the 14th value of the SALARRAY would be set equal to the new value of the salary. After that, execution resumes with the main program. If there is no match at line 27, that is, if the DO loop continues until it is satisfied, this indicates that the employee number from the punched card is that of a new employee, and hence that number and the associated salary must be added to the existing list of employee data.

When the loop is exhausted, execution continues at line 29, where the TOTAL number of records is incremented by 1. If TTL was equal to 70, and the punched card record was that of a new employee, line 29 changes the value of TTL (TOTAL in the main program) to 71. Next, the 71st elements of the number and salary arrays are assigned values N and S, respectively. After that, a branch is made to the RETURN statement, which returns control to the main program.

By inserting a statement between lines 1 and 2 which reads: COMMON EMPLNUM,EMPLSAL,NUMB,SALARY,TOTAL, and another statement between lines 24 and 25 reading COMMON NUMAR-

RAY, SALARRAY, N,S,TTL, the argument lists in the subroutine call and definition statements (lines 10 and 23) could be eliminated. Line 10 would then read CALL UPDATE, and line 24 would read SUBROUTINE UPDATE. It is permissible to have part of the arguments in a common statement and the other part in the argument list.

Note that the records written onto the tape are unsorted. Without further comment, a brief sort routine is shown below, which would sort the records in order of employee number. If used, such a routine would be inserted prior to the output routine:

```
        N1=TOTAL-1
        DO 10 I=1,N1
        N2=I+1
        DO 10 J=N2,TOTAL
        IF(EMPLNUM(I).LT.EMPLNUM(J)) GO TO10
        K=EMPLNUM(I)
        EMPLNUM(I)=EMPLNUM(J)
        EMPLNUM(J)=K
        X=EMPLSAL(I)
        EMPLSAL(I)=EMPLSAL(J)
        EMPLSAL(J)=X
     10 CONTINUE
```

Those interested can try to interpret the logic of the sorting routine.

Although statement numbers were all three-digit and started with 100, this is not necessary. Any number between one and five digits may be used, and the numbers need not be in any sequence, but the same number may not be used for more than one statement. Statement numbers in subprograms bear no relation to statement numbers in the main program.

Conclusion

To learn computer programming requires a certain minimum course of study, but in interactive computer terminal operation only a minimum of instruction is needed. The fundamentals of the three most popular languages—BASIC, FORTRAN, and COBOL—have been illustrated here to give the reader who is not familiar with programming some idea of the essential concepts. Many specific books are available which deal with various programming techniques and with detailed applications. The bibliography presented here is by no means complete, but should serve as a starting point.

BIBLIOGRAPHY—CHAPTER 18

CHERRY, G., *ADA Programming Structures: With an Introduction to Structured Concurrent Programming*. Reston, VA, Reston Publishing Co., Inc., 1981.

GENN, R. C. and GENN, E. L., *The Complete Microcomputer Handbook with Tested BASIC Programs*. Englewood Cliffs, NJ, Prentice-Hall, Inc., 1984.

GRAUER, R., *Cobol: A Vehicle for Information Systems*. Englewood Cliffs, NJ, Prentice-Hall, Inc., 1981.

HUME, J., and R. HOLT, *Better BASIC For the IBM PC*. Englewood Cliffs, NJ, Prentice-Hall, Inc., 1984.

KASSAB, V., *Basic Programming for the IBM Personal Computer with Technical Applications*. Englewood Cliffs, NJ, Prentice-Hall, Inc., 1984.

KHAILANY, A., *Business Programming in Fortran V1 and ANSI Fortran 77: A Structured Approach*. Englewood Cliffs, NJ, Prentice-Hall, Inc., 1981.

SMITH, J. M., and N. M. KOKOTOVICH, *Documenting an Organization's Computer Requirements*. New York, NY, John Wiley & Sons, Inc., 1985.

TENNENT, R., *Principles of Programming Languages*. Englewood Cliffs, NJ, Prentice-Hall, Inc., 1981.

TURNER, R., *Logics for Artificial Intelligence*. New York, NY, John Wiley & Sons, Inc., 1984.

CHAPTER 19

OPTO-ELECTRONIC DEVICES

The field of optics is related to the rest of electronics by the fact that light is a form of electromagnetic energy, just as radio waves, but with a much higher frequency and shorter wavelength. This chapter covers the fundamentals of optics, light emission devices, light sensing devices, and the combination of the two in the form of optical couplers. The most recent developments in fiber optics, optical communication, and optical computing systems are also included.

19.1 FUNDAMENTALS OF OPTICS

The visible spectrum and its place in the overall spectrum of electromagnetic radiation is shown in Figure 19-1. Note that it ranges from about 10^{14} to about 10^{15} Hertz or, in wavelengths, the center of the visible spectrum is approximately 10^3 millimicrons. The infrared region

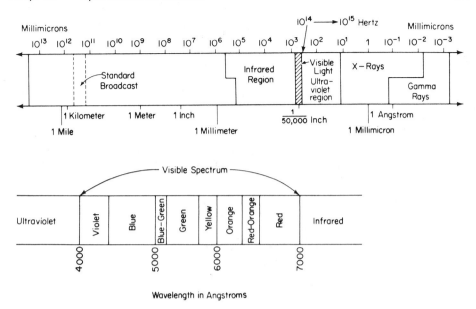

Figure 19-1.   The visible spectrum.

(Gerald L. Hansen, *Introduction to Solid-State Television Systems: Color and Black and White*, © 1969. Reprinted by permission of Prentice-Hall, Inc.)

is lower in frequency and longer in wavelength than visible light, while the ultraviolet region is higher in frequency and shorter in wavelength. X rays and gamma rays have still higher frequencies.

The visible spectrum is usually measured in Angstroms (Å); one Angstrom equals 10^{-9} meters. The colors that we can see range from the lowest frequency, red, to the highest, violet. As illustrated in Figure 19-1, these are the spectrum, saturated or pure colors in that each consists only of a single frequency. These are the colors visible in a rainbow, an oil slick, or when white light is diffracted through a prism. The colors most frequently seen by the human observer are not pure colors but so-called pastels, which means that they are made up of mixtures of pure colors. Colorimetry, the study of visible colors, is covered in Chapter 15 in connection with color television.

Chapter 11 dealt with RF and microwave fundamentals and introduced the effects of electromagnetic waves when they are reflected or refracted. The electromagnetic waves comprising visible light are much higher in frequency and have much shorter wavelengths and are therefore subject to a much more pronounced interaction effect when they encounter different materials. These interaction characteristics are generally classified as reflectance, absorbtance, transmittance, and refraction.

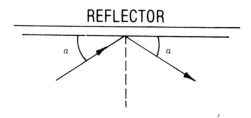

a) SPECULAR REFLECTANCE

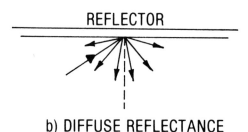

b) DIFFUSE REFLECTANCE

Figure 19-2. Diffuse and specular reflectance.

Reflectance

By definition, reflectance determines the amount of light that is reflected away from the surface it strikes. A mirror is an example of a device providing high reflectance, while a matte black surface would be an example providing minimum reflectance. The reflectance may be either diffuse, specular, or both. Diffuse reflectance is caused by a surface that is rough as compared to the wavelength, while a specular reflector has a smooth surface. A good mirror provides a large amount of specular reflection, while a beaded motion picture screen, for example, provides diffuse reflection. Figure 19-2 illustrates both specular and diffuse reflection. Note that in the case of specular reflection the angle of incidence is equal to the angle of reflectance.

Absorbtance

This describes the amount of light that is absorbed by the material it strikes. Absorbtance varies with different substances. Many methods of chemical analysis use absorbtance measurements because certain chemicals absorb specific wavelengths of the visible spectrum more than others. Absorbtance also enters into the efficiency of lens assemblies as described in Chapter 15 with regard to TV cameras.

Transmittance

The amount of electromagnetic radiation transmitted through a substance is another important characteristic. Transmittance becomes an essential characteristic when light is transmitted over long distances, as in the case of fiber optic devices and complex lens systems.

The sum of the reflectance, absorbtance, and transmittance must add up to 1.00 for any substance. This simply means that the total of light radiated toward a substance consists of some reflectance, some absorbtance, and some transmittance.

Refraction

Any electromagnetic wave changes direction at the interface of two mediums if the angle of incidence is not 90°. The index of refraction for light is the sine of the angle of incidence divided by the sine of the angle of refraction. The refractive index is also a function of wavelength or, in the case of visible light, of color. The refractive index of air is nominally 1.000.

Because the refractive index varies with the frequency or wavelength, a beam of white light can be diffracted by a prism or similar device to produce a "rainbow" display, the visible spectrum from red to blue and violet.

Polarization

The same principle of polarization applies to visible light as described in Chapter 10 for antennas. Ordinary daylight, diffused and traveling at many different angles, is polarized in many different ways. Just as specific antennas radiate and receive most efficiently the waves polarized in their direction, light filters can provide selective passage of polarized light. If, for example, a beam of light is passed through a vertically polarized filter, only the vertically polarized components will pass. If this light is then passed through a filter with horizontal polarization, none of the vertically polarized light can pass. The characteristics of polarized light are used in certain electronic light emission devices (liquid crystal displays), in optical communications systems, and, of course, for the reduction of solar radiation to the human eye (sunglasses).

Radiometry and Photometry

Radiometry is concerned with optical measurements, regardless of wavelength or color. Photometry, on the other hand, is concerned with visible light with respect to the spectral response of the human eye.

Definition	Radiometric		Photometric	
	Name	Unit (SI)	Name	Unit (SI*)
Energy	radiant energy	joule	luminous energy	lumen-sec
Energy per unit time = power = flux	radiant flux	watt	luminous flux	lumen
Power input per unit area	irradiance	W/m²	illuminance	lm/m² lux
Power per unit area	radiant exitance	W/m²	luminous intensity	lm/m²
Power per unit solid angle	radiant intensity	W/steradian	luminous intensity	candela
Power per unit solid angle per unit projected	radiance	W/m² steradian	luminance	candela/m²

Table 19-1. Radiometric and photometric units.

(M. Grossman, *Technician's Guide to Solid-State Electronics,* © 1976. Reprinted by permission of Prentice-Hall, Inc.)

The characteristics of human vision are covered in Chapter 15, Television Systems. Table 19-1 shows a comparison of radiometric and photometric units.

Figures 19-3a and b illustrate some of the essential concepts inherent in Table 19-1. A theoretical point source radiates uniformly over a 4π steradians space. At a distance of one foot, the illumination from a 1 candela or 12.6 lumen light source will be 1 lumen per square foot. At a distance of 1 meter from the point source, the spherical surface will be 1 square meter and will be illuminated by 1 lumen per square meter.

The radiometric and photometric units described in Table 19-1 can, of course, be related to each other. Table 19-2 provides illumination and luminance conversion factors to allow conversion between different units. These different units are either based on the English system (foot and inch) or the metric system (meter) as well as such optical units as candles, candelas, and lamberts. Photometric specialists use the lambert to describe the brightness of area type sources. The lumen is a measurement of the power of the visible electromagnetic radiation. For a point source the intensity of the source is measured in lumens per steradian, and this unit of intensity is called the candela. The intensity of point source light-emitting diodes is generally specified in candelas or millicandelas, but sources of light which represent an area rather than a point source are described as brightness by the

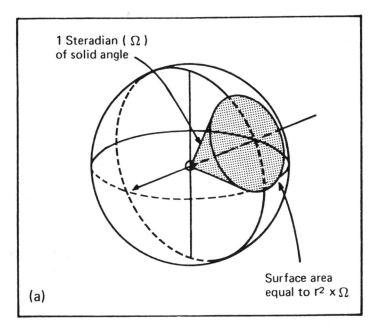

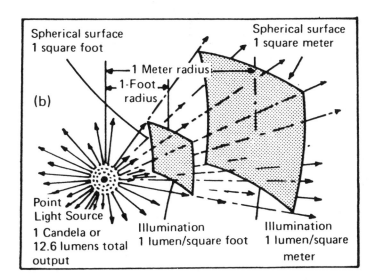

Figure 19-3.   Concept of illumination by a point source.

(M. Grossman, *Technician's Guide to Solid-State Electronics*, © 1976. Reprinted by permission of Prentice-Hall, Inc.)

ILLUMINATION CONVERSION FACTORS				
1 lumen - 1/680 lightwatt (at 555 nm) 1 lumen-hour = 60 lumen-minutes 1 footcandle = 1 lumen/ft²		1-watt-second = 1 joule = 10^7 ergs 1 phot = 1 lumen/cm² 1 lux = 1 lumen/m²*		
Number of ⟶ Multiplied by ⟶ Equals number of	Footcandles	Lux*	Phots	Milliphots
Footcandles	1	0.0929	929	0.929
Lux*	10.76	1	10,000	10
Phots	0.00108	0.0001	1	0.001
Milliphots	1.076	0.1	1,000	1

LUMINANCE CONVERSION FACTORS						
1 nit = 1 candela/m²* 1 stilb = 1 candela/cm² 1 apostilb (international) = 0.1 millilambert = 1 blondel 1 lambert = 1,000 millilamberts						
Number of ⟶ Multiplied by ⟶ Equals number of	Foot- lamberts	Candelas /m²	Milli- lamberts	Candelas /in.²	Candelas /ft.²	Stilbs
Footlamberts	1	0.2919	0.929	452	3.142	2,919
Candelas/m²*	3.426	1	3.183	1,550	10.76	10,000
Millilamberts	1.076	0.3142	1	487	3.382	3,142
Candelas/in.²	0.00221	0.000645	0.00205	1	0.00694	6.45
Candelas/ft.²	0.3183	0.0929	0.2957	144	1	929
Stilbs	0.00034	0.0001	0.00032	0.155	0.00108	1
*International System of Metric Units—recommended standard						

Table 19-2. Conversion tables for illumination and luminance.

(M. Grossman, *Technician's Guide to Solid-State Electronics*, © 1976. Reprinted by permission of Prentice-Hall, Inc.)

unit *lambert*. A number of additional units such as the *lux* and the *phot* have been used in optical specifications. Table 19-2 permits conversion between these various units.

Lenses, Prisms, Mirrors, and Optical Filters

These devices operate on the principles of reflectance, absorbtance, and refractance as described above. Lens systems are usually a combination of the two basic lenses shown in Figure 19-4. The planoconvex lens has a diameter D and a focal length f, indicating the distance from the center of the lens to the point of convergence. The planoconcave lens in Figure 19-4b has, in effect, the opposite action to the convex lens, but its focal length is based on an imaginary point at the front of the lens.

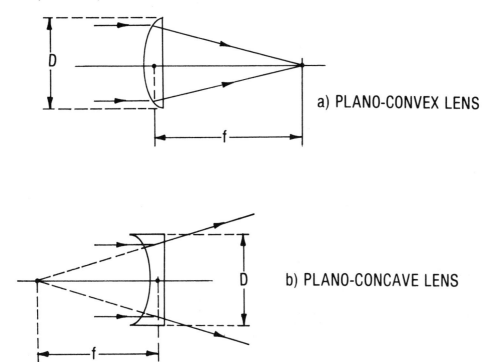

a) PLANO-CONVEX LENS

b) PLANO-CONCAVE LENS

Figure 19-4. Basic lenses.

Figure 19-5 illustrates the next degree of complexity in the bi-convex and biconcave lens. For the same radius of curvature as a pla-noconvex and planoconcave lens, the biconvex and biconcave lenses have a much shorter focal length due to the double action of the extra curvature. In all other respects these lenses operate in the same manner.

Meniscus lenses can be combinations of the curvature of convex and concave lenses. If both the front and rear surfaces are curved in the same direction, as illustrated in Figure 19-6a, it is considered to be a positive meniscus lens and operates similar to a convex lens. If the curvatures at the front and rear of the lens are basically concave, this is considered a negative meniscus lens. Lenses are available with a concave front, concave back, a convex front, and a convex back as well. A large variety of lens combinations can be assembled in optical systems to provide the required optical function.

Prisms are also available in a large variety of different shapes and angles. In determining the shape of a prism, the rules of refractance dominate the function. Depending on the angle of incidence with the entering and the exiting prism surfaces, different effects will be obtained.

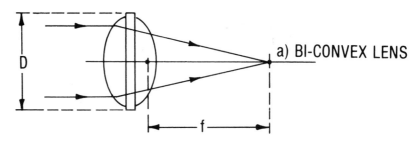

a) BI-CONVEX LENS

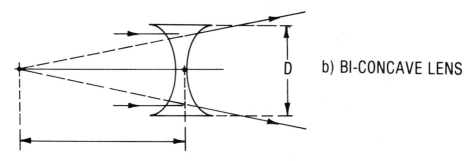

b) BI-CONCAVE LENS

Figure 19-5.   Double-action lenses.

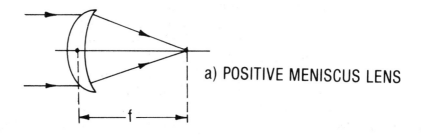

a) POSITIVE MENISCUS LENS

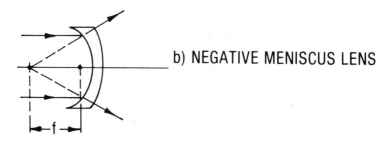

b) NEGATIVE MENISCUS LENS

Figure 19-6.   Meniscus lenses.

The simple household mirror can be compared to the reflector used with antennas in microwave transmission. Mirrors can be curved spherically or paraboloidally to provide focusing onto a point or a parallel beam. A special form of mirror is the half-silvered type. It has a very thin layer of reflecting material so that some of the light is reflected and some is passed through. Two such mirrors, placed at right angles, are used in color TV cameras to provide identical images to three or four separate photosensitive surfaces.

Filters consist basically of colored glass. The color of the filter indicates the color of light that it permits to pass. In electronic terms such filters correspond to band pass filters because they permit one narrow band of frequencies to pass through but reject other frequencies. Optical band stop type filters are also available but are rarely used. Such filters are usually a combination of band pass filters in which one particular band is omitted. It is also possible to obtain filters that will allow more infrared or more ultraviolet light to pass, corresponding to electronic low- and high-pass filters. The manufacture, specification, and selection of optical filters is a highly specialized field usually found within the discipline of astronomy and professional color graphics.

19.2 ELECTRONIC LIGHT EMISSION DEVICES

Incandescent Light Sources

The ordinary light bulb is the oldest incandescent electric source of light. The intensity of the light emitted from an incandescent lamp depends on the temperature of the tungsten wire; this, in turn, depends on the amount of electric current flowing through it. Any increase in brightness is accompanied by a corresponding increase in temperature and power consumption. Incandescent displays can take the simple form of pilot lights or they can be as specialized as numerical readouts. In numerical readouts, a series of filaments is arranged in the standard seven-segment format illustrated in Figure 19-7. The entire assembly is mounted in an evacuated glass envelope, and the required segments are connected to the power supply to display the selected numeral.

Electroluminescent Displays

The cathode-ray tube (CRT) is the most widely used application of electroluminescence. Light is emitted from the phosphor coating where a high-intensity electron beam strikes it. See Chapter 15, Television Systems, for further details on cathode-ray tubes. The principle of elec-

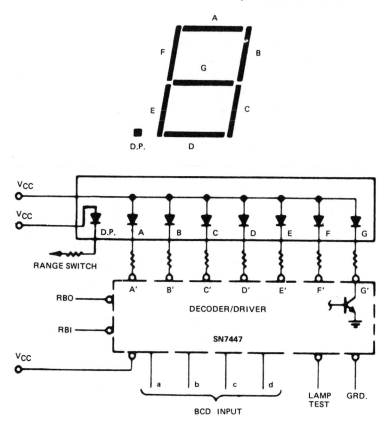

Figure 19-7. Seven-segment LED circuit.

(Walter H. Buchsbaum, *Complete Guide to Digital Test Equipment,* © 1977. Reprinted by permission of Prentice-Hall, Inc.)

troluminescence is also used in small vacuum tube indicators in which a phosphor-coated anode strip is bombarded by electrons from the cathode. Numerals can be displayed by arranging these anode strips in the form of a seven-segment display and using a common cathode. Electroluminescent displays require very low current and relatively high voltage; in most instances, the cathode is heated by a filament.

Multiple electron beam technology is emerging as a way to provide considerably more information by use of cathode-ray tube displays. Multiple-beam CRTs are desirable for medical imaging, military intelligence, analysis of the earth's resources, and computer-aided design and engineering. High-end applications such as these may have a 12-beam direct-viewing CRT. However, eight-beam CRTs are being used to obtain a multiple-beam raster-scan system. The eight beams are deflected in a bundle in order to create something like a paintbrush to paint the CRT screen with eight lines each sweep. The beams sweeping the CRT create 60 frames, as in standard television, except there is no interlace.

Gas Plasma Displays

The ordinary neon bulb is the simplest example of a gas plasma display. When a sufficiently high voltage is applied between the cathode and anode, the gas is ionized and an orange-red glow will appear at the positive electrode, the anode. This principle was used in the well-known "Nixie" displays where the numerals consisted of a series of specially shaped anodes, one mounted behind the next. A common cathode was used and, depending on which anode was connected to the positive voltage, that numeral glowed.

Much better visibility is obtained with segmented or dot matrix type gas plasma displays. Gas plasma displays operate at relatively high voltage (300 volts) and low current (1 ma or less per seven-segment number). Where several numerals are displayed, some economies can be achieved by operating all of them in a multiplex mode. Multiplexing provides power saving but requires that each numeral be illuminated at least 30 times a second to avoid flicker.

In dot matrix panel displays, vertical and horizontal strip electrodes, cathode and anode, are energized to generate a glowing dot at their intersection. The display circuitry for a dot matrix panel can be quite complex, but with the use of LSI-ICs fairly elaborate neon dot matrix displays have become competitive with other display systems.

Light-Emitting Diodes (LED)

The operation of this light source is based on the emission of photons when a current passes through a gallium arsenide silicon diode junction. A variety of different dopants have been developed to produce light of different colors, but the most common is a deep red. Special purpose LEDs are available in which different voltage levels produce different color illumination. These devices, however, are not in widespread commercial use. The familiar seven-segment display, illustrated in Figure 19-7, consists of individual diodes connected in parallel. The decoder-driver IC enables those diodes, representing individual segments, which make up the desired numeral. Note that external current limiting resistors connect the IC with each individual diode segment.

The most widely used LEDs operate on a 5-volt power supply and, in a typical unit, draw about 16 ma per segment for standard brightness. Power supply regulation is very important to LED operation. This is illustrated by comparing the power requirements for four 7-segment LEDs indicating the number 1111 and the number 8888. In the first case, a total of eight segments are illuminated, drawing 128 ma. In the second case, all 28 segments are illuminated and the power supply must deliver 448 ma. When multiplexing is used, different segments are illuminated at different, but brief, periods; this eases the power supply requirements somewhat. Multiplexing, however, requires that each segment be illuminated at least 30 times per second to eliminate flicker.

LED displays are also produced in the form of a 5 × 7 dot matrix. This permits the display of the letters of the alphabet as well as numerals, but requires more complex decoder-driver and multiplex circuits.

In the field of optical communications and in such devices as optoisolators, the most common LED is constructed using infrared-emitting materials (GaAs and GaAs:Si). These materials are more efficient than ordinary LEDs that emit visible light. Those doped with Si may be rated 10 mW at 100 mA. Similar devices are used extensively in fiber-optical communications systems. These are actually *near* infrared devices (1.3 to 1.6 μm wavelength) because operation in the near infrared spectrum results in lower fiber losses.

Liquid Crystal Displays (LCD)

This device is a light valve rather than a source of illumination. Liquid crystal displays get their name from a transparent liquid containing crystals which change their optical characteristics when an electric field is applied. The liquid-crystal state is called *nematic*.

When a voltage is applied between the two electrodes in Figure 19-8 the light transmission characteristics of the liquid change. In the dynamic scattering method the liquid is normally transparent, but when an electric field is applied, crystals form in a random fashion between the two electrodes. The reflectance of these crystals scatters the light and produces the effect of making the space between the two electrodes opaque. In some LCD displays, both electrodes are transparent, but in most cases, such as in digital watches, one electrode is a mirror. When the illumination comes from the front the electrodes forming the seven-segment numerals are transparent. In some digital wristwatches a tiny light bulb is mounted at the edge of the display for viewing in the dark.

The so-called field effect LCD appears to operate the same way, but works on a different principle. In this system the crystals are always

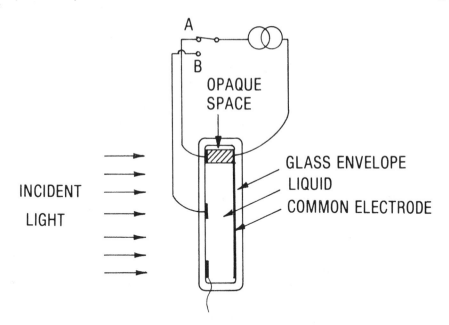

Figure 19-8.   Liquid crystal display (LCD) principles.

present in the liquid, and the application of an electric field causes the crystals to polarize the light passing through. When we add a polarizing filter on the front of the display, we control the polarization of the light passing through the liquid. If the filter permits only vertically polarized light waves to pass and the liquid crystals permit only horizontally polarized light to pass, we will observe the same effect as in the dynamic scattering type of LCD described above. Either type of LCD effectively controls the passage of light, but the field effect type of LCD requires polarized filters. Dynamic scattering type LCDs are usually available in fairly large sizes, but the field effect LCDs are most widely used in digital watches. Continuous application of a dc voltage damages the liquid material; therefore, only ac driving signals are used. Typical signals for driving LCDs are in the 25- to 100-Hz square wave range, at 20 to 120 volts. The current, however, is in the order of a few microamperes per segment. LCDs represent a high dc resistance shunted by a small capacity. A typical seven-segment LCD may be 200 megohms shunted by 300 pf.

A recent development based on liquid crystal technology is the Tektronix liquid crystal shutter. The Textronix shutter polarizes coherent light (from red and green filters) to create color images from monochromatic light.

Lasers

This acronym stands for *light amplification by stimulated emission of radiation.* A highly specialized field, the subject of many technical books, laser technology is based on quantum electronics. Atomic physics and chemistry are the key disciplines in the design and construction of lasers. In the application of lasers, however, the field of optics is sometimes of primary importance. This book presents a brief overview of those laser characteristics that are important to electronics professionals. In terms of electronic circuitry, we can say that a laser differs from other light sources in the same way as a noise generator or relaxation oscillator differs from a sine wave oscillator. Lasers generate a single frequency or a very narrow band of frequencies. What is even more important, however, is that these light signals, like the output of sine wave oscillators, are continuously coherent. It is this characteristic—continuously coherent single frequency optical radiation—which makes it possible to concentrate relatively large amounts of light power onto a very small area. Laser surgery and industrial applications use this characteristic of lasers.

Sine wave oscillators require a sharply tuned, resonant circuit or a crystal. Lasers, similarly, use an optical cavity, properly placed mirrors, rotating mirrors, polished and reflectively coated crystal surfaces, etc., to provide a single-frequency, coherent output. Any of the conventional modulation methods—AM, FM, pulse, and phase—can be employed with lasers. Pulsed lasers are used for distance measuring or the location of reflective objects, just as in radar technology.

Most commercially available lasers can be categorized as either solid state or gas devices. Gas lasers were developed first and they generally are capable of higher-power output, but their operation is also more complex. Just as in an ordinary neon tube, the gas is ionized by the passage of electric current and forms a plasma. This plasma can be concentrated by a magnetic field to generate the required atomic effects within the plasma to cause laser action. While there is a variety of methods to generate the lasing action of the ionized gas, they all rely on the principle of changing the states of energy within individual atoms. Helium, neon, argon, and xenon are the most frequently used gases for lasers. In many instances these gases are mixed.

Solid state lasers are either of the crystal type or the light-emitting diode (LED) type. Ruby crystals, especially, can be excited into lasing action by "optical pumping." In this method, a strong light excites the crystal structure of the ruby itself. An LED can be converted into a laser by special mechanical construction, specially ground and polished surfaces, and strong forward biasing. Figure 19-9 illustrates the construc-

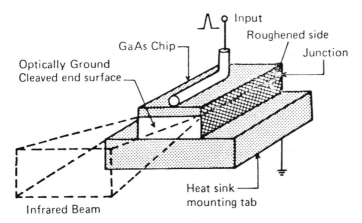

Figure 19-9.   LED laser.

(M. Grossman, *Technician's Guide to Solid-State Electronics.* © 1976. Reprinted by permission of Prentice-Hall, Inc.)

tion of an LED laser. Lasing action requires very high currents within the LED structure. This means that only pulsed operation is practical and cooling must be supplied where the power dissipation reaches the LED's limit.

Modulation of Lasers

Lasers can be modulated in a number of different ways. Where optical pumping is used to initiate and maintain lasing action, the amplitude or pulsing of this optical pumping can be used to provide amplitude or pulse modulation. It is more difficult to vary the phase and the frequency of the laser output signal. A GaAs laser can be modulated with X-band information by placing it in an X-band cavity. In LED lasers it is possible to use the forward-biasing current as the modulating parameter. In many high-powered lasers, particularly those that use optical cavities, the modulation may be electromechanical in the form of rotating mirrors. Liquid crystal displays (LCDs) which operate as light valves, as described earlier in this chapter, have been used to modulate the output of low-power lasers.

19.3 ELECTRONIC LIGHT-SENSING DEVICES

The basic photo transistor has been briefly described in Chapter 8, and in this chapter solid state light-sensing devices will be discussed further. These devices can be generally divided into photoresistive (or conductive) devices and those that generate voltages, the photovoltaic devices.

Photoresistive Devices

Photoresistive or conductive devices were used long before semiconductors were known. These devices, cadmium-sulphate, cadmium-selenium, and lead-sulphide, have a high resistance as the amount of light increases. These devices are often referred to as bulk photoconductive cells; their response time to changes in illumination is relatively slow. Another limitation of bulk photoconductive cells is that their sensitivity is greatly reduced when they are exposed to strong light for long periods. One of their advantages is that they can pass substantial currents and can often be used to drive relays directly. PN junctions or diodes can also be photosensitive; their operation provides many advantages over the bulk photoconductive cells. Figure 19-10a illustrates the principles of PN junction photo diodes. With a voltage E applied, and resistor R in series to limit the current, a plot of the reverse-bias current of a photosensitive diode shows that the reverse saturation current increases in accordance with the level of illumination (11, 12, and 13). In Figure 19-10b the current-voltage relationship of a photo diode is shown. This device can be either photovoltaic or photoconductive, depending on biasing. This means that, in the fourth quadrant of the plot, the diode actually generates a voltage in proportion to the amount of illumination. Most photo diodes are either designed for conduction or photovoltaic operation. Solar cells are silicon diodes which convert sunlight into electric power, and are one example of photovoltaic diodes. Their special design involves very large pn junctions for maximum exposure to sunlight.

Photoconductive diodes provide the best high-frequency response. They are available in a variety of design configurations, with very low dark currents, very fast response to pulsed light, and very high sensitivities. An example of these is the silicon PIN photo diode.

As described in Chapter 8, photo transistors are essentially junction transistors that are reverse biased and exposed to illumination. As in a photo-diode, electron-hole pairs are created in a base-collector junction and the beta of the transistor multiplies the base current. For this reason, photo transistors are more sensitive than photo diodes. In addition to ordinary junction photo transistors, FET photo transistors have been developed. These devices have very high input impedance and can provide a much higher output current than a junction photo transistor.

Silicon-controlled rectifiers, SCRs, that are activated by illumina-

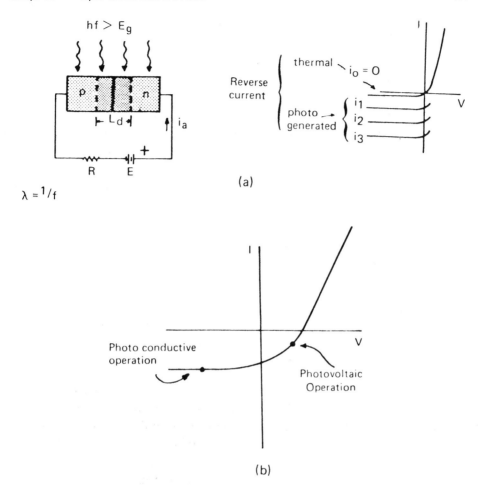

Figure 19-10.   Photo-diode principles.

(M. Grossman, *Technician's Guide to Solid-State Electronics.* © 1976. Reprinted by permission of Prentice-Hall, Inc.)

tion are also available. This device operates just like a regular SCR except that the gate signal is provided by the source of illumination instead of by the trigger voltage. Once the SCR is triggered, it conducts until the voltage across it is disconnected or its polarity is reversed. One drawback of photo-activated SCRs is their increased sensitivity to heat.

Photovoltaic Devices

As illustrated in Figure 19-10, a photo diode can be operated without bias voltage so that it generates electric power. Not all photo diodes operate in this manner. Early photographic light meters used selenium cells to generate a voltage sufficient to move a sensitive meter. Almost all present-day photovoltaic cells are of silicon and have the following common features:

Constant current output—current is relatively constant over a fair range of output voltage.

Output voltage does not change with cell area for a given type of construction and material.

Across the junction, dc resistance of the cell itself is as small as possible.

Junction surface is made as large as possible by interdigitated junction design.

Most cells generate less than 1 volt and a few ma.

Frequency response of a photovoltaic cell tends to be low. Photo cells are connected in series and parallel arrays to obtain maximum output.

19.4 OPTICAL COUPLERS

These devices combine the characteristics of light-emitting diodes (LEDs) and photoconductive diodes to convert electrical signals into optical signals and back again into electrical signals. The most common application of optical couplers is to provide an isolation barrier between two electrical entities. This makes it possible to greatly reduce or eliminate electromagnetic interference, leakage currents, ground currents, and other electrical or electromagnetic linkages between two circuits. One important application of optical couplers is in biomedical instrumentation, where they are used to provide electrical isolation between the patient and the instrument. Another use is in industrial control systems, where they isolate the driving circuit for a motor and the circuit that senses electrical motor characteristics. Optical couplers are available with a variety of amplitude, frequency, and other parameters.

In typical LED/phototransistor optoisolators, a transparent glass window called a *dielectric channel* isolates and insulates an infrared-emitting diode from the phototransistor. By this means, an isolation of 25,000–75,000, etc., volts is readily achieved.

All silicon photosensors (phototransistors, etc.) respond to the entire visible radiation range as well as to infrared. In fact, all opto-electronic devices such as diodes, transistors, Darlingtons, triacs, etc., have the same basic radiation frequency response (which actually peaks in the infrared range).

19.5 FIBER OPTICS

Fiber optic systems are finding wide use in telepone and data communications. Although glass fibers are referred to as *waveguides* (which
they are), the comparison between the transmission of light through
fibers of glass with the transmission of microwave signals through
waveguides is not exactly true, but in both cases electromagnetic radiation is transmitted along a fixed path. In both cases there are losses
due to the transmission path and in both cases the dimensions and
geometry of the transmission path are important.

Glass fibers used for transmitting light are generally circular in
diameter; their operation is based on the physical fundamentals of optics described at the beginning of this chapter. The most important of
these are reflection, refraction, absorption, and transmission. Figure
19-11(a) shows the single-mode transmission fiber. Only the light that

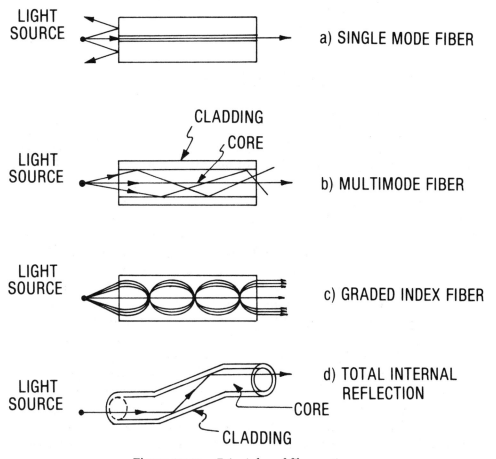

Figure 19-11.   Principles of fiber optics.

enters perpendicular to the input surface at the center of the fiber gets transmitted; all other light beams are reflected by the input surface. A much more widely used kind of fiber transmission is the multimode shown in Figure 19-11(b). Like a coaxial cable, the multimode fiber consists of a core material and a cladding which surrounds it. Because the index of refraction between the core and the cladding is very high, all the light beams which strike the cladding at an angle are refracted back into the core. The core material is usually a high-quality glass, and the cladding may either be a different kind of glass, plastic, or some other material.

The graded index fiber shown in Figure 19-11(c) provides a gradual transition between the core and the cladding, arranged in such a way that the refractive index bends the light rays as illustrated. Only the graded index fiber and the single-mode fiber are suitable for the transmission of phase-modulated coherent signals. In the multimode fiber transmission different light rays arrive at the destination with different delays. Figure 19-11(d) illustrates how total internal reflection makes it possible to bend optical fibers.

19.6 OPTICAL TRANSMISSION SYSTEMS

The availability of very long optical fibers and inexpensive coherent light sources and detectors makes it possible to construct long distance communications systems using light waves instead of radio waves. The major advantages of an optical communications system are the very high information capacity, the relatively small size, and the total immunity to many of the interference sources which plague existing communications systems. Optical communications are immune to cross talk, or interference due to lightning or other electromagnetic sources, and to the electromagnetic effects induced by nuclear explosions.

In recent years, advances by original equipment manufacturers have significantly enhanced the capabilities of optical communications systems. High-data-rate systems using optic fiber have been developed for free-space-related applications as well as here on earth.

The basic elements of an optical transmission system are shown in the block diagram of Figure 19-12. The light source corresponds to the RF oscillator and transmitter in a radio transmission system. The modulator accepts the voice, video, and data signals and modulates the light source in any of the four basic modulation methods. The light source and the modulator must be designed to work with each other. If an LED laser is used as light source, its modulation efficiency depends

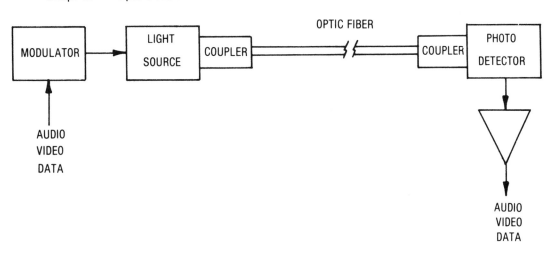

Figure 19-12.   Basic optical transmission system.

on the type of modulation used. It is also possible, however, to provide
other light sources, such as an incandescent lamp, and to modulate its
output by means of a liquid crystal device (LCD), an electrically variable
polarized filter, or an electromechanical mirror arrangement. The modulated light must then be coupled to the optical fiber, as illustrated in
Figure 19-12, to provide maximum efficiency and minimum insertion
loss. A variety of special techniques has been developed for this purpose, including a design in which the optical fiber itself is part of the
LED light source.

   At the receiving end, a photo detector produces an electrical signal
in accordance with the variations of the light it receives. Many of the
different photo detectors described earlier in this chapter can be used
for this purpose, but PIN diodes and other high-performance silicon
photo diodes are preferred.

   In the basic block diagram of Figure 19-12 only a one-way communication systems using, theoretically, a single fiber has been shown.
In practice, a fiber optic cable consists of hundreds of individual fibers.
Half of the fibers carry information in one direction, while the other
half carries information in the opposite direction, providing a large
number of two-way channels. In one experimental system, repeaters
are used at two-kilometer intervals. Each repeater consists of a photo
detector, amplifier, modulator, and new light source. The signal is, in
effect, regenerated at each repeater.

   An optical fiber replacement for standard parallel computer-to-
computer interfaces that will permit using fiber optic transmission technology over distances of up to 3000 feet has been introduced. Using

optical transmission allov 3 printers to operate at up to about 3100 lines per minute (typically, wire lines limit the system to about 2000 lines a minute at shorter distances).

BIBLIOGRAPHY—CHAPTER 19

Hewlett-Packard, *Opto-electronics/Fiber-optics Applications Manual*. New York, NY, McGraw-Hill.

PINSON, L., *Electro-optics*. New York, NY, John Wiley & Sons, Inc., 1985.

Reference Data For Engineers: Radio, Electronics, Computer, and Communications (7th Ed.). Indianapolis, IN, Howard W. Sams & Co., Inc., 1985.

SEIPPEL, R., *Optoelectronics*. Reston, VA, Reston Publishing Company, Inc., 1980.

VERDEYEN, J. T., *Laser Electronics*. Englewood Cliffs, NJ, Prentice-Hall, Inc., 1981.

YU, F., *White-light Optical Signal Processing*. New York, NY, John Wiley & Sons, Inc., 1985.

CHAPTER 20

POWER SUPPLIES

20.0 INTRODUCTION

Power supplies change electric power from a primary source into forms needed by electronic circuits. In the past linear power supplies have been the heaviest, largest, and most expensive part of electronic equipment. They were also a major source of heat and the site of large dielectric stresses. Today's switching-mode supplies are more reliable, smaller, and have significantly greater power densities than their predecessors.

Linears do have important advantages over switching-mode supplies, however. Linear supplies are superior to switches in load and line regulation. Also, output ripple amplitude from switches is considerably higher, making linears the power supplies of choice for applications that cannot tolerate a series of variations in output voltage. This chapter covers the essential aspects of both linear and switching-mode supplies.

Classification of Power Supplies

Power supplies are classified by their input and output characteristics. The characteristics usually specified for the input are:

Voltage and its variation
Phases if ac
Frequency if ac
Waveform if ac

The characteristics usually specified for the output are:

Voltage and allowable variation
Maximum and minimum load current
Frequency and allowable variation if ac
Waveform if ac
Allowable ripple and noise

The common types of supplies convert voltage from ac to dc, dc to ac, or dc to ac to dc. They will be discussed under those headings.

Regulation, Ripple, and Noise

Changes in input voltage and output current cause changes in output voltage. When these changes are above tolerable limits, regulation is used. Regulation is a major concern in power supply design. Residual ac voltage on a dc output is also of great concern to power supply users. Referred to as *ripple voltage* when its origin is the ac line through the rectification process, residual voltage can be generated by extraneous sources. The latter is frequently called *noise*.

20.1 AC-TO-DC SUPPLIES

Because of the universal availability of ac power and the requirements of electronic circuits, the ac to dc type is the most commonly used power supply. This conversion is accomplished with rectifiers, devices that permit current to flow in only one direction. This characteristic is described in the voltage current curves shown in Figure 20-1. For the ideal diode or rectifier, there should be no current flow when the voltage is negative (to the left of the vertical axis) and infinite current flow just as soon as the voltage passes beyond the vertical axis. In practice,

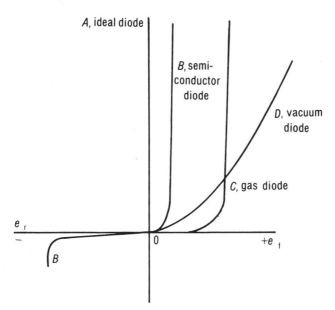

Figure 20-1.   Voltage-current curves of rectifiers.

(Lane K. Branson, *Introduction to Electronics*, © 1967. Reprinted by permission of Prentice-Hall, Inc.)

three different types of diodes or rectifiers respond with relatively different voltage-current curves, as shown in Figure 20-1. See Chapter 8, which contains additional data pertaining to diodes and rectifier components.

The voltage beyond the zero point that is required to start conduction of current is called the *off-set voltage* and is the least in the semiconductor diode and the most in the gas diode, where ionization potentials exist, depending on the type of gas used. Vacuum diodes exhibit a more gradual slope of the voltage-current curve than the other two types. Semiconductor diodes are described in detail in Chapter 8. Note that in semiconductor diodes there is a point when the voltage goes sufficiently negative so that current is conducted in the opposite direction. This reverse effect is used in a special type of diode, the zener, which is used in power supply regulation, as described later in this chapter.

Silicon zener diodes are used widely in rectifier networks where two or more diodes are connected in series to obtain a higher negative voltage than is available with a single diode. Diodes used in this manner are specifically designed to operate in the avalanche or zener condition. When the zener voltage of one diode multiplied by the number of such diodes in series is made larger than the inverse voltage applied to the series string, the inverse voltage on any single diode will not become excessive. No compensating networks are needed to insure safe operation. Silicon diodes with extremely high current ratings are available, so that parallel operation, which presents some problems, is seldom required. These developments have led to a widespread acceptance of silicon diodes in rectifier circuits.

Rectifier Networks

Rectifier networks are classified as half wave, full wave or bridge, and single-phase or multiphased. Figure 20-2 is a compilation of the essential characteristics of some of the common rectifier networks. The data presented with the circuits are based on sine wave input voltages, but the circuits may be used for other waveforms. The relationships assume ideal conditions, including zero source impedance (except where indicated otherwise), zero rectifier drop, and ideal transformers. Corrections for departures from these ideal conditions can be made when designing an actual circuit. The equivalent forward impedance of a diode can be approximated by a fixed voltage drop plus a small series resistance. In the case of silicon diodes, the voltage drop is about 0.6 volt per diode element and the series resistance is negligible. The transformer is usually the principal source impedance. Winding resistances are lumped with external circuit resistances to determine correct transformer ratios.

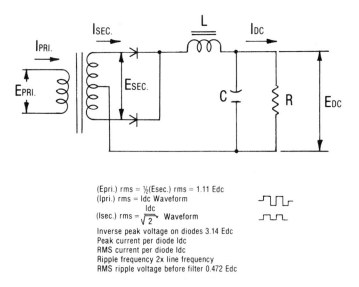

(Epri.) rms = ½(Esec.) rms = 1.11 Edc
(Ipri.) rms = Idc Waveform
(Isec.) rms = $\frac{Idc}{\sqrt{2}}$ Waveform

Inverse peak voltage on diodes 3.14 Edc
Peak current per diode Idc
RMS current per diode Idc
Ripple frequency 2x line frequency
RMS ripple voltage before filter 0.472 Edc

Figure 20-2a.   Single-phase full wave rectifier choke input circuit.

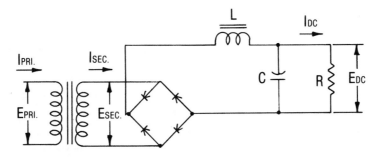

(Epri.) rms = (Esec.) rms = 1.11 Edc
(Ipri.) rms = (Isec.) rms = Idc Waveform
Inverse peak voltage on diodes 1.57 Edc
Peak current per diode Idc

RMS current per diode $\sqrt{\dfrac{Idc}{2}}$

Ripple frequency 2x line frequency
RMS ripple voltage before filter 0.472 Edc

Figure 20-2b.   Single-phase bridge rectifier choke input circuit.

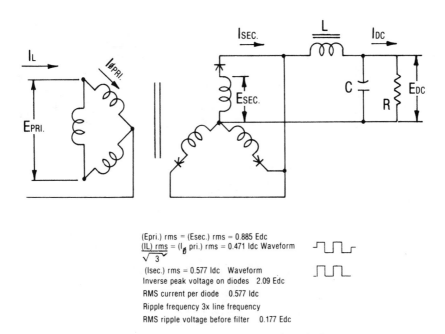

(Epri.) rms = (Esec.) rms = 0.885 Edc

$\dfrac{\text{(IL) rms}}{\sqrt{3}}$ = (I$_\emptyset$ pri.) rms = 0.471 Idc Waveform

(Isec.) rms = 0.577 Idc   Waveform
Inverse peak voltage on diodes   2.09 Edc
RMS current per diode   0.577 Idc
Ripple frequency 3x line frequency
RMS ripple voltage before filter   0.177 Edc

Figure 20-2c.   Three-phase half wave rectifier choke input circuit.

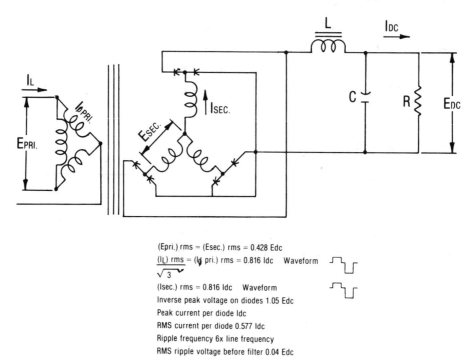

(Epri.) rms = (Esec.) rms = 0.428 Edc

$\dfrac{(I_L)\ rms}{\sqrt{3}}$ = (Iϕ pri.) rms = 0.816 Idc   Waveform

(Isec.) rms = 0.816 Idc   Waveform

Inverse peak voltage on diodes 1.05 Edc

Peak current per diode Idc

RMS current per diode 0.577 Idc

Ripple frequency 6x line frequency

RMS ripple voltage before filter 0.04 Edc

Figure 20-2d.   Three-phase full wave rectifier choke input circuit.

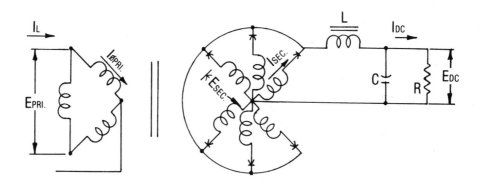

(Epri.) rms = (Esec.) rms = 0.74 Edc
(IL) rms = (Iϕpri.) rms = 0.577 Idc
(Isec.) rms = 0.408 Idc
Inverse peak voltage on diodes 2.09 Edc
Peak current per diode Idc
RMS current per diode 0.408 Idc
RMS ripple voltage before filter 0.04 Edc

Figure 20-2e.   Six-phase half wave rectifier choke input circuit.

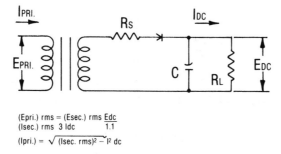

(Epri.) rms = (Esec.) rms Edc
(Isec.) rms 3 Idc   1.1

(Ipri.) = √ (Isec. rms)² − I² dc

Figure 20-2f. Single-phase half wave rectifier capacitor input circuit.

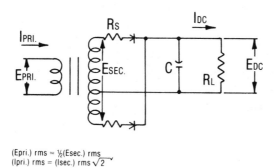

(Epri.) rms = ½(Esec.) rms
(Ipri.) rms = (Isec.) rms √2

Figure 20-2g. Single-phase full wave rectifier capacitor input circuit.

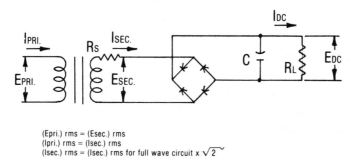

(Epri.) rms = (Esec.) rms
(Ipri.) rms = (Isec.) rms
(Isec.) rms = (Isec.) rms for full wave circuit x √2

Figure 20-2h. Single-phase bridge rectifier capacitor input circuit.

Ripple Filters

All the circuits in Figure 20-2 result in appreciable ripple voltage. The filters used to reduce this ripple voltage are of two types, choke input and capacitor input. Choke input filters give better performance, but are more costly and bulky than capacitor input filters.

For a choke input filter to work properly, the inductance of the choke must be above a critical value needed to maintain satisfactory regulation. The total current through the inductance is the sum of the dc current which flows through the load and the ac ripple current which flows through the capacitor. The critical inductance is that threshold value for which the peak ac ripple current is equal to the dc current. For any inductance less than this critical value, current flows in the inductance for less than 100% of the time. When this occurs, the capacitor tends to charge up to peak voltage during the interval when current flows. Since the reactance of the capacitor is small compared to that of the inductance, the ac ripple current is determined almost entirely by the inductance and the ripple voltage. The dc current is determined by the dc voltage and the load resistance. The critical value of inductance is directly proportional to the load resistance and inversely proportional to ripple or line frequency. The constant of proportionality contains a factor for the type of circuit used, accounting for the effect of the circuit on the magnitude of the ripple voltage. The critical inductance is:

$$L_c = KR_1/f$$

where

f = supply frequency
R_1 = dc load resistance
$K = 0.06$ for single-phase full wave circuits
$K = 0.0017$ for three-phase full wave circuits

Filter Attenuation

Attenuation of choke input filters is determined by the LC product of the filter section. The ripple reduction is given by the following approximate relationship:

$$a = \frac{1}{1 - \omega^2 LC}$$

where

  a = ratio of output ripple voltage to input ripple voltage
  ω = angular frequency of fundamental ripple voltage
  L = inductance of choke in henries
  C = capacitance of capacitor in farads

From this formula it can be seen that the term $\omega^2 LC$ must be large compared to 1 for effective filtering. As this value approaches 1, a resonant condition occurs which enhances the magnitude of the fundamental ripple frequency. This condition should be avoided. If two filter sections are used, the total attenuation is equal to the product of the two individual attenuation factors. Only the inductance of the input section need be above the critical value.

Swinging Chokes

Since the critical inductance depends upon the load, a disadvantageous condition exists when the load current is not constant. The inductance value of the choke is determined by the lowest dc current, and the dc current rating of the choke is determined by the highest dc current. This condition can be partially overcome through the use of a swinging choke. A swinging choke has two ratings, a large inductance for a small dc current and a smaller inductance for a larger dc current, in which case the choke is designed to partially saturate. The filter attenuation must be based on the smaller inductance. Reduced size and cost can be achieved through the use of swinging chokes, but circuit performance is not improved.

Capacitor Input Filters

Capacitor input filters are often preferred because of their smaller size and weight. The regulation of capacitor input circuits is poorer than for choke input circuits. A graphical analysis of several capacitor input circuits is given in Figures 20-3 through 20-6. Knowing the load resistance, the source resistance, and the allowable ripple voltage, one may obtain the required value of filter capacitor from Figure 20-3. Figure 20-4 or 20-5 may then be used to determine the relationship between ac and dc voltage. Figure 20-6 may be used to determine the relationship between ac and dc current. As can be seen from Figure 20-6, the RMS value of the current can become much higher than the dc current. This offsets some of the advantage of the capacitor input circuit.

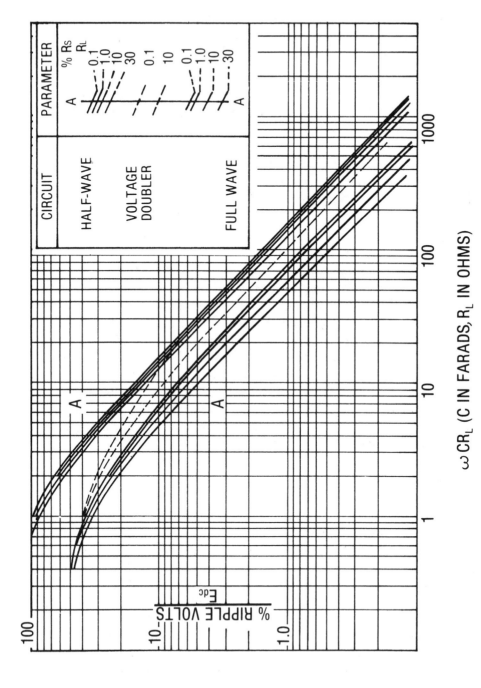

Figure 20-3. RMS ripple voltage of capacitor input filters.

(From O. H. Schade, Proc. *I.R.E.*, July 1943, p. 341.)

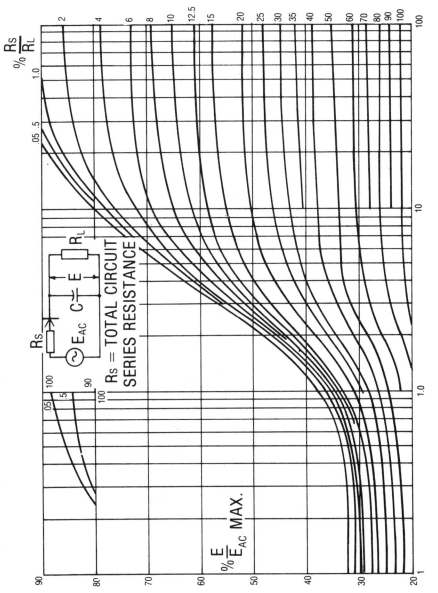

Figure 20-4. Relation of peak sine voltage to dc voltage in half wave capacitor input circuits.

(From O. H. Schade, Proc. *I.R.E.*, July 1943, p. 341.)

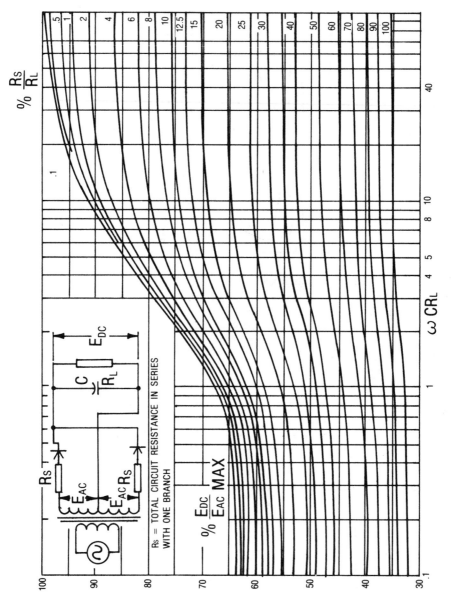

Figure 20-5. Relation of peak sine voltage to dc voltage in full wave capacitor input circuits.

(From O. H. Schade, *Proc. I.R.E.*, July 1943, p. 341.)

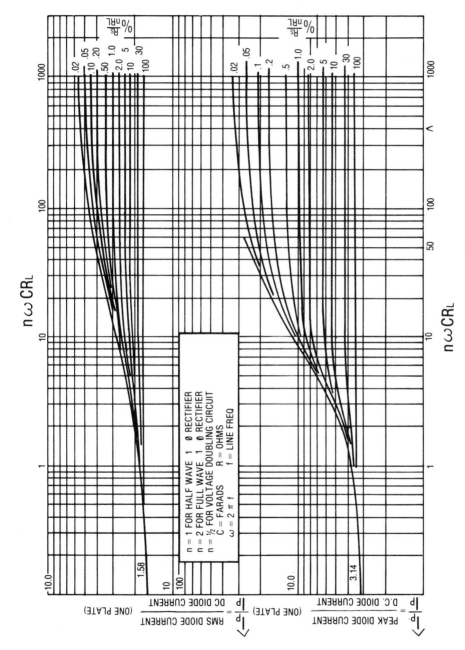

Figure 20-6. Relation of peak, average, and RMS diode current in capacitor input circuits.

(From O. H. Schade, Proc. I.R.E., July 1943, p. 341.)

Illustrative Problem

A dc source is needed to supply 200 volts at 0.050 amps. The RMS ripple voltage must not exceed 2% of the dc voltage. The power source available is 120 volts RMS 60 Hz. Minimum size and weight are desired. Determine the ratings for the capacitor, rectifiers, and transformer.

Solution: Minimum size and weight will be achieved with a capacitor input bridge circuit (refer to Figure 20-3). To use this and the other curves, the ratio of source to load resistance must be known. Since this value is unknown it must be estimated. A value of 4% is typical. Then, from the curve for the full wave circuit at the specified ripple, a value for ωCR_L of about 35 is obtained. Use 50 for safety. Then:

$$C = \frac{50}{\omega R_L} = \frac{50}{377 \times 4000} = 33 \mu F$$

The voltage rating on the capacitor must be high enough to take the peak voltage at no load. This value is determined below. Using the above values of ωCR_L and R_s/R_L, refer to Figure 20-5. The ratio of dc voltage to peak ac voltage is seen to be about 86%. Then the RMS secondary voltage will be:

$$E_{RMS} = \frac{200}{0.86 \times 1.414} = 165 \text{ V}$$

Refer to Figure 20-6. The parameter n is 2 for the circuit under consideration. From the curve the ratio of RMS diode current to average diode current is found to be about 2.5. The RMS diode current will be:

$$I_{diode, RMS} = 2.5 \times 0.025 = 0.063 \text{ A}$$

The ratio of peak diode current to average diode current is seen to be about 7. The peak diode current will be:

$$I_{diode, pk.} = 7 \times 0.025 = 0.175 \text{ A}$$

An inspection of the bridge circuit will show that the peak inverse voltage on each diode will be equal to the peak voltage across the transformer. The peak inverse diode voltage will be:

$$E_{diode, pk.} = 165 \times 1.414 = 233 \text{ V}$$

From Figure 20-2h the RMS current in the secondary of the transformer will be:

$$I_{sec., RMS} = 0.063 \times 1.414 = 0.089 \text{ A}$$

A summary of the ratings follows:

Capacitor $33\mu F$, 233 volts dc

Diodes, 4 required, Peak inverse voltage   233 V
$\qquad\qquad\qquad\qquad$ Peak current $\qquad\qquad$ 0.175 A
$\qquad\qquad\qquad\qquad$ Average current $\qquad$ 0.025 A
$\qquad\qquad\qquad\qquad$ RMS current $\qquad\quad$ 0.063 A

Transformer   Primary 120 volts RMS 60 Hz.
$\qquad\qquad\quad$ Secondary no load voltage 165 V RMS
$\qquad\qquad\qquad\quad$ Current 0.089 A RMS

Voltage Multipliers

Voltage multipliers are capacitor input circuits that charge capacitors in parallel and discharge them in series. Two types of voltage-doubling circuits are shown in Figure 20-7. A multiplier that can multiply the input voltage any number of times is shown in Figure 20-8. These

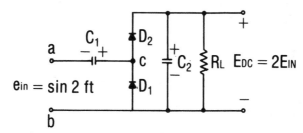

Figure 20-7a.   Half wave voltage doubler.

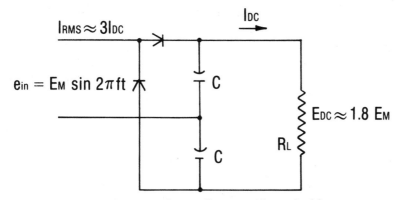

Figure 20-7b.   Full wave voltage doubler.

(Lane K. Branson, *Introduction to Electronics,* © 1967. Reprinted by permission of Prentice-Hall, Inc.)

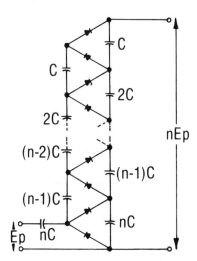

Figure 20-8. Circuit for high-order voltage multiplication.

circuits are often useful in high-voltage, low-current applications. The multiplying efficiency decreases with increasing load current. Voltage multipliers are often used with high-frequency generators, the high frequency improving the multiplying efficiency.

20.2 DC-TO-AC AND DC-TO-AC-TO-DC SUPPLIES

When the primary power source is dc, a source of ac voltage or a dc voltage at a different level is often needed. This need is met by the inverter, a switching device that converts dc voltage to a square wave ac voltage. This square wave ac voltage may then be transformed to the desired voltage level and rectified to obtain dc, or the square wave voltage may be used directly to operate ac devices. The square wave is preferred in inverters to the sine wave universally used in power distribution because of the much greater efficiency obtainable from the switching devices which the square wave allows. Sine wave generation requires a proportional control device that is wasteful of power. Many ac devices intended for operation from sine waves, including much rotating machinery, will operate satisfactorily from square waves. Other ac devices can be modified for square wave power. A third alternative is to remove the higher harmonics from the square wave with a low pass filter. Sophisticated techniques involving stepped or notched pulses are occasionally used to simplify filtering.

The inverter also enjoys wide use in applications where ac power is available. The magnetic components required in circuits operating at the usual power frequencies are often prohibitively large and heavy.

Inverters operating in the kilohertz range provide a substantial reduction in the size and weight of the power transformer. Filtering is greatly simplified. Usually all that is required is a small filter capacitor. Inverters are regularly used to supply power from milliwatts to many kilowatts. They cover the same voltage range as conventional supplies. The dc to operate inverters is usually obtained by rectifying the ac line voltage with a bridge rectifier, eliminating the need for a low-frequency power transformer. This artifice requires isolation of the inverter circuit from ground since the ac line is referenced to ground.

Inverter Operation

The inverter uses semiconductor switches, either transistors or SCRs, to apply a dc voltage with rapidly reversing polarity to the primary winding of a transformer. A simplified inverter circuit is shown in Figure 20-9. In this circuit two switching transistors are used in a push-pull arrangement to alternately apply a dc voltage to opposite halves of a center-tapped output transformer. The bases of these transistor switches are driven by a separate switching source through a second transformer. Power is delivered to the load through the secondary winding of the output transformer.

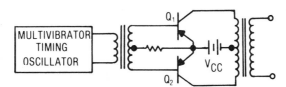

Figure 20-9.   Simple driven inverter.

The above method of obtaining the necessary switching action is used when it is desired to have the switching frequency synchronous with some other frequency. It has the advantage of simplified circuitry. The switches may also be driven by one of several feedback arrangements. This requires a tertiary winding on the output transformer as shown in Figure 20-10. The feedback circuit must have a short switching interval and must switch at the desired rate. The switching rate is controlled by having the transformer saturate on each half of the square wave. Assume, for example, that transistor Q1 is conducting. Exciting current will flow in one half of the primary winding of the transformer and increase linearly. The flux density in the core is porportional to the exciting current. When the flux density reaches saturation the inductance drops abruptly to a very small value. The transformer can no longer support the applied voltage, and the voltage across the feedback winding as well as the base current of Q1 drops. The time rate of change

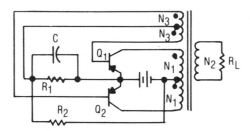

Figure 20-10. Inverter with transformer feedback windings and self-starting circuit, C and R_2.

of Q1 collector current reverses sign, causing the induced voltage in the transformer to reverse polarity. Base current in transistor Q2 begins to flow, switching on the collector current in that transistor, and the cycle is repeated. The conduction time interval is determined by the volt time integral applied to the transformer and the saturation flux density of its core.

An alternate arrangement uses two transformers. In this case the output transformer is linear and the second input transformer saturates. A two-transformer circuit is illustrated in Figure 20-11. This circuit is useful for high-power circuits where a saturating output transformer becomes unwieldy. The operation is similar to the single-transformer circuit.

A primary consideration in inverter design is efficiency. Semiconductor switches have low collector-to-emitter voltage so that the losses are low during the conduction interval. Most of the power loss occurs during the switching interval. This interval is kept small through the use of fast switching transistors and by careful transformer design. While the power losses are largely dissipated in the transistors, the causes for these losses are found in the operation of the circuit as a whole. Some power is lost in the windings and core of the transformer. A more troublesome source of inefficiency originating with the trans-

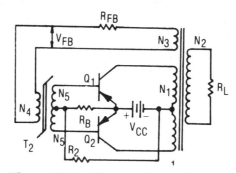

Figure 20-11. Two-transformer inverter.

former is the reactive energy stored there which must be dissipated by the transistors. This energy is stored in the shunt inductance of the transformer through which exciting current flows and in the leakage inductance through which load current flows. The magnetizing energy can be kept small by making the shunt inductance large, but the leakage inductance is a real problem. Various winding geometries can be used to reduce the leakage inductance to an acceptable minimum. The overt effect of leakage inductance is to increase switching time.

The above explanation of the operation of inverters assumes that oscillation is in progress. Actually this does not occur with great ease, particularly with start-up under load. To insure start-up, a bias may be applied to the bases of the transistors to insure initial conduction. Figure 20-10 shows a self-starting circuit consisting of bias resistor R_2 and speed-up capacitor C which accelerates switching.

20.3 REGULATION

Automatic control of power supply output is used when variations due to line voltage or load current, or both, cannot be tolerated. There are many different ways to solve this problem. A convenient way to examine the possibilities is to divide regulators into ac and dc output.

AC Regulators

An early type of ac regulator was developed for electric utilities in which an output voltage sensing circuit is used to control the selection of taps on a transformer. This system is still in use. Its appeal is due to the fact that there is very little limitation to its power-handling capability. Obvious shortcomings of this regulator include slow response and poor resolution. A slightly more sophisticated version of this regulator is widely used in industrial electronics where the power demands are more modest. In this system a toroidally wound variable transformer is used. A carbon brush wipes the entire winding of the transformer. The resolution is as good as the number of turns on the winding. The carbon brush can engage two turns simultaneously without causing a dead short. Transition from turn to turn is smoother than with a tap-changing switch. The brush is driven by a servo motor, and a sensing circuit on the output supplies the signal to the servo amplifier that operates the motor. This system is sometimes used as a preregulator followed by an electronic regulator.

The ferroresonant transformer is a useful regulator for line voltage and constant loads. Its principle of operation can be understood with the aid of the equivalent circuit in Figure 10-12. The shunt inductance,

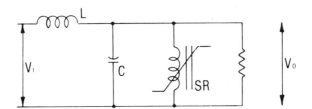

Figure 20-12. Equivalent circuit of the ferroresonant transformer.

SR, is designed to saturate below the lowest operating voltage. At the onset of saturation this inductance is resonant with the capacitor, C. At input voltages above saturation, the voltage across the resonant circuit tends to remain constant and the series inductance, L, absorbs most of the voltage between line and output. The limitations of the ferroresonant transformer illustrate the problems of ac regulation. Regulation approaching 1% for a 20% line voltage change is achievable. This claim must be immediately qualified. Since the method of regulation is based on ferromagnetic saturation, the average value of the output is regulated closely, but not the RMS value. The waveform is distorted, presenting a flat-topped appearance. The saturation point is frequency dependent, so the nominal output voltage varies with frequency. If the output is to be rectified, if the line frequency is constant, and if the load current is nearly constant, then the device works well. Compared to the mechanical regulator, the ferroresonant transformer is fast acting, within one cycle of power frequency. The functions of regulation and voltage transformation can be combined into a single component with little increase in size and weight over the transformer alone.

The problem of waveform distortion and the associated problem of sensing extend into more sophisticated types of ac regulators. The saturable reactor and its more elegant cousin, the magnetic amplifier, are frequently used as series ac regulators (see Sections 6.3 and 6.6). In this application the saturable reactor gate windings are placed in series with the load. The reactor is gated to allow that amount of current to flow which will maintain a predetermined voltage across the load. An output sensing circuit supplies an error voltage to an amplifier which drives the control windings of the saturable reactor. Waveform distortion is inherent in the operation of saturable reactors and magnetic amplifiers. The question is whether we want to regulate peak, average, or RMS values. Peak voltages usually should be limited to a safe value. This can be done with simple limiting circuits, which often operate in conjunction with regulator systems. Zener diodes across the input are sometimes used. Gas discharge tubes serve the same purpose. High voltage supplies frequently use spark gaps to limit peak voltages.

The average value is of concern when the output is to be rectified. Sensing of average value can be done with rectifier circuits, a fast and convenient method. Average value sensing can be used when RMS control is desired if the distortion is kept low. With average sensing a zener diode is used as a voltage reference. The rectified ac voltage is divided with an adjustable pot and compared with the zener. The difference signal is amplified and applied to the control windings of the saturable reactor. The phasing is arranged so that if the reference voltage is greater than the divider voltage, the control winding current increases, increasing the conduction angle of the reactor. If the divider voltage is greater than the reference voltage, the control winding current decreases, and the conduction angle decreases.

If true RMS control over a distorted output is required, this can be achieved through the use of a thermocouple sensor which, however, is relatively slow. In this method the sampled signal is fed to a resistor, the heat from which raises the temperature of the hot junction of the thermocouple. The thermocouple output is a dc current proportional to RMS input current. The thermocouple current is converted to a voltage and compared to the reference voltage across a zener diode. The difference signal controls the regulator amplifier.

Silicon controlled rectifiers (SCRs) may be used as series control devices. Their operation is similar to magnetic amplifiers. The SCR requires lower control power and switches faster. The faster switching causes transients and spurious oscillations. SCRs are widely used in nonsensitive applications. They are valuable in handling large amounts of power.

DC Regulators

There is a growing demand for regulated dc power, and a sophisticated technology has developed to meet this demand. A simple regulator circuit using a zener diode is shown in Figure 20-13. The series resistor, R, limits the current to both the load and the zener. Once the zener

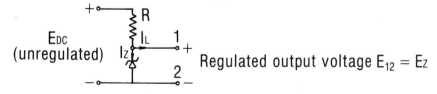

Figure 20-13.   Basic zener circuit.

(Lane K. Branson, *Introduction to Electronics*, © 1967. Reprinted by permission of Prentice-Hall, Inc.)

firing voltage is reached, current increases rapidly through the zener diode. Any further increase in the unregulated voltage, E_{dc}, will result in an increase in current through the zener diode but will not substantially change the voltage across it. If the load current increases, some of the current which otherwise would go through the diode will be available for the load. If the load current is decreased, more current will be taken by the zener diode.

A more useful regulator is shown in Figure 20-14. This is a simple form of the series regulator. The zener diode, D, is used as a reference, and the output is sampled by means of a voltage divider. The divider is adjustable to provide for varying the output and for calibration. Once set, the ratio of the divider must remain constant. Lack of stability in the divider is just as serious as it is in the voltage reference. This is sometimes a problem in high impedance dividers, which are sensitive to both temperature and voltage changes. The voltage from the zener and the divider are compared, and the difference signal provides base current for transistor Q2. The base-to-emitter voltage of Q1 is small, and the voltage difference between input and output will therefore appear across R3. If there is an increase in input voltage the divider voltage will tend to increase, causing an increase in Q2 base current and collector current. The increase in collector current will flow through R3, increasing the voltage drop across R3 and tending to maintain the output voltage constant. A decrease in load resistance will have the opposite effect.

The variation of output voltage with varying line voltage and load current can be reduced in a regulator by increasing the gain of the regulator amplifier with additional stages. However, the regulator circuit is subject to the same bandwidth and stability limitations of other feedback amplifiers. The regulator of Figure 20-14 can be refined by replacing R3 with a constant current preregulator as shown in Figure 20-15. The zener diode 1N746 maintains a constant voltage across the 1.0-K resistor, and therefore a constant emitter current in transistor 2N1131 through control of its base current. Through the comparison circuit, transistor 2N338 collector current takes a lesser or larger percentage of the current from the constant current source, thus controlling the base current of transistor 2N656 more effectively. Transistor 2N656 with 2N1049 form a Darlington pair (see Section 8.4), providing greater gain than a single-series transistor. A differential amplifier is sometimes used to form the comparison circuit in which case the zener voltage forms one input and the voltage divider forms the other input to the differential amplifier.

The transistor used as the series element limits the regulator by its

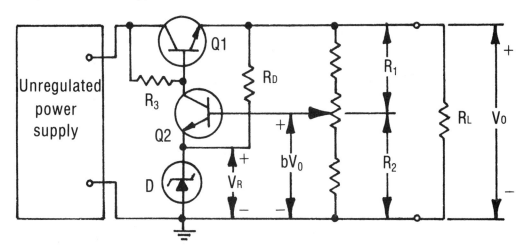

Figure 20-14.   A semiconductor regulator using a series pass transistor.

collector current rating, its emitter-to-collector voltage rating, and its power dissipation capability. Transistors are available with a wide range of current ratings, but there is a low ceiling on their voltage ratings. High-voltage regulators still use vacuum tubes for the series control. In high-voltage regulators precautions must be taken to provide for isolation of the control circuitry; otherwise the operation is similar to transistor regulators.

Filtering and Noise

Filtering for the removal of ripple voltage is a simple procedure in theory. When filter attenuation requirements are modest, networks developed from this theory yield satisfactory results. With stringent attenuation requirements other factors come into play. For example, the low amplitude and high frequency of the ripple voltage from three-phase rectifiers make these circuits a popular choice for low-ripple dc supplies. But the published values of ripple voltage and frequency for three phase rectifer circuits are based on a perfectly balanced three-phase line. A small imbalance in the line causes a troublesome increase in ripple amplitude and the introduction of a frequency component one third of that for a perfectly balanced line. The filter must attenuate the ripple components resulting from the line imbalance.

Preregulators are often used in the ac line of power supplies providing low-ripple dc. Magnetic amplifiers and SCRs in these preregulators introduce waveform distortion that causes high-ripple voltage.

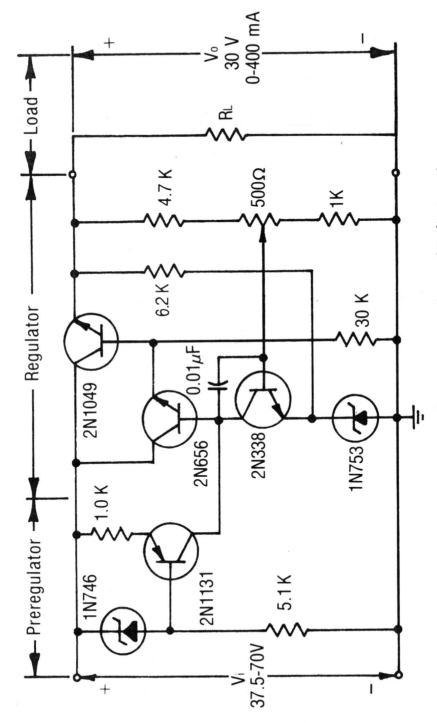

Figure 20-15. Series regulator using preregulator and Darlington pair.

(Courtesy of Texas Instruments, Inc.)

Also a larger filter choke is needed to assure continuous current flow during nonconducting intervals.

Rectifier circuitry often suffers from high-frequency resonance. Commonly the leakage inductance of the power transformer may resonate with the capacitance of a capacitor input filter or with the stray capacitance of the power transformer itself. Diode switching excites these circuits into damped oscillation. Fast switching aggravates the situation, which can sometimes be alleviated by using slower switching diodes. High-frequency traps are sometimes used to attenuate these oscillations.

Inverters and switching regulators are common sources of noise. The most that can be expected is to reduce this noise to an acceptable level. Often the switching frequency of the inverter or regulator is chosen to lie in a range which will be the least troublesome to sensitive circuits.

The power line can be either a source of noise introduced into the supply or an effective conductor of noise from the supply to other equipment. To reduce conducted noise on power lines, high-frequency line filters are sometimes used. These filters must be rated to pass power frequency with low insertion loss as well as attenuate high-frequency noise. Electrostatic shields in the power transformers reduce conducted noise by reducing the capacitive coupling between primary and secondary.

Magnetic components in the power supply have stray magnetic fields. These fields can cause hum in high gain circuits. The field is highly directional. Sometimes reorienting the component will help. Magnetic shielding may be necessary. The carbon steels used in chassis construction make good shields. Several thin partitions separated by air spaces are much more effective than a single thicker section. Shields constructed of exotic high-permeability steels are an expensive last resort.

Very high voltage circuits are seldom completely free of ionization discharge noise. These discharges increase the ambient electrical noise level and cause premature insulation failure. The requirement for low ambient noise is a more stringent requirement than adequate insulation life. Ionization discharges take place most commonly between electrodes with small radii separated by two different dielectric materials, one dielectric material being air. When a high ac voltage is applied between electrodes with this geometry, the gases in the air space are subjected to severe dielectric stress. Increasing the radii of electrodes and eliminating air by the use of solid or liquid dielectric materials will help to reduce this problem. It is difficult to attenuate this type of noise by filtering.

Dynamic Characteristics

The dynamic characteristics of a power supply are of interest because of their relationship to ripple reduction as well as to sudden changes in line voltage or load current. Consider the condition where there is a sudden increase in load current. Assuming regulators have capacitors across their outputs, if the load current momentarily increases, the output capacitor must supply the additional current until the regulator responds. The output voltage will change in accordance with the following relationship:

$$\frac{\Delta V}{\Delta t} = \frac{i_c}{C}$$

where C is the capacitance across the output. This relationship is a consideration in selecting the value of the capacitor across the load. A second consideration is that the time constant of the output circuit should be large compared to the reciprocal of the ripple frequency.

Now consider the case where there is a sudden change in line voltage. The regulator requires a finite time to react to that change. However, the filter that precedes the regulator has a time constant much longer than the regulator circuit. The voltage across the filter capacitor changes at a rate sufficiently slow to allow the regulator time to respond.

Regulators may be used in programming. The filter preceding the regulator precludes switching in the circuit prior to the regulator, and the time constant of the load circuit limits the rate of change of output voltage when switching is done in the regulator. When rapid rise of output voltage is needed, the output capacitance should be made as small as possible, consistent with ripple requirements.

An ac voltage across the output of the regulator causes a voltage to develop across the regulator in opposition to the ac voltage on output. The regulator is thus a good ripple filter. This opposition voltage is approximately equal to the ripple voltage times the voltage divider ratio times the gain of the voltage gain of the regulator. A further improvement in ripple reduction can be achieved by increasing the ac gain of the regulator with a capacitor placed across the long leg of the divider. Without affecting the dc regulating properties of the circuit, a large improvement in ripple reduction is obtained. The capacitor also has the undesirable effect of increasing the tendency to oscillate. The value of this capacitor must be chosen judiciously to accomplish its purpose without creating an unstable condition. Capacitor C3 in Figure 20-17 is such a capacitor.

The regulator is a negative feedback amplifier. It is subject to instability if sufficient phase shift exists in the feedback circuit. The low cutoff frequency in the output circuit inhibits oscillation. Instability problems arise when the ac gain of the regulator is increased to improve ripple reduction. Since the circuits that discriminate between ac and dc must be reactive, they introduce phase shift. A compromise between high ac gain with good ripple reduction and instability must be made. One solution is to provide greater filtering before the regulator.

In the circuits described above, regulation is achieved by maintaining a controlled voltage drop across a series dissipative element. The output voltage is slightly less than the minimum input voltage, and the series element must absorb a voltage slightly greater than the difference between the maximum and minimum input voltages. The power that the element dissipates is equal to the difference between input and output voltage times the load current. This type of regulator is a very inefficient device.

The switching regulator achieves much greater efficiency at the expense of increased complexity. In the switching regulator the series control element is a switch that dissipates little power. Regulation is achieved by controlling the percentage of time that the switch is closed. Energy storage elements are used to provide voltage during the intervals when the switch is off.

The operation of the switching regulator can be understood with the aid of the simplified circuit in Figure 20-16. Q1 is a transistor switch. The base drive for this transistor is a square current pulse, the width of which varies with the commands from the comparison circuit. D1 is a free-wheeling diode which allows the continued flow of dc current during the interval when Q1 is off. The voltage across inductance L is on and is equal to the output voltages when Q1 is off. In order to maintain these voltages across the inductor and to satisfy the relationship $V = L di/dt$, the rate of change of current has two values for these

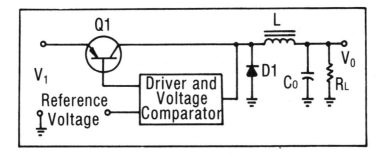

Figure 20-16. A simple switching voltage regulator circuit.

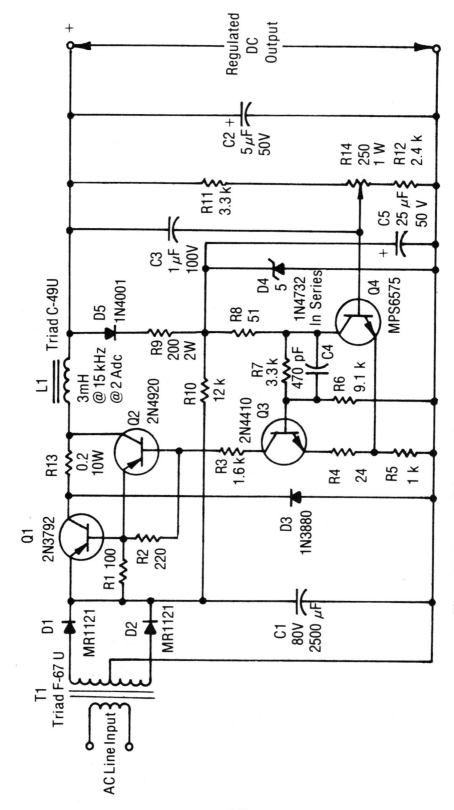

Figure 20-17. A switching regulator using a Schmitt trigger as the driver.

two conditions. When Q1 is on, diode D1 is biased off, and the current in L in excess of the load current changes capacitor C_0. When Q1 switches off, diode D1 is biased on. The current in L starts to decrease, reversing the voltage across L. Capacitor C_0 tends to maintain that voltage constant and the current in the inductor decreases linearly until Q1 switches on again. The difference between the load current and the current in the inductor is supplied by C_0. The resulting drop in voltage appears as ripple voltage on the output.

A more complete circuit for a switching regulator is shown in Figure 20-17. The control circuit contains the usual zener diode and voltage divider. Transistors Q3 and Q4 form a Schmitt trigger circuit. This is a form of bistable multivibrator which remains in one mode as long as the input signal is above a threshold value and switches to the opposite mode when the input signal is below that value. In this case the zener diode determines that threshold signal. Transistor Q3 is on when switch Q1 is on, and transitor Q4 is on when Q1 is off. Capacitor C3 provides the ac path for the switching signal to the base of Q4. Resistor R10 provides the initial current to start oscillation. The switching frequency is controlled largely by the values of L1 and C2.

Current Limiting and Short Circuit Protection

Voltage regulators can be overloaded easily. It is customary to provide current-limiting circuits to prevent damage to the regulator and load. The current-limiting circuit is often placed in the ground return, separating it from the regulator. In the circuit of Figure 20-18, diodes D1 and D2 are nonconducting as long as the voltage across them is below their forward threshold voltage. When the voltage across R_s increases due to increased load current to the point where the diode threshold voltage is reached, then base current is diverted from Q1 in such a way that the voltage across R_s is limited to a maximum value determined by the diode threshold voltage. The short circuit current will be:

$$I_s = \frac{V_i}{R_3} + \frac{V_{D1} + V_{D2} - V_{BE}}{R_s}$$

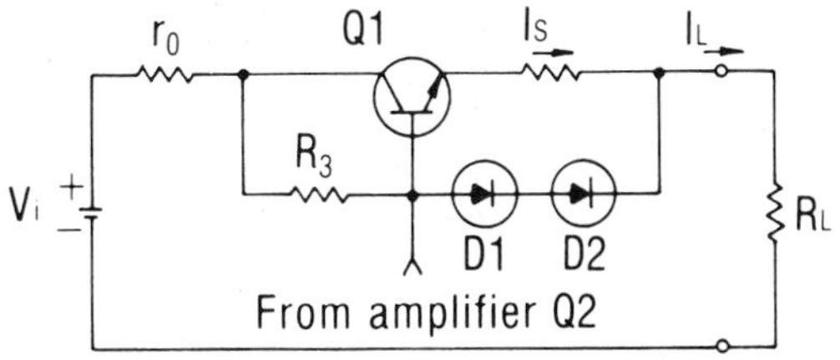

Figure 20-18. A short circuit overload protection circuit.

Crowbar Circuits

An electronic crowbar is a protective circuit used to guard against excessive voltage or currents. As the name implies, the circuit places a short across two terminals when established safe limits of operation have been exceeded. Crowbars are used to achieve faster operation than can be obtained from conventional circuit protection and to dissipate stored energy quickly when abnormal operating conditions require it. Crowbars vary in complexity and function. They can be simple clipping circuits; they can be automatically resetting or manually resetting.

A simple crowbar circuit is shown in Figure 20-19. This is a circuit to limit the peak voltage of an ac line. When the avalanche voltage of the zener diode is exceeded, one of the two SCRs fires, causing a high current to flow through the line impedance, which drops the voltage. When the ac voltage reverses polarity, the SCR that fired is reset. If the peak voltage of opposite polarity is also excessive then the other SCR will fire.

A more sophisticated crowbar is shown in Figure 20-20. This crow-

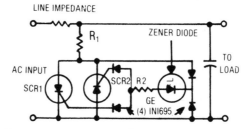

Figure 20-19. A crowbar circuit used to limit peak line voltage.

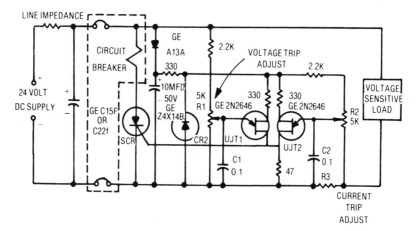

Figure 20-20. A crowbar circuit having overvoltage and overcurrent protection.

bar contains two control circuits. One functions when the voltage is excessive and the other functions when the current is excessive. Either circuit will cause the SCR to fire, shorting the dc output. The short removes the voltage from the load and accelerates the operation of a circuit breaker.

20.4 INTEGRATED CIRCUIT REGULATORS

Entire voltage regulator circuits are available on monolithic chips. One series of such chips will provide dc output voltages from 5 to 24 V at 1 A and will accommodate up to 40 V input when properly heat sinked. The regulator is contained in a standard TO-3 package. The circuit diagram of this regulator is shown in Figure 20-21. In addition to providing fixed output voltages, these regulators may be used in circuits that will provide voltage adjustment, positive or negative outputs, and higher current outputs. Typical circuits to achieve these objectives are shown in Figures 20-22, 20-23, and 20-24. Extensive application literature is available from the manufacturers of these devices.

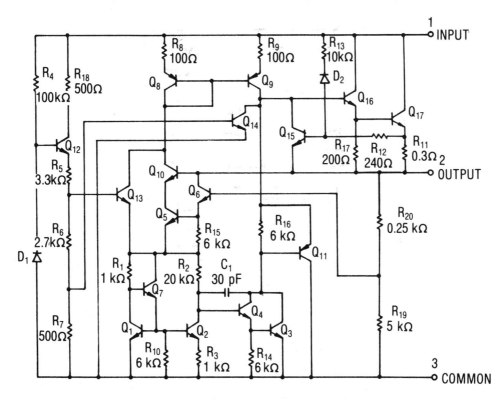

Figure 20-21.   An integrated circuit regulator providing 24 V dc at 1 A.

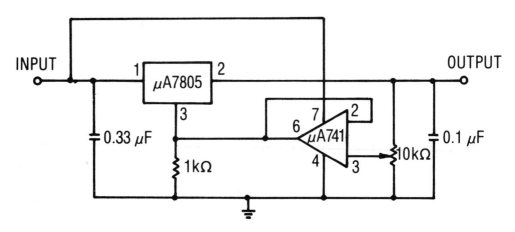

Figure 20-22. An adjustable output regulator using the fixed voltage regulator of Figure 20-21.

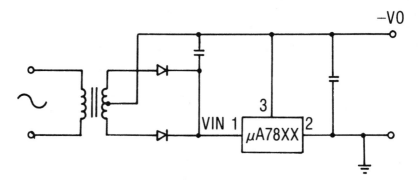

Figure 20-23. A negative output regulator circuit using the positive output regulator of Figure 20-21.

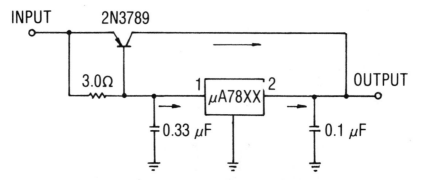

Figure 20-24. A high current regulator circuit using the 1-A regulator of Figure 20-21.

BIBLIOGRAPHY—CHAPTER 20

BUCHSBAUM, W., *Encyclopedia of Integrated Circuits: A Practical Handbook of Essential Reference Data*. Englewood Cliffs, NJ, Prentice-Hall, Inc., 1981.

HNATEK, E. R., *Design of Solid State Power Supplies* (2nd Ed.). New York, NY, Van Nostrand Reinhold Co., 1981.

Maxim Integrated Products, Inc., *Ac-to-dc Converter Chips, MAX610, 611, and 612 Series*. Sunnyvale, CA, 1985.

Motorola Semiconductor Products, Inc., *Rectifiers and Zener Diodes Data Book*. Phoenix, AZ, Motorola Semiconductor Products, Inc., 1982.

Reference Data for Engineers: Radio, Electronics, Computer, and Communications (7th Ed.). Indianapolis, IN, Howard W. Sams & Co., Inc., 1985.

CHAPTER 21

CONTROL DEVICES AND TRANSDUCERS

21.1 BASIC ELECTROMAGNETIC CONTROL DEVICES

Coils in Control Devices

Coils with iron cores find extensive applications in all branches of electronic controls. The types, shapes, sizes, and uses vary greatly, and only a few typical applications are presented in Figure 21-1. A representation of a basic relay is shown at (A). When current flows through the coil, an electromagnet is formed and the metal spring attached to an iron pole piece is pulled to the coil, thus closing the contact. Although only a single, normally open contact is shown in the example, relays in a variety of different contact arrangements are possible, as discussed more fully later on.

At (B), an iron coil is shown as a sensing device. When magnetic metal studs pass directly under the coil, a voltage pulse is induced into

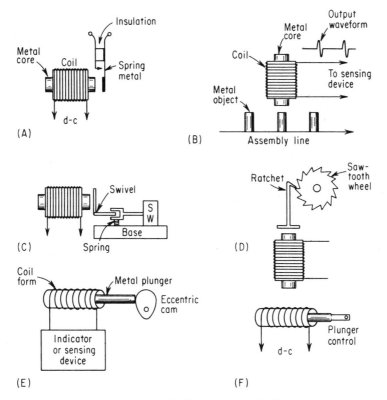

Figure 21-1.   Coils in control devices.

(Matthew Mandl, *Industrial Control Electronics*, © 1961. Reprinted by permission of Prentice-Hall, Inc.)

the coil. This signal can be used for counting, indexing, speed measurement, iron (nail) detecting, etc. This is really a transducer, the subject of a later paragraph.

At (C), the coil is used to actuate a switch, or a hydraulic or pneumatic valve. Similar to the relay, a switch (normally open or normally closed, or any other possible combination) can control relatively large amounts of electric power, with a small amount of power applied to the coil. If a valve is actuated, large amounts of hydraulic or pneumatic power can be turned on and off.

At (D), the coil is represented as actuating a ratchet mechanism. Successive pulls of the coil on the ratchet would cause stepped rotation of the sawtooth wheel. This wheel can be used to drive any number of devices. When attached to a set of electrical contacts, it becomes the widely used multipoint stepping relay.

At (E), an iron plunger is moved into or out of a coil by an eccentric cam or other device. Change of inductance of the coil can be used to control the current of an ac circuit. This device is described in more detail in Section 21.5, Fundamentals of Transducers.

A very important application of the coil to industrial controls, commonly referred to as a *solenoid*, is shown in Figure 21-1(F). An iron plunger is drawn into the coil upon application of electricity to the coil. An opposing spring can be used to return the plunger when the current is interrupted. This is the most basic way of converting electrical energy into mechanical motion, and can be applied to innumerable tasks in all types of equipment, where a relatively short, but powerful (dependent upon coil size, etc.) linear motion, electrically controlled, is required. Solenoids will be discussed in more detail later.

21.2 RELAYS

One of the most important qualities of a relay is the high degree of isolation between its control input (the coil) and its output (the contacts). A relay can be considered as an amplifier, since a small amount of power to its input coil can control a much higher amount of power at its contacts. Multiple contact arrangements permit complex control and sequencing actions that would require rather involved tube or semiconductor circuits. Relays are used in large quantities in industrial controls, despite the fact that they have moving parts, erodible contacts, and relatively slow operating speeds compared to solid state devices.

Although semiconductor devices such as thyristors and silicon-controlled rectifiers, etc., are often used as control devices in modern equipment (see Chapter 8), another popular device found in many applications is the opto-coupler/isolator described in Chapter 19.

There are many thousands of variations of so-called standard mechanical relay types to meet the special requirements of different applications. When specifying a relay, the following four areas must be considered: (1) the contact system, (2) the actuator system, (3) environmental conditions, (4) physical requirements. Table 21-1 provides a quick reference outline of the detailed characteristics that may apply in each area.

The versatility of the relay is determined to a great extent by the myriad contact arrangements possible. Figure 21-2 shows the basic ones; multiple combinations of each type—and mixed combinations, up to the capacity of the actuator system—may be put on one relay coil. Besides contact arrangement, careful consideration must be given to the contact structure and materials. The nature of the load (inductive, noninductive, lamps, ac, dc, etc.) and the amount of the load will de-

Contact System:	Actuator System:
Contact Arrangement	Type of Power Source
Load Details per Contact	Energy Available
Open Circuit Voltage	Nominal Voltage or Current
AC or DC	AC or DC
Frequency	Maximum Voltage or Current
Resistive	Battery or Line
Inductive	Pull-In
Motor	Drop-Out
Lamp	Differential
Maximum Surge	Tolerances
Current	Operating Speed
Normal or Steady State Current	Fast Operation
Duty Cycle	Time Delay
Life Requirement	Rectified AC
Circuit	Unusual Wave Forms
	Coil Resistance

Environmental Conditions:	Physical Requirements:
Normal Ambient Temperature	Space Available
Maximum Temperature	Size
Minimum Temperature	Shape
Military Specifications	Mounting
Underwriters' Laboratories Specifications	Termination
	Plug-In
Moisture	Printed Circuit
Humidity	Enclosure
Corrosive or Explosive Atmosphere	Dust Cover
Dust	Hermetically Sealed
Shock	Marking
Vibration	Finishes
Altitude	
Linear Acceleration	

Table 21-1.   Relay characteristics.

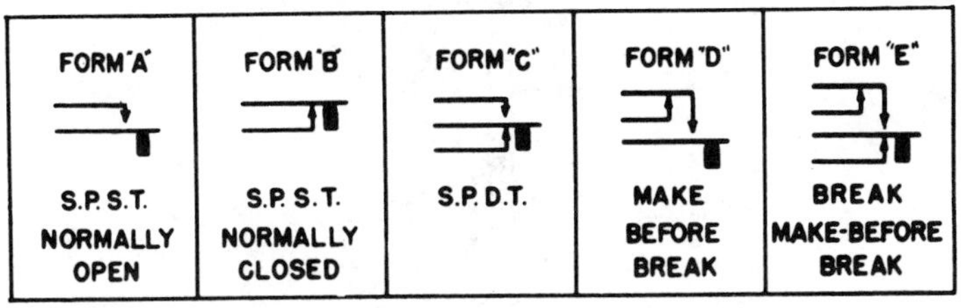

FORM "A"	FORM "B"	FORM "C"	FORM "D"	FORM "E"
S.P.S.T. NORMALLY OPEN	S.P.S.T. NORMALLY CLOSED	S.P.D.T.	MAKE BEFORE BREAK	BREAK MAKE-BEFORE BREAK

Figure 21-2.   Contact arrangements.

(Illustration courtesy of Allied Control Company, Inc.)

termine the material and spacing requirements of the contacts. Where open contacts are not suitable, sealed contacts of two basic types are available. One is the magnetic reed, in which the contacts themselves are sealed in glass in a pure, insulating gas atmosphere. A small magnet or iron piece is attached to one of the contact springs; when current is applied to the surrounding coil, the contact is opened or closed, depending on the contact arrangement.

The second sealed type uses a pool of mercury in a glass capsule to make the connection. This arrangement is usually dependent on a suitable mounting position.

Depending on environmental conditions and physical requirements, relays may be sealed in metal cans, housed in plastic enclosures, impregnated, encapsulated, etc.

21.3 SOLENOIDS

In the specification or selection of a solenoid, the following basic information is required:

1. Voltage and frequency.
2. Force required.
3. Type of motion—pull, push, push-pull, or rotary.
4. Length of stroke.
5. Ambient temperature.
6. Duty cycle—time on to time off.
7. Mounting requirements and other physical environmental conditions.

A typical solenoid application is illustrated in Figure 21-3. Manufacturers' catalogs list many different types of solenoids available as stock or off-the-shelf items. Force specifications must include such information as initial pull (or push) and holding force; in many appli-

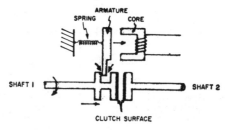

Figure 21-3. Solenoid example.

(Illustration courtesy of *Electronics World*.)

cations, the intermediate force is important. The force-stroke curves of solenoids are often available from the manufacturer, and provide important information of the variation of force with distance from the final seat of the plunger. The force usually increases very sharply near the final seating point, and care must be taken that sufficient excess force is available throughout the stroke to prevent stalling of the motion. In some solenoids, intermediate stroke points may have less force than the initial force.

The duty cycle of stock solenoids is often limited to the classifications of *continuous* and *intermittent*. While the term *continuous* is quite clear, the definition *intermittent* is not precise at all. The final criterion in duty cycle is the amount of heating that can be tolerated by the solenoid, without damage to itself or its surroundings, during its operating cycle. It is important to specify, completely, ambient temperature, number of solenoid strokes per minute, and percentage of time the solenoid coil is energized, then leave it to the manufacturer to provide the correct solenoid. The alternative would be to try the unit experimentally and check for excessive heating. It should be recognized that force availability drops off rapidly with rising coil temperature, and must be taken into account for proper performance of the mechanism to be driven.

Solenoids are essentially short-stroke devices, usually limited to a stroke of less than 4 in. When the stroke exceeds 2 in. and the force exceeds 50 lbs, dc solenoids are generally more efficient, since they require less iron and are therefore smaller and lighter. The work requirements, force times distance, can often be implemented best by the use of mechanical levers and linkages to the device to be actuated. Where long strokes and heavy forces are required, it is often advantageous to use air or hydraulic cylinders, which are discussed below.

The return of a solenoid can be attained by: (1) gravity, if the plunger is heavy enough and appropriately oriented; (2) springs (seldom supplied by the solenoid manufacturer), which must be accounted for when specifying the solenoid force requirements; or (3) a second return solenoid. This latter device must be actuated at precisely timed periods to fit in with the main solenoid operation.

21.4 SOLENOID-CONTROLLED PNEUMATIC AND HYDRAULIC DEVICES

Where the force and stroke requirements exceed the normal limitations of the electromagnetic solenoid, pneumatic and hydraulic cylinders are used to provide the needed push or pull. They become the "muscle"

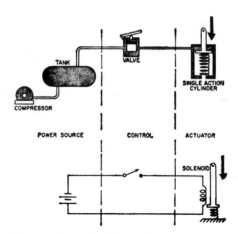

Figure 21-4.   Fluid vs. electric circuitry.

(Illustration courtesy of *Electronics World.*)

in many systems, and the solenoid is relegated to the job of actuating the valve mechanism that directs the flow of the fluid (compressed air or some suitable hydraulic oil under pressure) into and out of the cylinder.

The circuitry of fluid devices is readily understandable when the analogy to electrical circuits is recognized. Figure 21-4 makes this comparison. The compressor and tank are analogous to the battery or power supply, the switch and valve serve the same function of interrupting or redirecting the flow of the working medium (fluid or electricity), and the cylinder is roughly equivalent to a solenoid. The return motion is shown to be caused by a spring in each case. Upon opening the switch (or the valve), the working medium must be removed from the cylinder (or solenoid). If the fluid is air, the valve usually provides for venting to the atmosphere, relieving the pressure and permitting the spring to return the cylinder. In hydraulic systems, a sump tank is usually provided to recover the oil into a low-pressure region. It is then pressurized by the pump (compressor) and recirculated without loss. The solenoid usually dissipates the energy stored in its magnetic field through a shunting resistor.

Fluidic systems use pressure gages that measure pounds per square inch, comparable to electric meters. Pneumatic systems typically have pressures in the region of 150 to 300 lbs/in^2, and hydraulic systems can range upward of 2000 lbs/in^2. Similar to voltage regulators, pressure regulators are almost always present to keep the system pressure nearly constant at some preset value.

A cylinder with a piston surface of 3 in^2 can exert a force of 900

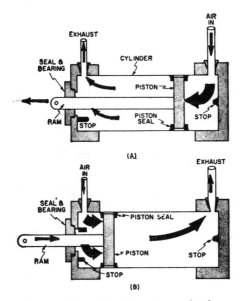

Figure 21-5.   Double-action cylinder.

(Illustration courtesy of *Electronics World.*)

lbs with a fluid pressure of 300 lbs/in². The size of the piping connecting the tank to the cylinder and the orifice size of the valves determine the speed of action of the cylinder to a large extent. This can be
compared to wire size and switch contact resistance in limiting current
flow.

Although a spring return cylinder was used in Figure 21-4 to illustrate the analogy to the solenoid, most often the cylinder is double
acting as shown in Figure 21-5. Here the return stroke, like the forward
stroke, is under complete control of the air. The variety of cylinders
available is probably greater than that of solenoids, with different specifications of seals, bearings, lubrication, end-stops, coupling methods,
etc. Manufacturers' literature should be consulted for the parameters.

As is the case for cylinders, valves also are available in prolific
varieties, when details of construction and operation are considered.
Figures 21-6 and 21-7 show the basic types, with their electrical switch
analogs.

Compressed air pneumatic systems furnish the muscle power for
many automated control systems in industry. Pneumatic cylinders are
found in operations involving hammering, riveting, die punching,
clamping and holding devices, and many other applications. The principles of pneumatic and hydraulic systems are the same and, as concerns the solenoid control of circuits, the same electrical considerations

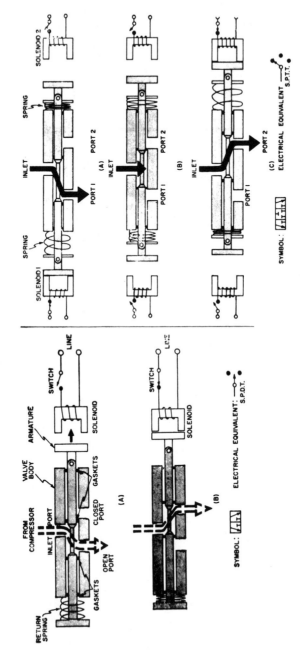

Figure 21-6. Basic valve types A.

(Illustration courtesy of *Electronics World*.)

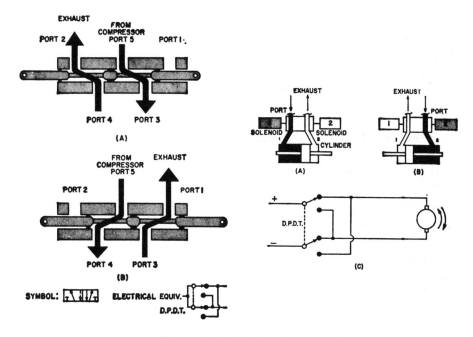

Figure 21-7.   Basic valve types B.

(Illustration courtesy of *Electronics World.*)

apply. Hydraulic cylinders are often found in machine tools for precision movement of slides and ways, chucking operations, tool indexing and movement, and where any powerful but precisely controllable motion is required, such as large antenna steering systems.

21.5 FUNDAMENTALS OF TRANSDUCERS

A transducer is a device that converts energy from one form into another. Thus, a microphone is a transducer because it converts sound energy (a form of mechanical energy) into electrical energy. A loudspeaker, on the other hand, reverses the process and converts electrical energy into sound waves.

The variety and complexity of transducers in use, in the broadest definition of the word, is so great that complete coverage would fill a library. This description is limited to those basic types that have mechanical, thermal, and luminous (within or close to the visible band) inputs and electrical outputs. Special devices working in the nuclear, X-ray, or chemical fields, etc., are not considered.

Strictly speaking, such devices as motors, solenoids, heaters, lamps, and other items with electrical inputs and mechanical, thermal, or light outputs are also transducers, but they are essentially output devices. The most frequently used "input" transducers or sensors in electronic systems can be divided into mechanical, thermal, photo, and fluid flow sensors. Basic types of each category are described below.

21.6 MECHANICAL TRANSDUCERS

Figure 21-8 illustrates a variety of transducers whose outputs are simply on-off switches. Even an ordinary light switch is a transducer which can be manually actuated or laterally moved by some piece of machinery, such as a limit switch to control the motion of a lathe, stopping or reversing the motion of the cutting tool arm. Figure 21-8b shows a diaphragm actuating a sensitive switch, an arrangement used widely for air or liquid control. Instructions are usually supplied by the manufacturer for calibrating the tripping pressure and adjusting the differential between the on and off points. The centrifugal switch shown in Figure 21-8a is used to switch over during the start-up of an ac motor. It senses the rotary speed and converts it to an electric control action. Liquid level floats of all shapes, sizes, and kinds actuate switches to

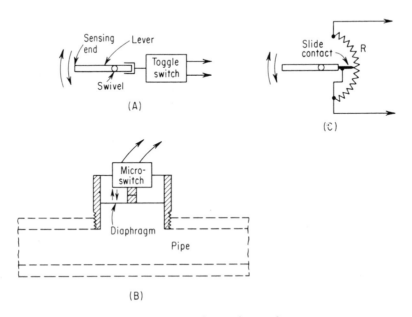

Figure 21-8. Mechanical transducers.

(Matthew Mandl, *Industrial Control Electronics*, © 1961. Reprinted by permission of Prentice-Hall, Inc.)

control pumps and sound alarms and to provide sequencing in industrial processes.

If the on-off switch in each of the switched devices illustrated in Figure 21-8 is replaced by a variable resistor, proportionate control can be obtained.

A basic resistance-type transducer is shown in the simplified diagram of Figure 21-8c. Here pressure moves a diaphragm linked to the arm of a potentiometer. The control circuit for this kind of transducer might be an amplifier whose plate current passes through the control winding of a saturable reactor. In this way, the resistance of the transducer controls the ac voltage delivered to the rectifier, and thence to the motor.

Next to the resistor type of displacement transducer, the inductive types are probably the most widely used. They range from a simple tuning-slug motion inside a coil to the linear differential transformer types. There are many different winding arrangements, but the principle of a differential transformer can be seen in Figure 21-9. Pressure on the diaphragm moves the core, changing the flux linkage between the primary and the two secondaries in such a way that one of the two triodes receives a larger grid signal than the other. The difference is amplified and can be measured by a meter connected between the plates, or amplified further and used as error voltage for a servo positioning system. By proper arrangement of the voltages, it is possible to obtain a difference in phase as well as amplitude, and thereby indicate the direction of diaphragm motion.

The change in capacitance as the distance between capacitor plates is changed is the principle used in the capacitive transducer shown in Figure 21-10b. The housing is part of the grounded outer plate, which may be in the form of a diaphragm, and the inner plate is mounted on an insulator. To increase the capacitive change, both plates sometimes contain concentric rings and grooves that mesh into each other. In con-

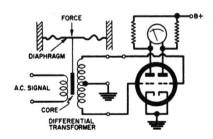

Figure 21-9.   Inductive transducers.

(Illustration courtesy of *Electronics World.*)

trol systems using capacitive transducers, the control element includes either an oscillator or a tuned amplifier. Here an oscillator is shown, which could be part of a telemetering system to transmit pressure changes as frequency modulation of a carrier signal.

In addition to the basic resistive, capacitive, and inductive pressure transducers, there are a few special types such as the magnetostrictive, electrokinetic, and piezoelectric. The first of these works on the principle that a physical distortion of certain nickel alloys causes changes in their magnetic properties. The second depends on the flow of a polar fluid (an electrically charged fluid, such as an electrolyte) through a porous membrane, which generates a potential across this membrane. The third type will be more familiar since, in a way, it is used in many phonographs and microphones. When a crystalline structure is distorted physically, it generates a small voltage in proportion to the distorting force, as occurs in phono cartridges or microphones using crystal or ceramic elements. This property is harnessed differently in the accelerometer circuit of Figure 21-10a.

Another type of dynamic transducer is shown in Figure 21-11. In the inductive vibration transducer, a small mass vibrates and, through a reed linkage, drives the magnetic core of a coil. When a magnet moves through a coil, a voltage is generated that varies with the speed of motion. An amplifier is usually used to drive a remote recorder. Transducers such as this can be used in automated factory systems and industrial processes where vibratory motion must be monitored.

The capacitive accelerometer is basically the same as the capacitive pressure transducer shown in simplified form in Figure 21-10, except that the inner plate (mass) moves, and its motion depends on the acceleration exerted on it. The change in capacitance is proportional to

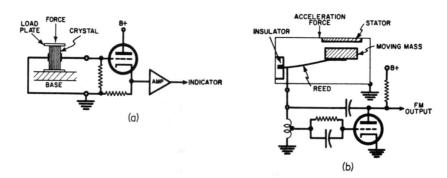

Figure 21-10. Capacitive and piezoelectric transducers.

(Illustration courtesy of *Electronics World.*)

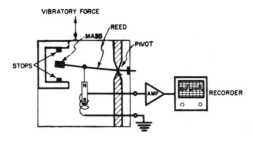

Figure 21-11.   Dynamic transducer.

(Illustration courtesy of *Electronics World*.)

acceleration, and the output will be in the form of a change in frequency.

The strain gage is a special category of mechanical transducer found in a wide variety of forms and applications. Strain gages are available as individual resistance units which can be cemented to any mechanical member whose strain (stretch or compression) is to be measured. Others are available as complex "rosette" arrangements that can distinguish direction of strain. Others are permanently built into so-called load cells for the measurement of force or as pressure transducers.

Because they are resistance-type transducers, they are usually connected in bridge-type circuits, which produce greater electrical outputs, and can be balanced for zero output at some reference (or zero strain). Typical of these are the strain gage arrangements shown in Figure 21-12. The opposite arms of the Wheatstone bridge network are bonded directly to the structural member whose stress or strain is being measured. The strain gage elements are made of fine wire that changes its resistance as it is stretched or compressed. These changes are very small, but, because they occur in opposite arms of the bridge, their effect is much more noticeable.

An unbonded type of strain gage where all four arms of the bridge

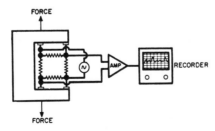

Figure 21-12.   Strain gage circuit.

(Illustration courtesy of *Electronics World*.)

form a free grid with opposite arms under tension, is used in electronic weighing systems where a small expansion of a calibrated stress member is measured. The bonded type of strain gage is used mostly for continuous measurements and tests of a single member, but can also be found as part of a separate pressure transducer.

21.7 TEMPERATURE-SENSITIVE TRANSDUCERS

There are a variety of heat-sensitive transducers, most of which are constructed with bimetal elements. The principle of bimetal temperature sensitivity is shown in Figure 21-13. Two dissimilar metals with different coefficients of expansion are bonded together. When the temperature changes, the difference in coefficient of expansion causes the unit to bend. Thus, it can be used to open a circuit for a rise in temperature or to close a circuit as shown in Figure 21-13.

A thermistor is a temperature-sensitive resistance frequently used in electronic circuits, such as the simple relay system of Figure 21-14.

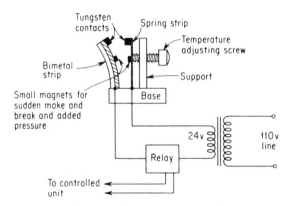

Figure 21-13. Thermostat system.

(Matthew Mandl, *Industrial Control Electronics*, © 1961. Reprinted by permission of Prentice-Hall, Inc.)

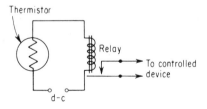

Figure 21-14. Thermistor circuit.

(Matthew Mandl, *Industrial Control Electronics*, © 1961. Reprinted by permission of Prentice-Hall, Inc.)

Besides the thermostat and the thermistor, a variety of other temperature-sensitive transducers are available.

21.8 PHOTOELECTRIC TRANSDUCER

Photoelectric devices are transducers because they convert luminous energy into electrical signals. A large variety of photoelectric units is employed in industry, some of which are described in more detail in Chapters 7 and 8. They are used for turning equipment on or off, counting objects, indexing, positioning, observing, or otherwise regulating electrical and mechanical devices. The large variety of photoelectric transducers available makes them extremely versatile devices. They are available for use over a wide electromagnetic spectral range from the infrared, through the visible, to the ultraviolet range, and beyond.

Applications range from opening doors when a light beam is interrupted to counting merchandise on a moving conveyor belt. Burglar alarms, color matching, and chemical analysis also find applications for the photoelectric device. Each application and each type of photocell has its own special requirements and characteristics. An example is the vision system used in industrial robotics.

21.9 HUMIDITY TRANSDUCER

To convert the effects of humidity to an electrical output, use is usually made of some porous material whose absorption of moisture either changes its physical property or its electrical resistance. The first type

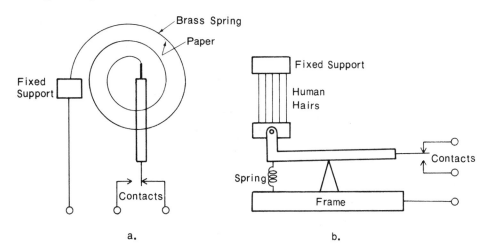

Figure 21-15. Humidity transducers.

is illustrated by Figure 21-15, where drying of the special paper or human hair causes mechanical shrinking, and this movement is used to actuate a switch. Special films change their resistance with moisture. These resistance changes can be converted to electrical signals in a Wheatstone bridge, just as for the case of the strain gage, shown previously.

21.10 FLOW TRANSDUCERS

Precise gaging of liquid and gaseous flow is essential in many industrial processes and in such diverse applications as the control of fuel flow to rocket motors and the flow of blood to heart-lung machines during surgery.

The all-mechanical flowmeter is used by the water and gas companies. Four types of flowmeters are described below.

1. Turbine type—a propeller assembly in the flow path rotates at a speed proportionate to the speed of flow of the fluid. A sensing coil outside the pipe picks up magnetic flux variations produced by the propeller, and the frequency of the signal is a direct indication of the speed of flow.
2. The Venturi and orifice flowmeter, where the cross section of pipe is reduced, thus developing a pressure differential between the wide and narrow sections. The magnitude of this differential pressure depends upon the flow rate. Figure 21-16 shows such an arrangement. A diaphragm and variable reluctance (moving iron

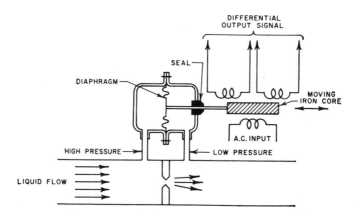

Figure 21-16.   Venturi flowmeter.

(Illustration courtesy of *Electronics World*.)

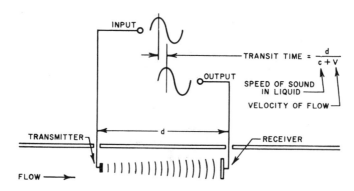

Figure 21-17.   Ultrasonic flowmeter.

(Illustration courtesy of *Electronics World.*)

core) differential transformer, previously shown in Figure 21-9, sense the difference in pressure and provide an electrical output.

3. The ultrasonic flowmeter is shown in simplified form in Figure 21-17. The change in sound frequency due to the Doppler effect (see Chapter 13) in the fluid is indicative of the velocity of fluid flow. More elaborate schemes using this principle are found in many applications.

4. The caloric flowmeter, where the cooling effect of fluid flow is measured by a thermocouple. A heater element is immersed in the fluid stream and operates at a higher temperature than the fluid. This element is cooled at a rate dependent upon the rate of flow. An array of thermocouples, to compensate for ambient temperature changes, measures the amount of cooling and provides an electrical output as a function of flow.

Again, there are many possible variations, and variations within variations, of the flowmeters mentioned above. Many technicians who work in industrial electronics will come in contact with particular types of flowmeters as part of control systems. For repair and maintenance of flowmeters, detailed manufacturers' information is essential.

BIBLIOGRAPHY—CHAPTER 21

Institute of Electrical and Electronic Engineers, Inc., *Standard Digital Interface for Programmable Instrumentation.* New York, NY, John Wiley & Sons, Inc., 1984.

Kosow, K., *Control of Electric Machines*. Englewood Cliffs, NJ, Prentice-Hall, Inc., 1973.

Kuo, B., *Automated Control Systems* (3rd Ed.). Englewood Cliffs, NJ, Prentice-Hall, Inc., 1975.

Motorola: The Complete MAP Solution—BR329. Phoenix, AZ, Motorola Semiconductor Products, Inc., 1986.

Sante, D., Automatic Control System Technology. Englewood Cliffs, NJ, Prentice-Hall, Inc., 1980.

Shimon, N., *Handbook of Industrial Robotics*. New York, NY, John Wiley & Sons, Inc., 1985.

CHAPTER 22

INDUSTRIAL ELECTRONICS

22.1 FUNDAMENTALS OF AUTOMATIC CONTROL

Every principle of automatic control is used in modern factory automation and industrial robots. For example, high-precision servo motors are essential for factory automation using industrial robots. Piezoelectric and electrostatic feedback mechanisms are used as sensors and must be sensitive to variations in grasping force. However, all components and systems used in manufacturing automation are primarily software driven.

At present, the problem is that computer vendors use numerous unique communications protocols, therefore one machine cannot communicate with another machine without custom-made interfaces. A new standard called *manufacturing automation protocol* (MAP) is currently receiving worldwide acceptance. The International Standards Organization's (ISO) "Open System Interconnect" forms the basis of the MAP standard.

The hardware used is dependent upon the specific nature of the control function to be performed, the accuracy and speed required, and the type of mechanisms chosen to implement the system. The underlying concepts, principles, and mathematics are all essentially the same. Control systems using pneumatic devices can be described by electrical analogs and vice versa. This chapter deals with electronic and electrical automatic control systems, but other disciplines will be mentioned for clarity.

Two basic control systems are the on-off or discontinuous type and the continuous variable type. The amount of energy required for closing or opening a switch, or adjusting a rheostat to control the electric current to a motor, bears little relationship to the amount of energy available at the motor shaft, but the timing of these control actions may be critical to the work done by the motor. If no indications are provided of the control action's result, the system is called *open cycle*. A typical example of an open cycle system is the usual traffic signal light controlled by a time clock. The lights change, and the cycle runs on, whether there are cars at the intersection or not. The control and results are independent of each other.

By adding some means to measure and evaluate the result of the control action against some criteria and using this information to adjust the controlling mechanism, the loop is closed; such a system is called a *closed-cycle* or *closed-loop system*. The part of the system that uses this evaluated information is called the *feedback path*.

Most home heating systems use a thermostat as the sensing, evaluating, and control switch functions in the feedback path to regulate the temperature of a room. The basics of the feedback system are illustrated in Figure 22-1. This is an example of a closed-cyle, on-off system, with no proportionate control. It is either full-on or full-off.

The difference between the desired performance and the actual performance, at any instance, is called the *error*. In some systems only the final result (or position) is of importance, with the path taken or

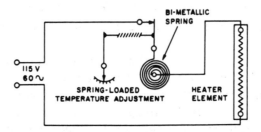

Figure 22-1. Thermostatic control system.

(Philco Technological Center, *Servomechanism Fundamentals and Experiments*, © 1964. Reprinted by permission of Prentice-Hall, Inc.)

the speed in getting there only secondarily relevant. Positioning machines (as for drilling operations, etc.) fall into this category. Other control systems such as industrial robots require that the path and even the speed and acceleration be carefully controlled at all times. An example of continuous control is the contour cutting machine (for cams, turbine blades, etc.)

22.2 PRINCIPLES OF SERVOMECHANISM

When the error or difference between the desired and the actual result is fed back to achieve the desired result, the term *servo system* or *servo operation* can be applied. The act of reading these lines requires sophisticated servo systems in the eye muscles and in the neck. Tracking a moving target visually or with a weapon is a good example of servo operation. The eyes follow the target, and the difference in eye and head motion causes the head to turn so that we can look at the target head on.

Figure 22-2 illustrates a block diagram of a generalized servo-control system. The requirement for some error, as a basic element of the servo system, explains the fact that most of servo theory is concerned with the control and reduction of this error signal. The various functional blocks in Figure 22-2 usually represent mechanical, fluidic, or electrical components. In a typical system, the controlled device may be the motor, rotating a robot's mechanical fingers. The error measuring device may be a comparator circuit, while the follow-up device may be a synchro follower or a tachometer. The input command may come

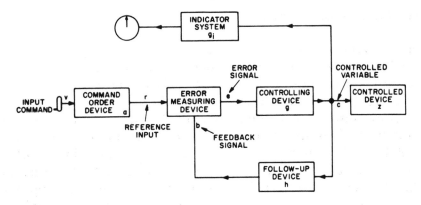

Figure 22-2.   Servo system block diagram.

(Philco Technological Center, *Servomechanism Fundamentals and Experiments*, © 1964. Reprinted by permission of Prentice-Hall, Inc.)

from a circuit, and the command order device may be a circuit converting timing pulses into a signal suitable for comparison with the feedback signal. As a controlling device, a high-power amplifier could be used, which furnishes driving power to the motor. In such a system the position is constantly changing, while in other systems, such as palletizing robots, one correct and final position is the result.

Error Analysis in Servo Systems

The less "friction" and "inertia" the "load" presents to the system, the easier it is for the servo system to provide a low error result. Quotation marks are used about the words load, friction, inertia because the control system is not always concerned only with physical loads, but the same theory applies to systems with "nonphysical" loads, such as traffic control problems.

Generally, lower friction will produce a smaller error, but too low a friction load is likely to overshoot the desired result. Excessive friction can cause a *lag*, and the servo system can undershoot or stop short of the correct result. Between these two extremes, there exists an optimum value which is attained when there is sufficient friction, also called *damping*, to prevent the uncontrolled buildup of oscillations, or hunting of the system. In the previous tracking example, excessive friction would mean that the head always moves too late so that the target is seen only out of the corners of the eyes. Too little friction would mean a back and forth motion of the head. The point of optimum friction is called *critical damping*. Friction beyond this point will increase the lag error.

Servo Stabilization Methods

When a sudden command change occurs in a servo system, the drive unit will run at a speed proportionate to the command change, after overcoming initial transient variations. This steady state speed is called the *slewing speed*. As the system approaches its final destination, the speed slows down proportionately and, depending upon the friction and inertia, will either undershoot, overshoot, hunt, dither, or stop accurately at the desired point. The drive unit must have sufficient power to overcome all expected friction, and an adjustable damping mechanism is then used to adjust for the best performance compromise. Dependent upon the particular application, a "best" compromise will include a small amount of controlled hunting, which often produces a more accurate stopping point than a critically damped system. This

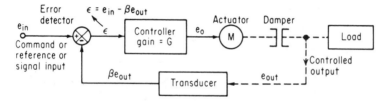

Figure 22-3. Generalized servo system with viscous damping.

(I. L. Kosow, *Electric Machinery and Control,* © 1964. Reprinted by permission of Prentice-Hall, Inc.)

method of servo system stabilization is called *viscous output damping,* and a system block diagram using a fluid mechanism is shown in Figure 22-3.

Figure 22-4 shows how viscous damping can be accomplished by electromagnetic means. Where the controlled hunting method is not sufficient, the error rate damping method can be used. By introducing a signal proportional to the rate of change of the error into the system, in addition to the error signal itself, it is possible to adjust the stabilization of hunting due to sudden starts and stops or transients. This method has no effect on the steady state error that occurs during steady slewing, but, combined with viscous damping, provides an independent means of controlling hunting. Figure 22-5 shows the method of introducing error rate damping into a viscously damped dc system.

Output rate damping is an alternative method of performing transient stabilization. Mathematically, and in performance, it behaves very much like the error rate method of damping. A signal, proportionate to the speed of the output, is added to the steady state error signal instead of using a signal derived from the rate of change of the error

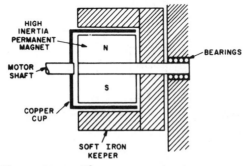

Figure 22-4. Electromagnetic viscous damping.

(Philco Technological Center, *Servomechanism Fundamentals and Experiments,* © 1964. Reprinted by permission of Prentice-Hall, Inc.)

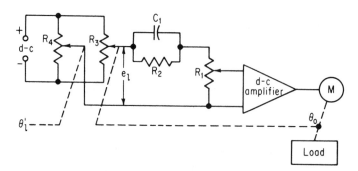

Figure 22-5. Error rate damping.

(I. L. Kosow, *Electric Machinery and Control*, © 1964. Reprinted by permission of Prentice-Hall, Inc.)

itself. Figure 22-6 illustrates the method of implementing this by using a tachometer. The electrical output of a tachometer is proportionate to shaft speed and, during hunting, provides negative feedback to dampen this oscillation. This method is also often called *tachometer stabilization* and is a very common and satisfactory way of stabilizing high-gain servo systems.

While the two methods above control hunting, and thereby only indirectly affect the steady state error, the integral control damping method can act directly to reduce any steady state error.

When an error tends to persist for a period of time, such as in steady state slewing, the integration or accumulation of error in a so-called lag circuit will gradually increase the error signal and cause the system to reduce the error. Figure 22-7 is a diagram illustrating the

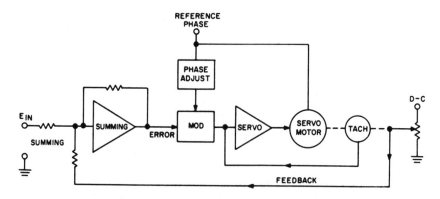

Figure 22-6. Tachometer stabilization.

(Philco Technological Center, *Servomechanism Fundamentals and Experiments*, © 1964. Reprinted by permission of Prentice-Hall, Inc.)

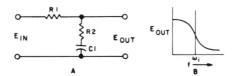

Figure 22-7. DC integral stabilizing network.

(Philco Technological 'Center, *Servomechanism Fundamentals and Experiments,* © 1964. Reprinted by permission of Prentice-Hall, Inc.)

principles of this method. As the error signal is integrated, its corrective ability is limited only by the maximum drive power of the system. Theoretically, the error can be reduced to zero, but rapidly changing or transient errors are not followed. Only the net result of errors integrated over a period of time is effective in the integral control method.

Precision servo systems will usually combine viscous damping with either error rate or tachometer damping, as well as integration damping.

System Implementation

Servo systems have thus far been described as block diagrams and according to basic functions. The design of servo systems is a highly specialized field utilizing integral and differential mathematics. From a practical point of view, a qualitative knowledge is sufficient for the testing, adjusting, and maintaining of servo systems. Actual components represented by the various blocks in the block diagrams above are discussed in succeeding sections.

Some types of components, such as synchro units, potentiometers, amplifiers, tachometers, comparators, etc., either electrical, mechanical, or electromechanical in nature, are usually of the analog or continuous type. Digital servo systems use the same fundamentals but depend on digital components and circuits, which are covered in Chapter 16.

Systems can be of the all-analog type, all-digital type, or can combine some of the best features of each. Such hybrid systems require analog-to-digital and digital-to-analog interfacing conversions as well as modifications in software, if the systems are software driven.

22.3 FUNDAMENTALS OF ELECTRIC MOTORS AND SYNCHRO UNITS

In motors, as in solenoids, the transformation of electrical energy into mechanical energy is accomplished through the attractive and repulsive action of the electromagnetic field. Figure 22-8 illustrates the basic mo-

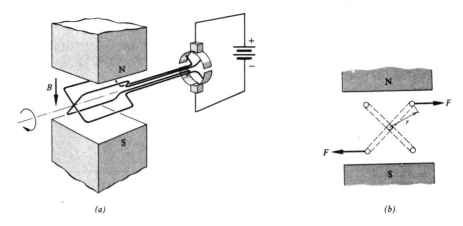

Figure 22-8. Basic motor action.

(Lane K. Branson, *Introduction to Electronics*, © 1967. Reprinted by permission of Prentice-Hall, Inc.)

tor action common to all motors, ac or dc. If, at any instance, the polarities of the current and the main magnetic fields are as shown, the magnetic fields generated by the current in the stator and the rotor are in opposition. As a result, the rotor field tends to align itself with the stator field, thus producing a turning action. In a dc motor, a commutator is provided to reverse the polarity of the rotor current to the rotating coil at the right instant so that rotation continues. In ac motors, a variety of devices and arrangements is employed to keep the rotor and stator fields in the proper phase (polarity) relationship to sustain rotary motion.

There are a tremendous number of specialized motor types, both ac and dc, but an understanding of the basic types is essential to the application of any of the large number of variations and combinations the electric motor industry has to offer.

Figure 22-9 shows a breakdown of the different types of electric motors and their subcategories. Each will be discussed in the following paragraphs.

Basic DC Motors

The speed of a dc motor depends upon the strength of the magnetic field between the stator and rotor, and the speed is increased as the field strength increases. The torque or turning force is proportional to the current in the rotor coils, called the armature, and the magnetic field strength. Therefore, for a given motor load, output torque, the speed can be controlled by varying the current in the coils of the stator.

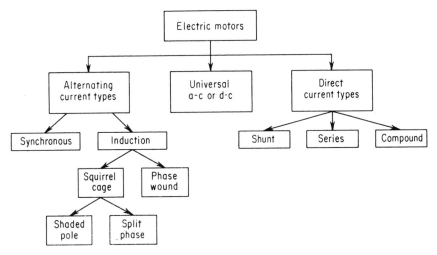

Figure 22-9.   Breakdown of electric motors.

(Matthew Mandl, *Industrial Control Electronics*, © 1961. Reprinted by permission of Prentice-Hall, Inc.)

When more load is imposed upon the motor, it tends to slow down somewhat and increase the armature current to produce the torque called for. The three basic dc motor types are the shunt, series, and compound wound motors. These names refer to the manner in which the stator field coils are connected with reference to the armature.

The shunt wound motor has a high resistance, many turns of small wire, with the coil for the stator, called the field coil, connected in parallel across the line with the armature. A variable resistor in series with this field coil provides a simple and convenient means for varying the field strength, and thereby the motor speed, as shown in Figure 22-10. Speed changes very little with wide changes in load and, because the field is quite steady, the torque is directly proportional to the armature current. The shunt motor is a general purpose type with medium starting torque and good speed regulation, and finds use in such applications as pumps, fans, blowers, printing presses, etc.

The series motor has a low resistance (large wire of few turns) field coil, and this field coil is connected in series with the armature across the power line. Since the same current flows through the armature and the field, strength and armature current are both high at start-up; this results in a high starting torque, because torque is proportional to the product of field strength and armature current. Mathematically expressed, the torque is proportional to the square of the armature current. In this type of motor the speed is highly variable, because the field strength changes with armature current, which in turn changes with

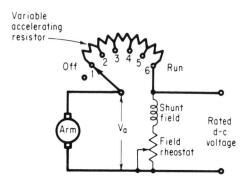

Figure 22-10.   Shunt motor starting circuit.

(I. L. Kosow, *Electric Machinery and Control*, © 1964. Reprinted by permission of Prentice-Hall, Inc.)

the load. The motor speed drops with increasing load and speeds up when the load is reduced. These type motors must not be operated without any load, because it is possible for them to speed up so fast as to damage themselves (run away).

Applications that involve frequent starts and stops, require high starting torques, and involve wide speed variations are suited to the series motor. They include such uses as trolley cars, golf caddies, automobile starter motors, elevators, industrial cranes, and other hoisting devices.

Small motors used in many domestic home appliances, such as vacuum cleaners, mixing machines, hand drills, etc., are of the so-called universal type. These motors are basically dc series wound motors, but because of the series connection, the same phase (or polarity) of current must flow in the armature and field coils. For this reason, these motors can also be used on ac power and are called *universal*. The slightly lower efficiency, when used on ac, is usually no serious handicap in the small motor sizes.

The compound motor, which has a series field winding aiding its shunt field winding, provides high torque at a given armature current. Increases in load increase the armature current, which also flows through a compounding series winding, thereby increasing the field strength. This counteracts the tendency of the armature current to rise, to increase the torque to match the load, because the field strength has been increased. By proper design, a correct proportioning of the number of turns in the shunt and series field will result in the desired speed-load characteristic. The series field is also of help in increasing starting torque as compared to a simple shunt motor. Differentially compounded motors are seldom used, because of their inherent instability and generally undesirable qualities. While it is possible to attain ex-

cellent speed constancy under changing load with differential compounding, the torque characteristics are very poor (low torque) at starting. Unless very great care is taken in proportioning the differential
field, the speed may have a tendency to rise with load and cause a
"runaway" condition. Figure 22-11 shows the comparative characteristics of the shunt, series, and compound wound dc motor.

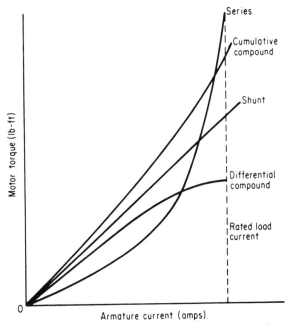

Figure 22-11.   Comparison of motor characteristics.

(I. L. Kosow, *Electric Machinery and Control*, © 1964. Reprinted by permission of Prentice-Hall,
Inc.)

The dc motor may be reversed by reversing the polarity of the
power to the armature relative to the field coil. The switching is usually
done in the armature because it has less inductance and will cause less
arcing at the switch contacts. In larger motors, resistors are inserted
temporarily into the armature circuit while starting, to limit the current
to a safe value. After coming up to speed, these starter resistors are
removed. Figure 22-10 shows a typical control circuit for a shunt wound
motor.

Basic AC Motors

The operation of an ac motor depends on an electrically rotating field
generated by the stator. The generation of a rotating field depends upon
the use of a polyphase (two-phase, three-phase, or more) source of ac.

The vector addition of these "time-spaced" electric currents that are also physically spaced about the rotor, by proper location of the field coils in the stator slots, results in this rotating magnetic field.

There is a whole category of motors that operate from single-phase ac. The major difference among them is the manner in which the second phase is derived from the single-phase power line. Various methods of obtaining the second phase are described below. This second phase is then used to create the rotating field necessary for ac motor operation, as described above.

The nature of the rotor then basically determines in which of the various ac categories the motor falls, as indicated in Figure 22-9.

In the synchronous motor, the rotor uses a dc field. This dc field, created with power from a dc generator in large motors, or rectified from the ac, or supplied by permanent magnets in small fractional horsepower motors, "locks in" and follows the rotating field of the stator. The speed of the synchronous motor is dependent upon the frequency of the ac power and the number of poles (which is determined by the physical construction and arrangement of the field coil windings), and both together determine the speed of the rotating field. Increasing the load on a synchronous motor causes the rotor to lag behind the rotating stator field by some angle dependent upon the load. This lag angle increases with increasing load up to the maximum available torque from the motor. Beyond that, the rotor will slip out of synchronism. This is not desirable in large motors since it can cause damage. Because of the erratic behavior of the synchronous motor at other than synchronous speed, various special schemes are used to start such motors. Besides the obvious use in clocks, etc., where speed synchronous with the power lines is required, the synchronous motor has the important characteristic of behaving as a large capacitor when its rotor field current is raised. This is often used to correct poor load power factors in large industrial plants.

The induction motor depends for its operation on a certain amount of slip between the speed of rotation of the rotor and the rotating magnetic field of the stator. When the rotating field is faster than the rotor, this field is in effect cutting across the conductors in the rotor and inducing currents in them. These induced currents generate a magnetic field, which interacts with the rotating field, thus dragging the rotor around with it at a slightly slower speed. In small motors the rotor may consist of bars of conductors, shorted at their ends, embedded in the slotted iron of the rotor as indicated in Figure 22-12. These are so-called squirrel cage motors. The nature and resistance of the rotor conductor bars determine the amount of slip and torque characteristics. Higher-

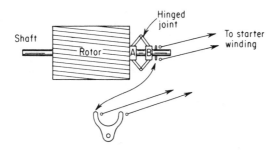

Figure 22-12.   Single-phase induction motor.

(Matthew Mandl, *Industrial Control Electronics*, © 1961. Reprinted by permission of Prentice-Hall, Inc.)

resistance bars in the rotor tend to provide higher starting torques but more slip at normal running speeds. Figure 22-13 is a graph illustrating the slip-torque characteristics for different types of rotor construction and resistance.

Where greater control of the torque slip characteristics and higher torques are needed for starting without excess slip, the wound rotor is used. Instead of shorted conducting bars, a winding similar to the construction of a dc motor is used. The coils are usually arranged in a three-phase circuit, using Y connections. In starting the motor, resistors are inserted to increase the starting torque. When the motor comes up to proper speed, the resistors are shorted out, and the motor behaves

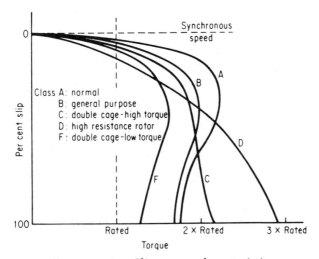

Figure 22-13.   Slip torque characteristics.

(I. L. Kosow, *Electric Machinery and Control*, © 1964. Reprinted by permission of Prentice-Hall, Inc.)

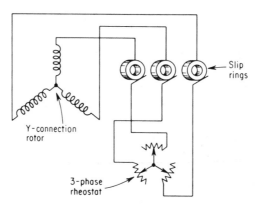

Figure 22-14. Circuit of wound rotor induction
motor.

(Matthew Mandl, *Industrial Control Electronics,* © 1961. Reprinted by permission of Prentice-
Hall, Inc.)

very much like a squirrel cage type. Figure 22-14 shows the circuit
arrangement for starting a wound rotor motor.

As mentioned previously, single-phase motors must provide some
means of phase splitting to create the second phase, so that a rotating
stator field is generated. Some motors use a high inductance starter
winding which causes the current through it to lag in phase. This
current is then of a different phase than the main winding, and the
combination creates a rotating field. Other methods include a starting
capacitor in conjunction with the starter winding, to obtain a better
phase relationship (more nearly 90° out of phase) and higher starting
torque. In small motors used for fans, phonograph turntables, etc., for
sizes normally under 1/4 hp, the shaded pole method is used. One
pole has a shorted turn of heavy copper wire imbedded in a portion of
the pole face of the motor. Currents induced in the shorted turn of wire
cause the magnetic flux, in the portion of the pole face surrounded (or
shaded) by the turn of wire, to lag behind the rest of the field, and it
behaves as if it were another pole of a different phase. Again, the combination of field phases and their physical spacing combine to produce
a rotating field. Figure 22-15 shows the physical outline of such a small
shaded pole motor.

Once a single-phase motor is started, it will continue to run without the need for the starter winding action, because of the interaction
between the stator and rotor fields. In the larger, single-phase motors
(1/4 to 1 hp) centrifugal switches (or other means) are provided to
disconnect the starter windings, which are designed for intermittent
duty, heavy current use.

Motor manufacturers very often classify motors by their power rat-

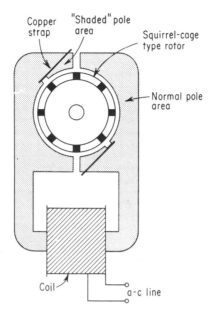

Figure 22-15.   Shaded pole motor.

(Matthew Mandl, *Industrial Control Electronics*, © 1961. Reprinted by permission of Prentice-Hall, Inc.)

ing. Fractional horsepower motors (less than 1 hp) are generally meant when reference is made to a "small" motor. Integral horsepower motors are those of 1 hp and larger.

Figure 22-16 is a schematic representation and table summarizing the various electric motors discussed and their applications.

Basic Synchro Units

Synchro units are included under electric motors because they use the same principles and the same construction techniques as ac motors. The synchro unit, however, is essentially a single-phase device, and the rotating magnetic field is usually generated by a rotating device elsewhere. The stator of the synchro unit generally has three windings, spaced 120° (electrical degrees) per pole. If the windings are supplied with single-phase ac voltages of the proper amplitude relationships, a magnetic field is generated whose angular position is the vector sum of the fields from each coil. When the amplitudes of the voltages on the stator coils change, the vector sum changes, and the resultant stator field change causes the rotor to move or rotate continuously. Figure 22-17 illustrates this concept. The rotor of TX is mechanically rotated and TR follows.

There are seven general types of synchro units, each classified

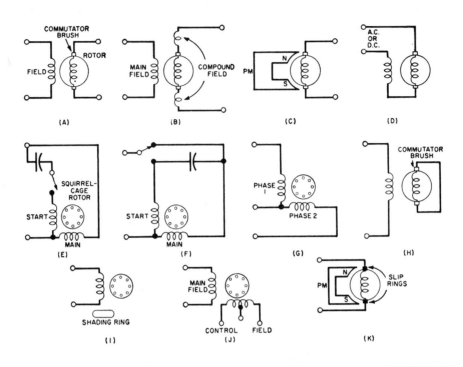

TYPE & FIG. REFERENCE	STARTING TORQUE	SPEED CONTROL	OPERATING FEATURES	TYPICAL APPLICATIONS
D.c. shunt (A)	Medium	Thyratron or voltage control	Adjustable speed; constant torque or constant power	Pumps, conveyors, wire and paper winding
D.c. compound (B)	High	Usually not used	Speed adjustable over small range; high but varying torque	Flywheel drive, shears, punch presses, hoists
D.c.-PM field (C)	Low	Power tubes or transistors		Fans, blowers, battery-operated devices
Universal series d.c. or a.c. (D)	Very high	Thyratron, saturable reactor, series resistor	High speed; high efficiency	Hoists, cranes, vehicles, hand tools, appliances, general utility
Capacitor start a.c. (E)	Very high	Saturable reactor	Limited range of speed control as torque drops with voltage	Compressors, pumps, blowers
Capacitor running (reversible) (F)	Low	Usually not used	Speed varies greatly with load	Fans, blowers, centrifugal pumps
Squirrel-cage induction (poly-phase) (G)	Depends on type used	Saturable reactor, resistors	Available in six classes of performance characteristics	General-purpose industrial motor used as main power source for heavy machinery
Repulsion-start, induction-run (H)	Very high	Usually not used	High starting-current surge	Pumps, compressors, conveyors
Shaded pole (I)	Very low	Usually not used	Relatively inefficient, but low in cost	Fans, blowers, heaters, phonographs
Servo (J)	High	Power amplifier, saturable reactor	Accurate control through special control winding	Positioning systems, computers
Synchronous (K)	Low	None	Constant speed depends on number of poles and line frequency	Clocks, timers, blowers, fans, compressors

Figure 22-16. Table of motor types.

(Illustration courtesy of *Electronics World*.)

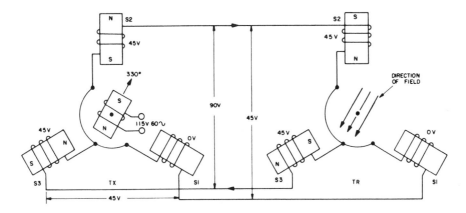

Figure 22-17.   Basic synchro transmitter and receiver.

(Philco Technological Center, *Servomechanism Fundamentals and Experiments*, © 1964. Reprinted by permission of Prentice-Hall, Inc.)

according to function. These include the transmitter, receiver, differential transmitter, differential receiver, control transformer, and resolver. The synchro types are further classified as a control unit or a torque unit, which can exert substantial power.

Synchro Transmitter.   The synchro transmitter is sometimes referred to as a *synchro generator* or *torque transmitter.* Its rotor, which is excited from an ac source, is usually mechanically coupled to the controlling shaft and is usually so restrained that it cannot turn except under the influence of the controlling shaft. The voltages induced in the stator windings, as a result of the field set up by the rotor windings, are representative of the angular position of the rotor at any instant.

Synchro Receiver.   The synchro receiver, also known as a synchro motor, torque receiver, or repeater, is similar electrically to the transmitter. The rotor, which is free to turn, usually positions a light load—a pointer dial or some other indicating device—directly or through a light gear train. The angular position that the rotor assumes depends upon the voltages received from the synchro transmitter.

Synchro Torque Differential Transmitter.   The synchro torque differential transmitter has three separate windings on the rotor, displaced by 120 electrical degrees. It is used to compensate for errors existing in various parts of the system. For example, if the voltages from a synchro transmitter are applied to one winding of the differential transmitter, and the other winding is then connected to the stator of a

synchro receiver, the voltages appearing at the receiver may be modified by the angular position of the differential rotor. The angular position of the receiver rotor, with respect to the transmitter rotor, may
be adjusted by the setting of the differential rotor.

Synchro Torque Differential Receiver. The synchro torque differential receiver is similar in design to the torque differential transmitter, but its rotor is free to turn. This unit is used to interpret directly
the sum or difference of two angular motions. When the windings of
the differential receiver are connected to two synchro transmitters, the
differential receiver rotor assumes a position corresponding to the angular sum of difference of the two transmitted motions.

Synchro Control Differential Transmitter. The synchro control
differential transmitter is similar in design to the other synchro differential units, but its rotor is mechanically driven. This unit is used to
transmit electrically the algebraic sum or difference of two angles. The
stator voltages of the control differential are supplied by a synchro
transmitter, and the rotor voltage is transmitted to a synchro control
transformer.

Synchro Control Transformer. The synchro control transformer,
while generally similar to a synchro receiver, has physical and electrical features that are different. The rotor is positioned mechanically.
The control transformer is used to obtain a signal indicative of angular
position. Voltages from a synchro transmitter or control differential
transmitter produce a voltage in the rotor of the control transformer,
which represents the angular displacement of the transmitter rotor.

Synchro Resolver. The most common type of resolver has two
stator windings displaced by 90 degrees, and two rotor windings also
displaced by 90 degrees. The stators act as transformer primary windings, while the rotors act as transformer secondary windings. The synchro resolver is used as a triangle, or vector, solver, and its operation
is somewhat analogous to the synchro control transformer in that voltages, rather than torque or rotation, are produced.

22.4 AMPLIFIERS FOR INDUSTRIAL CONTROLS

Amplifier principles are discussed in Chapter 8 (Semiconductors, Transistors, and Circuits), and this section only deals with special amplifiers
as they apply to industrial controls. Industrial amplifiers do not usually

require the bandwidth and linearity used in communications, but must be rugged, trouble-free, and usually capable of supplying substantial power. Where possible, dc amplifiers are avoided, and, because of the narrow bandwidth requirements, modulator-demodulator systems are used instead. Dependent on the input and output signal requirements in particular systems, different amplifier arrangements are in use. Servo amplifiers, probably the most widely used type in industrial applications, can be classified as shown in Figure 22-18.

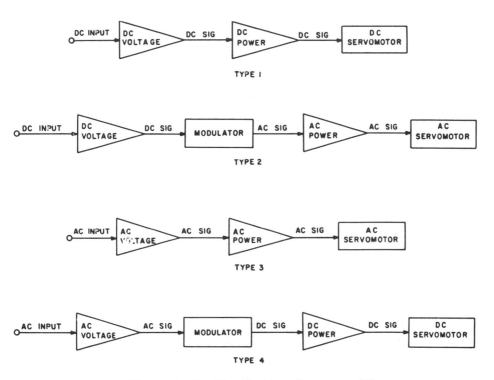

Figure 22-18.   Classification of servo amplifiers.

(Philco Technological Center, *Servomechanism Fundamentals and Experiments,* © 1964. Reprinted by permission of Prentice-Hall, Inc.)

In industrial amplifiers, it is important that they do not alter the phase or introduce excessive hum and noise. Phase distortion is particularly troublesome in servo systems because of the quadrature components introduced. The results can be sluggish operation, unstable behavior such as dithering, positioning errors, or combinations of all three.

Vacuum Tube and Solid State Amplifiers

Figure 22-19 is an example of a vacuum tube amplifier using feedback in its circuit to stabilize performance, as employed in a low-power output servomechanism. Such an amplifier can also be designed using solid state transistors instead of the vacuum tubes.

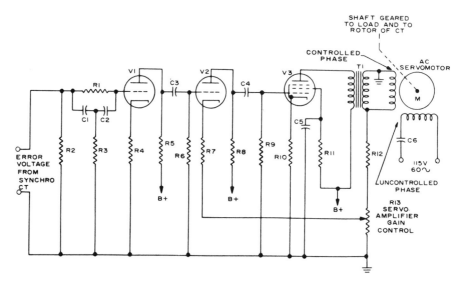

Figure 22-19.   Vacuum tube feedback amplifier.

(Philco Technological Center, *Servomechanism Fundamentals and Experiments,* © 1964. Reprinted by permission of Prentice-Hall, Inc.)

SCR and Thyratron Amplifiers.   Tube or transistor amplifiers that control the operation of motors are quite satisfactory in the low and medium power ranges, up to about 50 W. When higher power outputs are required, however, thyratrons and their solid state equivalents, SCRs, provide more efficiency, lower cost, and smaller size.

These devices are clearly not linear amplifiers, and their output is determined by either amplitude or phase-shift control. Amplitude control is obtained by varying the bias on the input control element, which sets the voltage at which the device fires. Figure 22-20 illustrates bias amplitude control for a thyratron, but the SCR behaves in much the same fashion. The shaded portion denotes the output current pulse, and it is obvious that bias amplitude variations cannot control the output current beyond 90° (1/4 cycle), since no current flows beyond this point.

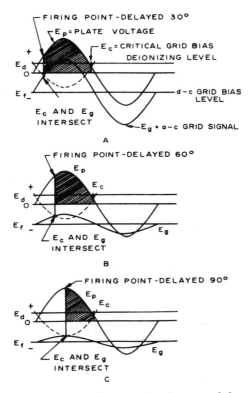

Figure 22-20.   Bias amplitude control for thyratron.

(Philco Technological Center, *Servomechanism Fundamentals and Experiments*, © 1964. Reprinted by permission of Prentice-Hall, Inc.)

A more effective control is obtained if the phase of the control signal is shifted. This can be done by adding an ac bias signal, usually 90° out of phase with the control signal. As shown in Figure 22-21, this ac bias (E_r), the dc bias (E_f), and the control signal (E_s) combine to form the final signal (E_g). Since the control signal E_s is either in phase or 180° out of phase with the plate (or collector), complete 180° control of the output current wave is obtained.

Magnetic Amplifiers.   Magnetic amplifiers have the advantages of ruggedness, long life, and low cost, as well as high power-handling capacity. Their basic operation is described in Chapter 6. Where high distortion levels and limited response speed are not important, as in motor drive controls, magnetic amplifiers are often used. The 60-Hz power line frequency is widely used, but faster response can be obtained with power supply frequencies up to 2000 Hz. Magnetic ampli-

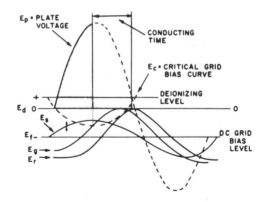

Figure 22-21. Bias phase control for thyratron.

(Philco Technological Center, *Servomechanism Fundamentals and Experiments*, © 1964. Reprinted by permission of Prentice-Hall, Inc.)

fiers can operate over a wide temperature range of -50 to $+100°C$ and are resistant to high nuclear radiation environments. With power inputs of as little as 2 to 4 W, it is possible to control up to 10,000 W.

The magnetic amplifier lends itself to use in servomechanism motor drives, and a typical circuit is illustrated in Figure 22-22. In this arrangement, each saturable reactor has three windings. A dc error signal is necessary; in addition, dc bias must be used. With no error present, reactors X1 and X2 are equally and partially saturated by the dc bias current, resulting in equal reactance values for the two ac coils, and the servomotor is not energized.

When a dc error signal is present, the current through the X2 con-

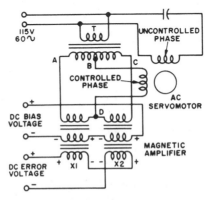

Figure 22-22. Magnetic amplifier servo control.

(Philco Technological Center, *Servomechanism Fundamentals and Experiments*, © 1964. Reprinted by permission of Prentice-Hall, Inc.)

trol winding acts to further saturate X2, decreasing the reactance of the X2 ac winding. Conversely, the same current through the X1 control winding, tends to cancel the effect of the bias current, thereby decreasing the saturation of X1 and increasing the reactance of the X1 ac winding. Hence, point D is in effect connected to point C, causing motor rotation. Reversal of the error signal polarity causes the system to operate in the same manner but in the opposite direction.

Electromechanical Rotational Amplifiers

A dc generator, which is driven from some prime mover or an electric motor, can be used as a dc amplifier, because a small change in the current in the field of a dc generator causes a much larger change in its output armature current. Current change ratios of 50 to 100 times can be obtained in a simple arrangement. 0.2 A driving a small generator, connected to the field coil of a much larger one, as shown in Figure 22-23, can produce 100 A output. Commercially packaged systems that work on this principle are available

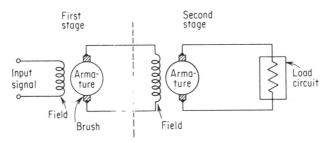

Figure 22-23.   Cascaded rotating amplifiers.

(Matthew Mandl, *Industrial Control Electronics*, © 1961. Reprinted by permission of Prentice-Hall, Inc.)

Working on somewhat similar principles, the Amplidyne can produce power gains of 25,000 or more and work with inputs on the order of 1 W. A two-generator cascade effect is obtained with a single armature and field coil, mechanically driven by an electric motor. Figure 22-24 shows that the two brushes normally used in a generator are short circuited. As a result, a large current flows in this winding, which causes armature reaction in the generator. Armature reaction is the magnetic field due to the armature current, usually an undesired effect. This armature reaction field is at right angles to the main field; its generated voltage is picked up by two additional brushes, added at right angles to the shorted ones. The output load is connected to these two

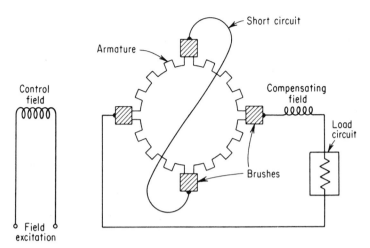

Figure 22-24. Amplidyne circuit.

(Matthew Mandl, *Industrial Control Electronics*, © 1961. Reprinted by permission of Prentice-Hall, Inc.)

brushes. To compensate for the effects of the load current reaction field, a special compensating field coil is included in series with the load current.

Because of the high inductance involved in electromechanical amplifiers, the response speed is relatively low, usually in the order of 0.1 second. When the inputs to these electromechanical amplifiers are from solid state or vacuum tube amplifiers, very small signals can control very large amounts of power.

22.5 INDUSTRIAL ELECTRONICS APPLICATIONS

Induction Heating Systems

Wherever metals must be heated, induction heating provides a rapid, efficient, and easily controllable source of heat. The heat can be applied exactly where required, it can be precisely controlled, and the workpiece and the surrounding area remain clean and uncontaminated by combustion products.

As the name implies, the principle of operating depends on the induction of electric currents by electromagnetic induction (transformer action) into the workpiece to be heated. The flow of these currents, called *eddy currents*, develops heat because of the ohmic resistance of the material. Where magnetic materials, such as iron-based alloys, etc., are involved, the hysteresis effects of the material contribute to the heating. When a relatively large workpiece is heated, lower frequencies are employed, and transformers are often used to simply couple the work coil and the workpiece to the 60-cycle ac power. Where substantial power is required and a good impedance match is important, rotating motor generator sets are used to deliver higher frequencies, usually in the range of 960, 3000, and 10,000 Hz.

As the frequency of the applied induced current rises, the so-called skin effect comes into play. Because the induced current also generates a magnetic field which tends to counter the current flow, at very high frequencies current only flows in the "skin" or outer surfaces of the workpiece. This effect is extremely useful in controlling the depth of penetration of the heating effect by selecting the proper frequency for the material, size, and shape of the workpiece. The graph of Figure 22-25 shows the depth of penetration of the heating effect versus frequency and material.

The design of a suitable work coil to couple the power source to the workpiece is of great importance. Work coils must accommodate the shape, size, frequency, and particular area of heat application, and they are therefore tailor made for each application. Figure 22-26 shows some typical shapes for various jobs of hardening, brazing, and soldering.

High frequencies are obtained from RF generators which are, in

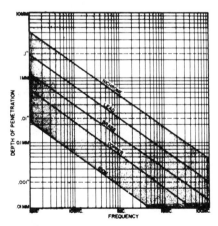

Figure 22-25.   Graph of penetration vs. frequency.

(Illustration courtesy of *Electronics World.*)

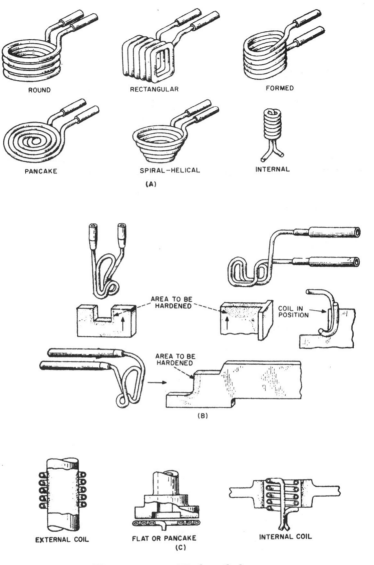

Figure 22-26. Work coil shapes.

(Illustration courtesy of *Electronics World*.)

effect, high-power radio transmitters. Care must be taken that RF energy is not radiated into space, because of interference with communications channels, etc.; and all RF generators must conform to the applicable FCC regulations. Energy lost to radiation reduces the efficiency of the unit, and industrial RF generators must therefore be carefully constructed, shielded, and tested to conform to these regulations.

MATERIAL	EXCELLENT	GOOD	POOR	LITTLE EFFECT	MATERIAL	EXCELLENT	GOOD	POOR	LITTLE EFFECT
Acetate Film		x			Polyvinyl Acetate		x		
Butyrate Film		x			Polyvinyl Chloride (Plastisol)	x			
Epoxy		x			Resorcinol Formaldehyde Resin		x		
Kel-F			x		Rubber	x			
Micalex			x		Saran	x			
Melamine Formaldehyde Resin		x			Silicone Materials			x	
Nylon, Dacron, Orlon, Dynel (Mylar)		x			Teflon				x
Plexiglass, Lucite		x			Urea Formaldehyde Resins		x		
Polyester Resins		x			Vinyl Film	x			
Polyethylene				x	Vinyl Foam	x			
Polystyrene, styrene				x	Water	x			
Polyurethane Foam		x			Wood	x			

Figure 22-27.   Table of dielectric heated materials.

(Illustration courtesy of *Electronics World.*)

Dielectric Heating Systems

While induction heating is produced by causing the flow of current in a conducting material by magnetic flux induction, there are many materials and processes that are not conductors of electricity. Many of these materials lend themselves to electronic heating by a high-frequency electrostatic field. These materials are essentially insulators (or dielectrics) and vary from plywood to plastics. The table of Figure 22-27 is a compilation of such dielectrically heatable materials and their relative susceptibility to this method.

Induction heating uses a transformer effect to induce energy into the workpiece; dielectric heating uses a capacitor effect to couple the energy into the material. Figure 22-28 shows schematically how the dielectric is placed between the plates of a capacitor. Materials most suitable to this type of heating are ''lossy'' type dielectrics, because the RF energy (similar to the hysteresis effect in iron) causes heating in the material by agitating the molecules and atoms in pulling and distorting them from their normal positions as the polarity of the plates changes.

The choice of frequency and power is determined, to a great extent, by the type of material to be heated, but the thickness of material, the

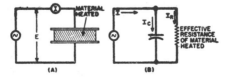

Figure 22-28.   Dielectric heating principles.

(Illustration courtesy of *Electronics World.*)

area to be heated, and the temperature to be attained are also important. Most dielectric heating is done between 5 and 100 megacycles.

Dielectric heating has found important applications in the manufacture of raincoats, seat covers, plastic bags, wading pools, inflated toys, shower caps, etc., made from thermoplastic materials such as vinyl, nylon, orlon, saran, etc. This process is called *heat sealing*.

Applications of dielectric heating are also found in such diverse areas as thawing frozen bales of wool in textile mills, curing of foam rubber articles on a conveyer belt, drying of building materials, etc. Cooking by the microwave oven method is also a form of dielectric heating, and is a very popular application.

RF power sources for dielectric heating are basically similar to those used for RF induction heating, with, of course, different methods of coupling to the workpiece. FCC regulations as to RF shielding and radiation limiting also apply here.

Electronic Cooling

Although the principle of cooling by the use of the *Peltier effect* has been known for over 100 years, only the new semiconductor materials have made electronic cooling practical. The Peltier effect is the reversible transformation of heat into electrical energy or electrical energy into heat at the junction of dissimilar conducting materials. That is, if an electric current flows across a junction of two different materials at the same temperature, the junction will either absorb or emit heat. Figure 22-29 shows such an arrangement. For greatest Peltier effect, p-type semiconductor materials in contact with n-type materials are selected.

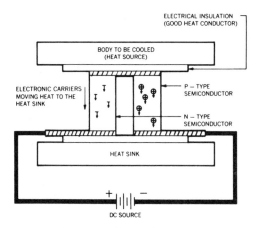

Figure 22-29.   Peltier effect principles.

(Illustration courtesy of Borg-Warner Thermoelectric Prod.)

The amount of heat absorbed or emitted is directly proportional to the current and the so-called Peltier coefficient, which depends on the particular materials. For electronic cooling, the two materials should have a high Peltier coefficient, good conductivity for electricity, and poor conductivity for heat, because the ohmic (or joule) heat generated by the current could swamp out the cooling effect. The most commonly used material is bismuth telluride, with various dopants (impurities) to enhance thermoelectric properties, and make p- and n-types of conductors. The Peltier effect for cooling can be considered as a heat pump. The heat-pumping capacity increases with current until ohmic heat (I^2R) and the thermally conducted heat to the cooling junction equal and surpass the Peltier cooling effect. Larger heat-pumping capacities can be obtained by using many couples (junctions), thermally in parallel and electrically in series.

Practical applications of electronic cooling are still confined to special situations and relatively low heat-pumping requirements, where other means would take too much room or otherwise be inconvenient. Small portable refrigerators, spot cooling of electronic components or infrared detectors etc., make up the bulk of the uses.

Welding

While the actual electric welding current does its job simply through generating heat, its time duration and magnitude must be carefully controlled to avoid burning oxidation, melting, or otherwise damaging the work. The three basic welding processes are spot, seam, and arc welding.

1. *Spot welding*—The basic circuit setup is as shown in Figure 22-30a. This is a type of resistance welding, because the two metal plates to be welded provide sufficient surface-contact resistance to generate heat when passing current from the welding electrodes. The heat causes a fusion of the sheets at the spot to be welded. In large welders, the contacting electrodes are cooled by a continuously running stream of water through the electrode piping, as shown in Figure 22-30b. Both electrodes also apply pressure during the fusion process.

As shown in a, the line voltage (through the gates thyratron circuits) is stepped down by a transformer to deliver approximately 2 to 10 V at a current ranging anywhere from 1000 to 50,000 A. Precise control must be maintained to assure proper welding without burn-through or the making of a bond of insufficient holding characteristics.

2. *Seam welding*—is another form of spot welding, though instead of point electrodes, rollers are used. As the material passes between the

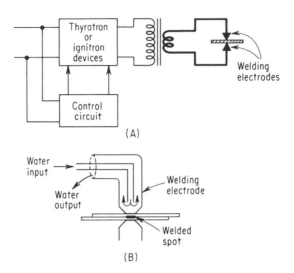

Figure 22-30. Spot welding.

(Matthew Mandl, *Industrial Control Electronics*, © 1961. Reprinted by permission of Prentice-Hall, Inc.)

rollers, the control circuits apply bursts of current to the roller electrodes, and a series of spot welds is formed in a running seam. Again, thyratron-controlled circuits are employed to apply the proper amount of power at the right intervals for just the right degree of welding.

3. *Arc welding*—consists of producing a concentrated high-intensity arc between two electrodes and applying the heat so produced to the metal to be welded. The electrodes are made up of special welding rods which melt during the arc, so that a portion of the molten electrode material can flow onto the metal being welded. For arc welding, either dc or ac can be employed.

22.6 FACTORY AUTOMATION

Although factory automation can be found in almost every major manufacturing facility in the United States and Japan, it still has many unsolved problems. Even though the problems exist (for example, no manufacturing automation protocol), automation is now creating an important niche in heavy industrial assembly. Makers of semiconductor-processing equipment are developing a host of new technologies, and improved coordination now permits the simultaneous use of two robot arms for a single task.

Automation minimizes product variance (improving quality control) during high-precision parts machining. New developments in programmable-controller software allow the manufacturer to specify more precisely the dimensions of a certain part. Improved software also permits a broader use of numerical controllers, which can be reprogrammed to work with any design changes almost as soon as the specifications are received.

Numerical Control of Machine Tools

The term *numerical control* (N/C) is essentially self-explanatory—control of a machine tool, such as a lathe or drill, by numbers.

Two different systems are in general use. The first is called a *point-to-point positioning system* because it involves separate, automatic positioning steps of the tool or workpiece. Between steps, the tool is withdrawn from the workpiece. The exact path of motion is not closely controlled between cuts, since the tool is not in contact with the workpiece while moving from point to point. These machines are relatively simple and inexpensive and constitute the bulk of present-day N/C.

The second system is the more complex and expensive *contour* or *continuous-path system*, used with the more sophisticated, multi-axis machine tools. Two or more tool motions are controlled simultaneously, while the tool is in contact with the workpiece, to generate a contour or profile.

Point-to-point systems are used for such machines as drilling, boring, tapping, etc., that perform operations at specific stations. Profile mills, contouring lathes, and multidimensional contouring mills are in the second category.

In addition to positioning the tool and the workpiece, N/C can also be used for sequencing of a machine. N/C circuits can control the selection of tools, the kind of operation to be performed, the axis of motion, speeds, feeds, turret positions, control coolants, etc.

For the majority of N/C systems, information and instructions are presented to the machine in the form of punched-paper, eight-hole tape, now standard with the computer and communications industry. These tapes are prepared by trained programmers, often with the aid of a computer, especially in the case of contour machines, which require large amounts of positional data. A few machines work from punched cards, but these require bulkier card readers and can easily get out of sequence. Digital magnetic tape is increasing in popularity because it has a very high information density and can be erased readily and used over, or altered, as required.

The programmed information is fed into the machine's reader unit, where the codes for the numbers or instructions are decoded and used

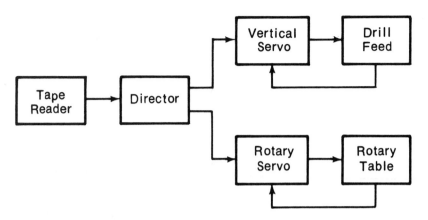

Figure 22-31.   Block diagram of N/C machine.

to control the machine's switching circuitry that translates the information into the required machine movements.

Most N/C systems work on the closed loop system, described in Section 22.2, and incorporate a transducer to indicate the actual position of the machine. The electrical output of this transducer is compared to the desired position as obtained from the tape (or other) input, and the error signal is used to direct the machine to the correct position.

Very few N/C machines operate on the open loop system, which is simpler but requires highly reliable systems, because it depends on accurate calibration that will not change with time, temperature, use, etc.—an obvious drawback.

Figure 22-31 is a block diagram of a two-axis N/C machine, one rotary and the other linear, used for boring workpieces mounted on a rotary table. In this system the tape reader feeds data to the director, which controls the two axis feedback loops as well as the auxiliary functions.

Electronic Weighing

Practically all raw and many finished materials of modern industry are weighed at one time or another. Weighing is done before mixing in quality control and to establish the quantity of the packaged product.

All of the principles of controls and servo systems, discussed in this chapter, find application in one or another of the weighing systems. Electronics helps to automate the weighing process by detecting when the "scales" are in balance, what the weight is, and indicating or recording the weight. Some systems are of the closed loop variety and are subject to all the principles as outlined for feedback servomecha-

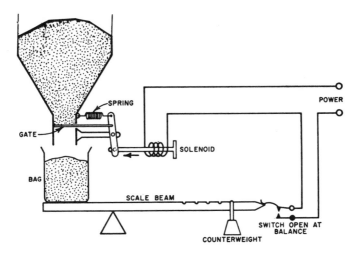

Figure 22-32.   Balance sensing system.

(Illustration courtesy of *Electronics World.*)

nisms. Figure 22-32 is a simplified system of balance sensing, using a switch to indicate the balance of a scale beam. While most industrial systems are more complex, this example illustrates the basic elements. As the material pours into the bag, the scale beam tilts upward until, at the correct weight, the switch is tripped. This opens the holding circuit for the solenoid, which controls the flow of material, and the shutoff gate across the loading funnel is closed. Replacing the full bag with an empty one causes the scale indicator to move down, closing the switch and actuating the solenoid again, repeating the whole process.

Industrial scales use a variety of motion or balance sensors. They include almost all of the mechanical input transducers described in Chapter 21, such as differential transformers, variable inductors, photocells, magnetically operated switches, etc. The differential transformer is probably the most widely used device for balance detecting or load displacement.

Figure 22-33 is a basic circuit for null balancing with a differential transformer. Note that the two windings of each differential transformer are connected in series, with voltages adding. T_1 acts as the load-sensing unit, and its displacement against a stiff spring member (not shown) is directly proportionate to the weight. A matching differential transformer, T_2, serves as a comparator, and the difference signal is applied to the input of a servoamplifier. The servomotor will drive the linkage-

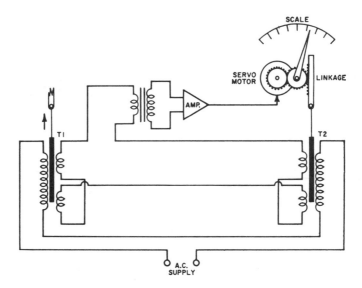

Figure 22-33.   Differential transformer in weighing.

(Illustration courtesy of *Electronics World.*)

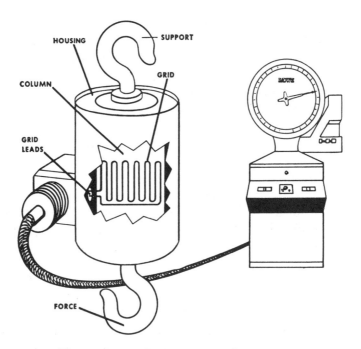

Figure 22-34.   Strain gage weighing system.

(Illustration courtesy of *Electronics World.*)

gear mechanism, including the dial indicator, and T_2, until the difference signal is zero, thus indicating the weight.

If the dial indicator of Figure 22-33 is replaced by a counter mechanism, the weight can be displayed in numerals. A similar numeral wheel assembly can be used for printing the weight onto a paper tape, ticket, or card.

Figure 22-34 is an example of an open loop weighing system that uses a strain-gage load cell. Strain gages and their accompanying circuitry are discussed in Chapter 21, and the application for weighing depends on the strain exerted by the weight on a supporting member.

BIBLIOGRAPHY—CHAPTER 22

CHAMBERS, H., and C. CHACEY, *Drafting and Manual Programming for Numerical Control*. Englewood Cliffs, NJ, Prentice-Hall, Inc., 1980.

GROVER, M., *Automation, Production Systems, and Computer-aided Manufacturing*. Englewood Cliffs, NJ, Prentice-Hall, Inc., 1980.

KUO, B., *Automatic Control Systems* (4th Ed.). Englewood Cliffs, NJ, Prentice-Hall, Inc., 1982.

KUO, B., *Digital Control Systems*.   New York, NY, Holt, Rinehart & Winston, 1980.

MILLER, R., *Servomechanisms: Devices and Fundamentals*. Reston, VA, Reston Publishing Company, Inc., 1977.

NOF, S., *Handbook of Industrial Robotics*. New York, NY, John Wiley & Sons, Inc. 1985.

Reference Data for Engineers: Radio, Electronics, Computer, and Communications (7th Ed.). Indianapolis, IN, Howard W. Sams & Co., Inc., 1985.

RICHARDSON, D., *Handbook of Rotating Electric Machinery*. Reston, VA, Reston Publishing Company, Inc., 1980.

CHAPTER 23

MATHEMATICAL FORMULAS

The mathematical formulas presented in this chapter are particularly useful if the reader is familiar with the mathematical operations involved, but needs to look up specific formulas. If the reader has never studied calculus, for example, he should refer to the reference books listed at the end of this chapter before attempting to use strange formulas.

The notations used in the following pages are uniform within each numbered paragraph, but are not uniform throughout the chapter. For example, the letter n is used to indicate any number, but in 23.9, under the headings Taylor's Series and Maclaurin's Series, where n is part of the series involving f', f'', etc., the letter n indicates a higher-order derivative. In all cases, the notation corresponds to the notation used in practical applications, rather than that used in specialized textbooks.

Chapter 23 is arranged in the approximate order of complexity of the mathematics required. This implies that a reader having difficulties with matrix algebra may find it helpful to refer to the fundamentals of algebra, for example.

23.1 FUNDAMENTALS OF ALGEBRA

Signs

$$A + A = 2A$$

$$A - A = 0$$

$$A \cdot A = (+A) \cdot (+B) = +AB$$

$$(+A) \cdot (-B) = -AB$$

$$(-A) \cdot (-B) = +AB$$

Name	Capital	Lower case	Designates
Alpha	A	α	Angles
Beta	B	β	Angles, flux density
Gamma	Γ	γ	Conductivity
Delta	Δ	δ	Variation of a quantity, increment
Epsilon	E	ϵ	Base of natural logarithms (2.71828)
Zeta	Z	ζ	Impedance, coefficients, coordinates
Eta	H	η	Hysteresis coefficient, efficiency
Theta	Θ	θ	Phase angle
Iota	I	ι	
Kappa	K	κ	Dielectric constant, coupling coefficient, susceptibility
Lambda	Λ	λ	Wavelength
Mu	M	μ	Permeability, micro, amplification factor
Nu	N	ν	Reluctivity
Xi	Ξ	ξ	
Omicron	O	o	
Pi	Π	π	3.1416
Rho	P	ρ	Resistivity
Sigma	Σ	σ	
Tau	T	τ	Time constant, time-phase displacement
Upsilon	Υ	υ	
Phi	Φ	ϕ	Angles, magnetic flux
Chi	X	χ	
Psi	Ψ	ψ	Dielectric flux, phase difference
Omega	Ω	ω	Ohms (capital), angular velocity ($2\pi f$)

Figure 23-1.   Greek alphabet.

(Harry E. Thomas, *Handbook for Electronic Engineers and Technicians*, © 1965. Reprinted by permission of Prentice-Hall, Inc.)

Exponents

$$A^x = A \cdot A \cdot A \cdot \text{etc.} \to A, \text{ x times}$$

$$A^x \cdot A^y = A^{x+y}$$

$$A^x \cdot A^{-y} = \frac{A^x}{A^y} = A^{x-y}$$

$$A^x \cdot B^x = (AB)^x$$

$$\frac{A^x}{B^x} = \left(\frac{A}{B}\right)^x; \quad (A^x)^y = A^{xy}$$

Commutative Law

$$A + B = B + A$$

$$A \cdot B = B \cdot A$$

Associative Law

$$(A + B) + C = A + (B + C)$$

$$(AB) \cdot C = A \cdot (BC)$$

Distributive Law

$$A \cdot (B + C) = AB + AC$$

Equations

1 unknown:
(1st order)

$$ax + b = c$$

$$ax = c - b$$

$$x = \frac{c - b}{a}$$

2 unknowns:
(2nd order)

$$a_1 x + b_1 y = c_1 \quad (1)$$

$$\underline{a_2 x + b_2 y = c_2 \quad (2)}$$

multiply (1) by b_2, multiply (2) by $-b_1$, add (1) + (2)

$$a_1 b_2 x + b_1 b_2 y = b_2 c_1$$

$$\frac{-a_2 b_1 x - b_1 b_2 y = -b_1 c_2}{a_1 b_2 x - a_2 b_1 x = b_2 c_1 - b_1 c_2}$$

$$x = \frac{b_2 c_1 - b_1 c_2}{a_1 b_2 - a_2 b_1} = \frac{D_x}{D}$$

repeat for a_1, $-a_2$ to obtain $y = \dfrac{a_1c_2 - a_2c_1}{a_1b_2 - a_2b_1} = \dfrac{D_y}{D}$

D_x = Determinant of x

D_y = Determinant of y

D = Determinant of common denominator

In Matrix Form

$$x = \frac{\begin{vmatrix} c_1 & b_1 \\ c_2 & b_2 \\ a_1 & b_1 \\ a_2 & b_2 \end{vmatrix}}{} = \frac{c_1b_2 - c_2b_1}{a_1b_2 - a_2b_1}$$

$$y = \frac{\begin{vmatrix} a_1 & c_1 \\ a_2 & c_2 \\ a_1 & b_1 \\ a_2 & b_2 \end{vmatrix}}{} = \frac{a_1c_2 - a_2c_1}{a_1b_2 - a_2b_1}$$

3 unknowns:
(3rd order)

$$a_1x + b_1y + c_1z = k_1$$
$$a_2x + b_2y + c_2z = k_2$$
$$a_3x + b_3y + c_3z = k_3$$

In Matrix Form

$$x = \frac{\begin{vmatrix} k_1 & b_1 & c_1 \\ k_2 & b_2 & c_2 \\ k_3 & b_3 & c_3 \end{vmatrix}}{\begin{vmatrix} a_1 & b_1 & c_1 \\ a_2 & b_2 & c_2 \\ a_3 & b_3 & c_3 \end{vmatrix}} = \frac{D_x}{D}$$

$$y = \frac{\begin{vmatrix} a_1 & k_1 & c_1 \\ a_2 & k_2 & c_2 \\ a_3 & k_3 & c_3 \end{vmatrix}}{D} = \frac{D_y}{D}$$

$$z = \frac{\begin{vmatrix} a_1 & b_1 & k_1 \\ a_2 & b_2 & k_2 \\ a_3 & b_3 & k_3 \end{vmatrix}}{D} = \frac{D_z}{D}$$

$$x = \frac{k_1 b_2 c_3 + b_1 c_2 k_3 + c_1 k_2 b_3 - k_3 b_2 c_1 - b_3 c_2 k_1 - c_3 k_2 b_1}{a_1 b_2 c_3 + b_1 c_2 a_3 + c_1 a_2 b_3 - a_3 b_2 c_1 - b_3 c_2 a_1 - c_3 a_2 b_1}$$

For details of matrix operation see section 23.5.

Quadratic Equations

2nd degree

$$ax^2 + bx + c = 0$$

$$x = \frac{-b \pm \sqrt{b^2 - 4ac}}{2a}$$

Complete squares: $x^2 + x(a + b) + ab = (x + a)(x + b)$

$$x^2 + x(a - b) - ab = (x + a)(x - b)$$

$$x^2 - x(a + b) + ab = (x - a)(x - b)$$

$$x^2 - a^2 = (x + a)(x - a)$$

Higher-Order Equations

$$x^n + a_1 x^{n-1} + a_2 x^{n-2} \ldots + c = 0$$

Complete squares: $(x + a)^n = x^n + nx^{n-1}a + \dfrac{n(n-1)}{2!} x^{n-2}a^2 +$

$$\ldots \frac{n! x^{n-k} a^k}{(n-k)!\, k!} \ldots a^2 < x^2$$

by logarithms: $x^n = n \log x$

(see logarithms to base 10 on page 577) $a_1 x^{n-1} = \log a_1 + (n-1) \log x$

$$c = \log(10^c)$$

Progressions

Arithmetic $A_1 + A_2 + A_3 + \ldots A_n = \Sigma_n$

$$A_n = a + (n-1)\,d; \qquad a = \text{1st term}$$

$$\Sigma_n = \frac{n}{2}(a + A_n) \qquad d = \text{common difference}$$

$$(A_{n-m} - A_{n-m-1})$$

Geometric $A_1 + A_2 + A_3 + \ldots A_n = \Sigma_n$

$$A_n = ar^{n-1}; \qquad a = \text{1st term}$$

$$\Sigma_n = \frac{a(r^n - 1)}{r - 1} \qquad r = \text{common ratio}$$

$$(A_{n-m}/A_{n-m-1})$$

Factorials

$$x! = 1 \cdot 2 \cdot 3 \cdot \text{etc.}$$

x	x!
1	1
2	2
3	6
4	24
5	120
6	720
7	5040
8	40,320
9	362,880
10	3,628,800

For $x > 10$, $x! = x^x e^{-x}\sqrt{2\pi x}$
(Less than 1% error)
$\log(x!) = (x + \tfrac{1}{2})\log x - 0.434x + 0.399$

Binomial Theorem

$$(a \pm b)^n = a^n \pm na^{n-1}b + \frac{n(n-1)}{2!}a^{n-2}b^2$$

$$\pm \frac{n(n-1)(n-2)}{3!}a^{n-3}b^3 + \ldots$$

If n is a positive integer: $n + 1$ terms, finite

If n is not a positive integer: Converges for $[b/a] < 1$

Diverges for $[b/a] > 1$.

Complex Quantities

Imaginary roots and numbers

$$\sqrt{-1} = i \text{ or } j; \quad i^2 = -1, \quad j^2 = -1$$

j is used to indicate reactive component in electrical circuit (see Chapter 4); i is used in nonelectrical work.

$$(a + jb) + (c + jd) = (a + c)$$

$$+ j(b + d) = (ac - bd) + j(bc + ad)$$

$$a + jb = \sqrt{a^2 + b^2} \cdot \epsilon^{j\theta}, \text{ where } \sqrt{a^2 + b^2} > 0$$

$$a + jb = \sqrt{a^2 + b^2}\,[\cos \theta + j \sin \theta], \quad \sin \theta = \frac{b}{\sqrt{a^2 + b^2}}$$

$$\sin \theta = \frac{\epsilon^{j\theta} - \epsilon^{-j\theta}}{2j} = I_m[\epsilon^{j\theta}] \quad \cos \theta = \frac{a}{\sqrt{a^2 + b^2}}$$

$$\cos \theta = \frac{\epsilon^{j\theta} + \epsilon^{-j\theta}}{2} = \text{Real } [\epsilon^{j\theta}]$$

$$(a + jb)(c + jd) = ac + j^2bd + j(ad + bc)$$

$$= ac - bd + j(ad + bc)$$

Trigonometric Identities

$$\sin A = \frac{\cos A}{\cot A} = \frac{1}{\csc A} = \cos A \tan A = \pm\sqrt{1 - \cos^2 A}$$

$$\cos A = \frac{\sin A}{\tan A} = \frac{1}{\sec A} = \sin A \cot A = \pm\sqrt{1 - \sin^2 A}$$

$$\sin (A \pm B) = \sin A \cos B \pm \cos A \sin B$$

$$\cos (A \pm B) = \cos A \cos B \mp \sin A \sin B$$

$$\sin A = \frac{\epsilon^{jA} - \epsilon^{-jA}}{2j}$$

$$\cos A = \frac{\epsilon^{jA} + \epsilon^{-jA}}{2}$$

$$\sin A + \sin B = 2 \sin \tfrac{1}{2}(A + B) \cos \tfrac{1}{2}(A - B)$$

$$\sin A - \sin B = 2 \cos \tfrac{1}{2}(A + B) \sin \tfrac{1}{2}(A - B)$$

$$\cos A + \cos B = 2 \cos \tfrac{1}{2}(A + B) \cos \tfrac{1}{2}(A - B)$$

$$\cos B - \cos A = 2 \sin \tfrac{1}{2}(A + B) \sin \tfrac{1}{2}(A - B)$$

$$\sin 2A = 2 \sin A \cos A$$

$$\cos 2A = \cos^2 A - \sin^2 A$$

$$\sin \tfrac{1}{2} A = \pm \sqrt{\frac{1 - \cos A}{2}}$$

$$\cos \tfrac{1}{2} A = \pm \sqrt{\frac{1 + \cos A}{2}}$$

Logarithms to Base 10

$$\text{let } N = 10^n$$

$$\log N = n \qquad (\text{see Table 23-2})$$

$$\log 10 = 1 \qquad\qquad \log 2 = 0.3010$$

$$\log 100 = 2 \qquad\qquad \log 20 = 1.3010$$

$$\log 1 = 0 \qquad\qquad \log 200 = 2.3010$$

$$\log 0.1 = -1 \qquad\qquad \log 0.2 = -1.3010$$

$$\log 0.01 = -2 \qquad \log 0.02 = -2.3010$$

$$\log a + \log b = 10^a \cdot 10^b = \log (a \cdot b)$$

$$\log a - \log b = \frac{10^a}{10^b} = \log (a - b)$$

$$\log (-a) = -\log (+a)$$

Logarithms to Base ϵ (NATURAL LOGARITHMS)

$$\text{let } N = \epsilon^n; \ \epsilon = 2.71828$$

$$\ln N = n \quad \text{see Table 24-3}$$

$$\ln 1 = 0$$

$$\ln 2.71828 = 1$$

Angle	Sin A	Cos A	Tan A	Cot A	Sec A	Csc A
0°	0	1	0	Inf	1	Inf
30°	0.50u0	0.8660	0.5774	1.73	1.15	2
45°	0.7071	0.7071	1	1	1.41	1.41
60°	0.8660	0.5000	1.73	0.5774	2	1.15
90°	1	0	Inf	0	Inf	1
120°	0.8660	−0.5000	−1.73	−0.5774	−2	1.15
180°	0	−1	0	Inf	−1	Inf
270°	−1	0	Inf	0	Inf	−1
360°	0	1	0	Inf	1	Inf

Table 23-1.   Values of frequently used angles

Harry E. Thomas, *Handbook for Electronic Engineers and Technicians,* © 1965. Reprinted by permission of Prentice-Hall, Inc.

Number	Number^2	$\sqrt{Number}$	$\sqrt{10 \times Number}$	Number^3
1	1	1.000000	3.162278	1
2	4	1.414214	4.472136	8
3	9	1.732051	5.477226	27
4	16	2.000000	6.324555	64
5	25	2.236068	7.071068	125
6	36	2.449490	7.745967	216
7	49	2.645751	8.366600	343
8	64	2.828427	8.944272	512
9	81	3.000000	9.486833	729
10	100	3.162278	10.00000	1,000
11	121	3.316625	10.48809	1,331
12	144	3.464102	10.95445	1,728
13	169	3.605551	11.40175	2,197
14	196	3.741657	11.83216	2,744
15	225	3.872983	12.24745	3,375
16	256	4.000000	12.64911	4,096
17	289	4.123106	13.03840	4,913
18	324	4.242641	13.41641	5,832
19	361	4.358899	13.78405	6,859
20	400	4.472136	14.14214	8,000
21	441	4.582576	14.49138	9,261
22	484	4.690416	14.83240	10,648
23	529	4.795832	15.16575	12,167
24	576	4.898979	15.49193	13,824
25	625	5.000000	15.81139	15,625

Table 23-2.   Number functions.

(John D. Lenk, *Practical Semiconductor Data Book for Electronic Engineers and Technicians,* © 1970. Reprinted by permission of Prentice-Hall, Inc.)

Number	Number^2	$\sqrt{Number}$	$\sqrt{10 \times Number}$	Number^3
26	676	5.099020	16.12452	17,576
27	729	5.196152	16.43168	19,683
28	784	5.291503	16.73320	21,952
29	841	5.385165	17.02939	24,389
30	900	5.477226	17.32051	27,000
31	961	5.567764	17.60682	29,791
32	1,024	5.656854	17.88854	32,768
33	1,089	5.744563	18.16590	35,937
34	1,156	5.830952	18.43909	39,304
35	1,225	5.916080	18.70829	42,875
36	1,296	6.000000	18.97367	46,656
37	1,369	6.082763	19.23538	50,653
38	1,444	6.164414	19.49359	54,872
39	1,521	6.244998	19.74842	59,319
40	1,600	6.324555	20.00000	64,000
41	1,681	6.403124	20.24846	68,921
42	1,764	6.480741	20.49390	74,088
43	1,849	6.557439	20.73644	79,507
44	1,936	6.633250	20.97618	85,184
45	2,025	6.708204	21.21320	91,125
46	2,116	6.782330	21.44761	97,336
47	2,209	6.855655	21.67948	103,823
48	2,304	6.928203	21.90890	110,592
49	2,401	7.000000	22.13594	117,649
50	2,500	7.071680	22.36068	125,000
51	2,601	7.141428	22.58318	132,651
52	2,704	7.211103	22.80351	140,608
53	2,809	7.280110	23.02173	148,877
54	2,916	7.348469	23.23790	157,464
55	3,025	7.416198	23.45208	166,375
56	3,136	7.483315	23.66432	175,616
57	3,249	7.549834	23.87467	185,193
58	3,364	7.615773	24.06319	194,112
59	3,481	7.681146	24.28992	205,379
60	3,600	7.745967	24.49490	216,000
61	3,721	7.810250	24.69818	226,981
62	3,844	7.874008	24.89980	238,047
63	3,969	7.937254	25.09980	250,047
64	4,096	8.000000	25.29822	262,144
65	4,225	8.062258	25.49510	274,625
66	4,356	8.124038	25.69047	287,496
67	4,489	8.185353	25.88436	300,763
68	4,624	8.246211	26.07681	314,432
69	4,761	8.306624	26.26785	328,509
70	4,900	8.366600	26.45751	343,000

Table 23-2. Continued.

Number	Number^2	$\sqrt{Number}$	$\sqrt{10 \times Number}$	Number^3
71	5,041	8.426150	26.64583	357,911
72	5,184	8.485281	26.83282	373,248
73	5,329	8.544004	27.01851	389,017
74	5,476	8.602325	27.20294	405,224
75	5,625	8.660254	27.38613	421,875
76	5,776	8.717798	27.56810	438,976
77	5,929	8.774964	27.74887	456,533
78	6,084	8.831761	27.92848	474,552
79	6,241	8.888194	28.10694	493,039
80	6,400	8.944272	28.28427	512,000
81	6,561	9.000000	28.46050	531,441
82	6,724	9.055385	28.63564	551,368
83	6,889	9.110434	28.80972	571,787
84	7,056	9.165151	28.98275	592,704
85	7,225	9.219544	29.15476	614,125
86	7,396	9.273618	29.32576	636,056
87	7,569	9.327379	29.49576	658,503
88	7,744	9.380832	29.66479	681,472
89	7,921	9.433981	29.83287	704,969
90	8,100	9.486833	30.00000	729,000
91	8,281	9.539392	30.16621	753,571
92	8,464	9.591663	30.33150	778,688
93	8,649	9.643651	30.49590	804,357
94	8,836	9.695360	30.65942	830,584
95	9,025	9.746794	30.82207	857,375
96	9,216	9.797959	30.98387	884,736
97	9,409	9.848858	31.14482	912,673
98	9,604	9.899495	31.30495	941,192
99	9,801	9.949874	31.46427	970,299
100	10,000	10.00000	31.62278	1,000,000

Number	$\sqrt[3]{Number}$	$\sqrt[3]{10 \times Number}$	$\sqrt[3]{100 \times Number}$
1	1.000000	2.154435	4.641589
2	1.259921	2.714418	5.848035
3	1.442250	3.107233	6.694330
4	1.587401	3.419952	7.368063
5	1.709976	3.684031	7.937005
6	1.817121	3.914868	8.434327
7	1.912931	4.121285	8.879040
8	2.000000	4.308869	9.283178
9	2.080084	4.481405	9.654894
10	2.154435	4.641589	10.00000

Table 23-2. Continued.

Number	$\sqrt[3]{Number}$	$\sqrt[3]{10 \times Number}$	$\sqrt[3]{100 \times Number}$
11	2.223980	4.791420	10.32280
12	2.289428	4.932424	10.62659
13	2.351335	5.065797	10.91393
14	2.410142	5.192494	11.18689
15	2.466212	5.313293	11.44714
16	2.519842	5.428835	11.69607
17	2.571282	5.539658	11.93483
18	2.620741	5.646216	12.16440
19	2.668402	5.748897	12.38562
20	2.714418	5.848035	12.59921
21	2.758924	5.943922	12.80579
22	2.802039	6.036811	13.00591
23	2.843867	6.126926	15.20006
24	2.884499	6.214465	13.38866
25	2.924018	6.299605	13.57209
26	2.962496	6.382504	13.75069
27	3.000000	6.463304	13.92477
28	3.036589	6.542133	14.09460
29	3.072317	6.619106	14.26043
30	3.107233	6.694330	14.42250
31	3.141381	6.767899	14.58100
32	3.174802	6.839904	14.73613
33	3.207534	6.910423	14.88806
34	3.239612	6.979532	15.03695
35	3.271066	7.047299	15.18294
36	3.301927	7.113787	15.32619
37	3.332222	7.179054	15.46680
38	3.361975	7.243156	15.60491
39	3.391211	7.306144	15.74061
40	3.419952	7.368063	15.87401
41	3.448217	7.428959	16.00521
42	3.476027	7.488872	16.13429
43	3.503398	7.547842	16.26133
44	3.530348	7.605905	16.38643
45	3.556893	7.663094	16.50964
46	3.583048	7.719443	16.63103
47	3.608826	7.774980	16.75069
48	3.634241	7.829735	16.86865
49	3.659306	7.883735	16.98499
50	3.684031	7.937005	17.09976
51	3.708430	7.989570	17.21301
52	3.732511	8.041452	17.32478
53	3.756286	8.092672	17.43513
54	3.779763	8.143253	17.54411
55	3.802952	8.193213	17.65174

Table 23-2. Continued.

Number	$\sqrt[3]{Number}$	$\sqrt[3]{10 \times Number}$	$\sqrt[3]{100 \times Number}$
56	3.825862	8.242571	17.75808
57	3.848501	8.291344	17.86316
58	3.870877	8.339551	17.96702
59	3.892996	8.387207	18.06969
60	3.914868	8.434327	18.17121
61	3.936497	8.480926	18.27160
62	3.957892	8.527019	18.37091
63	3.979057	8.572619	18.46915
64	4.000000	8.617739	18.56636
65	4.020726	8.662391	18.66256
66	4.041240	8.706588	18.75777
67	4.061548	8.750340	18.85204
68	4.081655	8.793659	18.94536
69	4.101566	8.836556	19.03778
70	4.121285	8.879040	19.12931
71	4.140818	8.921121	19.21997
72	4.160168	8.962809	19.30979
73	4.179339	9.004113	19.39877
74	4.198336	9.045042	19.48695
75	4.217163	9.085603	19.57434
76	4.235824	9.125805	19.66095
77	4.254321	9.165656	19.74681
78	4.272659	9.205164	19.83192
79	4.290840	9.244335	19.91632
80	4.308869	9.283178	20.00000
81	4.326749	9.321698	20.08299
82	4.344481	9.359902	20.16530
83	4.362071	9.397796	20.24694
84	4.379519	9.435388	20.32793
85	4.396830	9.472682	20.40828
86	4.414005	9.509685	20.48800
87	4.431048	9.546403	20.56710
88	4.447960	9.582840	20.64560
89	4.464745	9.619002	20.72351
90	4.481405	9.654894	20.80084
91	4.497941	9.690521	20.87759
92	4.514357	9.725888	20.95379
93	4.530655	9.761000	21.02944
94	4.546836	9.795861	21.10454
95	4.562903	9.830476	21.17912
96	4.578857	9.864848	21.25317
97	4.594701	9.898983	21.32671
98	4.610436	9.932884	21.39975
99	4.626065	9.966555	21.47229
100	4.641589	10.00000	21.54435

Table 23-2. Continued.

N	0	1	2	3	4	5	6	7	8	9	u. d.
10	0000	0043	0086	0128	0170	0212	0253	0294	0334	0374	4.2
11	0414	0453	0492	0531	0569	0607	0645	0682	0719	0755	3.8
12	0792	0828	0864	0899	0934	0969	1004	1038	1072	1106	3.5
13	1139	1173	1206	1239	1271	1303	1335	1367	1399	1430	3.2
14	1461	1492	1523	1553	1584	1614	1644	1673	1703	1732	3.0
15	1761	1790	1818	1847	1875	1903	1931	1959	1987	2014	2.8
16	2041	2068	2095	2122	2148	2175	2201	2227	2253	2279	2.6
17	2304	2330	2355	2380	2405	2430	2455	2480	2504	2529	2.5
18	2553	2577	2601	2625	2648	2672	2695	2718	2742	2765	2.4
19	2788	2810	2833	2856	2878	2900	2923	2945	2967	2989	2.2
20	3010	3032	3054	3075	3096	3118	3139	3160	3181	3201	2.1
21	3222	3243	3263	3284	3304	3324	3345	3365	3385	3404	2.0
22	3424	3444	3464	3483	3502	3522	3541	3560	3579	3598	1.9
23	3617	3636	3655	3674	3692	3711	3729	3747	3766	3784	1.8
24	3802	3820	3838	3856	3874	3892	3909	3927	3945	3962	1.8
25	3979	3997	4014	4031	4048	4065	4082	4099	4116	4133	1.7
26	4150	4166	4183	4200	4216	4232	4249	4265	4281	4298	1.6
27	4314	4330	4346	4362	4378	4393	4409	4425	4440	4456	1.6
28	4472	4487	4502	4518	4533	4548	4564	4579	4594	4609	1.5
29	4624	4639	4654	4669	4683	4698	4713	4728	4742	4757	1.5
30	4771	4786	4800	4814	4829	4843	4857	4871	4886	4900	1.4
31	4914	4928	4942	4955	4969	4983	4997	5011	5024	5038	1.4
32	5051	5065	5079	5092	5105	5119	5132	5145	5159	5172	1.3
33	5185	5198	5211	5224	5237	5250	5263	5276	5289	5302	1.3
34	5315	5328	5340	5353	5366	5378	5391	5403	5416	5428	1.3
35	5441	5453	5465	5478	5490	5502	5514	5527	5539	5551	1.2
36	5563	5575	5587	5599	5611	5623	5635	5647	5658	5670	1.2
37	5682	5694	5705	5717	5729	5740	5752	5763	5775	5786	1.2
38	5798	5809	5821	5832	5843	5855	5866	5877	5888	5899	1.1
39	5911	5922	5933	5944	5955	5966	5977	5988	5999	6010	1.1
40	6021	6031	6042	6053	6064	6075	6085	6096	6107	6117	1.1
41	6128	6138	6149	6160	6170	6180	6191	6201	6212	6222	1.0
42	6232	6243	6253	6263	6274	6284	6294	6304	6314	6325	1.0
43	6335	6345	6355	6365	6375	6385	6395	6405	6415	6425	1.0
44	6435	6444	6454	6464	6474	6484	6493	6503	6513	6522	1.0
45	6532	6542	6551	6561	6571	6580	6590	6599	6609	6618	1.0
46	6628	6637	6646	6656	6665	6675	6684	6693	6702	6712	.9
47	6721	6730	6739	6749	6758	6767	6776	6785	6794	6803	.9
48	6812	6821	6830	6839	6848	6857	6866	6875	6884	6893	.9
49	6902	6911	6920	6928	6937	6946	6955	6964	6972	6981	.9
50	6990	6998	7007	7016	7024	7033	7042	7050	7059	7067	.9
51	7076	7084	7093	7101	7110	7118	7126	7135	7143	7152	.8
52	7160	7168	7177	7185	7193	7202	7210	7218	7226	7235	.8
53	7243	7251	7259	7267	7275	7284	7292	7300	7308	7316	.8
54	7324	7332	7340	7348	7356	7364	7372	7380	7388	7396	.8

Table 23-3.   Four-place log tables.

(Harry E. Thomas, *Handbook for Electronic Engineers and Technicians*, © 1965. Reprinted by permission of Prentice-Hall, Inc.)

N	0	1	2	3	4	5	6	7	8	9	u. d.
55	7404	7412	7419	7427	7435	7443	7451	7459	7466	7474	.8
56	7482	7490	7497	7505	7513	7520	7528	7536	7543	7551	.8
57	7559	7566	7574	7582	7589	7597	7604	7612	7619	7627	.8
58	7634	7642	7649	7657	7664	7672	7679	7686	7694	7701	.7
59	7709	7716	7723	7731	7738	7745	7752	7760	7767	7774	.7
60	7782	7789	7796	7803	7810	7818	7825	7832	7839	7846	.7
61	7853	7860	7868	7875	7882	7889	7896	7903	7910	7917	.7
62	7924	7931	7938	7945	7952	7959	7966	7973	7980	7987	.7
63	7993	8000	8007	8014	8021	8028	8035	8041	8048	8055	.7
64	8062	8069	8075	8082	8089	8096	8102	8109	8116	8122	.7
65	8129	8136	8142	8149	8156	8162	8169	8176	8182	8189	.7
66	8195	8202	8209	8215	8222	8228	8235	8241	8248	8254	.7
67	8261	8267	8274	8280	8287	8293	8299	8306	8312	8319	.6
68	8325	8331	8338	8344	8351	8357	8363	8370	8376	8382	.6
69	8388	8395	8401	8407	8414	8420	8426	8432	8439	8445	.6
70	8451	8457	8463	8470	8476	8482	8488	8494	8500	8506	.6
71	8513	8519	8525	8531	8537	8543	8549	8555	8561	8567	.6
72	8573	8579	8585	8591	8597	8603	8609	8615	8621	8627	.6
73	8633	8639	8645	8651	8657	8663	8669	8675	8681	8686	.6
74	8692	8698	8704	8710	8716	8722	8727	8733	8739	8745	.6
75	8751	8756	8762	8768	8774	8779	8785	8791	8797	8802	.6
76	8808	8814	8820	8825	8831	8837	8842	8848	8854	8859	.6
77	8865	8871	8876	8882	8887	8893	8899	8904	8910	8915	.6
78	8921	8927	8932	8938	8943	8949	8954	8960	8965	8971	.6
79	8976	8982	8987	8993	8998	9004	9009	9015	9020	9025	.5
80	9031	9036	9042	9047	9053	9058	9063	9069	9074	9079	.5
81	9085	9090	9096	9101	9106	9112	9117	9122	9128	9133	.5
82	9138	9143	9149	9154	9159	9165	9170	9175	9180	9186	.5
83	9191	9196	9201	9206	9212	9217	9222	9227	9232	9238	.5
84	9243	9248	9253	9258	9263	9269	9274	9279	9284	9289	.5
85	9294	9299	9304	9309	9315	9320	9325	9330	9335	9340	.5
86	9345	9350	9355	9360	9365	9370	9375	9380	9385	9390	.5
87	9395	9400	9405	9410	9415	9420	9425	9430	9435	9440	.5
88	9445	9450	9455	9460	9465	9469	9474	9479	9484	9489	.5
89	9494	9499	9504	9509	9513	9518	9523	9528	9533	9538	.5
90	9542	9547	9552	9557	9562	9566	9571	9576	9581	9586	.5
91	9590	9595	9600	9605	9609	9614	9619	9624	9628	9633	.5
92	9638	9643	9647	9652	9657	9661	9666	9671	9675	9680	.5
93	9685	9689	9694	9699	9703	9708	9713	9717	9722	9727	.5
94	9731	9736	9741	9745	9750	9754	9759	9763	9768	9773	.5
95	9777	9782	9786	9791	9795	9800	9805	9809	9814	9818	.5
96	9823	9827	9832	9836	9841	9845	9850	9854	9859	9863	.5
97	9868	9872	9877	9881	9886	9890	9894	9899	9903	9908	.4
98	9912	9917	9921	9926	9930	9934	9939	9943	9948	9952	.4
99	9956	9961	9965	9969	9974	9978	9983	9987	9991	9996	.4

Table 23-3. Continued.

N	0	1	2	3	4	5	6	7	8	9
1.0	0.0 0000	0995	1980	2956	3922	4879	5827	6766	7696	8618
1.1	9531	*0436	*1333	*2222	*3103	*3976	*4842	*5700	*6551	*7395
1.2	0.1 8232	9062	9885	*0701	*1511	*2314	*3111	*3902	*4686	*5464
1.3	0.2 6236	7003	7763	8518	9267	*0010	*0748	*1481	*2208	*2930
1.4	0.3 3647	4359	5066	5767	6464	7156	7844	8526	9204	9878
1.5	0.4 0547	1211	1871	2527	3178	3825	4469	5108	5742	6373
1.6	7000	7623	8243	8858	9470	*0078	*0672	*1282	*1879	*2473
1.7	0.5 3063	3649	4232	4812	5389	5962	6531	7098	7661	8222
1.8	8779	9333	9884	*0432	*0977	*1519	*2058	*2594	*3127	*3658
1.9	0.6 4185	4710	5233	5752	6269	6783	7294	7803	8310	8813
2.0	9315	9813	*0310	*0804	*1295	*1784	*2271	*2755	*3237	*3716
2.1	0.7 4194	4669	5142	5612	6081	6547	7011	7473	7932	8390
2.2	8846	9299	9751	*0200	*0648	*1093	*1536	*1978	*2418	*2855
2.3	0.8 3291	3725	4157	4587	5015	5442	5866,	6289	6710	7129
2.4	7547	7963	8377	8789	9200	9609	*0016	*0422	*0826	*1228
2.5	0.9 1629	2028	2426	2822	3216	3609	4001	4391	4779	5166
2.6	5551	5935	6317	6698	7078	7456	7833	8208	8582	8954
2.7	9325	9695	*0063	*0430	*0796	*1160	*1523	*1885	*2245	*2604
2.8	1.0 2962	3318	3674	4028	4380	4732	5082	5431	5779	6126
2.9	6471	6815	7158	7500	7841	8181	8519	8856	9192	9527
3.0	9861	*0194	*0526	*0856	*1186	*1514	*1841	*2168	*2493	*2817
3.1	1.1 3140	3462	3783	4103	4422	4740	5057	5373	5688	6002
3.2	6315	6627	6938	7248	7557	7865	8173	8479	8784	9089
3.3	9392	9695	9996	*0297	*0597	*0896	*1194	*1491	*1788	*2083
3.4	1.2 2378	2671	2964	3256	3547	3837	4127	4415	4703	4990
3.5	5276	5562	5846	6130	6413	6695	6976	7257	7536	7815
3.6	8093	8371	8647	8923	9198	9473	9746	*0019	*0291	*0563
3.7	1.3 0833	1103	1372	1641	1909	2176	2442	2708	2972	3237
3.8	3500	3763	4025	4286	4547	4807	5067	5325	5584	5841
3.9	6098	6354	6609	6864	7118	7372	7624	7877	8128	8379
4.0	8629	8879	9128	9377	9624	9872	*0118	*0364	*0610	*0854
4.1	1.4 1099	1342	1585	1828	2070	2311	2552	2792	3031	3270
4.2	3508	3746	3984	4220	4456	4692	4927	5161	5395	5629
4.3	5862	6094	6326	6557	6787	7018	7247	7476	7705	7933
4.4	8160	8387	8614	8840	9065	9290	9515	9739	9962	*0185
4.5	1.5 0408	0630	0851	1072	1293	1513	1732	1951	2170	2388
4.6	2606	2823	3039	3256	3471	3687	3902	4116	4330	4543
4.7	4756	4969	5181	5393	5604	5814	6025	6235	6444	6653
4.8	6862	7070	7277	7485	7691	7898	8104	8309	8515	8719
4.9	8924	9127	9331	9534	9737	9939	*0141	*0342	*0543	*0744
5.0	1.6 0944	1144	1343	1542	1741	1939	2137	2334	2531	2728
5.1	2924	3120	3315	3511	3705	3900	4094	4287	4481	4673
5.2	4866	5058	5250	5441	5632	5823	6013	6203	6393	6582
5.3	6771	6959	7147	7335	7523	7710	7896	8083	8269	8455
5.4	8640	8825	9010	9194	9378	9562	9745	9928	*0111	*0293
5.5	1.7 0475	0656	0838	1019	1199	1380	1560	1740	1919	2098
5.6	2277	2455	2633	2811	2988	3166	3342	3519	3695	3871
5.7	4047	4222	4397	4572	4746	4920	5094	5267	5440	5613
5.8	5786	5958	6130	6302	6473	6644	6815	6985	7156	7326
5.9	7495	7665	7834	8002	8171	8339	8507	8675	8842	9009
6.0	1.7 9176	9342	9509	9675	9840	*0006	*0171	*0336	*0500	*0665
6.1	1.8 0829	0993	1156	1319	1482	1645	1808	1970	2132	2294
6.2	2455	2616	2777	2938	3098	3258	3418	3578	3737	3896
6.3	4055	4214	4372	4530	4688	4845	5003	5160	5317	5473
6.4	5630	5786	5942	6097	6253	6408	6563	6718	6872	7026
6.5	7180	7334	7487	7641	7794	7947	8099	8251	8403	8555
6.6	8707	8858	9010	9160	9311	9462	9612	9762	9912	*0061
6.7	1.9 0211	0360	0509	0658	0806	0954	1102	1250	1398	1545
6.8	1692	1839	1986	2132	2279	2425	2571	2716	2862	3007
6.9	3152	3297	3442	3586	3730	3874	4018	4162	4305	4448
7.0	4591	4734	4876	5019	5161	5303	5445	5586	5727	5869
7.1	6009	6150	6291	6431	6571	6711	6851	6991	7130	7269
7.2	7408	7547	7685	7824	7962	8100	8238	8376	8513	8650
7.3	8787	8924	9061	9198	9334	9470	9606	9742	9877	*0013
7.4	2.0 0148	0283	0418	0553	0687	0821	0956	1089	1223	1357
7.5	1490	1624	1757	1890	2022	2155	2287	2419	2551	2683
7.6	2815	2946	3078	3209	3340	3471	3601	3732	3862	3992
7.7	4122	4252	4381	4511	4640	4769	4898	5027	5156	5284
7.8	5412	5540	5668	5796	5924	6051	6179	6306	6433	6560
7.9	6686	6813	6939	7065	7191	7317	7443	7568	7694	7819

Table 23-4.   Natural (ln) logarithm tables.

(J. Westlake and G. Noden, *Applied Mathematics for Electronics*, © 1968. Reprinted by permission of Prentice-Hall, Inc.)

N	0	1	2	3	4	5	6	7	8	9
8.0	7944	8069	8194	8318	8443	8567	8691	8815	8939	9063
8.1	9186	9310	9433	9556	9679	9802	9924	*0047	*0169	*0291
8.2	2.1 0413	0535	0657	0779	0900	1021	1142	1263	1384	1505
8.3	1626	1746	1866	1986	2106	2226	2346	2465	2585	2704
8.4	2823	2942	3061	3180	3298	3417	3535	3653	3771	3889
8.5	4007	4124	4242	4359	4476	4593	4710	4827	4943	5060
8.6	5176	5292	5409	5524	5640	5756	5871	5987	6102	6217
8.7	6332	6447	6562	6677	6791	6905	7020	7134	7248	7361
8.8	7475	7589	7702	7816	7929	8042	8155	8267	8380	8493
8.9	8605	8717	8830	8942	9054	9165	9277	9389	9500	9611
9.0	9722	9834	9944	*0055	*0166	*0276	*0387	*0497	*0607	*0717
9.1	2.2 0827	0937	1047	1157	1266	1375	1485	1594	1703	1812
9.2	1920	2029	2138	2246	2354	2462	2570	2678	2786	2894
9.3	3001	3109	3216	3324	3431	3538	3645	3751	3858	3965
9.4	4071	4177	4284	4390	4496	4601	4707	4813	4918	5024
9.5	5129	5234	5339	5444	5549	5654	5759	5863	5968	6072
9.6	6176	6280	6384	6488	6592	6696	6799	6903	7006	7109
9.7	7213	7316	7419	7521	7624	7727	7829	7932	8034	8136
9.8	8238	8340	8442	8544	8646	8747	8849	8950	9051	9152
9.9	9253	9354	9455	9556	9657	9757	9858	9958	*0058	*0158
10.0	2.3 0259	0358	0458	0558	0658	0757	0857	0956	1055	1154
N	0	1	2	3	4	5	6	7	8	9

Table 23-4.   Continued.

Conversion Between LOG and LN

$$10 = \epsilon^{2.3026}; \quad \ln 10 = 2.3026$$

$$\therefore \log x = \ln 2.3026x$$

$$\epsilon = 10^{0.4343}; \quad \log \epsilon = 0.4343$$

$$\therefore \ln x = \log 0.4343x$$

Properties of ϵ

$$\epsilon = 1 + \frac{1}{1!} + \frac{1}{2!} + \frac{1}{3!} + \ldots = 2.71828$$

$$\frac{1}{\epsilon} = 0.36787$$

$$\epsilon^{+jx} = \cos x + j \sin x$$

$$\epsilon^{-jx} = \cos x - j \sin x$$

$$\epsilon^x = 1 + x + \frac{x^2}{2!} + \frac{x^3}{3!} + \ldots, \left| x \right| < \infty$$

23.2 FUNDAMENTALS OF TRIGONOMETRY

Right Triangle Properties

The trigonometric functions are defined, in terms of the sides of a right triangle, by

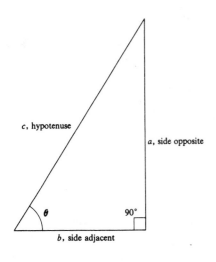

c, hypotenuse

a, side opposite

90°

θ

b, side adjacent

$$\text{sine } \theta = \sin \theta = \frac{\text{side opposite}}{\text{hypotenuse}} = \frac{a}{c}$$

$$\text{cosine } \theta = \cos \theta = \frac{\text{side adjacent}}{\text{hypotenuse}} = \frac{b}{c}$$

$$\text{tangent } \theta = \tan \theta = \frac{\text{side opposite}}{\text{side adjacent}} = \frac{a}{b}$$

The inverses of these functions are defined as

$$1/\sin \theta = \text{cosecant } \theta = \csc \theta;$$

$$1/\cos \theta = \text{secant } \theta = \sec \theta$$

$$1/\tan \theta = \text{cotangent } \theta = \cot \theta$$

If two sides of a triangle (or one side and the angle) are known, the above functions allow the determination of the other side and the angle (or the other two sides). It follows straightaway that

$$a = c \sin \theta \qquad \text{Area} = \frac{ab}{2}$$

$$b = c \cos \theta$$

Since

$$a^2 + b^2 = c^2 \text{ (Pythagorean theorem)}$$

we have

$$\sin^2 \theta + \cos^2 \theta = 1$$

$$\sin^2 A = \frac{1 - \cos 2A}{2}$$

(Lane K. Branson, *Introduction to Electronics*, © 1967. Reprinted by permission of Prentice-Hall, Inc.)

Angle	Radians	Sine	Cosine	Tangent	Angle	Radians	Sine	Cosine	Tangent
0°	.0000	.0000	1.0000	.0000	45°	.7854	.7071	.7071	1.0000
1	.0175	.0175	.9998	.0175	46	.8029	.7193	.6947	1.0355
2	.0349	.0349	.9994	.0349	47	.8203	.7314	.6820	1.0724
3	.0524	.0523	.9986	.0524	48	.8378	.7431	.6691	1.1106
4	.0698	.0698	.9976	.0699	49	.8552	.7547	.6561	1.1504
5	.0873	.0872	.9962	.0875	50	.8727	.7660	.6428	1.1918
6	.1047	.1045	.9945	.1051	51	.8901	.7771	.6293	1.2349
7	.1222	.1219	.9925	.1228	52	.9076	.7880	.6157	1.2799
8	.1396	.1392	.9903	.1405	53	.9250	.7986	.6018	1.3270
9	.1571	.1564	.9877	.1584	54	.9425	.8090	.5878	1.3764
10	.1745	.1736	.9848	.1763	55	.9599	.8192	.5736	1.4281
11	.1920	.1908	.9816	.1944	56	.9774	.8290	.5592	1.4826
12	.2094	.2079	.9781	.2126	57	.9948	.8387	.5446	1.5399
13	.2269	.2250	.9744	.2309	58	1.0123	.8480	.5299	1.6003
14	.2443	.2419	.9703	.2493	59	1.0297	.8572	.5150	1.6643
15	.2618	.2588	.9659	.2679	60	1.0472	.8660	.5000	1.7321
16	.2793	.2756	.9613	.2867	61	1.0647	.8746	.4848	1.8040
17	.2967	.2924	.9563	.3057	62	1.0821	.8829	.4695	1.8807
18	.3142	.3090	.9511	.3249	63	1.0996	.8910	.4540	1.9626
19	.3316	.3256	.9455	.3443	64	1.1170	.8988	.4384	2.0503
20	.3491	.3420	.9397	.3640	65	1.1345	.9063	.4226	2.1445
21	.3665	.3584	.9336	.3839	66	1.1519	.9135	.4067	2.2460
22	.3840	.3746	.9272	.4040	67	1.1694	.9205	.3907	2.3559
23	.4014	.3907	.9205	.4245	68	1.1868	.9272	.3746	2.4751
24	.4189	.4067	.9135	.4452	69	1.2043	.9336	.3584	2.6051
25	.4363	.4226	.9063	.4663	70	1.2217	.9397	.3420	2.7475
26	.4538	.4384	.8988	.4877	71	1.2392	.9455	.3256	2.9042
27	.4712	.4540	.8910	.5095	72	1.2566	.9511	.3090	3.0777
28	.4887	.4695	.8829	.5317	73	1.2741	.9563	.2924	3.2709
29	.5061	.4848	.8746	.5543	74	1.2915	.9613	.2756	3.4874
30	.5236	.5000	.8660	.5774	75	1.3090	.9659	.2588	3.7321
31	.5411	.5150	.8572	.6009	76	1.3265	.9703	.2419	4.0108
32	.5585	.5299	.8480	.6249	77	1.3439	.9744	.2250	4.3315
33	.5760	.5446	.8387	.6494	78	1.3614	.9781	.2079	4.7046
34	.5934	.5592	.8290	.6745	79	1.3788	.9816	.1908	5.1446
35	.6109	.5736	.8192	.7002	80	1.3963	.9848	.1736	5.6713
36	.6283	.5878	.8090	.7265	81	1.4137	.9877	.1564	6.3138
37	.6458	.6018	.7986	.7536	82	1.4312	.9903	.1392	7.1154
38	.6632	.6157	.7880	.7813	83	1.4486	.9925	.1219	8.1443
39	.6807	.6293	.7771	.8098	84	1.4661	.9945	.1045	9.5144
40	.6981	.6428	.7660	.8391	85	1.4835	.9962	.0872	11.43
41	.7156	.6561	.7547	.8693	86	1.5010	.9976	.0698	14.30
42	.7330	.6691	.7431	.9004	87	1.5184	.9986	.0523	19.08
43	.7505	.6820	.7314	.9325	88	1.5359	.9994	.0349	28.64
44	.7679	.6947	.7193	.9657	89	1.5533	.9998	.0175	57.29

Table 23-5. Trigonometric tables.

(Harry E. Thomas, *Handbook for Electronic Engineers and Technicians*, © 1965. Reprinted by permission of Prentice-Hall, Inc.)

$$\cos^2 A = \frac{1 + \cos 2A}{2}$$

$$\sin A \sin B = \tfrac{1}{2}\left[\cos(A - B) - \cos(A + B)\right]$$

$$\cos A \cos B = \tfrac{1}{2}\left[\cos(A + B) + \cos(A - B)\right]$$

$$\sin A \cos B = \tfrac{1}{2}\left[\sin(A + B) + \sin(A - B)\right]$$

$$\sin A = A - \frac{A^3}{3!} + \frac{A^5}{5!} - \frac{A^7}{7!} + \ldots$$

$$\cos A = 1 - \frac{A^2}{2!} + \frac{A^4}{4!} - \frac{A^6}{6!} + \ldots$$

Oblique Triangles

Angles A, B, C $A + B + C = 180°$

sizes a, b, c

Height $= h = b \sin A = a \sin B$

Sine Law: $\dfrac{a}{\sin A} = \dfrac{b}{\sin B} = \dfrac{c}{\sin C}$

Cosine Law: $a^2 = b^2 + c^2 - 2bc \cos A$

$$b^2 = a^2 + c^2 - 2ac \cos B$$

$$c^2 = a^2 + b^2 - 2ab \cos C$$

Area $= \tfrac{1}{2} bc \sin A = \tfrac{1}{2} ac \sin B = \tfrac{1}{2} ab \sin C$

Spherical Trigonometry

Right spherical triangle

Curved sides: a, b, c

Angles: A, B $C = 90°$

$$\sin a = \sin c \sin A = \tan b \cot B$$

$$\sin b = \sin c \sin B = \tan a \cot A$$

$$\cos c = \cos a \cos B = \cot A \cot B$$

$$\cos A = \sin B \cos a = \tan b \cot c$$

$$\cos B = \sin A \cos b = \tan a \cot c$$

Oblique spherical triangle

Sine Law: $\dfrac{\sin a}{\sin A} = \dfrac{\sin b}{\sin B} = \dfrac{\sin c}{\sin C}$

Cosine Law: $\cos a = \cos b \cos c + \sin b \sin c \cos A$

$$\cos b = \cos c \cos a + \sin c \sin a \cos B$$

$$\cos c = \cos a \cos b + \sin a \sin b \cos C$$

$$\cos A = -\cos B \cos C + \sin B \sin C \cos a$$

$$\cos B = -\cos A \cos C + \sin A \sin C \cos b$$

$$\cos C = -\cos A \cos B + \sin A \sin B \cos c$$

Area: Sphere radius = 1

$$\text{Area} = A + B + C - \pi$$

Hyperbolic Trigonometry

Applies to triangles drawn in non-Euclidean space; for example, reflection charts used in transmission line and microwave work. Hyperbolic sine, cosine, etc. = sinh, cosh, etc.

Right hyperbolic triangle:

Curved Sides: a, b, c

Angles: A, B, C, where $C = 90°$

$$\cosh C = \cosh a \cosh b = \cot A \cot B$$

$$\cos A = \sin B \cosh a = \tanh b \cosh c$$

$$\cos B = \sin A \cosh b = \tanh a \cosh c$$

Oblique hyperbolic triangle:

Sine Law: $\dfrac{\sinh a}{\sin A} = \dfrac{\sinh b}{\sin B} = \dfrac{\sinh c}{\sin C}$

Cosine Law: $\cosh a = \cosh b \cosh c - \sinh b \sinh c$

$$\cos A + \text{permutations}$$

$$\cos A = -\cos B \cos C +$$

$$\sin B \sin C \cosh a + \text{permutations}$$

$$\text{Area} = \pi - (A + B + C)$$

Hyperbolic Functions

Hyperbolic functions are defined by:

$$\sinh x = \frac{\epsilon^x - \epsilon^{-x}}{2}; \quad \cosh x = \frac{\epsilon^x + \epsilon^{-x}}{2}$$

Hyperbolic sine (sinh) and cosine (cosh) combine trigonometric and exponential algebraic functions.

$$\cosh^2 x - \sinh^2 x = 1$$

$$\tanh x = \frac{\sinh x}{\cosh x} = \frac{1}{\coth x}$$

$$\coth x = \frac{\cosh x}{\sinh x} = \frac{1}{\tanh x}$$

$$\sinh (-x) = -\sinh (x)$$

$$\cosh (-x) = \cosh x$$

$$\sinh (jx) = j \sin x$$

$$\cosh (jx) = \cos x$$

$$\sinh 2x = 2 \sinh x \cosh x$$

$$\cosh 2x = \cosh^2 x \times \sinh^2 x$$

$$\sinh (x \pm jy) = \sinh x \cosh y \pm j \cosh x \sinh y$$

$$\cosh (x \pm jy) = \cosh x \cosh x \pm j \sinh x \sinh y$$

$$\text{If } x = \ln \left[\tan \left(\frac{\pi}{4} + \frac{y}{2} \right) \right] = \text{gudermannian of } x$$

$$\text{then: } \sinh x = \tan y$$

$$\cosh x = \sec y$$

$$\tanh x = \sin y$$

23.3 FUNDAMENTALS OF GEOMETRY

Plane Analytic Geometry

Rectangular coordinates: x = horizontal axis

y = vertical axis

Straight line: $ax + by = c$

for $a = 0$ line crosses y at $\dfrac{c}{b}$

for $b = 0$ line crosses x at $\dfrac{c}{a}$

Intercept of 2 lines occurs when both x and both y values are equal

Slope of a line $s = \tan \theta = \dfrac{\sin \theta}{\cos \theta} = \dfrac{y}{x}$

Distance from point (x_1, y_1) to a line

$$= \frac{a}{\pm \sqrt{a^2 + b^2}} x_1 + \frac{b}{\pm \sqrt{a^2 + b^2}} y^2 + \frac{c}{\pm \sqrt{a^2 + b^2}}$$

Circle: $x^2 + y^2 = r^2$, $r =$ radius, center at origin

for center at m_1, n_1: $(x - m)^2 + (y - n)^2 = r^2$

tangent line at x_1, y_1: $y - y_1 = -\dfrac{x_1 - m}{y_1 - n} (x - x_1)$

normal line at x_1, y_1: $y - y_1 = \dfrac{y_1 - n}{x_1 - m} (x - x_1)$

Parabola: x_1, y_1 are coordinates of vertex
   $r =$ distance to focal point

open at right: $(y - y_1)^2 = 2r(x - x_1)$

open at left: $(y - y_1)^2 = -2r(x - x_1)$

open above: $(x - x_1)^2 = 2r(y - y_1)$

open below: $(x - x_1)^2 = -2r(y - y_1)$

Hyperbola: Foci at F_1, F_2, directrices D_1, D_2 eccentricity $e > 1$

x-hyperbola:
(centered at origin) $\dfrac{x^2}{a^2} - \dfrac{y^2}{b^2} = 1$; $b^2 = a^2(e^2 - 1)$

y-hyperbola:
(centered at origin) $\dfrac{y^2}{b^2} - \dfrac{x^2}{a^2} = 1$

Asymptotes: $y = \pm\dfrac{b}{a} x$

Ellipse: Foci at F_1, F_2, directrices D_1, D_2 eccentricity $e > 1$;

$$1 - e^2 = \frac{b^2}{a^2}$$

Major axis $= 2a$

Minor axis $= 2b$

Focal distance $= 2c$

$$\frac{x^2}{a^2} + \frac{y^2}{b^2} = 1; \text{ centered at origin}$$

Solid Analytic Geometry

Rectangular coordinates $-x =$ horizontal axis

$y =$ horizontal axis

$z =$ vertical axis

$x - y$ plane $=$ horizontal

$x - z$ plane $=$ vertical

$y - z$ plane $=$ vertical

Distance between points x_1, y_1, z_1, and x_2, y_2, z_2

$$d = \sqrt{(x_1 - x_2)^2 + (y_1 - y_2)^2 + (z_1 - z_2)^2}$$

Plane: $\quad \frac{x}{a} + \frac{y}{b} + \frac{z}{c} = 1$; a, b, c are x, y, z axes intercepts

Prolate spheroid: $\quad a^2(y^2 + z^2) + b^2 x^2 = a^2 b^2$

where $a > b$, revolution about x-axis

Oblate spheroid: $\quad b^2(x^2 + z^2) + a^2 y^2 = a^2 b^2$

where $a > b$, revolution about y-axis

Paraboloid: $\quad x^2 + z^2 = 2rx$, revolution about x-axis

Hyperboloid: $\quad a^2(y^2 + z^2) - b^2 x^2 = -a^2 b^2$, about x-axis

$\quad b^2(x^2 + z^2) - a^2 y^2 = a^2 b^2$, about y-axis

Ellipsoid: $\quad \frac{x^2}{a^2} + \frac{y^2}{b^2} + \frac{z^2}{c^2} = 1$

where a, b, c are semi-axes, or intercepts on the x, y, z axes.

Mensuration Formulas

Triangle: Angles $A + B + C = 180°$; Area = ½ base × height

Rectangle: Circumference = 2(height + width)

Area = height × width

Parallelogram: Circumference = 2(side a + side b)

Area = base × height

Trapezoid: Circumference = Sum of base + sides + top

Area = ½ height (base + top)

Circle: Circumference = 2π radius

Area = $\pi(\text{radius})^2$

Parabola: Area = ⅔ base × height

Ellipse: Area = π(½ major axis × minor axis)

Sphere: Surface = $4\pi(\text{radius})^2$

Volume = $\tfrac{4}{3}\pi(\text{radius})^3$

Cube: Surface = $6(\text{side})^2$

Volume = $(\text{side})^3$

Block: Dimensions a, b, c,

Surface = $2ab + 2ac + 2bc$

Volume = abc

Cylinder: Surface = 2π radius(radius + height)

Volume = $\pi(\text{radius})^2$ height

Cone: Surface = π radius $\sqrt{\text{radius}^2 + \text{height}^2} + \pi$ radius^2

Pyramid: Volume = ⅓ (base area) · height

Ellipsoid: Volume = 0.5231 (major axis)(minor axis)^2

Paraboloid: Volume = 1.5707 (base radius)^2 · height

23.4 FUNDAMENTALS OF VECTOR ANALYSIS

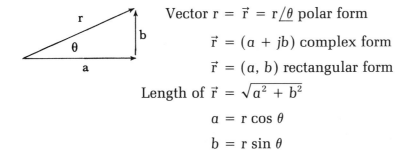

Vector $r = \vec{r} = r\underline{/\theta}$ polar form

$\vec{r} = (a + jb)$ complex form

$\vec{r} = (a, b)$ rectangular form

Length of $\vec{r} = \sqrt{a^2 + b^2}$

$a = r \cos \theta$

$b = r \sin \theta$

$$\tan \theta = \frac{b}{a}$$

Coordinate Transformation

Polar $\leftrightarrow$ rectangular: $r\underline{/\theta} = \underline{\sqrt{a^2 + b^2} \quad \tan^{-1}\left(\dfrac{b}{a}\right)}$

Polar $\leftrightarrow$ complex: $r\underline{/\theta} = a + jb = r \cos \theta + jr \sin \theta$

Example: $15\underline{/220°} = 15 \cos 220° + 15j \sin 220°$

$\cos 220° = -\cos 40°$; $\sin 220° = -\sin 40°$

$15\underline{/200°} = -15 \cos 40° - 15j \sin 40°$

$\cos 40° = 0.6428$; $\sin 40° = 0.7660$

$a = -11.5$; $b = -9.63$

$15\underline{/220°} = -11.5 - j\,9.63$

Answer: $15\underline{/220°} = (-11.5, -9.63) = -11.5 - j\,9.63$

Example: $-3 + j^2 = \underline{\sqrt{-3^2 + 2^2}\Big/\tan^{-1}\left(\dfrac{2}{-3}\right)}$

$-3 + j^2 = 3.6\underline{/\tan^{-1}(-0.6667)}$

$\tan^{-1}(-0.6667) = 33.7°$; $180° - 33.7° = 146.3°$

$-3 + j^2$ is in second quadrant

Answer: $-3 + j^2 = 3.6\underline{/146.3°}$

Vector Algebra

For addition and subtraction, vectors must be in complex form.

Addition: $(a + jb) + (c + jd) = (a + c) + j(b + d)$

Subtraction: $(a + jb) - (c + jd) = (a - c) + j(b - d)$

Multiplication: $(a + jb)(c + jd) = ac + j(bc + ad) - bd$

$(j^2 = -1)$

$$(r_1 \underline{/\theta_1})(r_2 \underline{/\theta_2}) = r_1 r_2 \underline{/(\theta_1 + \theta_2)}$$

Division: $$\frac{(a + jb)}{(c + jd)} = \frac{(a + jb)}{(c + jd)} \cdot \frac{(c - jd)}{(c - jd)}$$

$$= \frac{ac - j(bc + ad) + bd}{c^2 + d^2}$$

$$\frac{r_1 \underline{/\theta_1}}{r_2 \underline{/\theta_2}} = \frac{r_1}{r_2} \underline{/(\theta_1 - \theta_2)}$$

∇ (DEL) Operator

In rectangular coordinates, the differential vector operator ∇ is:

$$\nabla = \frac{\partial}{\partial x} \vec{x} + \frac{\partial}{\partial y} \vec{y} + \frac{\partial}{\partial z} \vec{z}$$

where $\vec{x}$, $\vec{y}$, $\vec{z}$ are the vector components in the three-dimensional co-ordinate system x, y, z.

If voltage V is a scalar (not a vector), then;

$$\nabla \cdot V = \frac{\partial v}{\partial x} \vec{x} + \frac{\partial v}{\partial y} \vec{y} + \frac{\partial v}{\partial z} \vec{z} \text{ and}$$

$\nabla \cdot V = $ gradient of V

If A is a vector $\vec{A}$, then;

$$\nabla \cdot \vec{A} = \frac{\partial A_x}{\partial x} + \frac{\partial A_y}{\partial y} + \frac{\partial A_z}{\partial z} \text{ and}$$

$\nabla \cdot \vec{A} = $ divergence of $\vec{A}$

$$\nabla_x \vec{A} = \left(\frac{\partial A_z}{\partial y} - \frac{\partial A_y}{\partial z}\right) \vec{x} + \left(\frac{\partial A_x}{\partial z} - \frac{\partial A_z}{\partial x}\right) \vec{y} + \left(\frac{\partial A_y}{\partial x} - \frac{\partial A_x}{\partial y}\right) \vec{Z}$$

or in matrix form:

$$\nabla \mathrm{x} A = \begin{bmatrix} \dfrac{\partial}{\partial x} & \dfrac{\partial}{\partial y} & \dfrac{\partial}{\partial z} \\[2mm] A_x & A_y & A_z \\[2mm] \vec{x} & \vec{y} & \vec{z} \end{bmatrix}$$

$$\nabla_x \vec{A} = \text{curl } A$$

Differential Vector Identities

$$\nabla^2 = \frac{\partial^2}{\partial_x^{\,2}} + \frac{\partial^2}{\partial_y^{\,2}} + \frac{\partial^2}{\partial_z^{\,2}}$$

div. curl $\vec{A} = \nabla \cdot (\nabla \times \vec{A}) = 0$

curl grad. $V = \nabla \mathrm{x}(\nabla \cdot V) = 0$

div. grad. $V = \nabla \cdot (\nabla \cdot V) = \nabla^2 V$

$$\nabla \mathrm{x} \nabla \mathrm{x} \vec{A} = \nabla(\nabla \cdot \vec{A}) - \nabla^2 \vec{A}$$

$$\nabla \cdot \vec{A}\mathrm{x}\vec{B} = \vec{B} \cdot \nabla \mathrm{x}\vec{A} - \vec{A} \cdot \nabla \mathrm{x}\vec{B}$$

$$\nabla(ab) = a\nabla b + b\nabla a$$

$$\nabla \cdot (a\vec{A}) = \vec{A} \cdot \nabla a + a\nabla \vec{A}$$

$$\nabla \mathrm{x}(a\vec{A}) = \nabla a \mathrm{x}\vec{A} + a\nabla \mathrm{x}\vec{A}$$

$$\nabla(\vec{A} \cdot \vec{B}) = (\vec{A} \cdot \nabla)\vec{B} + (\vec{B} \cdot \nabla)\vec{A} + \vec{A}\mathrm{x}(\nabla \mathrm{x}\vec{B}) + \vec{B}\mathrm{x}(\nabla \mathrm{x}\vec{A})$$

$$\nabla \mathrm{x}(\vec{A}\mathrm{x}\vec{B}) = \vec{A}\nabla \cdot \vec{B} - \vec{B}\nabla \cdot \vec{A} + (\vec{B} \cdot \nabla)\vec{A} - (\vec{A} \cdot \nabla)\vec{B}$$

23.5 FUNDAMENTALS OF MATRIX ALGEBRA

Matrix Notation and Properties

$$[A] = \begin{bmatrix} a_{11} & a_{12} & a_{13} & \ldots & a_{1n} \\ a_{21} & a_{22} & a_{23} & & \\ a_{31} & a_{32} & a_{33} & & \\ \vdots & & & & \\ a_{m1} & a_{m2} & a_{m3} & \ldots & a_{mn} \end{bmatrix} \begin{array}{l} \text{m rows} \\ \text{n columns} \end{array}$$

$$i = 1, 2, 3, \ldots, m$$

$$[A] = [a_{ij}]$$

$$j = 1, 2, 3, \ldots n$$

column matrix:
(order [3, 1]) $\quad [a_{ij}] \begin{array}{l} i = 1, 2, 3 \\ j = 1 \end{array} = \begin{vmatrix} a_{11} \\ a_{21} \\ a_{31} \end{vmatrix}$

row matrix:
(order [1, 3]) $\quad [a_{ij}] \begin{array}{l} i = 1 \\ j = 1, 2, 3 \end{array} = \begin{bmatrix} a_{11} & a_{12} & a_{13} \end{bmatrix}$

square matrix:
(order [2, 2]) $\quad [a_{ij}] \begin{array}{l} i = 1, 2 \\ j = 1, 2 \end{array} = \begin{bmatrix} a_{11} & a_{12} \\ a_{21} & a_{22} \end{bmatrix}$

zero matrix: $[0] = \begin{bmatrix} 0 & 0 & \ldots & 0 \\ 0 & 0 & \ldots & 0 \\ \vdots & & & \\ 0 & & \ldots & 0 \end{bmatrix}$

unit matrix: $[U] =$
(always square) $\begin{bmatrix} 1 & 0 & 0 & \ldots & 0 \\ 0 & 1 & 0 & & \vdots \\ 0 & 0 & 1 & & \vdots \\ \vdots & & & & 0 \\ 0 & \ldots & & 0 & 1 \end{bmatrix}$

Two matrices are equal only if they are of the same order and if all corresponding elements are equal.

Matrix Arithmetic and Transformations

Addition: Matrices must be of the same order

$$[A] + [B] = [a_{ij}]_n^m + [b_{ij}]_n^m = \begin{bmatrix} (a_{11} + b_{11})(a_{12} + b_{12}) & \ldots & (a_{1n} + b_{1n}) \\ (a_{21} + b_{21})(a_{22} + b_{22}) & & \vdots \\ \vdots & & \vdots \\ (a_{m1} + b_{m1}) & & (a_{mn} + b_{mn}) \end{bmatrix}$$

Subtraction: Matrices must be of the same order

$$[A] - [B] = [a_{ij}]_n^m - [b_{ij}]_n^m = \begin{bmatrix} (a_{11} - b_{11})(a_{12} - b_{12}) & \cdots & (a_{1n} + \mathbf{b}_{1n}) \\ (a_{21} - b_{21})(a_{22} - b_{22}) & & \vdots \\ \vdots & & \vdots \\ (a_{m1} - b_{m1}) & & (a_{mn} - b_{mn}) \end{bmatrix}$$

Multiplication:

by a scalar V: $V[A] = [Va_{ij}]_n^m = \begin{bmatrix} Va_{11} & Va_{12} & \cdots & Va_{1n} \\ Va_{21} & Va_{22} & \cdots & \\ \vdots & & & \\ Va_{m1} & \cdots & \cdots & Va_{mn} \end{bmatrix}$

of 2 matrices requires that one has as many rows as the other has columns.

$$[a_{ij}] \times [b_{jk}] = \begin{bmatrix} a_{11} & a_{12} \\ a_{21} & a_{22} \end{bmatrix} \times \begin{bmatrix} b_{11} \\ b_{21} \end{bmatrix} = \begin{bmatrix} a_{11}b_{11} + a_{12}b_{21} \\ a_{21}b_{11} + a_{22}b_{21} \end{bmatrix}$$

$$m \quad \boxed{[A]}^{\,k} \quad \times \quad \boxed{[B]}_{\,k}^{\,n} \quad = \quad \boxed{[C]}^{\,n} \quad m$$

Note: Matrix multiplication is not commutative

$$[A] \times [B] \neq [B] \times [A]$$

however, $[A] \times [U] = [U] \times [A] = [A]$

Division: multiply by inverse matrix

$$\frac{[A]}{[B]} = [A] \times [B]^{-1}$$

Inverse matrix $[A]^{-1} = \dfrac{\text{adj}\,[A]}{\det\,[A]}$

(reciprocal)

Determinant: $\det\,[A] = |A|$; for $[A] = \begin{bmatrix} a_{11} & a_{12} \\ a_{21} & a_{22} \end{bmatrix}$

$$\det\,[A] = a_{11}a_{22} - a_{12}a_{21}$$

For larger matrices, the use of cofactors is required as illustrated below

$$\det \begin{bmatrix} a_{11} & a_{12} & a_{13} \\ a_{21} & a_{22} & a_{23} \\ a_{31} & a_{32} & a_{33} \end{bmatrix} = a_{11} \underbrace{\begin{bmatrix} a_{22} & a_{23} \\ a_{32} & a_{33} \end{bmatrix}}_{\substack{\text{cofactor} \\ \text{of } a_{11}}} - a_{12} \underbrace{\begin{bmatrix} a_{21} & a_{23} \\ a_{31} & a_{33} \end{bmatrix}}_{\substack{\text{cofactor} \\ \text{of } a_{12}}} + a_{13} \underbrace{\begin{bmatrix} a_{21} & a_{22} \\ a_{31} & a_{32} \end{bmatrix}}_{\substack{\text{cofactor} \\ \text{of } a_{13}}}$$

Cofactor of a_{ij} has the sign $(-1)^{i+j}$.

Cofactor matrices can be reduced by determinants as shown above for 4-element matrix.

Adjoint: Transposed matrix with cofactors as illustrated below

$$[A] = \begin{bmatrix} a_{11} & a_{12} & a_{13} \\ a_{21} & a_{22} & a_{23} \\ a_{31} & a_{32} & a_{33} \end{bmatrix} ; \quad [A]^{T} = [A]' = \begin{bmatrix} a_{11} & a_{21} & a_{31} \\ a_{12} & a_{22} & a_{32} \\ a_{13} & a_{23} & a_{33} \end{bmatrix}$$

$$A_{11} = \text{cofactor of } a_{11} = + \begin{bmatrix} a_{22} & a_{32} \\ a_{23} & a_{33} \end{bmatrix}$$

$$A_{12} = \text{cofactor of } a_{12} = - \begin{bmatrix} a_{21} & a_{31} \\ a_{23} & a_{33} \end{bmatrix}$$

etc.

$$\text{adj} [A] = [|A|] = \begin{bmatrix} A_{11} & A_{21} & A_{31} \\ A_{12} & A_{22} & A_{32} \\ A_{13} & A_{23} & A_{33} \end{bmatrix}$$

Equivalence transformation:

$$[B] = [P][A][Q]$$

$$\begin{bmatrix} b_{11} & 0 & 0 & \ldots & 0 \\ 0 & b_{22} & 0 & & \\ 0 & 0 & b_{33} & & \\ \vdots & & & \ddots & \\ 0 & 0 & 0 & \ldots & b_{mn} \end{bmatrix} = [P]_{m}^{m} \begin{bmatrix} a_{11} & a_{12} & \ldots & a_{ln} \\ a_{21} & a_{22} & & \cdot \\ \cdot & & & \cdot \\ \cdot & & & \cdot \\ a_{ml} & \ldots & \ldots & a_{mn} \end{bmatrix} [Q]_{h}^{n}$$

Eigenvalues and Vectors (Similarity Transformation)

$$[Q]^{-1}[A][Q] = A; \ [A] \text{ is a square matrix; det } [A] \rightarrow 0$$

$$[Q] = \begin{bmatrix} x_{11} & x_{12} & \cdots & x_{lm} \\ x_{21} & x_{22} & & \\ \vdots & & & \\ x_{ml} & \cdots & \cdots & x_{mn} \end{bmatrix} ; \ x_{ij} = \text{eigenvector}$$

Basic equation: $[A][X] = \lambda[X]$

$$([A] - \lambda[U])[X] = 0$$

for $[A] = 2\text{x}2$:

$$\begin{bmatrix} a_{11} - \lambda_1 & a_{12} \\ a_{21} & a_{22} - \lambda_i \end{bmatrix} \begin{bmatrix} x_{1i} \\ x_{2i} \end{bmatrix} = 0 = \begin{bmatrix} \lambda_i x_{1i} \\ \lambda_i x_{2i} \end{bmatrix}$$

λ_1 = eigenvalues or characteristic roots of $[A]$, scalars

$$\begin{bmatrix} x_{11} & x_{12} \\ x_{21} & x_{22} \end{bmatrix}^{-1} \begin{bmatrix} a_{11} & a_{12} \\ a_{21} & a_{22} \end{bmatrix} \begin{bmatrix} x_{11} & x_{12} \\ x_{21} & x_{22} \end{bmatrix} = \begin{bmatrix} \lambda_1 & 0 \\ 0 & \lambda_2 \end{bmatrix}$$

23.6 FUNDAMENTALS OF CALCULUS

Differential Calculus

Limits:

$$\lim_{x \to a} x = a; \quad \lim_{x \to \infty} a^x = 0, \quad \text{if } 0 < a < 1$$

$$\lim_{x \to 0} (x + x^n) = \lim_{x \to 0} x + \lim_{x \to 0} x^n$$

$$\lim_{x \to 1} (a^x \cdot b^x) = (\lim_{x \to 1} a^x)(\lim_{x \to 1} b^x)$$

Derivatives:

$$\frac{df(x)}{dx} = \lim_{\Delta x \to 0} \frac{f(x + \Delta x) - f(x)}{\Delta x}; \quad \frac{d(x)}{dx} = 1$$

Examples: $y = x^2$

$$\frac{dy}{dx} = \lim_{\Delta x \to 0} \frac{(x + \Delta x)^2 - x^2}{\Delta x} = 2x$$

$$y = x^3 + \frac{1}{x}$$

$$\frac{dy}{dx} = 3x^2 - \frac{1}{x^2}$$

General: $\dfrac{d}{dx}(cx^n) = cnx^{n-1}$

$$\frac{d}{dx}(u + v + w) = \frac{du}{dx} + \frac{dv}{dx} + \frac{dw}{dx}$$

$$\frac{d(uv)}{dx} = u\frac{dv}{dx} + v\frac{du}{dx}$$

$$\frac{d}{dx}\left(\frac{u}{v}\right) = \frac{v\dfrac{du}{dx} - u\dfrac{dv}{dx}}{v^2}$$

$$\frac{d}{dx}\sin u = \cos u \frac{du}{dx}$$

$$\frac{d}{dx}\cos u = -\sin u \frac{du}{dx}$$

$$\frac{d}{dx}\tan u = \sec^2 u \frac{du}{dx}$$

$$\frac{d}{dx}\cot u = -\csc^2 u \frac{du}{dx}$$

$$\frac{d}{dx}\sin^{-1} u = \frac{1}{\sqrt{1 - u^2}}\frac{du}{dx}$$

$$\frac{d}{dx}\cos^{-1} u = \frac{-1}{\sqrt{1 - u^2}}\frac{du}{dx}$$

$$\frac{d}{dx}\tan^{-1} u = \frac{1}{1 + u^2}\frac{du}{dx}$$

$$\frac{d}{dx}\cot^{-1} u = \frac{-1}{1 + u^2}\frac{du}{dx}$$

$$\frac{d}{dx} \log u = \frac{1}{u} \frac{du}{dx}$$

$$\frac{d}{dx} c^u = c^u \ln c \frac{du}{dx}$$

$$\frac{d}{dx} \epsilon^u = \epsilon^u \frac{du}{dx}$$

$$\frac{d}{dx} v^u = uv^{u-1} \frac{dv}{dx} + (\ln v)v^u \frac{du}{dx}$$

Partial derivatives:

$$\frac{\partial}{\partial x} f(x, y, z) = \lim_{x \to 0} \frac{f(x + \Delta x, y, z) - f(x, y, z)}{\Delta x}$$

Example: $$f = x^3 + x^2 y + y^3$$

$$\frac{\partial f}{\partial x} = 3x^2 + 2xy; \quad y = \text{constant}$$

$$\frac{\partial f}{\partial y} = x^2 + 3y^2; \quad x = \text{constant}$$

Integral Calculus

$$\frac{dy}{dx} = f(x) \text{ and } y = \int f(x) \, dx$$

$$\frac{d}{dx} \left[\int f(x) \, dx \right] = f(x)$$

$$\int ax^n \, dx = \frac{ax^{n+1}}{n+1} + c$$

Examples: $$\int 6x^4 \, dx = \frac{6}{5} x^5 + c$$

$$\int 2\sqrt{x} \, dx = \frac{4}{3} x\sqrt{x} + c$$

$$\int \frac{3}{x^2} \, dx = -\frac{3}{x} + c$$

General: $$\int (u + v + w) \, dx = \int u \, dx + \int v \, dx + \int w \, dx$$

$$\int dx = x + c$$

$$\int \frac{dx}{x} = \log x + c$$

$$\int \epsilon^x \, dx = \epsilon^x + c$$

$$\int a^x \, dx = \frac{a^x}{\log a} + c$$

$$\int \cos x \, dx = \sin x + c$$

$$\int \sin x \, dx = -\cos x + c$$

$$\int \tan x \, dx = -\log \cos x + c$$

$$\int \cot x \, dx = \log \sin x + c$$

$$\int u dv = uv - \int v \, du$$

$$\int \log ax \, dx = x \log ax - x + c$$

$$\int (ax + b)^n \, dx = \frac{(ax + b)^{n+1}}{a(n + 1)} + c$$

$$\int \frac{dx}{ax + b} = \frac{1}{a} \log (ax + b) + c$$

$$\int \sin ax \, dx = -\frac{1}{a} \cos ax + c$$

$$\int \cos ax \, dx = \frac{1}{a} \sin ax + c$$

$$\int \tan ax \, dx = -\frac{1}{a} \log \cos ax + c$$

$$\int \cot ax \, dx = \frac{1}{a} \log \sin ax + c$$

$$\int \sin^2 ax \, dx = \frac{x}{2} - \frac{\sin 2ax}{4a} + c$$

$$\int \cos^2 ax\ dx = \frac{x}{2} + \frac{\sin 2ax}{4a} + c$$

$$\int \frac{dx}{\sin ax} = \frac{1}{a} \log \tan \frac{ax}{2} + c$$

$$\int \frac{dx}{\cos ax} = \frac{1}{a} \log \tan \left(\frac{ax}{2} + \frac{\pi}{4} \right) + c$$

$$\int \sin ax \sin bx\ dx = \frac{\sin (a - b)x}{2(a - b)} -$$
$$\frac{\sin (a + b)x}{2(a + b)} + c \quad \text{if } a^2 \neq b^2$$

$$\int x \sin ax\ dx = \frac{\sin ax}{a^2} - \frac{x \cos ax}{a} + c$$

$$\int x \cos ax\ dx = \frac{\cos ax}{a^2} + \frac{x \sin ax}{a} + c$$

$$\int \sin^{-1} ax\ dx = x \sin^{-1} ax + \frac{1}{a} \sqrt{1 - a^2 x^2} + c$$

$$\int \cos^{-1} ax\ dx = x \cos^{-1} ax - \frac{1}{a} \sqrt{1 - a^2 x^2} + c$$

$$\int \tan^{-1} ax\ dx = x \tan^{-1} ax - \frac{\log (1 + a^2 x^2)}{2a} + c$$

Differential Equations

If a linear differential equation can be reduced to:

$$\frac{dy}{dx} + Py = Q;\ P, Q = f(x)\ \text{only}$$

its solution is:

$$y \epsilon^{\int P dx} = \int Q \epsilon^{\int P dx}\ dx + c$$

$$\text{Example:}\quad x \frac{dy}{dx} + 2y = x^3$$

$$y \epsilon^{\int \frac{2}{x} dx} = \int x^2 \epsilon^{\int \frac{2}{x} dx}\ dx + c$$

$$yx^2 = \frac{x^5}{5} + c$$

Higher-order linear differential equations:

$$a_o \frac{d^n y}{dx^n} + a_1 \frac{d^{n-1}y}{dx^{n-1}} + a_2 \frac{d^{n-2}y}{dx^{n-2}} + \ldots a_{n-1} \frac{dy}{dx} + a_n y = 0$$

form of general solution is:

$$y = c_1 \epsilon^{r1x} + c_2 \epsilon^{r2x} + \ldots c_n \epsilon^{rnx}$$

where: $a_o r^n + a_1 r^{n-1} + a_2 r^{n-2} \ldots + a_n = 0$ and if there are n roots in this equation.

Example:

$$\frac{d^4 y}{dx^4} - \frac{d^3 y}{dx^3} - 4\frac{d^2 y}{dx^2} + 4\frac{dy}{dx} = 0$$

$$r^4 - r^3 - 4r^2 + 4r = 0$$

roots are 1, 2, −2, 0

Solution: $y = c_1 \epsilon^x + c_2 \epsilon^{2x} + c_3 \epsilon^{-2x} + c_4$

23.7 FUNDAMENTALS OF FOURIER SERIES AND TRANSFORMS

Fourier Series

Real form, period 2π:

$$f(x) = \frac{a_o}{2} + a_1 \cos x + a_2 \cos 2x + \ldots a_n \cos nx$$

$$+ \ldots + b_1 \sin x + b_2 \sin 2x + \ldots b_n \sin nx$$

$$f(x) = \frac{a_o}{2} + \sum_{n=1}^{\infty} (a_n \cos nx + b_n \sin nx)$$

$$f(x) = \frac{a_o}{2} + \sum_{n=1}^{\infty} C_n \cos(nx - \phi_n)$$

$$C_n = \sqrt{a_n^2 + b_n^2}; \quad \tan \phi_n = \frac{b_n}{a_n}$$

$$a_o = \frac{1}{\pi} \int_{-\pi}^{\pi} f(x) \, dx = \frac{1}{\pi} \int_{0}^{2\pi} f(x) \, dx$$

$$a_n = \frac{1}{\pi} \int_{-\pi}^{\pi} f(x) \cos nx \, dx; \quad b_n = \frac{1}{\pi} \int_{-\pi}^{\pi} f(x) \sin nx \, dx$$

Period T:

$$f(x) = \frac{a_o}{2} + \sum_{n}^{\infty} \left(a_n \cos \frac{2n\pi x}{T} + b_n \sin \frac{2n\pi x}{T} \right)$$

$$a_n = \frac{2}{T} \int_{0}^{T} f(x) \cos \frac{2n\pi x}{T} \, dx$$

$$b_n = \frac{2}{T} \int_{0}^{T} f(x) \sin \frac{2n\pi x}{T} \, dx$$

Complex form, period 2π:

$$f(x) = \sum_{n=-\infty}^{\infty} d_n \epsilon^{jnx}; \quad d_n = \frac{1}{2\pi} \int_{-\pi}^{\pi} f(x) \epsilon^{-jnx} \, dx$$

For real functions:

$$d_n = \frac{1}{2} (a_n - jb_n) = \frac{1}{2} c_n \epsilon^{j\phi n}$$

$$d_{-n} = \frac{1}{2} (a_n + jb_n) = \frac{1}{2} c_n \epsilon^{-j\phi n}$$

$$d_o = \frac{1}{2} a_o = \frac{1}{2} c_o$$

Period T:

$$f(x) = \sum_{n=-\infty}^{n=\infty} d_n \epsilon^{j \frac{2n\pi x}{T}}; \quad d_n = \frac{1}{T} \int_{0}^{T} f(x) \epsilon^{-j \frac{2n\pi x}{T}} \, dx$$

Average power of $f(x)$:

$$\frac{1}{T} \int_{0}^{T} |f(x)|^2 \, dx = \sum_{n=-\infty}^{\infty} |d_n|^2$$

$$= \frac{c_o^2}{4} + \frac{1}{2} \sum_{n=1}^{\infty} c_n^2$$

$$= \frac{a_o^2}{4} + \frac{1}{2} \sum_{n=1}^{\infty} (a_n^2 + b_n^2)$$

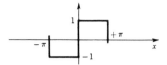

Square wave:

$$\frac{4}{\pi} \sum_{n=0}^{\infty} \frac{1}{2n+1} \sin (2n+1)x$$

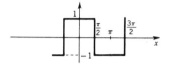

Square wave:

$$\frac{4}{\pi} \sum_{n=0}^{\infty} (-1)^n \frac{1}{2n+1} \cos (2n+1)x$$

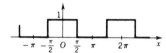

Square wave:

$$\frac{1}{2} + \frac{2}{\pi} \sum_{n=1}^{\infty} \frac{(-1)^n}{2n+1} \cos (2n+1)x$$

Square wave:

$$\frac{1}{2} + \frac{1}{\pi} \sum_{n=1}^{\infty} \frac{1-(-1)^n}{n} \sin nx$$

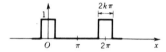

Rectangular wave:

$$k + \frac{2}{\pi} \sum_{n=1}^{\infty} \frac{1}{n} \sin nk\pi \cos nx$$

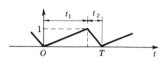

Sawtooth wave: $\omega = \dfrac{2\pi}{T}$, $T = t_1 + t_2$

$$\frac{1}{2} + \frac{4}{t_1 t_2 \omega^2} \sum_{n=1}^{\infty} \frac{1}{n^2} \sin \frac{nt_1\omega}{2} \sin n\omega \left(t - \frac{t_1}{2} \right)$$

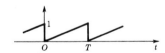

Rectangular sawtooth wave:

$$\frac{1}{2} - \frac{1}{\pi} \sum_{n=1}^{\infty} \frac{\sin n\omega t}{n}$$

Figure 23-2.   Fourier series.

(Roger Legros and A.V.J. Martin, *Transform Calculus for Electrical Engineers*, © 1961. Reprinted by permission of Prentice-Hall, Inc.)

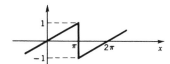

Rectangular sawtooth wave:

$$\frac{2}{\pi} \sum_{n=1}^{\infty} (-1)^{n+1} \frac{1}{n} \sin nx$$

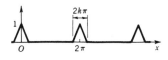

Triangular wave:

$$\frac{8}{\pi^2} \sum_{n=0}^{\infty} \frac{1}{(2n+1)^2} \cos(2n+1)x$$

Truncated triangular wave:

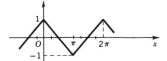

$$\frac{k}{2} + \frac{2}{\pi} \sum_{n=1}^{\infty} \frac{1}{n} \cos nx \left[\sin nk\pi \right.$$
$$\left. - \frac{1}{nk\pi} \left(nk\pi \sin nk\pi - 2 \sin^2 \frac{nk\pi}{2} \right) \right]$$

Absolute value cosine wave:

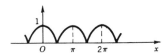

$$\frac{2}{\pi} + \frac{4}{\pi} \sum_{n=1}^{\infty} (-1)^{n+1} \frac{1}{4n^2 - 1} \cos 2nx$$

Half cosine wave:

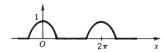

$$\frac{1}{\pi} + \frac{1}{2} \cos x - \frac{2}{\pi} \sum_{n=1}^{\infty} \frac{(-1)^n}{4n^2 - 1} \cos 2nx$$

Absolute value sine wave:

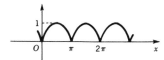

$$\frac{2}{\pi} - \frac{4}{\pi} \sum_{n=1}^{\infty} \frac{1}{4n^2 - 1} \cos 2nx$$

Half sine wave:

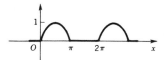

$$\frac{1}{\pi} + \frac{1}{2} \sin x - \frac{2}{\pi} \sum_{n=1}^{\infty} \frac{1}{4n^2 - 1} \cos 2nx$$

Figure 23-2. Continued.

Harmonics:

$$f(x) = \frac{a_o}{2} + \underbrace{\sum_{n=1}^{\infty} a_n \cos nx}_{\text{even}} + \underbrace{\sum_{n=1}^{\infty} b_n \sin nx}_{\text{odd}}$$

$$f(x) = \overbrace{\frac{f(x) + f(x + \pi)}{2}}^{} + \overbrace{\frac{f(x) - f(x + \pi)}{2}}^{}$$

Fourier Integral

Bilateral form: $f(t) = \dfrac{1}{\pi} \displaystyle\int_0^{\infty} d\omega \int_{-\infty}^{\infty} f(\tau) \cos \omega(t - \tau)\, d\tau$

Unilateral form: $f(t) = \dfrac{1}{2\pi} \displaystyle\int_{-\infty}^{\infty} d\omega \int_0^{\infty} f(\tau) \epsilon^{j\omega(t - \tau)}\, d\tau$

where $\omega = 2\pi f = $ radians per second

Trigonometric form: $f(t) = \displaystyle\int_0^{\infty} c(\omega) \cos\left[\omega t - \phi(\omega)\right] d\omega$

$$c(\omega) = \sqrt{a^2(\omega) + b^2(\omega)}$$

$$a(\omega) = \frac{1}{\pi} \int_{-\infty}^{\infty} f(\tau) \cos \omega\tau\, d\tau$$

$$b(\omega) = \frac{1}{\pi} \int_{-\infty}^{\infty} f(\tau) \sin \omega\tau\, d\tau$$

$$\tan \phi(\omega) = \frac{b(\omega)}{a(\omega)}$$

Fourier Transform

$$F(\omega) = A \int_{-\infty}^{\infty} f(x) \epsilon^{-j\omega x}\, dx = \mathfrak{F}[f(x)]$$

inverse: $f(x) = \dfrac{1}{A2\pi} \displaystyle\int_{-\infty}^{\infty} F(\omega) \epsilon^{j\omega x}\, d\omega = \mathfrak{F}^{-1}[f(x)]$

Change of time scale: If a is a positive number

then $\mathcal{L}\left[f\left(\dfrac{t}{a}\right)\right] = aF(as)$

Differentiation in the real (t) domain:

a) $f(t)$, $\dfrac{df(t)}{dt}$, etc. = exponential + continuous

$$\mathcal{L}\left[\frac{d^n f(t)}{dt^n}\right] = S^n F(s) + \sum_{p=1}^{n} S^{n-p} f^{(p-1)}(0+)$$

b) $f(t)$ is piecewise continuous; simple

discontinuities at $t = t_p$, $p = 0, 1, 2, \ldots$

$$\mathcal{L}\left[\frac{df(t)}{dt}\right] = sF(s) - f(0+) + \sum_{p=0}^{n} [f(t_p -) - f(t_p +)]\epsilon^{-stp}$$

Differentiation in complex(s) domain:

$$\frac{d^n F(s)}{ds^n} = \mathcal{L}[(-1)^n t^n f(t)], \; n = 0, 1, 2 \ldots$$

Integration in the real (t) domain:

$$\mathcal{L}\left[\int f_{(t)}\, dt\right] = \frac{1}{s} F(s) + \frac{1}{s} f^{-1}(0+)$$

$$\mathcal{L}\left[\int_0^t f(\tau)\, d\tau\right] = \frac{1}{s} F(s)$$

23.8 FUNDAMENTALS OF LAPLACE TRANSFORMS

Basic Laplace Equations

$$F(s) = \int_0^\infty \epsilon^{-st} f(t)\, dt; \quad t = \text{real, usually time}$$

$$s = \text{complex, sometimes}$$

$$s = j\omega; \; \omega = 2\pi f$$

$$F(s) = \mathcal{L}[f(t)]; \; f(t) = \mathcal{L}^{-1}[F(s)]$$

$\mathcal{L}$ = Laplace transform $\mathcal{L}^{-1}$ = Inverse $\mathcal{L}$

$\mathcal{L}$ has meaning only if $\displaystyle\int_0^\infty \epsilon^{-st} f(t)\, dt$ converges.

$\mathcal{L}$ is linear if:

$F(s) = \mathcal{L}[f(t)], G(s) = \mathcal{L}[g(t)]$ then

$\mathcal{L}[af(t) + bg(t)] = aF(s) + bG(s)$

$\mathcal{L}^{-1}[a'F(s) + b'G(s)] = a'f(t) + b'g(t)$

   where a, a', b, b' are independent coefficients

Translation of time axis:

   If $f(t) = 0$ for $t < 0$ and if $f(t - \tau) = 0$, $t < \tau$

   then $\mathcal{F}[f(t - \tau)] = \epsilon^{-s}\tau\mathcal{L}[f(t)]$

Translation in the complex plane:

$$F(s + a) = \mathcal{L}[\epsilon^{-at}f(t)]$$

Integration in the complex (s) domain:

   $f(t)$ is continuous, $\dfrac{f(t)}{t}$ exists

$$\int_z^\infty F(z)\,dz = \mathcal{L}\left[\frac{f(t)}{t}\right]$$

Periodic functions of (t):

function starts at $t = 0$, period $= T$

$f(t_1) = f(t)$ during T_1, but is zero elsewhere

$$F(s) = \frac{F_1(s)}{1 - \epsilon^{-Ts}}$$

$F(s) = \mathcal{L}[f(t)]$	$f(t) = \mathcal{L}^{-1}[F(s)]$	Remarks
1	$\delta(t)$	Unit impulse
s	$\delta'(t)$	Doublet impulse
s^n	$\delta^{(n)}(t)$	Impulse of $(n+1)$th order

Figure 23-3.   Laplace transform pairs.

(Roger Legros and A.V.J. Martin, *Transform Calculus for Electrical Engineers*, © 1961. Reprinted by permission of Prentice-Hall, Inc.)

$F(s) = \mathscr{L}[f(t)]$	$f(t) = \mathscr{L}^{-1}[F(s)]$	Remarks
s^{-1}	$u(t)$	Unit step
s^{-2}	t	Unit ramp
s^{-n}	$\dfrac{t^{n-1}}{(n-1)!}$	n positive integer
s^{-n}	$\dfrac{t^{n-1}}{\Gamma(n)}$	Any n, $\mathrm{Re}(n) > 0$ Γ: gamma function
$s^{-1/2}$	$\dfrac{1}{(\pi t)^{1/2}}$	
$s^{-(n+1/2)}$	$\dfrac{2^n t^{n-1/2}}{1\cdot 3\cdot 5 \ldots (2n-1)\pi^{1/2}}$	n positive integer
$\dfrac{1}{s+a}$	ϵ^{-at}	
$\dfrac{1}{(s+a)^2}$	$t\epsilon^{-at}$	
$\dfrac{1}{s^2+a^2}$	$\dfrac{1}{a}\sin at$	
$\dfrac{1}{s^2-a^2}$	$\dfrac{1}{a}\sinh at$	
$\dfrac{1}{(s^2+a^2)^2}$	$\dfrac{1}{2a^3}(\sin at - at\cos at)$	
$\dfrac{1}{(s^2+a^2)^n}$	$\dfrac{\pi^{1/2}}{\Gamma(n)}\left(\dfrac{t}{2a}\right)^{n-1/2}J_{n-1/2}(at)$	$n > 0$ J: Bessel function
$\dfrac{1}{(s^2-a^2)^n}$	$\dfrac{\pi^{1/2}}{\Gamma(n)}\left(\dfrac{t}{2a}\right)^{n-1/2}I_{n-1/2}(at)$	$n > 0$ I: modified Bessel function
$\dfrac{1}{(s+a)(s+b)}$	$\dfrac{\epsilon^{-at}-\epsilon^{-bt}}{b-a}$	
$\dfrac{1}{(s+a)(s+b)(s+c)}$	$-\dfrac{(c-b)\epsilon^{-at}+(a-c)\epsilon^{-bt}+(b-a)\epsilon^{-ct}}{(b-a)(c-b)(a-c)}$	
$\dfrac{1}{a^2+(s+b)^2}$	$\dfrac{1}{a}\epsilon^{-bt}\sin at$	
$\dfrac{1}{s^2(s+a)}$	$\dfrac{1}{a^2}(\epsilon^{-at}+at-1)$	
$\dfrac{1}{s(s+a)^2}$	$\dfrac{1}{a^2}[1-(1+at)\epsilon^{-at}]$	
$\dfrac{1}{(s+a)^2(s+b)}$	$\dfrac{1}{(b-a)^2}\epsilon^{-bt}+\dfrac{(b-a)t-1}{(b-a)^2}\epsilon^{-at}$	

Figure 23-3.   Continued.

$F(s) = \mathscr{L}[f(t)]$	$f(t) = \mathscr{L}^{-1}[F(s)]$	Remarks
$\dfrac{1}{s(s+a)(s+b)}$	$\dfrac{1}{ab} + \dfrac{b\epsilon^{-at} - a\epsilon^{-bt}}{ab(b-a)}$	
$\dfrac{1}{s(s^2+a^2)}$	$\dfrac{1}{a^2}(1 - \cos at)$	
$\dfrac{1}{s^2(s^2+a^2)}$	$\dfrac{1}{a^3}(at - \sin at)$	
$\dfrac{1}{s^3+a^3}$	$\dfrac{1}{3a^2}\left[\epsilon^{-at} - \epsilon^{-at/2}\left(\cos at\dfrac{\sqrt{3}}{2} - \sqrt{3}\sin at\dfrac{\sqrt{3}}{2}\right)\right]$	
$\dfrac{1}{s^4+4a^4}$	$\dfrac{1}{4a^3}(\sin at \cosh at - \cos at \sinh at)$	
$\dfrac{1}{s^4-a^4}$	$\dfrac{1}{2a^3}(\sinh at - \sin at)$	
$\dfrac{1}{s^{1/2}+a}$	$\dfrac{1}{(\pi t)^{1/2}} - a\epsilon^{a^2 t}[1 - \Theta(at^{1/2})]$	Θ: theta function also erf: error function
$\dfrac{1}{s^{1/2}(s^{1/2}+a)}$	$\epsilon^{a^2 t}[1 - \Theta(at^{1/2})]$	$[1 - \Theta(x)]$ is also erfc(x): complementary error function
$\dfrac{1}{(s^2+a^2)^{1/2}}$	$J_0(at)$	
$\dfrac{1}{(s^2-a^2)^{1/2}}$	$I_0(at)$	
$\dfrac{2}{s(s^2+4)}$	$\sin^2 t$	
$\dfrac{1}{s^{1/2}(s-a^2)}$	$\dfrac{1}{a}\epsilon^{a^2 t}\Theta(at^{1/2})$	
$\dfrac{1}{(s+a)(s+b)^{1/2}}$	$\dfrac{1}{(b-a)^{1/2}}\epsilon^{-at}\Theta[(b-a)^{1/2}t^{1/2}]$	
$\dfrac{s}{(s+a)(s+b)}$	$\dfrac{1}{b-a}(b\epsilon^{-bt} - a\epsilon^{-at})$	
$\dfrac{s}{s^2+a^2}$	$\cos at$	
$\dfrac{s}{s^2-a^2}$	$\cosh at$	
$\dfrac{s+b}{s^2+a^2}$	$\dfrac{1}{a}(b+a^2)^{1/2}\sin(at+\varphi)$	$\tan\varphi = \dfrac{a}{b}$

Figure 23-3. Continued.

DISCONTINUOUS FUNCTIONS

Function	$f(t)$	$F(s)$
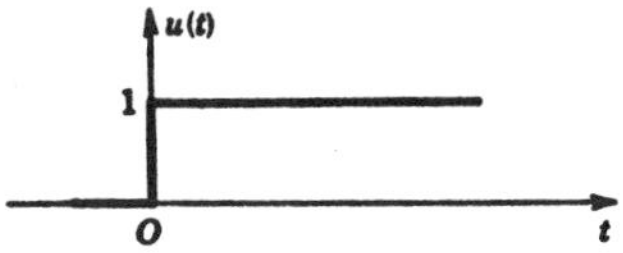 Unit step $u(t)$	Unit step $u(t)$	$\dfrac{1}{s}$
Unit impulse $\delta(t)$	Unit impulse $\delta(t)$	1
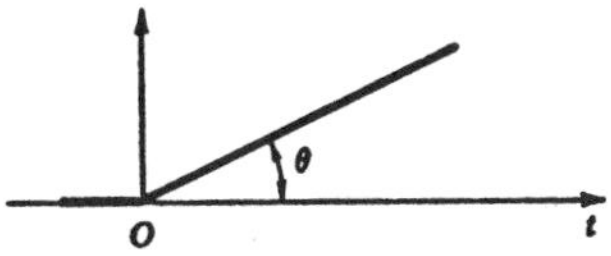 Ramp at, with $a = \tan\theta$	Ramp at, with $a = \tan\theta$	$\dfrac{a}{s^2}$
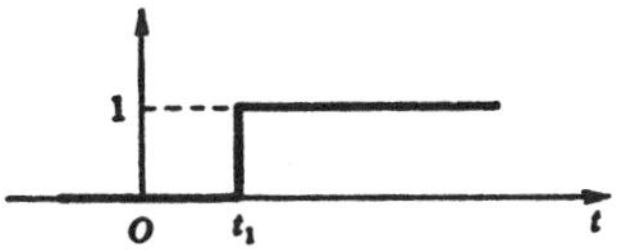 Shifted unit step $u(t - t_1)$	Shifted unit step $u(t - t_1)$	$\dfrac{1}{s}\epsilon^{-st_1}$
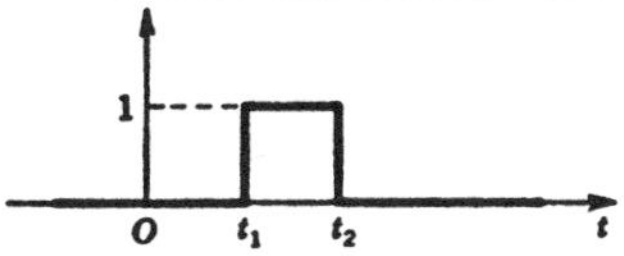 Rectangular pulse. Positive and negative unit steps $u(t - t_1) - u(t - t_2)$	Rectangular pulse. Positive and negative unit steps $u(t - t_1) - u(t - t_2)$	$\dfrac{1}{s}(\epsilon^{-st_1} - \epsilon^{-st_2})$
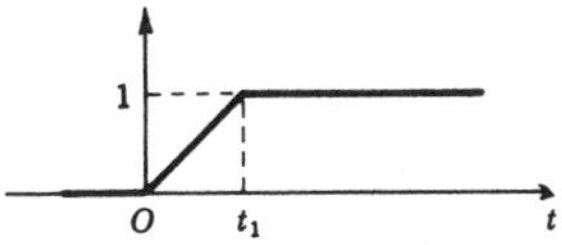 Truncated ramp $\dfrac{t}{t_1}$ for $0 < t < t_1$, 1 for $t > t_1$	Truncated ramp $\dfrac{t}{t_1}$ for $0 < t < t_1$, 1 for $t > t_1$	$\dfrac{1 - \epsilon^{-st_1}}{s^2 t_1}$
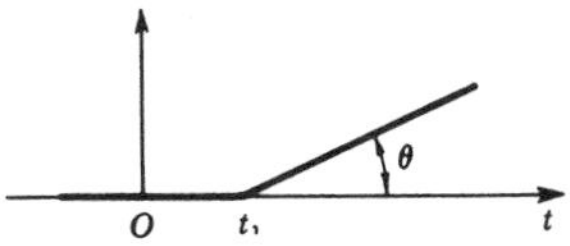 Shifted ramp $a(t - t_1)u(t - t_1)$	Shifted ramp $a(t - t_1)u(t - t_1)$	$\dfrac{a}{s^2}\epsilon^{-st_1}$
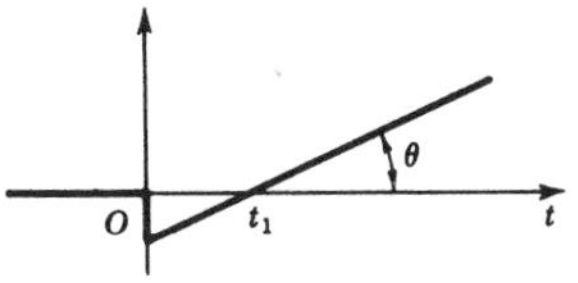 Down-shifted ramp $a(t - t_1)u(t)$	Down-shifted ramp $a(t - t_1)u(t)$	$a\left(\dfrac{1}{s^2} - \dfrac{t_1}{s}\right)$

Figure 23-3.   Continued.

Function	$f(t)$	$F(s)$
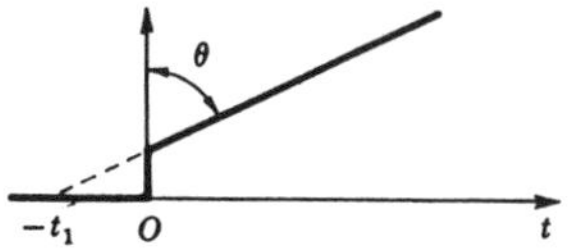	Up-shifted ramp $$a(t + t_1)u(t)$$	$a\left(\dfrac{1}{s^2} + \dfrac{t_1}{s}\right)$
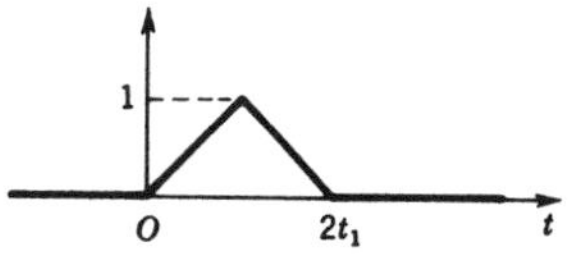	Isosceles triangle pulse $\dfrac{t}{t_1}$ for $0 < t < t_1$ $2 - \dfrac{t}{t_1}$ for $t_1 < t < 2t_1$ 0 for $t > 2t_1$	$\dfrac{(1 - \epsilon^{-st_1})^2}{s^2 t_1}$
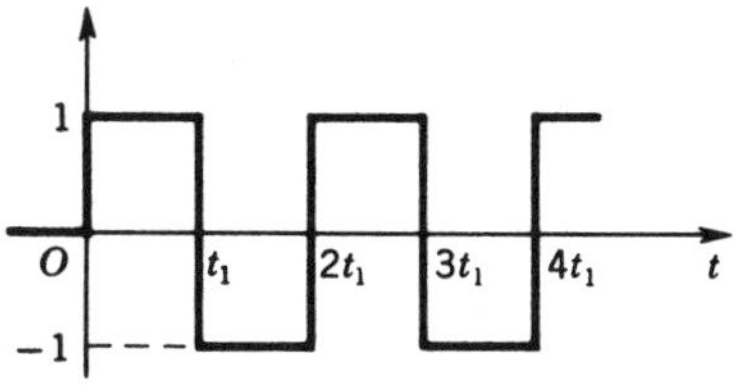	Symmetrical square wave $$u(t) + 2 \sum_{n=1}^{\infty} (-1)^n u(t - nt_1)$$	$\dfrac{1}{s}\tanh\dfrac{st_1}{2}$
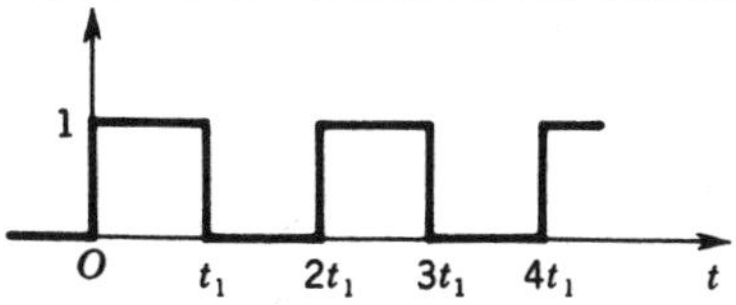	Nonsymmetrical square wave $$\sum_{n=0}^{\infty} (-1)^n u(t - nt_1)$$	$\dfrac{1}{s(1 + \epsilon^{-st_1})}$
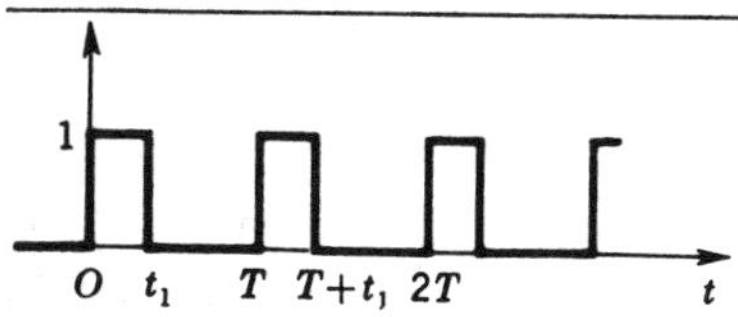	Rectangular wave $$\sum_{n=0}^{\infty} [u(t - nT) - u(t - nT - t_1)]$$	$\dfrac{1 - \epsilon^{-st_1}}{s(1 - \epsilon^{-sT})}$
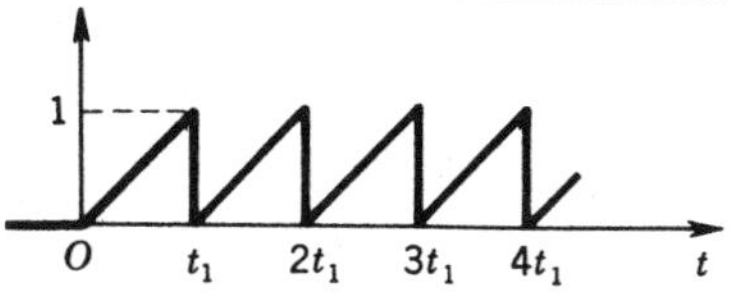	Rectangular sawtooth wave $$\dfrac{1}{t_1}\left[tu(t) - \sum_{n=1}^{\infty} u(t - nt_1)\right]$$	$\dfrac{\epsilon^{st_1} - st_1 - 1}{s^2 t_1(\epsilon^{st_1} - 1)}$
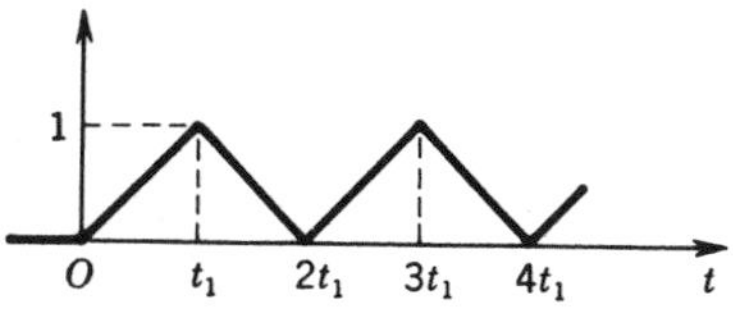	Isosceles triangular wave $$\dfrac{t}{t_1}u(t) + \dfrac{2}{t_1}\sum_{n=1}^{\infty} [(-1)^n(t - nt_1)$$ $$\times\, u(t - nt_1)]$$	$\dfrac{t_1}{s^2}\tanh\dfrac{st_1}{2}$
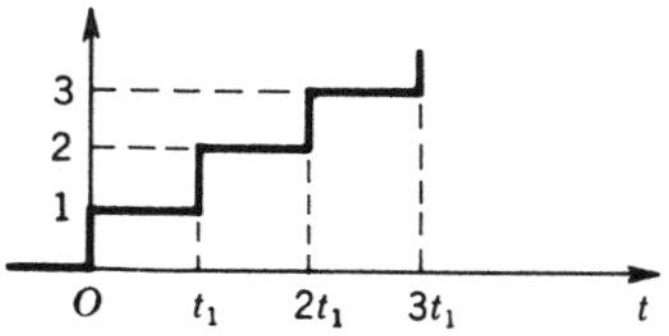	Staircase (unit steps) $$\sum_{n=0}^{\infty} u(t - nt_1)$$	$\dfrac{1}{s(1 - \epsilon^{-st_1})}$

Figure 23-3. Continued.

Function	$f(t)$	$F(s)$
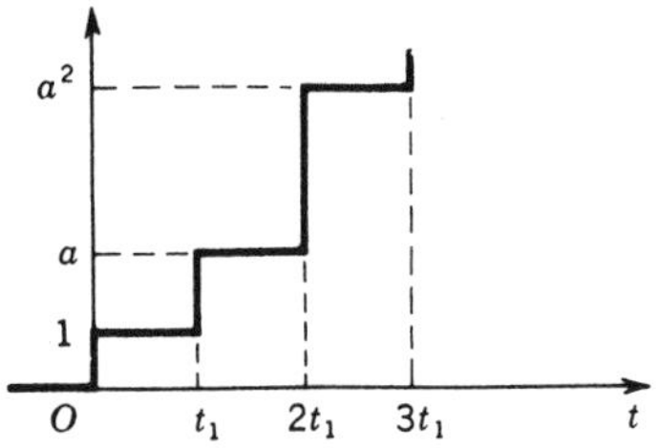	Geometrical staircase $u(t) + \dfrac{a-1}{a} \displaystyle\sum_{n=1}^{\infty} [a^n u(t - nt_1)]$	$\dfrac{1}{s}\left[1 + \dfrac{a-1}{\epsilon^{st_1} - a}\right]$ $\mathrm{Re}(s) > \dfrac{\ln a}{t_1}$
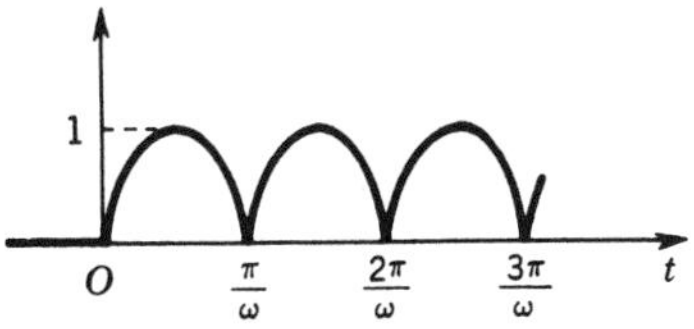	Absolute value sine wave $|\sin \omega t|$	$\dfrac{\omega}{s^2 + \omega^2} \coth \dfrac{s\pi}{2\omega}$
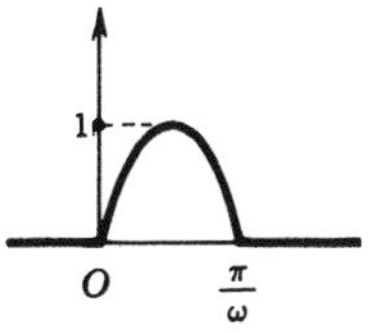	Half sine pulse $\sin \omega t \quad \text{for} \quad 0 < t < \dfrac{\pi}{\omega}$ $0 \qquad \text{for} \quad t > \dfrac{\pi}{\omega}$	$\dfrac{\omega}{s^2 + \omega^2}(1 + \epsilon^{-s\pi/\omega})$

Figure 23-3.   Continued.

23.9 SERIES AND EXPANSIONS

Taylor's Series

$$f(x) = f(x_o) + f'(x_o)(x - x_o) + \frac{f''(x_o)}{2!}(x - x_o)^2 + \ldots$$

$$f(x + a) = f(x) + f'(x)a + \frac{f''(x)}{2!}a^2 + \ldots + \frac{f^n(x)}{n!}a^n + \ldots$$

$$f'(x) = \frac{df(x)}{dx}; f''(x) = \frac{d^2f(x)}{dx^2}, \text{ etc.}$$

Maclaurin's Series

$$f(x) = f(o) + xf'(o) + \frac{x^2}{2!}f''(o) + \ldots + \frac{x^n}{n!}f^n(o) + \ldots$$

Expansion Series

$$(1 \pm x)^n = 1 \pm nx + \frac{n(n-1)}{2!}x^2 \pm \frac{n(n-1)(n-3)}{3!}x^3 + \ldots$$

$$\epsilon^x = 1 + x + \frac{x^2}{2!} + \frac{x^3}{3!} \ldots , \; |x| < \infty$$

$$\sin x = x - \frac{x^3}{3!} + \frac{x^5}{5!} - \frac{x^7}{7!} + \ldots , \; |x| < \infty, \; x = \text{radians}$$

$$\cos x = 1 - \frac{x^2}{2!} + \frac{x^4}{4!} - \frac{x^6}{6!} + \ldots , \; |x| < \infty, \; x = \text{radians}$$

$$\sin^{-1} x = x + \frac{1}{2}\frac{x^3}{3} + \frac{1 \cdot 3}{2 \cdot 4}\frac{x^5}{5} + \frac{1 \cdot 3 \cdot 5}{2 \cdot 4 \cdot 6}\frac{x^7}{7} + \ldots ; \; |x| < 1$$

$$\tan x = x + \frac{x^3}{3} + \frac{2x^5}{15} + \frac{17x^7}{315} + \frac{62x^9}{2835} + \ldots , \; |x| < \frac{\pi}{2}$$

23.10 FUNDAMENTALS OF PROBABILITY AND STATISTICS

Definitions and Notation

Discrete variates $x_1, x_2, \ldots x_n$; p_k is the probability of obtaining x_k in one trial

$$0 \le p_k \le 1; \sum_{\text{all } k} p_k = 1$$

cumulative probability $P(x) = \sum_{xk \le x} P_k$

Continuous variates $x, (x + dx), (x + 2\,dx) \ldots (x^f n\,dx)$; $px = p$ is the probability density function of one trial

Cumulative distribution function $P(x) = \displaystyle\int_{-\infty}^{x} p(s)\,ds$

$P(x)$ is the probability that variate is less than x

$$P_{(-\infty)} = 0; \; P_{(+\infty)} = \int_{-\infty}^{+\infty} p(s)\,ds = 1$$

For continuous variates $\displaystyle\int$ is used

For discrete variates Σ is used

Expected value for $y = f(x)$:

$$E_{(y)} = \sum_{\text{all k}} f(x_k)\, p_k = \int_{-\infty}^{+\infty} f(x)\, p(x)\, dx$$

Average or Mean: $E(x) = \mu$

$$\mu = \sum_{\text{all k}} p_k x_k = \int_{-\infty}^{+\infty} x p(x)\, dx$$

Root Mean Square (RMS): $r = \sqrt{E(x^2)}$

$$r = \sqrt{\sum_{\text{all k}} p_k x_k^2} = \sqrt{\int_{-\infty}^{+\infty} x^2 p(x)\, dx}$$

Moment of order r about the origin: $V_r = E(x^r)$

$$V_r = \sum_{\text{all k}} p_k x_k^r = \int_{-\infty}^{+\infty} x^r p(x)\, dx$$

Moment of order r about the mean: $\mu_r = E_{(x-\mu)}^{\ r}$

$$\mu_r = \sum_{\text{all k}} p_k (x_k - \mu)^r = \int_{-\infty}^{+\infty} (x - \mu)^r\, p(x)\, dx$$

Variance: $\sigma^2 = \mu^2 = E_{(x-\mu)}^{\ 2}$

$$\sigma^2 = \sum_{\text{all k}} p_k (x_k - \mu)^2 = \int_{-\infty}^{+\infty} (x - \mu)^2\, p(x)\, dx$$

Standard or RMS deviation from the mean: $\sigma = \sqrt{E(x - \mu)^2}$

$$\sigma = \sqrt{\sum_{\text{all k}} p_k (x_k - \mu)^2} = \sqrt{\int_{-\infty}^{+\infty} (x - \mu)^2\, p(x)\, dx}$$

Median: m such that variate x_k has equal probability of being larger or smaller than m.

$$\int_{-\infty}^{m} p(x)\, dx = \int_{m}^{+\infty} p(x)\, dx = 1/2$$

Combinations and Permutations

Number of possible combination of n items, taken m at a time is

$$C_m^n = \frac{n!}{m!(n - m)!} = \frac{n(n - 1)(n - 2) \ldots (n - m + 1)}{1 \times 2 \times 3 \times \ldots \times m}$$

Example: $n = 4$ people; $m = 2$, (1 pair)

$$C_2^4 = \frac{24}{4} = 6 \text{ possible combinations}$$

Number of possible permutations of n items, taken m at a time is

$$P_m^n = \frac{n!}{(n-m)!} = n(n-1)(n-2)\ldots(n-m+1)$$

when $n = m$ then $P_n^n = n!$

Characteristic Function

For discrete variants and their probability function P_k, the characteristic function is

$$C_{(y)} = \sum p_k \epsilon^{jyx^k} = \sum p_k^{\exp jyx}$$

For continuous variants, the characteristic is

$$C_{(y)} = E[\epsilon^{jyx}] = \int \epsilon^{jyx}\, dP_{(x)} = \int \epsilon^{jyx} p(x)\, dx$$

$$C_{(o)} = 1; \quad |C_y| \le 1$$

Distribution Functions

Binomial: $p = $ probability of one of two possible events

$q = 1 - p$; $n = $ number of trials

$k = $ number of "good" events

$$P_{(x)_k} = C_k^n p^k (1-p)^{n-k}$$

$$P_{(x)_k} = \frac{n!}{k!(n-k)!} p^k (1-p)^{n-k}$$

Example: 15% of a population is tall. Find the probability of 3 tall persons in a group of 10.

$$P_{(x)_3} = \frac{10 \times 9 \times 8}{1 \times 2 \times 3} 15^3 \times 85^7 \times 10^{-20}$$

$$P_{(x)_3} = 13\%$$

to find one short person among any three, $1 - (0.15)^3 = 99.7\%$

Expected value, $E = np$

Variance, $\sigma^2 = np(1-p)$

Poisson: Probability that K events will occur during time T

$$\mu = \text{aver. events during } T$$

$$P_{(x)_k} = \frac{\mu^k}{k!}\, \epsilon^{-\mu}$$

Expected value, $E = \mu$

Variance, $\sigma^2 = \mu$

Exponential: $f_x(y) = \dfrac{1}{\theta}\, \epsilon^{-y/\theta}$

θ = controlling parameter, any positive constant

Expected value, $E = \theta$

Variance, $\sigma^2 = \theta^2$

Normal or Gaussian:

$$f_x(y) = \frac{1}{\sqrt{2\pi}\,\sigma}\, \epsilon^{\frac{-(y-\mu)^2}{2\sigma^2}}$$

μ = mean, σ = standard deviation

Expected value, $E = \mu$

Variance, $\sigma^2 = \sigma^2$

BIBLIOGRAPHY—CHAPTER 23

DAHLQUIST, G. and A. BJORCK, *Numerical Methods*. Englewood Cliffs, NJ, Prentice-Hall, Inc., 1974.

FRANKLIN, J. N., *Matrix Theory*. Englewood Cliffs, NJ, Prentice-Hall, Inc., 1968.

HANSON, E. R., *A Table of Series and Products*. Englewood Cliffs, NJ, Prentice-Hall, Inc., 1967.

JUSZLI, F., et al., *Elementary Technical Mathematics with Calculus* (2nd Ed.). Englewood Cliffs, NJ, Prentice-Hall, Inc., 1980.

JUSZLI, F., et al., *Elementary Technical Mathematics* (3rd Ed.). Englewood Cliffs, NJ, Prentice-Hall, Inc. 1980.

INDEX